Statistical Thermodynamics

Donald A. McQuarrie
Department of Chemistry
University of California, Davis

University Science Books
Mill Valley, California

University Science Books
20 Edgehill Road
Mill Valley, CA 94941

Library of Congress Catalog Number 84-052641
ISBN 0-935702-18-0

Printed in the United States of America

CONTENTS

Chapter 4 BOLTZMANN STATISTICS, FERMI-DIRAC STATISTICS, AND BOSE-EINSTEIN STATISTICS 68

Chapter 5 IDEAL MONATOMIC GAS 81

Chapter 6 IDEAL DIATOMIC GAS 91

Chapter 7 CLASSICAL STATISTICAL MECHANICS 113

Chapter 8 IDEAL POLYATOMIC GAS 129

PREFACE

Statistical Thermodynamics represents the expansion of notes for a one-semester, first-year graduate course in statistical mechanics. The present volume comprises the first 15 chapters of a larger work entitled *Statistical Mechanics* that is to be published in 1974. There is intentionally more material in the present volume than should be necessary for a single semester; additional topics have been included in an attempt to recognize each instructor's personal prejudices. It is assumed that students have had a standard course in physical chemistry but have not previously had a serious course in quantum mechanics. The mathematical prerequisite of the book is, at most, a knowledge of linear differential equations with constant coefficients. All of the necessary mathematical techniques have been developed within the book itself, either in the main text, in problems, or in Appendix B.

There is no doubt that the term "statistical thermodynamics" or "statistical mechanics" either frightens or brings forth scorn from many people. To eradicate this attitude, a serious attempt has been made to write this book with the experimentalist in mind, whether he be instructor or student. There are extensive applications of the formal equations to experimental data both in the text itself and in the numerous applied problems. For example, a large number of figures and tables in the book compare theoretical predictions to experimental results for a variety of real systems. This has, I hope, not been done at the expense of a sensible degree of rigor or logical clarity; I have tried to achieve a balance in this respect. What I ask of the student is to make a serious effort to conceptualize and abstract for the first few chapters, because from then on he will find that practical results and rewards come quite rapidly and easily.

The first chapter is called a review, but no doubt much of the material will be new to many students. The topics included here are particularly those points of classical mechanics, quantum mechanics, and mathematics that play a central role in the development of statistical mechanics. Chapter 2 introduces the fundamental postulates and concepts of statistical mechanics with particular reference to the so-called canonical ensemble, in other words, to thermodynamic (macroscopic) systems with fixed volume, temperature, and number of particles. This chapter is fairly formal, but it concludes with problems that anticipate the utility of the formal equations derived in the chapter.

The approach used to develop statistical mechanics can be credited (at least in the textbook literature) to T. L. Hill's *Introduction to Statistical Thermodynamics* (1960). There are basically two reasons for this choice: It is the most appealing approach from both a logical and pedagogical point of view, but probably more to the point, I was a student of Hill shortly after the publication of his book and grew up statistical-mechanically with his approach.

Chapter 3 extends the formal development of Chapter 2, and Chapter 4 illustrates how the equations of Chapters 2 and 3 may be reduced to very practical formulas that are applicable to an amazing variety of systems. These applications, in fact, largely constitute the balance of this volume. The common feature of the systems discussed in Chapters 5 through 14 is this: Atoms or molecules making up these systems are either independent or "effectively" independent. This results in equations that are not only easy to use but lead to comparisons with experiment that are usually quite decent and rewarding.

The final chapter investigates imperfect gases, the simplest system in which atoms or molecules cannot be treated as independent or even effectively independent. Here the intermolecular potential begins to emerge as a major complicating factor. A principal goal of this chapter is the introduction of one of the more advanced techniques that can be used to treat complicated systems. Chapters 16 through 20 of the larger volume *Statistical Mechanics* (which might be covered in the first weeks of a second semester) treat other systems in which the intermolecular potential plays a central role. The inclusion of Chapter 15 on imperfect gases in the present volume is to assure the skeptical student that statistical mechanics is quite capable of treating general nonindependent systems—and possibly to entice him into a more challenging and much more exciting second semester when he can, in fifteen weeks, be brought close to a "literature level" in statistical mechanics.

Although this book has evolved from teaching a course in statistical mechanics for several years, much effort has been made to produce a book that is equally suitable for self-study. (I have also kept in mind the apprehensive instructor who is not a professional theoretical chemist.) To this purpose, there is a fairly extensive and specific bibliography available at the end of each chapter to direct either student or instructor to other treatments of the material covered. Obviously no book of this type is written in a vacuum, and the bibliography represents those sources that were particularly useful to me.

One of the most important and useful features of this book is the extensive collection of problems presented after each chapter. The only way to learn statistical mechanics, or any other physical chemical subject, is to work a great many problems. Consequently, there are over 450 problems in the book, ranging in difficulty from "plug in the number and get an answer" to those that require little writing but much ingenuity. A few of the problems fill in steps leading from one equation to another, while others extend material to research papers in the literature. Most of the problems, however, are intended as applications of the equations and concepts presented in chapters. It is primarily the problems that dictate the level of a course such as statistical mechanics, and the variety and number included here allow an instructor to present a straightforward application-oriented course that should be perfectly reasonable for seniors or a fairly demanding introductory course for theoretical graduate students.

I should like to acknowledge the people who have helped me in putting this book together: Donald DuPré and Stuart Rice for reviewing the entire manuscript; all of my students, particularly Steve Brenner, Jim Ely, Dennis Isbister, and Allen Nelson, who had to suffer through several earlier versions of the final manuscript; Wilmer Olivares for reading all of the page proofs; and Lynn Johnson, who typed the final manuscript and is more than happy to be finished. After reading many prefaces recently, it would seem appropriate at this point to thank my wife, Carole, for "encouragement and understanding during the writing of this book." It so happens, however, that she never did encourage me nor was she especially understanding, but she actually did something far better. As a biochemistry graduate student, she took both semesters of the course here (*not* from me) and found out for herself that statistical mechanics can be a rich and exciting field that has important applications not only in chemistry and physics but in biology as well. I hope that this book starts its readers along the same route.

Donald A. McQuarrie

STATISTICAL THERMODYNAMICS

CHAPTER 1

INTRODUCTION AND REVIEW

1–1 INTRODUCTION

Statistical mechanics is that branch of physics which studies macroscopic systems from a microscopic or molecular point of view. The goal of statistical mechanics is the understanding and prediction of macroscopic phenomena and the calculation of macroscopic properties from the properties of the individual molecules making up the system.

Present-day research in statistical mechanics varies from mathematically sophisticated discussions of general theorems to almost empirical calculations based upon simple, but nevertheless useful, molecular models. An example of the first type of research is the investigation of the question of whether statistical mechanics, as it is formulated today, is even capable of predicting the existence of a first-order phase transition. General questions of this kind are by their nature mathematically involved and are generally beyond the level of this book. We shall, however, discuss such questions to some extent later on. On the other hand, for many scientists statistical mechanics merely provides a recipe or prescription which allows them to calculate the properties of the physical systems which they are studying.

The techniques of statistical mechanics have been used in attacking a wide variety of physical problems. A quick glance through this text will show that statistical mechanics has been applied to gases, liquids, solutions, electrolytic solutions, polymers, adsorption, metals, spectroscopy, transport theory, the helix-coil transition of DNA, the electrical properties of matter, and cell membranes, among others.

Statistical mechanics may be broadly classified into two parts, one dealing with systems in equilibrium and the other with systems not in equilibrium. The treatment of systems in equilibrium is usually referred to as *statistical thermodynamics*, since it forms a bridge between thermodynamics (often called classical thermodynamics) and molecular physics.

Thermodynamics provides us with mathematical relations between the various experimental properties of macroscopic systems in equilibrium. An example of such a thermodynamic relation is that between the molar heat capacities at constant pressure and at constant volume,

$$C_p - C_v = \left[p + \left(\frac{\partial E}{\partial V}\right)_T\right]\left(\frac{\partial V}{\partial T}\right)_p \tag{1-1}$$

Another, and one that we shall use in the next chapter, is

$$\left(\frac{\partial E}{\partial V}\right)_{N,T} - T\left(\frac{\partial p}{\partial T}\right)_{N,V} = -p \tag{1-2}$$

Note that thermodynamics provides connections between many properties, but does not supply information concerning the magnitude of any one. Neither does it attempt to base any relation on molecular models or interpretations. This, in fact, is both the power and weakness of thermodynamics. It is a general discipline which does not need to recognize or rely upon the existence of atoms and molecules. Its many relations would remain valid even if matter were continuous. In addition, there are many systems (such as biological systems) which are too complicated to be described by an acceptable molecular theory, but here again the relations given by thermodynamics are exact. This great generality, however, is paid for by its inability to calculate physical properties separately or to supply physical interpretations of its equations. When one seeks a molecular theory which can do just this, one then enters the field of statistical thermodynamics. Thus thermodynamics and statistical thermodynamics treat the same systems. Thermodynamics provides general relations without the need of ever considering the ultimate constitution of matter, while statistical thermodynamics, on the other hand, assumes the existence of atoms and molecules to calculate and interpret thermodynamic quantities from a molecular point of view.

Statistical thermodynamics itself may be further divided into two areas: first, the study of systems of molecules in which molecular interactions may be neglected (such as dilute gases), and second, the study of systems in which the molecular interactions are of prime importance (such as liquids). We shall see that the neglect of intermolecular interactions enormously simplifies our problem. Chapters 4 through 14 of the book are devoted to the treatment of systems in which these interactions either may be ignored or highly simplified. This is the kind of statistical thermodynamics to which most undergraduates have been exposed and, to some extent, represents typical statistical thermodynamical research done in the 1930s. The more interesting and challenging problems, however, concern systems in which these molecular interactions cannot be neglected; Chapters 15 through 20 of the expanded volume, *Statistical Mechanics*, are devoted to the study of such systems. It is in this latter area that a great deal of the research of the 1940s, 1950s, and 1960s was carried out. There are, of course, many important problems of this sort still awaiting attack. The theory of concentrated electrolyte solutions and the proof for the existence of first-order phase transitions are just two examples.

The most difficult branch of statistical mechanics, both mathematically and conceptually, is the study of systems not in equilibrium. This field is often referred to as *nonequilibrium statistical mechanics*. This is presently a very active area of research.

There are still some important unsolved conceptual problems in nonequilibrium statistical mechanics. Nevertheless, in the 1950s great strides were made toward the establishment of a firm basis for nonequilibrium statistical mechanics, commensurate with that of equilibrium statistical mechanics, or what we have called statistical thermodynamics. Chapters 21 through 29 of *Statistical Mechanics* present an introduction to some of the more elementary of these fairly new and useful concepts and techniques.

In Chapter 2 we shall introduce and discuss the basic concepts and assumptions of statistical thermodynamics. We shall present these ideas in terms of quantum mechanical properties such as energy states, wave functions, and degeneracy. Although it may appear at this point that quantum mechanics is a prerequisite for statistical thermodynamics, it will turn out that a satisfactory version of statistical thermodynamics can be presented by using only a few quantum mechanical ideas and results. We assume that the student is familiar with only the amount of quantum mechanics taught in most present-day physical chemistry courses. About the only requirement of the first few chapters is an understanding that the Schrödinger equation determines the possible energy values E_j available to the system and that these may have a degeneracy associated with them which we denote by $\Omega(E_j)$.

Before discussing the principles, however, we shall present in this chapter a discussion of some of the terms or concepts that are particularly useful in statistical thermodynamics. In Section 1–2 we shall treat classical mechanics, including an introduction to the Lagrangian and Hamiltonian formalisms. In Section 1–3 we shall briefly review the main features of quantum mechanics and give the solutions of the Schrödinger equation for some important systems. The only new material in this section to most students probably will be the discussion of the eigenvalues or energy levels of a many-body system. Then in Section 1–4 we shall review thermodynamics briefly, since it is assumed that the reader is familiar with the three laws of thermodynamics and the tedious manipulations of partial derivatives. Two important topics that are not usually discussed in elementary physical chemistry texts are introduced, however. These two topics are the Legendre transformation and Euler's theorem, both of which are useful in studying statistical thermodynamics. Finally, in Section 1–5 we shall discuss some mathematical techniques and results that are particularly useful in statistical thermodynamics. Much of this section may be new material to the reader.

1–2 CLASSICAL MECHANICS

NEWTONIAN APPROACH

Everyone knows the equation $F = ma$. What this equation really says is that the rate of change of momentum is equal to the applied force. If we denote the momentum by $\mathbf{p}$, we have then a more general version of Newton's second law, namely,

$$\frac{d\mathbf{p}}{dt} \equiv \dot{\mathbf{p}} = \mathbf{F} \tag{1–3}$$

If the mass is independent of time, then $d\mathbf{p}/dt = m d\dot{\mathbf{r}}/dt = m\ddot{\mathbf{r}} = m\mathbf{a}$. If $\mathbf{F}$ is given as a function of position $\mathbf{F}(x, y, z)$, then Eq. (1–3) represents a set of second-order differential equations in x, y, and z whose solutions give x, y, and z as a function of time if some initial conditions are known. Thus Eq. (1–3) is called an equation of motion. We shall consider three applications of this equation.

Example 1. Solve the equation of motion of a body of mass m shot vertically upward with an initial velocity v_0 in a gravitational field.

If we choose the x-axis (positive in the upward direction) to be the height of the body, then we have

$$m\ddot{x} = -mg$$

where mg is the magnitude of the force. The negative sign indicates that the force is acting in a downward direction. The solution to this differential equation for x is then

$$x(t) = -\tfrac{1}{2}gt^2 + v_0 t + x_0 \tag{1-4}$$

with x_0 equal to $x(0)$, which in our case is 0. This then gives the position of the body at any time after it was projected. The extension of this problem to two dimensions (i.e., a shell shot out of a cannon) and the inclusion of viscous drag on the body are discussed in Problems 1–1 and 1–2.

Example 2. Set up and solve the equation of motion of a simple harmonic oscillator.

Let x_0 be the length of the unstrained spring. Hooke's law says that the force on the mass attached to the end of the spring is $F = -k(x - x_0)$. If we let $\xi = x - x_0$, we can write

$$\frac{d^2\xi}{dt^2} + \frac{k}{m}\xi = 0 \tag{1-5}$$

whose solution is

$$\xi(t) = A \sin \omega t + B \cos \omega t \tag{1-6}$$

The quantity

$$\omega = (k/m)^{1/2} \tag{1-7}$$

is the natural vibrational frequency of the system. Equation (1–6) can be written in an alternative form (see Problem 1–5)

$$\xi(t) = C \sin(\omega t + \phi) \tag{1-8}$$

This shows more clearly that the mass undergoes simple harmonic motion with frequency ω. Problems 1–3 through 1–5 illustrate some of the basic features of simple harmonic motion.

Example 3. Two-dimensional motion of a body under coulombic attraction to a fixed center.

In this case the force is $\mathbf{F} = -K\mathbf{r}/r^3$, that is, it is of magnitude $-K/r^2$ and directed radially. Newton's equations become

$$m\ddot{x} = F_x = -\frac{Kx}{(x^2 + y^2)^{3/2}}$$

$$m\ddot{y} = F_y = -\frac{Ky}{(x^2 + y^2)^{3/2}} \tag{1-9}$$

Unlike our previous examples, these two equations are difficult to solve. Since the force depends, in a natural way, on the polar coordinates r and θ, it is more convenient for us to set the problem up in a polar coordinate system. Using then

$$x = r \cos \theta$$
$$y = r \sin \theta$$

and some straightforward differentiation, we get

$$\left\{m(\ddot{r} - \dot{\theta}^2 r) + \frac{K}{r^2}\right\}\cos \theta - m(r\ddot{\theta} + 2\dot{\theta}\dot{r})\sin \theta = 0 \tag{1-10a}$$

$$\left\{m(\ddot{r} - \dot{\theta}^2 r) + \frac{K}{r^2}\right\}\sin \theta + m(r\ddot{\theta} + 2\dot{\theta}\dot{r})\cos \theta = 0 \tag{1-10b}$$

By multiplying the first of these equations by cos θ and the second by sin θ and then adding the two, one gets

$$m(\ddot{r} - \dot{\theta}^2 r) + \frac{K}{r^2} = 0 \tag{1–11}$$

But this is just the term in braces in Eq. (1–10), which leads us to the result that

$$m(r\ddot{\theta} + 2\dot{\theta}\dot{r}) = 0 \tag{1–12}$$

as well. Equation (1–12) can be written in the form

$$\frac{1}{r}\frac{d}{dt}(mr^2\dot{\theta}) = 0 \tag{1–13}$$

which implies that

$$mr^2\dot{\theta} = \text{constant} \tag{1–14}$$

This quantity, $mr^2\dot{\theta}$, which maintains a fixed value during the motion of the particle, is called the angular momentum of the particle and is denoted by l. The angular momentum is always conserved if the force is central, that is, directed along $\mathbf{r}$ (see Problem 1–10).

Equation (1–14) can be used to eliminate $\dot{\theta}$ from Eq. (1–11) to give an equation in r alone, called the radial equation:

$$m\ddot{r} - \frac{l^2}{mr^3} + \frac{K}{r^2} = 0 \tag{1–15}$$

This equation can be solved (at least numerically) to give $r(t)$, which together with Eq. (1–14) gives $\theta(t)$.

Even though the solution in this example is somewhat involved using polar coordinates, it is nevertheless much easier than if we had used Cartesian coordinates. This is just one example of many possibilities, which show that it is advantageous to recognize the symmetry of the problem by using the appropriate coordinate system.

This example was introduced, however, to illustrate another important point. Notice that Eq. (1–15) for r can be written as a Newtonian equation (i.e., in the form $F = ma$)

$$m\ddot{r} = -\frac{K}{r^2} + \frac{l^2}{mr^3}$$

if we interpret the term l^2/mr^3 as a force. This force is the well-known centrifugal force and must be introduced into the equation for $m\ddot{r}$.

This constitutes the main disadvantage of the Newtonian approach. The form of the equation $m\ddot{\eta} = F_\eta$ (where η is some general coordinate) is useful only in Cartesian systems, unless we are prepared to define additional forces, such as the centrifugal force in the above example. At times these necessary additional forces are fairly obscure.

There exist more convenient formulations of classical mechanics which are not tied to any one coordinate system. The two formulations that we are about to introduce are, in fact, independent of the coordinate system employed. These are the Lagrangian and the Hamiltonian formulations.

LAGRANGIAN APPROACH

Let K be the kinetic energy of a particle. In Cartesian coordinates

$$K(\dot{x}, \dot{y}, \dot{z}) = \frac{m}{2}(\dot{x}^2 + \dot{y}^2 + \dot{z}^2)$$

Let the potential energy be U. In many problems U is a function of position only, and so we write $U(x, y, z)$. Newton's equations are

$$m\ddot{x} = -\frac{\partial U}{\partial x}$$

with similar equations for y and z. Now introduce a new function

$$L(x, y, z, \dot{x}, \dot{y}, \dot{z}) \equiv K(\dot{x}, \dot{y}, \dot{z}) - U(x, y, z)$$

This function is called the Lagrangian of the system. In terms of L, we have

$$\frac{\partial L}{\partial \dot{x}} = \frac{\partial K}{\partial \dot{x}} = m\dot{x}$$

$$\frac{\partial L}{\partial x} = -\frac{\partial U}{\partial x}$$

$$\cdots$$

and we can write Newton's equations in the form

$$\frac{d}{dt}\left(\frac{\partial L}{\partial \dot{x}}\right) = \frac{\partial L}{\partial x} \tag{1-16}$$

with similar equations for y and z. These are Lagrange's equations of motion in Cartesian coordinates. The remarkable and useful property of Lagrange's equations is that they have the same form in any coordinate system. If the x, y, z are transformed into any other system, say q_1, q_2, q_3, Lagrange's equations take the form

$$\frac{d}{dt}\left(\frac{\partial L}{\partial \dot{q}_j}\right) = \frac{\partial L}{\partial q_j} \qquad j = 1, 2, 3 \tag{1-17}$$

This can be proved by writing $x = x(q_1, q_2, q_3)$, $y = y(q_1, q_2, q_3)$, and $z = z(q_1, q_2, q_3)$ and then transforming Eq. (1–16) into Eq. (1–17). (See Problem 1–13.)

Lagrange's equations are more useful than Newton's equations in many problems because it is usually much easier to write down an expression for the potential energy in some appropriate coordinate system than it is to recognize all the various forces. The Lagrangian formalism is based on the potential energy of the system, whereas the Newtonian approach is based on the forces acting on the system.

To illustrate the utility of the Lagrangian approach, we shall redo Example 3, the two-dimensional motion of a particle in a coulombic force field.

Example 3′. The kinetic energy is

$$K = \frac{m}{2}(\dot{x}^2 + \dot{y}^2) = \frac{m}{2}(\dot{r}^2 + r^2\dot{\theta}^2)$$

and the potential energy is $U = -K/r$. The Lagrangian, then, is

$$L(r, \theta, \dot{r}, \dot{\theta}) = \frac{m}{2}(\dot{r}^2 + r^2\dot{\theta}^2) + \frac{K}{r} \tag{1-18}$$

The two Lagrangian equations of motion are

$$\frac{d}{dt}\left(\frac{\partial L}{\partial \dot{r}}\right) = \frac{\partial L}{\partial r}$$

$$\frac{d}{dt}\left(\frac{\partial L}{\partial \dot{\theta}}\right) = \frac{\partial L}{\partial \theta}$$

or using Eq. (1–18) for L,

$$\frac{d}{dt}(m\dot{r}) = mr\dot{\theta}^2 - \frac{K}{r^2}$$

$$\frac{d}{dt}(mr^2\dot{\theta}) = 0$$

These two equations are just Eqs. (1–11) and (1–13). Note, however, that they were obtained in a much more straightforward manner than were Eqs. (1–11) and (1–13). Problems 1–10 through 1–12 further illustrate the utility of the Lagrangian formulation. Other problems involve the motion of one and two particles in central force fields.

Equations (1–17) are three second-order ordinary differential equations. To completely specify the solutions, we need three initial velocities $\dot{q}_1(0)$, $\dot{q}_2(0)$, $\dot{q}_3(0)$, and three initial positions $q_1(0)$, $q_2(0)$, $q_3(0)$. These six initial conditions along with Lagrange's equations completely determine the future (and past) trajectory of the system. If there were N particles in the system, there would be $3N$ Lagrange equations and $6N$ initial conditions.

There is another formulation of classical mechanics that involves $6N$ first-order differential equations. Although this formulation is not as convenient as Lagrange's for solving problems, it is more convenient from a theoretical point of view, particularly in quantum mechanics and statistical mechanics. This is the Hamiltonian formulation.

HAMILTONIAN APPROACH

We define a generalized momentum by

$$p_j = \frac{\partial L}{\partial \dot{q}_j} \qquad j = 1, 2, \ldots, 3N \tag{1–19}$$

This generalized momentum is said to be *conjugate* to q_j. Note that Eq. (1–19) is simply $p_x = m\dot{x}$, and so on, in Cartesian coordinates.

We now define the Hamiltonian function for a system containing just one particle (for simplicity) by

$$H(p_1, p_2, p_3, q_1, q_2, q_3) = \sum_{j=1}^{3} p_j \dot{q}_j - L(\dot{q}_1, \dot{q}_2, \dot{q}_3, q_1, q_2, q_3) \tag{1–20}$$

It is understood here that the $\dot{q}_j$'s have been eliminated in favor of the p_j's by means of Eq. (1–19).

An important difference between the Lagrangian approach and the Hamiltonian approach is that the Lagrangian is considered to be a function of the generalized velocities $\dot{q}_j$ and the generalized coordinates q_j, whereas the Hamiltonian is considered to be a function of the generalized momenta p_j and the conjugate generalized coordinates q_j. This may appear to be a fine distinction at this point, but it will turn out to be important later on. It also may seem, at this time, that the definition Eq. (1–20) is rather obscure, but we shall give a motivation for its form in Section 1–4. (See Problem 1–38.)

For the kinds of systems that we shall treat in this book, the kinetic energy is of the form

$$K = \sum_{j=1}^{3N} a_j(q_1, q_2, \ldots, q_{3N})\dot{q}_j^{\,2} \tag{1–21}$$

that is, a quadratic function of the generalized velocities. The coefficients a_j are, in general, functions of generalized coordinates but not an explicit function of time. If, furthermore, the potential energy is a function only of the generalized coordinates, then the p_j occurring in Eq. (1–20) are given by

$$p_j = \frac{\partial L}{\partial \dot{q}_j} = \frac{\partial K}{\partial \dot{q}_j} = 2a_j\,\dot{q}_j$$

where the last equality comes from Eq. (1–21). Substituting this into Eq. (1–20) gives the important result

$$H = K + U = \text{total energy} \tag{1–22}$$

We shall now show that if the Lagrangian is not an explicit function of time, then $dH/dt = 0$. We begin with the definition of H, that is, Eq. (1–20).

$$dH = \sum_j \dot{q}_j\,dp_j + \sum_j p_j\,d\dot{q}_j - \sum \frac{\partial L}{\partial \dot{q}_j}\,d\dot{q}_j - \sum \frac{\partial L}{\partial q_j}\,dq_j$$

But if we use Eqs. (1–17) and (1–19), we see that

$$dH = \sum \dot{q}_j\,dp_j - \sum \dot{p}_j\,dq_j \tag{1–23}$$

The total derivative of H is (assuming no explicit dependence on time)

$$dH = \sum \left(\frac{\partial H}{\partial p_j}\right) dp_j + \sum \left(\frac{\partial H}{\partial q_j}\right) dq_j \tag{1–24}$$

Comparing Eqs. (1–23) and (1–24), we get Hamilton's equations of motion:

$$\frac{\partial H}{\partial p_j} = \dot{q}_j \qquad \frac{\partial H}{\partial q_j} = -\dot{p}_j \qquad j = 1, 2, \ldots, 3N \tag{1–25}$$

Hamilton's equations are $6N$ first-order differential equations. It is easy to show from Eqs. (1–24) and (1–25) that $dH/dt = 0$. (See Problem 1–14.) This along with Eq. (1–22) says that energy is conserved in such systems.

Since the Hamiltonian is so closely related to the energy, and it is the total energy which is usually the prime quantity in quantum and statistical mechanics, the Hamiltonian formalism will turn out to be the most useful from a conceptual point of view. Fortunately, however, we shall never have to solve the equations of motion for macroscopic systems. The role of statistical mechanics is to avoid doing just that.

1–3 QUANTUM MECHANICS

In the previous section we have seen that a knowledge of the initial velocities and coordinates of a particle or a system of particles was sufficient to determine the future course of the system if the equations of motion, essentially the potential field that the system experiences, are known. If the state (its velocities and coordinates) of the system is known at time t_0, then classical mechanics provides us with a method of calculating the state of the system at any other time t_1.

By the 1920s it was realized that such a calculation was too detailed in principle. The Heisenberg uncertainty principle states that it is impossible to precisely specify both the momentum and position of a particle simultaneously. Consequently, the prescription given by classical mechanics had to be modified to include the principle of uncertainty. This modification resulted in the development of quantum mechanics.

There are a number of levels of introducing the central ideas of quantum mechanics, but for most of the material in this text, we need consider only the most elementary. A fundamental concept of quantum mechanics is the so-called wave function $\Psi(\mathbf{q}, t)$, where $\mathbf{q}$ represents the set of coordinates necessary to describe the system. The wave function is given the physical interpretation that the probability that at time t the system is found between q_1 and $q_1 + dq_1$, q_2 and $q_2 + dq_2$, and so on, is

$$\Psi^*(\mathbf{q}, t)\Psi(\mathbf{q}, t)\, dq_1\, dq_2 \cdots dq_{3N}$$

We shall often write $dq_1 \cdots dq_{3N}$ as $d\mathbf{q}$. The uncertainty principle dictates that $\Psi(\mathbf{q}, t)$ is the most complete description of the system that can be obtained. Since the system is sure to be somewhere, we have

$$\int \Psi^*(\mathbf{q}, t)\Psi(\mathbf{q}, t)\, d\mathbf{q} = 1 \tag{1–26}$$

If Eq. (1–26) is satisfied, Ψ is said to be normalized.

A central problem of quantum mechanics is the calculation of $\Psi(\mathbf{q}, t)$ for any system of interest. We denote the time-independent part of $\Psi(\mathbf{q}, t)$ by $\psi(\mathbf{q})$. The state of the system described by a particular $\psi(\mathbf{q})$ is said to be a stationary state. Throughout this book we shall deal with stationary states only.

For our purpose, the wave function ψ is given as the solution of the Schrödinger equation

$$\mathscr{H}\psi = E\psi \tag{1–27}$$

where $\mathscr{H}$ is the Hamiltonian operator, and E is a scalar quantity corresponding to the energy of the system. The Hamiltonian operator is

$$\begin{aligned} \mathscr{H} &= -\frac{\hbar^2}{2m}\left(\frac{\partial^2}{\partial x^2} + \frac{\partial^2}{\partial y^2} + \frac{\partial^2}{\partial z^2}\right) + U(x, y, z) \\ &= -\frac{\hbar^2}{2m}\nabla^2 + U(x, y, z) \end{aligned} \tag{1–28}$$

where $\hbar$ is $h/2\pi$, that is, Planck's constant divided by 2π. The first term here corresponds to the kinetic energy, and the second term is the potential energy. The Hamiltonian operator, then, corresponds to the total energy. There is a quantum mechanical operator and an equation similar to Eq. (1–27) corresponding to every quantity of classical mechanics, but we shall need only the one for the energy, namely, the Schrödinger equation.

Given certain physical boundary conditions of the system, a knowledge of $\mathscr{H}$ alone is sufficient to determine ψ and E. The wave function ψ is called an eigenfunction of the operator $\mathscr{H}$, and E is called an eigenvalue. There will usually be many ψ's and E's that satisfy Eq. (1–28), and this is indicated by labeling ψ and E with one or more subscripts. Generally, then, we have

$$\mathscr{H}\psi_j = E_j\psi_j \tag{1–29}$$

Equation (1–29) is a partial differential equation for ψ_j. The application of the boundary conditions often limits the values of E_j to only certain discrete values. Some simple examples are

1. a particle in a one-dimensional infinite well:

$$\mathscr{H} = -\frac{\hbar^2}{2m}\frac{\partial^2}{\partial x^2}$$

$$\varepsilon_n = \frac{h^2 n^2}{8ma^2} \qquad n = 1, 2, \ldots \tag{1–30}$$

2. a simple harmonic oscillator:

$$\mathscr{H} = -\frac{\hbar^2}{2m}\frac{\partial^2}{\partial x^2} + \frac{1}{2}kx^2$$

$$\varepsilon_n = (n + \tfrac{1}{2})\hbar\omega \qquad n = 0, 1, 2, \ldots \tag{1–31}$$

where $\omega = (k/m)^{1/2}$.

3. a rigid rotor (see Problem 1–21):

$$\mathscr{H} = -\frac{\hbar^2}{2I}\left\{\frac{1}{\sin\theta}\frac{\partial}{\partial\theta}\left(\sin\theta\frac{\partial}{\partial\theta}\right) + \frac{1}{\sin^2\theta}\frac{\partial^2}{\partial\phi^2}\right\}$$

$$\varepsilon_J = \frac{J(J+1)\hbar^2}{2I} \qquad J = 0, 1, 2, \ldots \tag{1–32}$$

Here I is the moment of inertia of the rotor (see Problem 1–15 for a treatment of the classical counterpart of this system).

The rigid rotor illustrates another important concept of quantum mechanics, namely, that of degeneracy. It happens that there may be a number of eigenfunctions or states of the system having the same eigenvalue or energy. The number of eigenfunctions having this energy is called the degeneracy of the system. For the rigid rotor, the degeneracy, ω_J, is $2J + 1$. The particle in a one-dimensional infinite well and the simple harmonic oscillator are nondegenerate, that is, the ω_n are unity. The concept of energy states and degeneracy plays an important role in statistical thermodynamics.

Consider the energy states of a particle in a three-dimensional infinite well. These are given by

$$\varepsilon_{n_x n_y n_z} = \frac{h^2}{8ma^2}(n_x^2 + n_y^2 + n_z^2) \qquad n_x, n_y, n_z = 1, 2, 3, \ldots \tag{1–33}$$

The degeneracy is given by the number of ways that the integer $M = 8ma^2\varepsilon/h^2$ can be written as the sum of the squares of three positive integers. In general, this is an erratic and discontinuous function of M (the number of ways will be zero for many values of M), but it becomes smooth for large M, and it is possible to derive a simple expression for it. Consider a three-dimensional space spanned by n_x, n_y, and n_z. There is a one-to-one correspondence between energy states given by Eq. (1–33) and the points in this n_x, n_y, n_z space with coordinates given by positive integers. Figure 1–1 shows a two-dimensional version of this space. Equation (1–33) is an equation for a sphere of radius $R = (8ma^2\varepsilon/h^2)^{1/2}$ in this space

$$n_x^2 + n_y^2 + n_z^2 = \frac{8ma^2\varepsilon}{h^2} = R^2$$

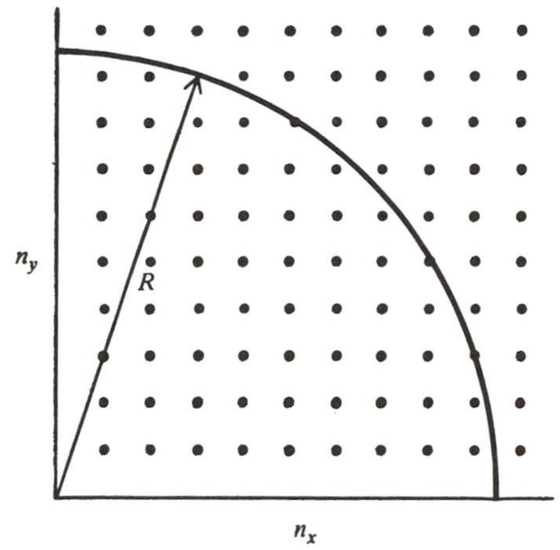

Figure 1-1. **A two-dimensional version of the (n_x, n_y, n_z) space, the space with the quantum numbers n_x, n_y, and n_z as axes.**

We wish to calculate the number of lattice points that are at some fixed distance from the origin in this space. In general, this is very difficult, but for large R we can proceed as follows. We treat R or ε as a continuous variable and ask for the number of lattice points between ε and $\varepsilon + \Delta\varepsilon$. To calculate this quantity, it is convenient to first calculate the number of lattice points consistent with an energy $\leq \varepsilon$. For large ε, it is an excellent approximation to equate the number of lattice points consistent with an energy $\leq \varepsilon$ with the volume of one octant of a sphere of radius R. We take only one octant, because n_x, n_y, and n_z are restricted to be positive integers. If we denote the number of such states by $\Phi(\varepsilon)$, we can write

$$\Phi(\varepsilon) = \frac{1}{8}\left(\frac{4\pi R^3}{3}\right) = \frac{\pi}{6}\left(\frac{8ma^2\varepsilon}{h^2}\right)^{3/2} \tag{1-34}$$

The number of states between ε and $\varepsilon + \Delta\varepsilon$ ($\Delta\varepsilon/\varepsilon \ll 1$) is

$$\begin{aligned}\omega(\varepsilon, \Delta\varepsilon) &= \Phi(\varepsilon + \Delta\varepsilon) - \Phi(\varepsilon) \\ &= \frac{\pi}{4}\left(\frac{8ma^2}{h^2}\right)^{3/2} \varepsilon^{1/2}\,\Delta\varepsilon + O((\Delta\varepsilon)^2)\end{aligned} \tag{1-35}$$

If we take $\varepsilon = 3kT/2$, $T = 300°\text{K}$, $m = 10^{-22}$ g, $a = 10$ cm, and $\Delta\varepsilon$ to be 0.01ε (in other words a 1% band around ε), then $\omega(\varepsilon, \Delta\varepsilon)$ is $O(10^{28})$.* So even for a system as simple as a particle-in-a-box, the degeneracy can be very large at room temperature.

For an N-particle system, the degeneracy is tremendously greater than $O(10^{28})$. To see this, consider a system of N noninteracting particles in a cube. The energy of this system is

$$E = \frac{h^2}{8ma^2}\sum_{j=1}^{N}(n_{xj}^2 + n_{yj}^2 + n_{zj}^2) = \frac{h^2}{8ma^2}\sum_{j=1}^{3N} s_j^2$$

where n_{xj}, n_{yj}, n_{zj}, and s_j are positive integers. The degeneracy of this system can be calculated by generalizing the above derivation for one particle. Using the volume of

* We use the notation $O(10^{28})$ say, to mean of the order of magnitude 10^{28}. This differs from standard mathematical notation, but there should be no confusion.

an N-dimensional sphere from Problem 1–24, the number of states with energy $\leq E$ is*

$$\Phi(E) = \frac{1}{\Gamma(N+1)\Gamma[(3N/2)+1]}\left(\frac{2\pi ma^2 E}{h^2}\right)^{3N/2} \tag{1–36}$$

where $\Gamma(n)$ here is the gamma function. (See Problem 1–58.) The number of states between E and $E + \Delta E$ is

$$\Omega(E, \Delta E) = \frac{1}{\Gamma(N+1)\Gamma(3N/2)}\left(\frac{2\pi ma^2}{h^2}\right)^{3N/2} E^{(3N/2-1)}\,\Delta E \tag{1–37}$$

In this case, $E = 3NkT/2$. If we take $T = 300°\text{K}$, $m = 10^{-22}$ g, $a = 10$ cm, $N = 6.02 \times 10^{23}$, and ΔE equal to $0.01E$, we get $\Omega(E, \Delta E)$ to be $0(10^N)$ (see Problem 1–23), an extremely large number. This shows that as the number of particles in the system increases, the quantum mechanical degeneracy becomes enormous. Although we have shown this only for a system of noninteracting particles confined to a cubical box, that is, an ideal gas, the result is generally true. We shall see in the next chapter that the concept of the degeneracy of a macroscopic system is very important.

There is another quantum mechanical result that we shall use later on. It often happens that the Hamiltonian of a many-body system can be written either exactly or approximately as a summation of one-particle or few-particle Hamiltonians, that is,

$$\mathscr{H} = \mathscr{H}_\alpha + \mathscr{H}_\beta + \mathscr{H}_\gamma + \cdots \tag{1–38}$$

Let the eigenvalues of $\mathscr{H}_j$ be ε_j, and the eigenfunctions be ψ_j, where $j = \alpha, \beta, \gamma, \ldots$. To solve the many-body Schrödinger equation, we let $\psi = \psi_\alpha \psi_\beta \psi_\gamma \ldots$. Then

$$\begin{aligned}
\mathscr{H}\psi &= (\mathscr{H}_\alpha + \mathscr{H}_\beta + \mathscr{H}_\gamma + \cdots)\psi_\alpha\psi_\beta\psi_\gamma\cdots \\
&= \psi_\beta\psi_\gamma\cdots\mathscr{H}_\alpha\psi_\alpha + \psi_\alpha\psi_\gamma\cdots\mathscr{H}_\beta\psi_\beta + \cdots \\
&= \psi_\beta\psi_\gamma\varepsilon_\alpha\psi_\alpha + \psi_\alpha\psi_\gamma\varepsilon_\beta\psi_\beta + \cdots \\
&= (\varepsilon_\alpha + \varepsilon_\beta + \cdots)\psi_\alpha\psi_\beta\psi_\gamma\cdots = E\psi
\end{aligned} \tag{1–39}$$

In other words, the energy of the entire system is the sum of the energies of the individual particles if they do not interact. This is a very important result and will allow us to reduce a many-body problem to a one-body problem if the interactions are weak enough to ignore, such as in the case of a dilute gas. We shall see a number of cases where, even though the interactions are too strong to be ignored (such as in a solid), it is possible to formally or mathematically write the Hamiltonian in the form of Eq. (1–38). This will lead to defining quasi-particles like phonons and photons.

The last quantum mechanical topic we shall discuss here is that of the symmetry of wave functions with respect to the interchange of identical particles. Consider a system of N identical particles, described by a wave function $\psi(1, 2, 3, \ldots, N)$, where 1 denotes the coordinates of particle 1, and so on. If we interchange the position of any two of the particles, say particles 1 and 2, the wave function must either remain the same or change sign. (See Problem 1–26.) Thus if we let P_{12} be an operator that exchanges the two identical particles 1 and 2, then

$$\begin{aligned}
P_{12}\psi(1, 2, 3, \ldots, N) &= \psi(2, 1, 3, \ldots, N) \\
&= \pm\psi(1, 2, 3, \ldots, N)
\end{aligned} \tag{1–40}$$

* The extra factor of $\Gamma(N+1)$ occurs here because of the indistinguishability of the N particles. This will be discussed fully in Chapter 4.

It turns out that whether the wave function remains the same or changes sign is a function of the nature of the two identical particles that are exchanged. For particles with an integral spin (such as the He-4 nucleus, photons, ...), the wave function remains the same. In this case the wave function is called symmetric, and such particles are called bosons. For particles with half-integral spin (such as electrons, ...), the wave function is called antisymmetric, and the particles are called fermions. Chapter 4 considers the consequences of this symmetry requirement of wave functions.

1–4 THERMODYNAMICS

In this section we shall not attempt to review thermodynamics, but shall simply state the three laws and briefly discuss their consequences. Problems 1–27 through 1–36 review some of the equations and manipulations that arise in thermodynamics. Two topics that are not often treated in elementary physical chemistry are presented here, namely, Legendre transformations and Euler's theorem. Both of these topics will be used later on.

The pressure-volume work done by a system on its surroundings in going from state A to state B is

$$w = \int_A^B p\,dV$$

where p is the pressure exerted by the surroundings on the system. The differential quantity δw is positive if dV is positive.

The heat absorbed by the system from the surroundings during the change of the system from state A to state B is

$$q = \int_A^B \delta q$$

The first law of thermodynamics states that even though w and q depend upon the path taken from A to B, their difference does not. Their difference, then, is a function only of the two states A and B, or, namely, is a state function. This function is called the internal energy or thermodynamic energy and is denoted by E.

The first law of thermodynamics is

$$\Delta E = E_B - E_A = q - w$$

$$= \int_A^B \delta q - \int_A^B p\,dV \tag{1–41}$$

For simplicity, we consider only p–V work.

A reversible change is one in which the driving force (a difference in pressure, a difference in temperature, and so on) is infinitesimal. Any other change is called irreversible or spontaneous. Problem 1–27 asks the reader to show that for an isothermal process, $w_{\text{rev}} > w_{\text{irrev}}$ and $q_{\text{rev}} > q_{\text{irrev}}$.

The first law of thermodynamics is nothing but a statement of the law of conservation of energy. The second law is somewhat more abstract and can be stated in a number of equivalent ways. One of them is: There is a quantity S, called entropy, which is a state function. In an irreversible process, the entropy of the system and its surroundings increases. In a reversible process, the entropy of the system and its surroundings remains constant. The entropy of the system and its surroundings never decreases. The system and its surroundings are often referred to as the universe.

The mathematical expression for the difference in entropy between states A and B of a system is given by

$$\Delta S = \int_A^B \frac{dq_{\text{rev}}}{T} \tag{1-42}$$

Note that the heat appearing here is that associated with a reversible process. To compute ΔS between two states A and B, we must take the system from A to B in a reversible manner.

Another statement of the second law is: Along any reversible path, there exists an integrating factor T, common to all systems such that

$$dS = \frac{dq_{\text{rev}}}{T} \tag{1-43}$$

is an exact differential, that is, that S is a state function. Thus

$$\Delta S = \int_A^B \frac{dq_{\text{rev}}}{T}$$

For all other processes

$$\Delta S > \int_A^B \frac{dq}{T}$$

where T is the temperature of the surroundings.

The third law of thermodynamics states: If the entropy of each element in some crystalline state be taken as zero at the absolute zero of temperature, every substance has a finite positive entropy, but at the absolute zero of temperature, the entropy may become zero, and does become so in the case of perfect crystalline substances.

The second law is concerned with only the difference in the entropy between two states. The third law allows us to calculate the absolute entropy of a substance by means of the expressions

$$S - S_0 = \int_0^T \frac{dq_{\text{rev}}}{T} \quad \text{and} \quad S_0 = 0 \tag{1-44}$$

Problem 1–36 asks you to calculate the absolute entropy of gaseous nitromethane at its boiling point.

For simple one-component systems, the first law can be written in the form

$$dE = T\,dS - p\,dV \tag{1-45}$$

This implies that

$$\left(\frac{\partial E}{\partial S}\right)_V = T \quad \text{and} \quad \left(\frac{\partial E}{\partial V}\right)_S = -p \tag{1-46}$$

The simplicity of these partial derivatives implies that E is a "natural" function of S and V. For example, if we were to consider E to be a function of V and T (see Problem 1–30), we would get

$$dE = \left[T\left(\frac{\partial p}{\partial T}\right)_V - p\right] dV + C_V\,dT$$

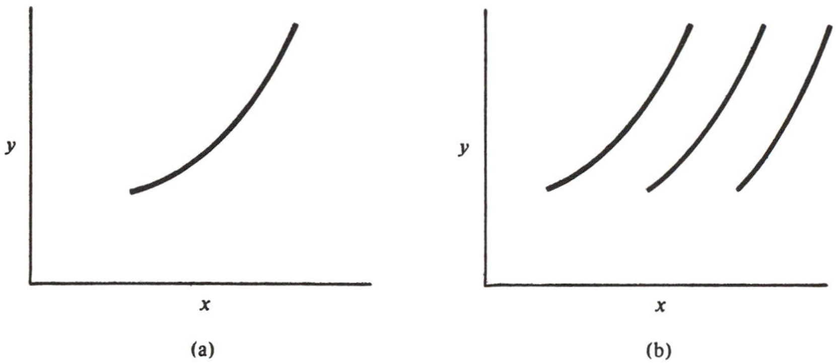

Figure 1–2. **(a) shows the function $y(x)$ and (b) shows a family of functions, all of which give the same value of y for any fixed value of p.**

Note that in this case the coefficients of dV and dT are not as simple as the coefficients of dV and dS obtained when E is expressed as a function of S and V. The "simplicity" of the expression $dE = T\,dS - p\,dV$ suggests that S and V are the "natural" variables for E. The quantities S and V (especially S) are difficult to control in the laboratory and consequently are not always the most desirable independent variables. A more useful pair might be (T, V) or (T, p). An important question that arises, then, is the existence of other thermodynamic state functions whose natural variables are (T, V) or (T, p), and so on. Furthermore, how would one find them if they do exist. This leads us to the topic of Legendre transformations.

We shall discuss a function of one variable in some detail and then simply present the generalization to a function of many variables. Consider a function $y = y(x)$, and let its slope be $p = p(x)$. We wish to describe the function $y(x)$ in terms of its slope. Figure 1–2, however, shows that the slope alone is not sufficient to completely specify $y(x)$. Figure 1–2(a) shows the curve $y(x)$, and Fig. 1–2(b) shows a family of curves, all of which give the same value of y for any one value of p. In order to uniquely describe the curve in Fig. 1–2(a), we must select one member of the family of curves in Fig. 1–2(b). We do this by specifying the intercepts of the tangent lines with the y-axis. Let the intercept be $\phi(p)$. Instead of describing the curve in Fig. 1–2(a) by y versus x, then, we can equally well represent it by specifying the slope at each point along with the intercept of the slope with the y-axis. Figure 1–3 shows these two representations. One sees that either representation can be used to describe the function. The relation between the two representations can be obtained by referring to Fig. 1–4. This figure shows that the slope p at any point is given by

$$p = \frac{y - \phi}{x - 0}$$

The result that we are after is

$$\phi(p) = y - px \tag{1–47}$$

The function $\phi(p)$ is the Legendre transformation of y. It is completely equivalent to $y(x)$, but considers p to be the independent variable instead of x. This may not be clear from the notation in Eq. (1–47), but it is understood there that y and x have been eliminated in favor of p by using the equations $y = y(x)$ and $x = x(p)$.

Let us apply this to the thermodynamic energy $E(S, V)$. We seek a function of T and V that is completely equivalent to E. Equation (1–46) shows that $T = (\partial E/\partial S)_V$,

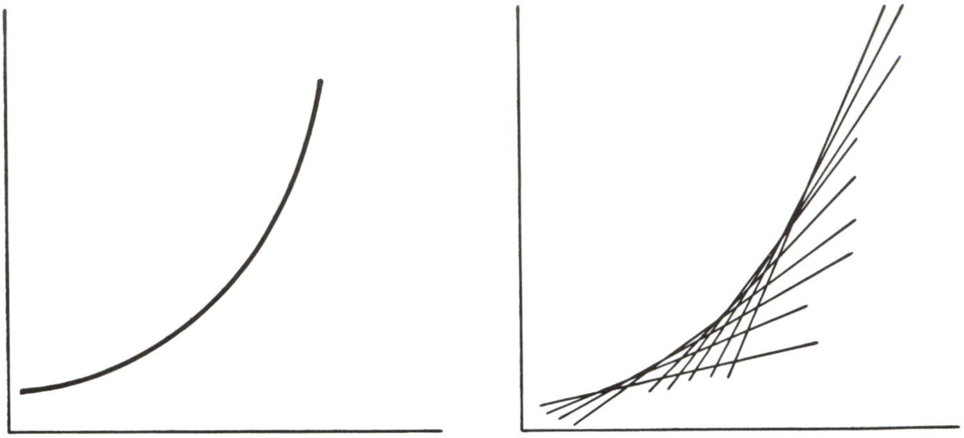

(a) y-x representation (b) ϕ-p representation

Figure 1–3. **In (a), the function is represented by the locus of points, y versus x. In (b), the same function is given by the envelope of its tangent curves.**

and so we are in a position to apply Eq. (1–47) directly. This can be treated as a one-variable problem, since V is held fixed throughout. Therefore the Legendre transformation of E that considers T and V to be the independent variables is $E - TS$. Of course, this is the Helmholtz free energy

$$A(T, V) = E - TS \tag{1–48}$$

whose differential form is

$$dA = -S\,dT - p\,dV \tag{1–49}$$

This shows that the natural variables of A are T and V. Another motivation for saying this is that the condition for equilibrium at constant T and V is that A assume its minimum value, or that $\Delta A \leq 0$ for a spontaneous process at constant T and V. To prove this, write

$$\begin{aligned} dA &= dE - T\,dS - S\,dT \\ &= \delta q - p\,dV - \delta q_{\text{rev}} - S\,dT \\ &= \delta q - \delta q_{\text{rev}} \end{aligned} \tag{1–50}$$

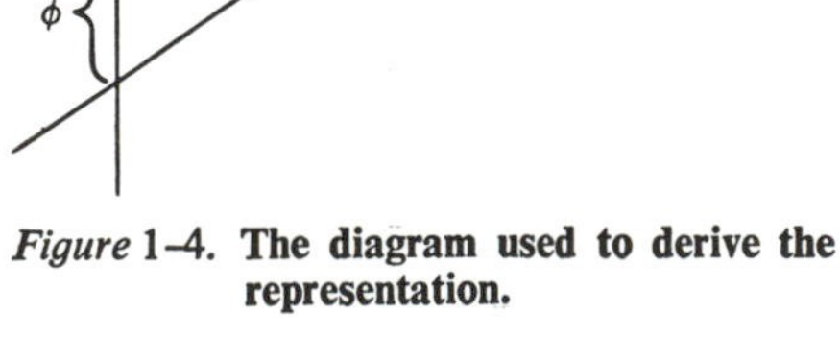

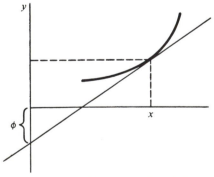

Figure 1–4. **The diagram used to derive the connection between the y–x representation and the ϕ–p representation.**

at constant T and V. But $\delta q \le \delta q_{\text{rev}}$ (see Problem 1–27), and so $\Delta A \le 0$ at constant T and V.

In elementary physical chemistry, the function $A = E - TS$ is often presented as an a priori definition. But it should be apparent now that this form is dictated by the Legendre transformation if one specifies T and V to be the independent variables.

A function, whose natural variables are S and p, can be obtained in the same manner. Equation (1–46) shows that $p = -(\partial E/\partial V)_S$, so Eq. (1–47) gives that $E + pV$ is a thermodynamic state function, whose natural variables are S and p. This function is, of course, the enthalpy.

The generalization of Eq. (1–47) to more than one variable is simply

$$\phi(p) = y - \sum_j p_j x_j \tag{1–51}$$

where the x_j's are the independent variables of y, and $p_j = (\partial y/\partial x_j)$. We can use Eq. (1–51) to construct a thermodynamic state function, whose natural variables are T and p. Using Eqs. (1–46) and Eq. (1–51), we see that such a function is $E - TS + pV$, the Gibb's free energy. Its differential form $dG = -S\,dT + V\,dp$ and the fact that $\Delta G \le 0$ for a spontaneous change at constant T and p suggest that T and p are the natural variables of G.

Up to this point we have considered only closed one-component systems. In general, E, H, A, and G depend upon the number of moles or molecules of each component. If we let N_j be the number of moles of component j, we have

$$dE = T\,dS - p\,dV + \sum_j \left(\frac{\partial E}{\partial N_j}\right)_{S,V,N_K,\,j\neq K} dN_j \tag{1–52}$$

$$= T\,dS - p\,dV + \sum_j \mu_j\,dN_j \tag{1–53}$$

where the second line defines μ_j. By adding $d(pV)$ to both sides of Eq. (1–53), we get

$$dH = T\,dS + V\,dp + \sum_j \mu_j\,dN_j \tag{1–54}$$

If we subtract $d(TS)$ from both sides of Eq. (1–53), we get

$$dA = -S\,dT - p\,dV + \sum_j \mu_j\,dN_j \tag{1–55}$$

Similar manipulations give

$$dG = -S\,dT + V\,dp + \sum_j \mu_j\,dN_j \tag{1–56}$$

Equations (1–52) through (1–56) show that

$$\mu_j = \left(\frac{\partial E}{\partial N_j}\right)_{S,V,\ldots} = \left(\frac{\partial H}{\partial N_j}\right)_{S,p,\ldots} = \left(\frac{\partial A}{\partial N_j}\right)_{V,T,\ldots} = \left(\frac{\partial G}{\partial N_j}\right)_{p,T,\ldots} \tag{1–57}$$

The quantity μ_j is called the chemical potential.

There is a mathematical theorem, called Euler's theorem, which is very useful in thermodynamics. Before discussing Euler's theorem, however, we must define extensive and intensive variables. Extensive properties are additive; their value for the whole system is equal to the sum of their values for the individual parts. Examples are the volume, mass, and entropy. Intensive properties are not additive. Examples are temperature and pressure. The temperature of any small part of a system in equilibrium is the same as the temperature of the whole system. Euler's theorem deals with extensive and intensive variables. If

$$f(\lambda x_1, \lambda x_2, \ldots, \lambda x_N) = \lambda^n f(x_1, x_2, \ldots, x_N) \tag{1-58}$$

f is said to be a homogeneous function of order n. The functions $f(x) = 3x^2$ and $f(x, y, z) = xy^2 + z^3 - 6x^4/y$ are homogeneous functions of degree 2 and 3, respectively, whereas $f(x) = x^2 + 2x - 3$ and $f(x, y) = xy - e^{xy}$ are not homogeneous. Euler's theorem states that if $f(x_1, \ldots, x_N)$ is a homogeneous function of order n, then

$$nf(x_1, \ldots, x_N) = x_1 \frac{\partial f}{\partial x_1} + x_2 \frac{\partial f}{\partial x_2} + \cdots + x_N \frac{\partial f}{\partial x_N} \tag{1-59}$$

The proof of Euler's theorem is simple. Differentiate Eq. (1–58) with respect to λ:

$$n\lambda^{n-1} f(x_1, x_2, \ldots, x_N) = \left(\frac{\partial f}{\partial \lambda x_1}\right)\left(\frac{\partial \lambda x_1}{\partial \lambda}\right) + \left(\frac{\partial f}{\partial \lambda x_2}\right)\left(\frac{\partial \lambda x_2}{\partial \lambda}\right) + \cdots + \left(\frac{\partial f}{\partial \lambda x_N}\right)\left(\frac{\partial \lambda x_N}{\partial \lambda}\right)$$

$$= x_1\left(\frac{\partial f}{\partial \lambda x_1}\right) + x_2\left(\frac{\partial f}{\partial \lambda x_2}\right) + \cdots + x_N\left(\frac{\partial f}{\partial \lambda x_N}\right)$$

Euler's theorem is proved by letting $\lambda = 1$.

Extensive thermodynamic variables are homogeneous of degree 1. Let us apply Euler's theorem to the Gibb's free energy.

$$G(T, p, \lambda N_1, \lambda N_2, \ldots) = \lambda G(T, p, N_1, N_2, \ldots)$$

The variables T and p here can be treated as constants. Equation (1–59) gives that

$$G = \sum_j N_j \left(\frac{\partial G}{\partial N_j}\right)_{T, p, \ldots} = \sum_j N_j \mu_j \tag{1-60}$$

Taking the derivative of this at constant T and p,

$$dG = \sum_j N_j \, d\mu_j + \sum_j \mu_j \, dN_j \qquad \text{(constant } T \text{ and } p\text{)}$$

But using Eq. (1–56) at constant T and p, we have

$$\sum_j N_j \, d\mu_j = 0 \qquad \text{(constant } T \text{ and } p\text{)} \tag{1-61}$$

This is called the Gibbs-Duhem equation and is very useful in the thermodynamic study of solutions. (See *Physical Chemistry*, 4th ed., by W. J. Moore, p. 235, under "Additional Reading," for a simple application of the Gibbs-Duhem equation.)

We shall conclude this section on thermodynamics with a brief discussion of the application of thermodynamics to chemical equilibria. Consider the general reaction

$$\nu_A A + \nu_B B + \cdots \rightleftharpoons \nu_D D + \nu_E E + \cdots \tag{1-62}$$

The capital letters represent the formulas of the compounds, and the ν_j represent stoichiometric coefficients. It is more convenient to write Eq. (1–62) mathematically as

$$\nu_D D + \nu_E E + \cdots - \nu_A A - \nu_B B - \cdots = 0 \tag{1-63}$$

Define the extent of reaction λ, such that $dN_j = \nu_j \, d\lambda$ for all j, where the ν's for products are positive, and those for reactants are negative.

At constant T and p, we have

$$dG = \sum_j \mu_j \, dN_j = \left(\sum_j \mu_j \nu_j\right) d\lambda \qquad \text{(constant } T \text{ and } p\text{)}$$

At equilibrium, G must be a minimum with respect to λ, so we write

$$\sum_j \mu_j \nu_j = \nu_D \mu_D + \nu_E \mu_E + \cdots - \nu_A \mu_A - \nu_B \mu_B - \cdots = 0 \tag{1-64}$$

at equilibrium. The equilibrium between phases can be considered to be a chemical reaction of the form $A \rightleftharpoons B$, and so Eq. (1–64) gives that $\mu_A = \mu_B$ from the equilibrium condition between two pure phases.

Now consider the application of Eq. (1–64) to a chemical reaction between gases dilute enough to be considered ideal. Let the reaction be $\nu_A A + \nu_B B \rightleftharpoons \nu_C C + \nu_D D$. At constant temperature,

$$dG = V\,dp \qquad \text{(constant } T\text{)}$$

and so

$$G - G^0 = \int_{p_0}^{p} V\,dp = \int_{p_0}^{p} \frac{NkT}{p}\,dp = NkT \ln \frac{p}{p_0} \tag{1-65}$$

In this equation G^0 is the standard free energy of the gas, the standard state being the gas at a pressure p_0. Usually p_0 is taken to be 1 atmosphere. If we take N to be 1 mole, then G and G^0 become μ and μ^0. Each component in the reactive gas mixture will have an equation of the form of Eq. (1–65), and so we have

$$\mu_j(T, p) = \mu_j{}^0(T) + RT \ln \frac{p_j}{p_{0j}} \tag{1-66}$$

The total free energy change is

$$\begin{aligned} \Delta\mu &= \nu_C \mu_C + \nu_D \mu_D - \nu_A \mu_A - \nu_B \mu_B \\ &= \Delta\mu^0 + RT \ln \frac{(p_C')^{\nu_C}(p_D')^{\nu_D}}{(p_A')^{\nu_A}(p_B')^{\nu_B}} \end{aligned} \tag{1-67}$$

In this equation the (p')'s are p/p_0, that is, they are the pressures relative to the standard states. These (p')'s are unitless. The argument of the logarithm here has the *form* of an equilibrium constant, but is not equal to the equilibrium constant unless the pressures are those which exist at chemical equilibrium. Equation (1–67) gives the change in free energy of the conversion of reactants at *arbitrary* pressures to products at *arbitrary* pressures.

At equilibrium, $\Delta\mu = 0$, and we have

$$\begin{aligned} \Delta\mu^0 &= -RT \ln \left[\frac{(p_C')^{\nu_C}(p_D')^{\nu_D}}{(p_A')^{\nu_A}(p_B')^{\nu_B}}\right]_{\text{equilibrium}} \\ &= -RT \ln K_p \end{aligned} \tag{1-68}$$

There are extensive tabulations of μ^0's, and so $\Delta\mu^0$ is a simple matter to calculate. We see that if $\Delta\mu^0 < 0$, then $K_p > 1$, that is, the conversion of reactants *in their standard states* to products *in their standard states* proceeds spontaneously. On the other hand, if $\Delta\mu^0 > 0$, then $K_p < 1$, and we conclude that the reaction does not proceed spontaneously. It should be understood, however, that this applies only to reactants and products in their standard states. In general, it is $\Delta\mu$ along with Eq. (1–67) that determines the extent of a chemical reaction. (See Problem 1–34.)

1-5 MATHEMATICS

In this section we shall discuss several mathematical techniques or results that are repeatedly used in statistical thermodynamics. The topics we shall discuss here are random variables and distribution functions, Stirling's approximation, the binomial and multinomial coefficients, the Lagrange method of undetermined multipliers, and the behavior of binomial and multinomial coefficients for large numbers.

PROBABILITY DISTRIBUTIONS

Let u be a variable which can assume the M discrete values $u_1, u_2, \ldots, u_M$ with corresponding probabilities $p(u_1), p(u_2), \ldots, p(u_M)$. The variable u is said to be a discrete random variable, and $p(u)$ is said to be a discrete distribution. The mean, or average, value of u is

$$\bar{u} = \frac{\sum_{j=1}^{M} u_j\, p(u_j)}{\sum_{j=1}^{M} p(u_j)}$$

Since $p(u_j)$ is a probability, $p(u_j)$ must be normalized, that is, the summation in the denominator must equal unity. The mean of any function of u, $f(u)$, is given by

$$\overline{f(u)} = \sum_{j=1}^{M} f(u_j)p(u_j) \tag{1-69}$$

If $f(u) = u^m$, $\overline{f(u)}$ is called the mth moment of the distribution $p(u)$. If $f(u) = (u - \bar{u})^m$, $\overline{f(u)}$ is called the mth central moment, that is, the mth moment about the mean. In particular, the mean of $(u - \bar{u})^2$ is called the variance, and is a measure of the spread of the distribution. The square root of the variance is the standard deviation.

A very commonly occurring and useful discrete distribution is the Poisson distribution:

$$P(m) = \frac{a^m e^{-a}}{m!} \qquad m = 0, 1, 2, \ldots \tag{1-70}$$

This distribution has been applied to shot noise in electron tubes, the distribution of galaxies in space, aerial search, and many others.* (See Problem 1–42.)

If the random variable U is continuous rather than discrete, then we interpret $p(u)\,du$ as the probability that the random variable U lies between the values u and $u + du$. The mean of any function of U is

$$\overline{f(u)} = \int f(u)p(u)\,du \tag{1-71}$$

The limits of the integral are over the entire range of U.

The most important continuous probability distribution is the Gaussian distribution:

$$p(x) = \frac{1}{(2\pi\sigma^2)^{1/2}} \exp\left\{-\frac{(x - \bar{x})^2}{2\sigma^2}\right\} \qquad -\infty \leq x \leq \infty \tag{1-72}$$

The quantity σ^2, which is the variance, controls the width of the Gaussian distribution. The smaller the σ, the narrower the Gaussian distribution becomes. In the limit $\sigma \to 0$, Eq. (1–72) becomes a delta function (this is one representation of a delta function of Appendix B). Problems 1–43 through 1–45 involve some important results based on Eq. (1–72).

* See *Modern Probability Theory and Its Applications* by E. Parzen (New York: Wiley, 1960).

STIRLING'S APPROXIMATION

In statistical thermodynamics we often encounter factorials of very large numbers, such as Avogadro's number. The calculation and mathematical manipulation of factorials become awkward for large N. Therefore it is desirable to find an approximation for $N!$ for large N. Problems of this sort occur often in mathematics and are called asymptotic approximations, that is, an approximation to a function which improves as the argument of that function increases. Since $N!$ is actually a product, it is convenient to deal with $\ln N!$ because this is a sum. The asymptotic approximation to $\ln N!$ is called Stirling's approximation, which we now derive.

Since $N! = N(N-1)(N-2)\cdots(2)(1)$, $\ln N!$ is

$$\ln N! = \sum_{m=1}^{N} \ln m \tag{1-73}$$

Figure 1–5 shows $\ln x$ plotted versus x. The sum of the areas under these rectangles up to N is $\ln N!$. Figure 1–5 also shows the continuous curve $\ln x$ plotted on the same graph. Thus $\ln x$ is seen to form an envelope to the rectangles, and this envelope becomes a steadily smoother approximation to the rectangles as x increases. We can approximate the area under these rectangles by the integral of $\ln x$. The area under $\ln x$ will poorly approximate the rectangles only in the beginning. If N is large enough (we are deriving an asymptotic expansion), this area will make a negligible contribution to the total area. We may write, then,

$$\ln N! = \sum_{m=1}^{N} \ln m \approx \int_1^N \ln x\, dx = N \ln N - N \qquad (N \text{ large}) \tag{1-74}$$

which is Stirling's approximation to $\ln N!$. The lower limit could just as well have been taken as 0 in Eq. (1–74), since N is large. (Remember that $x \ln x \to 0$ as $x \to 0$.)

A more refined derivation of Stirling's approximation gives $\ln N! \approx N \ln N - N + \ln(2\pi N)^{1/2}$, but this additional term is seldom necessary. (See Problem 1–59.)

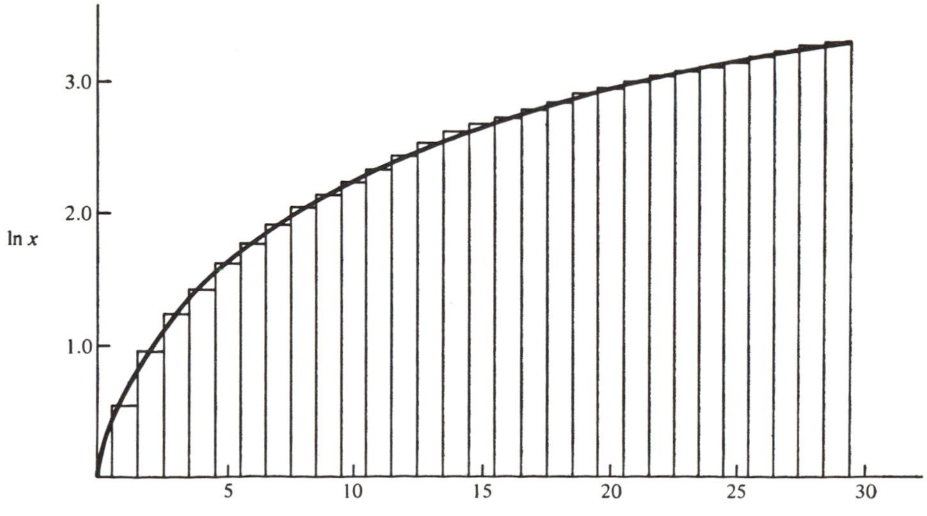

Figure 1–5. **A plot of $\ln x$ versus x, showing how the summation of $\ln m$ can be approximated by the integral of $\ln x$.**

BINOMIAL AND MULTINOMIAL DISTRIBUTION

During the course of our discussion of the canonical ensemble, we shall encounter the problem of determining how many ways it is possible to divide N distinguishable systems into groups such that there are n_1 systems in the first group, n_2 systems in the second group, and so on, and such that $n_1 + n_2 + \cdots = N$, that is, all the systems are accounted for. This is actually one of the easiest problems in combinatorial analysis. To solve this, we first calculate the number of permutations of N distinguishable objects, that is, the number of possible different arrangements or ways to order N distinguishable objects. Let us choose one of the N objects and place it in the first position, one of the $N-1$ remaining objects and place it in the second position, and so on, until all N objects are ordered. Clearly there are N choices for the first position, $N-1$ choices for the second position, and so on, until finally there is only one object left for the Nth position. The total number of ways of doing this is then the product of all the choices,

$$N(N-1)(N-2)\cdots(2)(1) \equiv N! \qquad \text{(distinguishable objects)}$$

Next we calculate the number of ways of dividing N distinguishable objects into two groups, one group containing N_1 objects, say, and the other containing the remaining $N - N_1$. There are $N(N-1)\cdots(N - N_1 + 1)$ ways to form the first group, and $N_2! = (N - N_1)!$ ways to form the second group. The total number is, then, the product

$$N(N-1)\cdots(N-N_1+1) \times (N-N_1)! = \frac{N!}{(N-N_1)!} \times (N-N_1)! = N!$$

But this has overcounted the situation drastically, since the order in which we place N_1 members in the first group and N_2 in the second group is immaterial to the problem as stated. All $N_1!$ orders of the first group and $N_2!$ orders of the second group correspond to just one division of N objects into N_1 objects and N_2 objects. Therefore the desired result is

$$\frac{N!}{N_1!(N-N_1)!} = \frac{N!}{N_1!N_2!} \tag{1–75}$$

Since the combination of factorials in Eq. (1–75) occurs in the binomial expansion,

$$(x+y)^N = \sum_{N_1=0}^{N} \frac{N!x^{N-N_1}y^{N_1}}{N_1!(N-N_1)!} = \sum_{N_1 N_2}^{*} \frac{N!x^{N_1}y^{N_2}}{N_1!N_2!} \tag{1–76}$$

$N!/N_1!(N-N_1)!$ is called a binomial coefficient. The asterisk on the second summation in Eq. (1–76) signifies the restriction $N_1 + N_2 = N$.

The generalization of Eq. (1–75) to the division of N into r groups, the first containing N_1, and so on, is easily seen to be

$$\frac{N!}{N_1!N_2!\cdots N_r!} = \frac{N!}{\prod_{j=1}^{r} N_j!} \tag{1–77}$$

where $N_1 + N_2 + \cdots + N_r = N$. This is known as a multinomial coefficient, since it occurs in the expansion

$$(x_1 + x_2 + \cdots + x_r)^N = \sum_{N_1=0}^{N} \sum_{N_2=0}^{N} \cdots \sum_{N_r=0}^{N}{}^{*} \frac{N!x_1^{N_1}\cdots x_r^{N_r}}{\prod_{j=1}^{r} N_j!} \tag{1–78}$$

where this time the asterisk signifies the restriction $N_1 + N_2 + \cdots + N_r = N$.

There are a number of other combinatorial formulas that are useful in statistical thermodynamics, but Eq. (1–77) is the most useful for our purposes. Combinatorial formulas can become rather demanding to derive. We refer to Appendix AVII of Mayer and Mayer* which contains a collection of formulas.

METHOD OF LAGRANGE MULTIPLIERS

It will be necessary, later, to maximize Eq. (1–77) with the constraint $N_1 + N_2 + \cdots + N_r = \text{constant}$. This brings us to the mathematical problem of maximizing a function of several (or many) variables $f(x_1, x_2, \ldots, x_r)$ when the variables are connected by other equations, say $g_1(x_1, \ldots, x_r) = 0$, $g_2(x_1, \ldots, x_r) = 0$, and so on. This type of problem is readily handled by the method of Lagrange undetermined multipliers.

If it were not for the constraints, $g_j(x_1, x_2, \ldots, x_r) = 0$, the maximum of $f(x_1, \ldots, x_r)$ would be given by

$$\delta f = \sum_{j=1}^{r} \left(\frac{\partial f}{\partial x_j}\right)_0 \delta x_j = 0 \tag{1–79}$$

where the zero subscript indicates that this equation equals zero only when the r partial derivatives are evaluated at the maximum (or minimum) of f. Denote these values of x_j by x_j^0. If there were no constraints, each of the δx_j would be able to be varied independently and arbitrarily, and so we would conclude that $(\partial f/\partial x_j) = 0$ for every j, since δf must equal zero. This would give r equations from which the values of the $r x_j^0$ could be obtained.

On the other hand, if there is some other relation between the x's, such as $g(x_1, x_2, \ldots, x_r) = 0$, we have the additional equation

$$\delta g = \sum_{j=1}^{r} \left(\frac{\partial g}{\partial x_j}\right)_0 \delta x_j = 0 \tag{1–80}$$

This equation serves as a constraint that the δx_j must satisfy, thus making one of them depend upon the other $r - 1$. In the Lagrange method, one multiplies Eq. (1–80) by some parameter, say λ, and adds the result to Eq. (1–79) to get

$$\sum_{j=1}^{r} \left(\frac{\partial f}{\partial x_j} - \lambda \frac{\partial g}{\partial x_j}\right)_0 \delta x_j = 0 \tag{1–81}$$

The δx_j are still not independent, because of Eq. (1–80), and so they cannot be varied independently. Equation (1–80), however, can be treated as an equation giving one of the δx_j in terms of the other $r - 1$ independent ones. Pick any one of the r δx_j as the dependent one. Let this be δx_μ.

The trick now is that we have not specified λ yet. We set it equal to $(\partial f/\partial x_\mu)_0/(\partial g/\partial x_\mu)_0$, making the coefficient of δx_μ in Eq. (1–81) vanish. The subscript zero here indicates that $(\partial f/\partial x_\mu)$ and $(\partial g/\partial x_\mu)$ are to be evaluated at values of the x_j such that f is at its maximum (or minimum) under the constraint of Eq. (1–80). Of course, we do not know these values of x_j yet, but we can nevertheless formally define λ in this manner. This leaves a sum of terms in Eq. (1–81) involving only the independent δx_j, which can be varied independently, yielding that

$$\left(\frac{\partial f}{\partial x_j}\right)_0 - \lambda \left(\frac{\partial g}{\partial x_j}\right)_0 = 0 \qquad j = 1, 2, \ldots, \mu - 1, \mu + 1, \ldots, r$$

* See Mayer and Mayer, *Statistical Mechanics* (New York: Wiley, 1940).

If we combine these $r-1$ equations with our choice for λ, we have

$$\left(\frac{\partial f}{\partial x_j}\right)_0 - \lambda\left(\frac{\partial g}{\partial x_j}\right)_0 = 0 \tag{1-82}$$

for all j.

As we said above, the choice of λ here is certainly formal, since both $(\partial f/\partial x_\mu)_0$ and $(\partial g/\partial x_\mu)_0$ must be evaluated at these values of x_j which maximizes f, but these are known from Eq. (1–82) only in terms of λ. But this presents no difficulty, since in practice λ is determined by physical requirements. Examples of this will occur in the next two chapters.

Lagrange's method becomes no more difficult in the case in which there are several constraints. Let $g_1(x_1, \ldots, x_r), g_2(x_1, \ldots, x_r), \ldots$ be a set of constraints. We introduce a Lagrange multiplier for each $g_i(x_1, \ldots, x_r)$ and proceed as above to get

$$\frac{\partial f}{\partial x_j} - \lambda_1 \frac{\partial g_1}{\partial x_j} - \lambda_2 \frac{\partial g_2}{\partial x_j} - \cdots = 0 \tag{1-83}$$

BINOMIAL DISTRIBUTION FOR LARGE NUMBERS

Lastly, there is one other mathematical observation we need here in order to facilitate the discussion in the next chapter. This observation concerns the shape of the multinomial coefficient [Eq. (1–78)] as a function of the N_j's, as the N_j's become very large. To simplify notation, we shall consider only the binomial coefficient, but this will not affect our conclusions. Let us first find the value of N_1 for which $f(N_1) = N!/N_1!(N-N_1)!$ reaches its maximum value. Since N_1 and N are both very large, we treat them as continuous variables. Also since $\ln x$ is a monotonic function of x, we can maximize $f(N_1)$ by maximizing $\ln f(N_1)$. This allows us to use Stirling's approximation. The maximum of $f(N_1)$ is found, then, from

$$\frac{d \ln f(N_1)}{dN_1} = 0$$

to be located at $N_1^* = N/2$. Let us now expand $\ln f(N_1)$ about this point. The Taylor expansion is

$$\ln f(N_1) = \ln f(N_1^*) + \frac{1}{2}\left(\frac{d^2 \ln f(N_1)}{dN_1^{\,2}}\right)_{N_1=N_1^*} (N_1 - N_1^*)^2 + \cdots \tag{1-84}$$

The linear term in $N_1 - N_1^*$ is missing, because the first derivative of $\ln f(N)$ is zero at $N_1 = N_1^*$. The second derivative appearing in Eq. (1–84) is equal to $-4/N$. Thus if we ignore higher-order terms (see Problem 1–53), Eq. (1–84) can be written in the form of a Gaussian curve

$$f(N_1) = f(N_1^*) \exp\left\{-\frac{2(N_1 - N_1^*)^2}{N}\right\} \tag{1-85}$$

Comparison of this with the standard form of the Gaussian function

$$f(x) = \frac{1}{(2\pi\sigma^2)^{1/2}} \exp\left\{-\frac{(x - x^*)^2}{2\sigma^2}\right\} \tag{1-86}$$

shows that the standard deviation is of the order of $N^{1/2}$. Equation (1–85) is, therefore, a bell-shaped function, centered at $N_1^* = N/2$ and having a width of a few multiples of $N^{1/2}$. Problem 1–43 establishes the well-known fact that a Gaussian function goes essentially to zero when x differs from x^* by a few σ's. Since we are interested only in large values of N_1 (or N), say numbers of the order of 10^{20}, we have a bell-shaped curve that is contained between $10^{20} \pm$ a few multiples of 10^{10}, which, if plotted, would for all practical purposes look like a delta function centered at $N_1^* = N/2$. Thus we have shown that the binomial coefficient peaks very strongly at the point $N_1 = N_2 = N/2$. This same behavior occurs for a multinomial coefficient as well. If there are s N_j's, the multinomial coefficient has a very sharp maximum at the point $N_1 = N_2 = \cdots = N_s = N/s$. (See Problem 1–50.) This peak becomes sharper as the N_j's become larger, and become a delta function in the limit $N_j \to \infty$ for all j.

MAXIMUM TERM METHOD

Another important result, which is a consequence of the large numbers encountered in statistical mechanics, is the *maximum-term method.* It says that under appropriate conditions the logarithm of a summation is essentially equal to the logarithm of the maximum term in the summation. To see how this goes, consider the sum

$$S = \sum_{N=1}^{M} T_N$$

where $T_N > 0$ for all N. Since all the terms are positive, the value of S must be greater than the value of the largest term, say $T_{\max}$, and less than the product of the number of terms and the value of the largest term. Thus we can write

$$T_{\max} \le S \le M T_{\max}$$

Taking logarithms gives

$$\ln T_{\max} \le \ln S \le \ln T_{\max} + \ln M$$

We shall see that it is often the case in statistical mechanics that $T_{\max}$ will be $O(e^M)$. Thus we have

$$O(M) \le \ln S \le O(M) + \ln M$$

For large M, $\ln M$ is negligible with respect to M itself, and so we see that $\ln S$ is bounded from above and below by $\ln T_{\max}$, and so

$$\ln S = \ln T_{\max}$$

This is a rather remarkable theorem, and like a number of other theorems used in statistical mechanics, its validity results from the large numbers involved.

ADDITIONAL READING

1–2

DAVIDSON, N. 1962. *Statistical mechanics.* New York: McGraw-Hill. Chapter 2.
GOLDSTEIN, H. 1950. *Classical mechanics.* Reading, Mass.: Addison-Wesley.
SYMON, K. R. 1960. *Mechanics.* Reading, Mass.: Addison-Wesley.
TOLMAN, R. C. 1938. *Statistical mechanics.* London: Oxford University Press. Chapters 1 and 2.

1–3

HAMEKA, H. F. 1967. *Introduction to quantum theory.* New York: Harper & Row.
HANNA, M. W. 1969. *Quantum mechanics in chemistry,* 2nd ed. New York: Benjamin.
HILL, T. L. 1956. *Statistical mechanics.* New York: McGraw-Hill. Section 8.
KARPLUS, M., and PORTER, R. N. 1970. *Atoms and molecules.* New York: Benjamin.
KESTIN, J., and DORFMAN, J. R. 1971. *A course in statistical thermodynamics.* New York: Academic. Chapter 3.
PILAR, F. L. 1968. *Elementary quantum chemistry.* New York: McGraw-Hill.
TOLMAN, R. C. 1938. *Statistical mechanics.* London: Oxford University Press. Chapters 7 and 8.

1–4

ANDREWS, F. C. 1971. *Thermodynamics*. New York: Wiley.
CALLEN, H. B. 1960. *Thermodynamics*. New York: Wiley.
MAHAN, B. H. 1963. *Elementary chemical thermodynamics*. New York: Benjamin.
MOORE, W. J. 1972. *Physical chemistry*, 4th ed. Englewood Cliffs, N.J.: Prentice-Hall.
REISS, H. 1965. *Methods of thermodynamics*. Boston, Mass.: Blaisdell.
WASER, J. 1966. *Basic chemical thermodynamics*. New York: Benjamin.

1–5

ABRAMOWITZ, M., and STEGUN, I. A. "Handbook of Mathematical Functions," *Natl. Bur. Stan. Appl. Math. Series*, **55**, 1964.
ARFKEN, G., 1970. *Mathematical methods for physicists*, 2nd ed. New York: Academic.
KESTIN, J., and DORFMAN, J. R. 1971. *A course in statistical thermodynamics*. New York: Academic. Chapter 4.
KREYSZIG, E. 1962. *Advanced engineering mathematics*, 2nd ed. New York, Wiley.
REIF, F. 1965. *Statistical and thermal physics*. New York: McGraw-Hill, Chapter 1.

PROBLEMS

1–1. Solve the equation of motion of a body of mass m dropped from a height h. Assume that there exists a viscous drag on the body that is proportional to and in the opposite direction to the velocity of the body. (Let the proportionality constant be γ.) Solve for the so-called terminal velocity, that is, the limiting velocity as $t \to \infty$.

1–2. Calculate the trajectory of a shell shot out of a cannon with velocity v_0, assuming no aerodynamic resistance and that the cannon makes an angle θ with the horizontal axis.

1–3. Remembering that the potential energy is given by

$$V(x) = -\int_0^x F(\xi)\, d\xi = \tfrac{1}{2}kx^2$$

for a simple harmonic oscillator, derive an expression for the total energy as a function of time. Discuss how the kinetic and potential energy behave as a function of time.

1–4. Solve the equation for a harmonic oscillator of mass m and force constant k that is driven by an external force of the form $F(t) = F_0 \cos \omega_0 t$.

1–5. Show that

$$\xi(t) = A \sin \omega t + B \cos \omega t$$

can be written as

$$\xi(t) = C \sin(\omega t + \phi)$$

1–6. Show that the total linear momentum is conserved for a system of N particles with an interaction potential which depends only on the distance between particles.

1–7. When does $p = \partial L/\partial \dot{q}$ but $\neq \partial K/\partial \dot{q}$?

1–8. Consider a system of two-point particles with masses m_1 and m_2 moving in two dimensions. It is very common for their potential of interaction to depend upon their relative coordinates $(x_1 - x_2, y_1 - y_2)$ only. Thus the total energy is

$$E = \frac{m_1}{2}(\dot{x}_1^2 + \dot{y}_1^2) + \frac{m_2}{2}(\dot{x}_2^2 + \dot{y}_2^2) + U(x_1 - x_2, y_1 - y_2)$$

Now introduce four new variables

$$X = \frac{m_1 x_1 + m_2 x_2}{m_1 + m_2} \qquad Y = \frac{m_1 y_1 + m_2 y_2}{m_1 + m_2}$$

$$x_{12} = x_1 - x_2 \qquad y_{12} = y_1 - y_2$$

and show that this two-body problem can be reduced to two one-body problems, one involving the center of mass of the system and one involving the *relative* motion of the two particles.

Give a physical interpretation of the ratio $m_1 m_2/(m_1 + m_2)$ that arises naturally in the relative motion. What is this quantity called? This result is easily extended to three dimensions.

1–9. Extend the development of Problem 1–8 to the case in which each particle also experiences an external potential energy, say $U(x_1, y_1, z_1)$ and $U(x_2, y_2, z_2)$. Interpret the resulting equations.

1–10. Derive Lagrange's equations for a particle moving in two dimensions under a central potential $u(r)$. Which of these equations illustrates the law of conservation of angular momentum? Is angular momentum conserved if the potential depends upon θ as well?

1–11. For a particle moving in three dimensions under the influence of a spherically symmetrical potential $U = U(r)$, write down the Lagrangian and the equations of motion in spherical coordinates (r, θ, ϕ). Show that $H = K + V$ from

$$H = \sum p_i \dot{q}_i - L$$

for this potential.

1–12. Solve the equation of motion of two masses m_1 and m_2 connected by a harmonic spring with force constant k.

1–13. Start with Lagrange's equations in Cartesian coordinates, that is,

$$\frac{d}{dt}\left(\frac{\partial L}{\partial \dot{x}}\right) = \frac{\partial L}{\partial x}$$

and so on. Now introduce three generalized coordinates q_1, q_2, and q_3 which are related to the Cartesian coordinates by $x = x(q_1, q_2, q_3)$, and so on. Show that by transforming Lagrange's equations from $x, \dot{x}, y, \dot{y}, z$, and $\dot{z}$ as independent variables to $q_1, \dot{q}_1, q_2, \dot{q}_2, q_3$, and $\dot{q}_3$ we get

$$\frac{d}{dt}\left(\frac{\partial L}{\partial \dot{q}_1}\right) = \frac{\partial L}{\partial q_1}$$

and so on.

1–14. If H, the classical Hamiltonian, does not depend explicitly on time, show that $dH/dt = 0$. What does this mean physically? Is this true if H does depend explicitly upon time?

1–15. Consider the rotation of a diatomic molecule with a fixed internuclear separation l and masses m_1 and m_2. By employing center of mass and relative coordinates, show that the rotational kinetic energy can be written in spherical coordinates as

$$\tfrac{1}{2}I(\dot{\theta}^2 + \dot{\phi}^2 \sin^2 \theta)$$

and from this derive the rotational Hamiltonian

$$H_{rot} = \frac{1}{2I}\left(p_\theta^{\,2} + \frac{p_\phi^{\,2}}{\sin^2 \theta}\right)$$

In these equations, $I = \mu l^2$, where μ is the reduced mass. This Hamiltonian is useful for studying the rotation of diatomic molecules.

1–16. Show that the motion of a particle under a central force law takes place entirely in a single plane.

1–17. What is the expectation (average) value for the linear momentum p_x of a particle in a one-dimensional box $p_x^{\,2}$? Briefly discuss your results.

1–18. Show that the energy eigenvalues of a free particle confined to a cube of length a are given by

$$\varepsilon = \frac{h^2}{8ma^2}(n_x^{\,2} + n_y^{\,2} + n_z^{\,2}) \qquad n_x, n_y, n_z = 1, 2, \ldots$$

1–19. Show that the energy eigenvalues of a free particle confined to a rectangular parallepiped of lengths a, b, and c are given by

$$\varepsilon = \frac{h^2}{8m}\left(\frac{n_x^2}{a^2} + \frac{n_y^2}{b^2} + \frac{n_z^2}{c^2}\right) \qquad n_x, n_y, n_z = 1, 2, \ldots$$

1–20. Calculate the energy eigenvalues of a particle confined to a ring of radius a.

1–21. Show that the Hamiltonian operator of a rigid rotor is given by Eq. (1–32).

1–22. Calculate the degeneracy of the first few levels of a free particle confined to a cube of length a.

1–23. Verify the calculation that follows Eq. (1–37) which shows that the quantum mechanical degeneracy of a macroscopic system is $O(10^N)$.

1–24. We need to know the volume of an N-dimensional sphere in order to derive Eq. (1–36). This can be determined by the following device. Consider the integral

$$I = \int_{-\infty}^{\infty} \cdots \int e^{-(x_1^2 + x_2^2 + \cdots + x_N^2)}\, dx_1\, dx_2 \cdots dx_N$$

First show that $I = \pi^{N/2}$. Now one can formally transform the volume element $dx_1\, dx_2 \cdots dx_N$ to N-dimensional spherical (hyperspherical) coordinates to get

$$\int_{\text{angles}} dx_1\, dx_2 \cdots dx_N \to r^{N-1} S_N\, dr$$

where S_N is the factor that arises upon integration over the angles. Show that $S_2 = 2\pi$ and $S_3 = 4\pi$. S_N can be determined for any N by writing I in hyperspherical coordinates:

$$I = \int_0^{\infty} e^{-r^2} r^{N-1} S_N\, dr$$

Show that $I = S_N \Gamma(N/2)/2$, where $\Gamma(x)$ is the gamma function (see Problem 1–58). Equate these two values for I to get

$$S_N = \frac{2\pi^{N/2}}{\Gamma(N/2)}$$

Show that this reduces correctly for $N = 2$ and 3. Lastly now, convince yourself that the volume of an N-dimensional sphere of radius a is given by

$$V_N = \int_0^a S_N r^{N-1}\, dr = \frac{\pi^{N/2}}{\Gamma\left(\frac{N}{2} + 1\right)} a^N$$

and show that this reduces correctly for $N = 2$ and 3.

1–25. Derive an expression for the density of translational quantum states for a two-dimensional ideal gas.

1–26. Prove that a many-body wave function must be either symmetric or antisymmetric under the interchange of any two particles. Hint: Apply the exchange operation twice.

1–27. Show for an isothermal process that $w_{\text{rev}} > w_{\text{irrev}}$ and $q_{\text{rev}} > q_{\text{irrev}}$.

1–28. Derive the thermodynamic equation

$$C_p - C_V = \left[p + \left(\frac{\partial E}{\partial V}\right)_T\right]\left(\frac{\partial V}{\partial T}\right)_p$$

and evaluate this difference for an ideal gas and a gas that obeys the van der Waals equation.

1–29. Derive the thermodynamic equation of state

$$\left(\frac{\partial E}{\partial V}\right)_T - T\left(\frac{\partial p}{\partial T}\right)_V = -p$$

1–30. Derive the equation

$$dE = \left[T\left(\frac{\partial p}{\partial T}\right)_V - p\right] dV + C_V\, dT$$

and from this show that $(\partial E/\partial V)_T = a/V^2$ for a van der Waals gas.

1–31. Show that

$$\left(\frac{\partial E}{\partial V}\right)_{\mu/T,\,1/T} + \frac{1}{T}\left(\frac{\partial p}{\partial(1/T)}\right)_{\mu/T,\,V} = -p$$

1–32. Derive an expression for $\partial \ln K/\partial T$ in terms of ΔH, the heat of reaction, and in terms of C_p, the heat capacity at constant pressure.

1–33. Consider the "water-gas" reaction

$$CO + H_2O(g) \rightarrow H_2 + CO_2$$

where

$$K_p = \frac{P_{H_2} P_{CO_2}}{P_{CO} P_{H_2O}}$$

and given the following data:

Substance	(kcal/mole)	a	$b \times 10^3$	$c \times 10^7$	ΔH^0_{298}(kcal/m)
CO	−32.81	6.42	1.67	1.96	−26.4157
$H_2O(g)$	−54.64	7.26	2.30	2.83	−57.7979
CO_2	−94.26	6.21	10.40	−35.45	−94.0518
H_2	0.00	6.95	−0.20	4.81	0.00

where the heat capacity of the gases in cal deg^{-1} $mole^{-1}$ is given by

$$C_p = a + bT + cT^2$$

Calculate K_p at 298°K and 800°K.

1–34. Calculate the free energy change at 700°C for the conversion of carbon monoxide at 10 atm and water vapor at 5 atm to carbon dioxide and hydrogen at partial pressures of 1.5 atm each. The equilibrium constant K_p for this reaction is 0.71. Is this process theoretically feasible?

1–35. It is illustrated in Chapter 17 that the speed of sound c_0 propagated through a gas is

$$c_0 = (m\rho\kappa_S)^{-1/2}$$

where κ_S is the adiabatic compressibility

$$\kappa_S = -\frac{1}{V}\left(\frac{\partial V}{\partial p}\right)_S$$

Show that this is equivalent to

$$c_0 = V\left\{-\frac{\gamma}{M}\left(\frac{\partial p}{\partial V}\right)_T\right\}^{1/2}$$

where $\gamma = C_p/C_V$, and M is the molecular weight of the gas. Using the above result, show that

$$c_0 = \left(\gamma \frac{RT}{M}\right)^{1/2}$$

for an ideal gas.

1–36. Jones and Giauque obtained the following values for C_p of nitromethane.

°K	15	20	30	40	50	60	70	80	90	100
C_p	0.89	2.07	4.59	6.90	8.53	9.76	10.70	11.47	12.10	12.62
°K	120	140	160	180	200	220	240	260	280	300
C_p	13.56	14.45	15.31	16.19	17.08	17.98	18.88	25.01	25.17	25.35

The melting point is 244.7°K, heat of fusion 2319 cal/mole. The vapor pressure of the liquid at 298.1°K is 3.666 cm. The heat of vaporization at 298.0°K is 9147 cal/mole. Calculate the third-law entropy of CH_2NO_2 gas at 298.1°K and 1 atm pressure (assuming ideal gas behavior).

1–37. Derive the Legendre transformation of E in which $S \to T$ and $N \to \mu$.

1–38. Apply a Legendre transformation to the Lagrangian $L(q_j, \dot{q}_j)$ to eliminate the generalized velocities in favor of generalized momenta, defined by $p_j \equiv \partial L/\partial \dot{q}_j$. What function does this turn out to be?

1–39. Find the natural function of V, E, and μ. Hint: Start with the natural function of V, E, and N, namely, S, and transform $N \to \mu$.

1–40. Derive the Legendre transformation of E in which $S \to T$, $N \to \mu$, and $V \to p$. What peculiar thing happens when all the extensive variables are transformed out?

1–41. Show that $\overline{(x-\bar{x})^2} = \overline{x^2} - \bar{x}^2$.

1–42. Show that the Poisson distribution $P(m) = a^m e^{-a}/m!$ is normalized. Calculate $\bar{m}$ and the variance. What is the significance of the parameter a?

1–43. Sketch the Gaussian distribution as σ (or even $\sigma/\bar{x}$) becomes smaller and smaller. To what type of distribution does a Gaussian go in the limit $\sigma \to 0$. Discuss the meaning of this distribution.

1–44. For the Gaussian distribution $p(x)$ show that

(a)

$$\int_{-\infty}^{\infty} p(x)\, dx = 1$$

(b) Calculate the nth central moment where $n = 0$, 1, 2, and 3.

(c) In the limit $\sigma \to 0$ what kind of distribution is approached where

$$p(x) = \frac{1}{\sigma\sqrt{2\pi}} \exp\left(-\frac{(x-\bar{x})^2}{2\sigma^2}\right)$$

1–45. The quantity $\overline{(x-\bar{x})^j}$ is called the jth central moment. Show that all odd central moments of a Gaussian vanish. What about the even ones? Relate the $j = 2$ central moment to the parameter σ.

1–46. Let $f(x, y)$ be a joint probability density, that is, $f(x, y)\, dx\, dy$ is the probability that X lies between x and $x + dx$ *and* Y lies between y and $y + dy$. If X and Y are *independent*, then

$$f(x, y)\, dx\, dy = f_1(x) f_2(y)\, dx\, dy$$

If X and Y are independent, show that the mean and variance of their sum is equal to the sum of the means and variances, respectively, of X and Y; that is, show that if $W = X + Y$, then

$$\bar{W} = \bar{X} + \bar{Y}$$

$$\overline{(W - \bar{W})^2} = \overline{(X - \bar{X})^2} + \overline{(Y - \bar{Y}^2)}$$

1–47. Let X be a random variable on the positive numbers, $0 \le x < \infty$, and let $p(x)$ be its probability density function. The function $\phi(s)$ defined by

$$\phi(s) = \int_0^{\infty} e^{-sx} p(x)\, dx$$

is called the characteristic function of $p(x)$. Find the relation between $\phi(s)$ and the moments of $p(x)$. Is knowledge of all the moments of $p(x)$ (assuming they exist) sufficient to specify $p(x)$ itself? Why or why not?

1–48. Show that the characteristic function of the density function of the sum of two independent random variables is the product of the characteristic functions of the densities of the two random variables themselves. What is the density function of $W = X + Y$?

1–49. Maximize

$$W(N_1, N_2, \ldots, N_M) = \frac{N!}{\prod_{j=1}^{M} N_j!}$$

with respect to each N_j under the constraints that

$$\sum N_j = N = \text{a fixed constant}$$
$$\sum E_j N_j = \mathscr{E} = \text{another fixed constant}$$

Hint: Consider the N_j's to be continuous, large enough to use Stirling's approximation of $N_j!$, and leave your answer in terms of the two undetermined multipliers.

1–50. Show that the maximum of a multinomial distribution is given when $N_1 = N_2 = \cdots = N_s = N/s$.

1–51. Use the method of undetermined multipliers to show that

$$-\sum_{j=1}^{N} P_j \ln P_j$$

subject to the condition

$$\sum_{j=1}^{N} P_j = 1$$

is a maximum when $P_j = \text{constant}$.

1–52. Consider the sum

$$\sum_{N=0}^{M} \frac{M! x^N}{N!(M-N)!}$$

where $x = O(1)$, and M and N are $O(10^{20})$. First show that $\ln \sum = M \ln(1 + x)$ *exactly*, and then calculate the logarithm of the maximum term. Hint: Remember the binomial expansion.

1–53. Show that the higher terms that were dropped in the expansion of $\ln f(N)$ in Eq. (1–84) are completely negligible for large values of N and M.

1–54. The Planck blackbody distribution law

$$\rho(\omega, T)\, d\omega = \frac{\hbar}{\pi^2 c^3} \frac{\omega^3\, d\omega}{\exp(\beta\hbar\omega) - 1}$$

gives the blackbody radiation energy density between frequencies ω and $\omega + d\omega$. ($\hbar \equiv h/2\pi$, $\omega \equiv 2\pi\nu$, and $\varepsilon = h\nu = \hbar\omega$.) Substitute this into

$$\frac{E}{V} = \int_0^\infty \rho(\omega, T)\, d\omega$$

to derive the temperature dependence of E/V. Do this by expressing your result as a group of factors multiplying a dimensionless integral. You do not need to evaluate this integral.

1–55. Show that $e^x/(1 \pm e^x)^2$ is an even function of x.

1–56. The heat capacity of the Einstein model of a crystal is given by

$$C_V = 3Nk \left(\frac{\Theta_E}{T}\right)^2 \frac{e^{\Theta_E/T}}{(e^{\Theta_E/T} - 1)^2}$$

where Θ_E is the "characteristic temperature" of the crystal. Determine both the high- and low-temperature limiting expressions for the heat capacity. Do the same thing for the Debye model of crystals, in which

$$C_V = 9Nk\left(\frac{T}{\Theta_D}\right)^3 \int_0^{\Theta_D/T} \frac{x^4 e^x\, dx}{(e^x - 1)^2}$$

where Θ_D is the Debye temperature of the crystal.

1–57. Recognizing it as a geometric series, sum the following series in closed form:

$$S = \sum_{n=0}^{\infty} e^{-\alpha n}$$

Compare this result to

$$I = \int_0^{\infty} e^{-\alpha n}\, dn$$

Under what conditions are these two results the same?

1–58. One often encounters the gamma function in statistical thermodynamics. It was introduced by Euler as a function of x, which is continuous for positive values of x and which reduces to $n!$ when $x = n$, an integer. The gamma function $\Gamma(x)$ is defined by

$$\Gamma(x) = \int_0^{\infty} e^{-t} t^{x-1}\, dt$$

First show by integrating by parts that

$$\Gamma(x+1) = x\Gamma(x)$$

Using this, show that $\Gamma(n+1) = n!$ for n an integer. Show that

$$\Gamma(\tfrac{1}{2}) = \sqrt{\pi}$$

Evaluate $\Gamma(\frac{3}{2})$ using the recurrence formula $\Gamma(x+1) = x\Gamma(x)$. Lastly show that

$$\Gamma\left(n + \frac{1}{2}\right) = \frac{1 \cdot 3 \cdots (2n-1)}{2^n}\,\Gamma\left(\frac{1}{2}\right)$$

$$= \frac{(2n)!}{2^{2n} n!}\sqrt{\pi}$$

For a discussion of the gamma function, see G. Arfken, *Mathematical Methods for Physicists*, 2nd ed. (New York: Academic, 1970).

1–59. We can derive Stirling's approximation from an asymptotic approximation to the gamma function $\Gamma(x)$. From the previous problem

$$\Gamma(N+1) = N! = \int_0^{\infty} e^{-x} x^N\, dx$$

$$= \int_0^{\infty} e^{Ng(x)}\, dx$$

where $g(x) = \ln x - x/N$. If $g(x)$ possesses a maximum at some point, say x_0, then for large N, $\exp(Ng(x))$ will be extremely sharply peaked at x_0. Under this condition, the integral for $N!$ will be dominated by the contribution of the integrand from the point x_0. First show that $g(x)$ does, in fact, possess at maximum at the point $x_0 = N$. Expand $g(x)$ about this point, keeping terms only up to and including $(x - N)^2$ to get

$$g(x) \approx g(N) - \frac{(x-N)^2}{2N^2} + \cdots$$

Why is there no linear term in $(x-N)$? Substitute this expression for $g(x)$ into the integral for $N!$ and derive the asymptotic formula

$$\ln N! \approx N \ln N - N + \ln(2\pi N)^{1/2}$$

1–60. Verify the energy conversion factors in Appendix A. (The one labeled "temperature" means that temperature required to give an energy equal to kT, where k is the Boltzmann constant.)

1–61. An integral that appears often in statistical mechanics and particularly in the kinetic theory of gases is

$$I_n = \int_0^\infty x^n e^{-ax^2}\, dx$$

This integral can be readily generated from two basic integrals. For even values of n, we first consider

$$I_0 = \int_0^\infty e^{-ax^2}\, dx$$

The standard trick to evaluate this integral is to square it, and then transform the variables into polar coordinates.

$$\begin{aligned} I_0^2 &= \int_0^\infty \int_0^\infty e^{-ax^2} e^{-ay^2}\, dx\, dy \\ &= \int_0^\infty \int_0^{\pi/2} e^{-ar^2} r dr d\theta \\ &= \frac{\pi}{4a} \\ I_0 &= \frac{1}{2}\left(\frac{\pi}{a}\right)^{1/2} \end{aligned}$$

Using this result, show that for even n

$$I_n = \frac{1 \cdot 3 \cdot 5 \cdots (n-1)}{2(2a)^{n/2}} \left(\frac{\pi}{a}\right)^{1/2} \qquad n \text{ even}$$

For odd values of n, the basic integral I_1 is easy. Using I_1, show that

$$I_n = \frac{\Gamma\left(\frac{n+1}{2}\right)}{2a^{(n+1)/2}} \qquad n \text{ odd}$$

1–62. Show that a Gaussian distribution is extremely small beyond a few multiples of σ.

1–63. Another function that occurs frequently in statistical mechanics is the Riemann zeta function, defined by

$$\zeta(s) = \sum_{k=1}^{\infty} k^{-s}$$

First show that $\zeta(1) = \infty$, but that $\zeta(s)$ is finite for $s > 1$. Show that another definition of $\zeta(s)$ is

$$\zeta(s) = \frac{1}{\Gamma(s)} \int_0^\infty \frac{x^{s-1}\, dx}{(e^x - 1)}$$

that is, show that this is identical to the first definition. In addition, show that

$$\eta(s) = \sum_{k=1}^{\infty} (-1)^{k-1} k^{-s} = (1 - 2^{1-s})\zeta(s)$$

$$\lambda(s) = \sum_{k=0}^{\infty} (2k+1)^{-s} = (1 - 2^{-s})\zeta(s)$$

The evaluation of $\zeta(s)$ for integral s can be done using Fourier series, and some results are $\zeta(2) = \pi^2/6$ and $\zeta(4) = \pi^4/90$.

For a discussion of the Riemann zeta function, see G. Arfken, *Mathematical Methods for Physicists*, 2nd ed. (New York: Academic, 1970).

CHAPTER 2

THE CANONICAL ENSEMBLE

In this chapter we shall introduce the basic concepts and assumptions of statistical thermodynamics, and then apply them to a system which has fixed values of V and N and is in thermal equilibrium with its environment. We shall derive the fundamental connection between the quantum mechanical energy levels available to an N-body system and its thermodynamic functions. This link is effected by a function, called the partition function, which is of central importance in statistical thermodynamics. In Section 2–4 we discuss the relevance of the statistical thermodynamic equations to the second and third laws of classical thermodynamics.

2–1 ENSEMBLE AVERAGES

Our goal is to calculate thermodynamic properties in terms of molecular properties. Given the structure of the individual molecules of our system and the form of the intermolecular potential, we wish to be able to calculate thermodynamic properties, such as entropy and free energy. We shall do this first with respect to mechanical properties (such as pressure, energy, volume), which are quantum mechanical or classical mechanical quantities, and then we shall bring nonmechanical thermodynamic variables (such as entropy, free energy) into our discussion by appealing to the equations of thermodynamics. One useful distinction between mechanical and nonmechanical properties is that mechanical properties are defined without appealing to the concept of temperature, whereas the definitions of nonmechanical properties involve the temperature.

Consider some macroscopic system of interest, such as a liter of water or a salt solution. From a macroscopic point of view, we can completely specify a system by a few parameters, say the volume, concentration or density, and temperature. Regardless of the complexity of the system, it requires only a small number of parameters to describe it. From a microscopic point of view, on the other hand, there will be an enormous number of quantum states consistent with the fixed macroscopic properties.

We saw in Chapter 1 that the degeneracy of an isolated N-body system is of the order of 10^N for all but the very lowest energies. This means that the liter of water or the salt solution could be in any one of the order of 10^N possible quantum states. It would be impossible for us to ever determine which of the order of 10^N possible states the system is in. The state of the system must be known, however, in order to calculate a mechanical thermodynamic property, such as the pressure, since the values of that property in each of the possible quantum states would, in general, be different. Thus we are faced with what appears to be an impossible task.

It is at this point that we appeal to the work of Maxwell, Boltzmann, and particularly Gibbs. The modern (postquantum) version of their approach is that in order to calculate the value of any mechanical thermodynamic property (say, the pressure), one calculates the value of that mechanical property in each and every one of the quantum states that is consistent with the few parameters necessary to specify the system in a macroscopic sense. The average of these mechanical properties is then taken, giving each possible quantum state the same weight. We then *postulate* that this average mechanical property corresponds to a parallel thermodynamic property. For example, we postulate that the average energy corresponds to the thermodynamic energy and that the average pressure corresponds to the thermodynamic pressure. It turns out that the calculation of a mechanical property averaged over all the consistent quantum states can be readily performed. Before doing this, however, we shall introduce some concepts that will make this procedure clearer.

We first discuss the concept of an ensemble of systems, first introduced by Gibbs. An ensemble is a (mental or virtual) collection of a very large number of systems, say $\mathscr{A}$, each constructed to be a replica on a thermodynamic (macroscopic) level of the particular thermodynamic system of interest. For example, suppose the system has a volume V, contains N molecules of a single component, and is known to have an energy E. That is, it is an isolated system with N, V, and E fixed. Then the ensemble would have a volume $\mathscr{A}V$, contain $\mathscr{A}N$ molecules, and have a total energy $\mathscr{E} = \mathscr{A}E$. Each of the systems in this ensemble is a quantum mechanical system of N interacting atoms or molecules in a container of volume V. The values of N and V, along with the force law between the molecules, are sufficient to determine the energy eigenvalues E_j of the Schrödinger equation along with their associated degeneracies $\Omega(E_j)$. These energies are the only energies available to the N-body system. Hence the fixed energy E must be one of these E_j's and, consequently, there is a degeneracy $\Omega(E)$. Note that there are $\Omega(E)$ different quantum states consistent with the only things we know about our macroscopic system of interest, namely, the values of N, V, and E. Although all the systems in the ensemble are identical from a thermodynamic point of view, they are not necessarily identical on a molecular level. So far we have said nothing about the distribution of the members of the ensemble with respect to the $\Omega(E)$ possible quantum states.

We shall further restrict our ensemble to obey the *principle of equal a priori probabilities*. That is to say, we require that each and every one of the $\Omega(E)$ quantum states is represented an equal number of times in the ensemble. Since we have no information to consider any one of the $\Omega(E)$ quantum states to be more important than any other, we must treat each of them equally, that is, we must utilize the principle of equal a priori probabilities. All of the $\Omega(E)$ quantum states are consistent with the given values of N, V, and E, the only information we have about the system. Clearly, the number of systems in the ensemble must be an integral multiple of $\Omega(E)$. The number of systems in an ensemble is a very large number and can be made arbitrarily large by simply doubling, tripling, and so on, the size of the ensemble. An alternative interpretation of the principle of equal a priori probabilities is that *an isolated system (N, V, and E fixed) is equally likely to be in any of its $\Omega(E)$ possible quantum states.*

We now define an ensemble average of a mechanical property as the average value of this property over all the members of the ensemble, utilizing the principle of equal a priori probabilities. We postulate that the ensemble average of a mechanical property can be equated to its corresponding thermodynamic property.

There are two complications in the above treatment that we should mention; neither of them, fortunately, is of any practical consequence. We have assumed that the isolated system that we have been using as an example has precisely the energy E. We know, however, from quantum mechanics that there always exists a small uncertainty ΔE in the value of E. For all thermodynamic purposes, this complication is completely inconsequential, and we shall therefore ignore it. The explanation of the other complication involves a greater knowledge of quantum mechanics than is generally required in this book. We have assumed that the systems of the ensemble are in one of the $\Omega(E)$ degenerate eigenstates having the eigenvalue E. The choice of these $\Omega(E)$ eigenfunctions, however, is somewhat arbitrary since any linear combination of these is also an eigenfunction with energy E. Moreover, a quantum mechanical system will, in general, not be in one of the $\Omega(E)$ selected states, but will be some linear combination of them. Thus we have tacitly assumed that any system with N, V, and E given will be a "pure state," whereas a system with N, V, and E given will most likely be in a "mixed state," that is, in a state described by a linear combination of the pure states we have chosen. In any event, this complication need not be considered, since the results do not differ appreciably from those obtained from the simpler and more naive point of view which we have presented above and now adopt.

Let us summarize this section by stating that we wish to calculate the ensemble average of some mechanical property, and then show that this can be set equal to the corresponding thermodynamic property. We have stated above that the calculation of the ensemble average is not difficult, and now we shall address ourselves to that problem. As Schrödinger says in his book:* "There is, essentially, only one problem in statistical thermodynamics, the distribution of a given amount of energy E over identical systems. Or perhaps better, to determine the distribution of an assembly of identical systems over the possible states in which the system can find itself, given that the energy of the assembly is a constant E."

So far, in this section, we have focused our attention on an ensemble whose members have N, V, and E fixed. This is called the *microcanonical ensemble* and is useful for theoretical discussions. For more practical applications, however, we consider not isolated systems, but those in which the temperature rather than the energy is fixed. The most commonly used ensemble in statistical thermodynamics is the *canonical ensemble*, in which the individual systems have N, V, and T fixed. The remainder of this chapter will deal with the canonical ensemble. There are many other types of ensembles, in fact, one for each set of thermodynamic variables that are used to specify an individual member of the ensemble. We shall discuss some of these other ensembles in the next chapter.

2–2 METHOD OF THE MOST PROBABLE DISTRIBUTION

Consider an experimental system with N, V, and T as its independent thermodynamic variables. We can mentally construct an ensemble of such systems in the following manner. We enclose each system in a container of volume V with walls that are heat

* E. Schrödinger, *Statistical Thermodynamics* (Cambridge: Cambridge University Press, 1952).

conducting but impermeable to the passage of molecules. The entire ensemble of systems is then placed in a very large heat bath at temperature T. When equilibrium is reached, the entire ensemble is at a uniform temperature T. Since the containing walls of each system are heat conducting, each and every system of the ensemble has the same fixed values of N, V, and T. Now, the entire ensemble is surrounded by thermal insulation, thus making the ensemble itself an isolated system with volume $\mathscr{A}V$, number of molecules $\mathscr{A}N$, and some total energy $\mathscr{E}$. (The actual value of $\mathscr{E}$ is not important.) Each of the $\mathscr{A}$ members of the canonical ensemble finds itself in a large heat bath at temperature T.

Because each of the systems of the canonical ensemble is not isolated but is at a fixed temperature, the energy of each system is not fixed at any set value. Thus we shall have to consider the entire spectrum of energy states for each member of the canonical ensemble. Let the energy eigenvalues of the quantum states of a system be $E_1(N, V)$, $E_2(N, V) \ldots$, ordered such that $E_{j+1} \geq E_j$. It is important to understand here that any particular energy, say E_i, is repeated according to its degeneracy, that is, occurs $\Omega(E_i)$ times. Any particular system might be found in any of these quantum states. We shall show later that the *average energy* or the *probability* that some system has a certain energy depends upon the temperature; however, any of the set of energies $\{E_j\}$ is possible, and so must be considered.

We can specify a state of the entire ensemble by saying that $a_1, a_2, a_3, \ldots$ of the systems are in states 1, 2, 3, ..., respectively, with energies $E_1, E_2, E_3, \ldots$. Thus we can describe any one state of the ensemble by writing

State No.	$1, \; 2, \; 3, \; \ldots, \; l \ldots$
Energy	$E_1, E_2, E_3, \ldots, E_l \ldots$
Occupation No.	$a_1, a_2, a_3, \ldots, a_l \ldots$

Occupation Number means the number of systems of the ensemble in that particular state. The set of occupation numbers is called a distribution. We shall often denote the set $\{a_j\}$ by $\mathbf{a}$.

Of course, the occupation numbers satisfy the two conditions:

$$\sum_j a_j = \mathscr{A} \tag{2–1}$$

$$\sum_j a_j E_j = \mathscr{E} \tag{2–2}$$

The first condition simply accounts for all the members of the ensemble, and the second represents the fact that the entire canonical ensemble is an isolated system, and hence has some fixed energy $\mathscr{E}$.

Since the canonical ensemble has been isolated from its surroundings by thermal insulation, we can apply the principle of equal a priori probabilities to this isolated system. In the form that we wish to use here, the principle of equal a priori probabilities says that every possible state of the canonical ensemble, that is, every distribution of occupation numbers $a_1, a_2, \ldots$, consistent with Eqs. (2–1) and (2–2) is equally probable and must be given equal weight in performing ensemble averages.

The number of ways $W(\mathbf{a}) \equiv W(a_1, a_2, a_3, \ldots)$ that any particular distribution of the a_j's can be realized is the number of ways that $\mathscr{A}$ *distinguishable* objects can be arranged into groups, such that a_1 are in the first group, a_2 in the second, and so on [see Eq. (1–77)]:

$$W(\mathbf{a}) = \frac{\mathscr{A}!}{a_1! a_2! a_3! \cdots} = \frac{\mathscr{A}!}{\prod_k a_k!} \tag{2–3}$$

The systems are distinguishable since they are macroscopic systems, which we could, in principle, furnish with labels.

In general, there are very many distributions which are consistent with Eqs. (2–1) and (2–2). In any particular distribution, $a_j/\mathscr{A}$ is the fraction of systems or members of the canonical ensemble in the jth energy state (with energy E_j). The overall probability P_j that a system is in the jth quantum state is obtained by averaging $a_j/\mathscr{A}$ over all the allowed distributions, giving equal weight to each one according to the principle of equal a priori probabilities. Thus P_j is given by

$$P_j = \frac{\bar{a}_j}{\mathscr{A}} = \frac{1}{\mathscr{A}} \frac{\sum_{\mathbf{a}} W(\mathbf{a}) a_j(\mathbf{a})}{\sum_{\mathbf{a}} W(\mathbf{a})} \tag{2–4}$$

In Eq. (2–4), the notation $a_j(\mathbf{a})$ signifies that the value of a_j depends upon the distribution, and the summations are over all distributions that satisfy Eqs. (2–1) and (2–2). We shall later let $\mathscr{A} \to \infty$, but the ratio $\overline{a_j}/\mathscr{A}$ will remain finite since $\bar{a}_j \to \infty$ as well.

Given the probability that a system with fixed values of N, V, and T is in the jth quantum state, one can calculate the canonical ensemble average of any mechanical property from

$$\overline{M} = \sum_j M_j P_j \tag{2–5}$$

where M_j is the value of M in the jth quantum state. Thus the prescription for calculating the ensemble average of any mechanical property is given by Eqs. (2–4) and (2–5) and is, in principle, complete. The summations involved in Eq. (2–4), however, are very difficult to perform mathematically, and thus in practice Eqs. (2–4) and (2–5) are too complicated to use.

The fact that we can let $\mathscr{A} \to \infty$, however, allows us to appeal to the results of Section 1–5. We have seen there that multinomial coefficients, such as $W(\mathbf{a})$, are extremely peaked about their maximum value if all the variables a_j are large. In Eqs. (2–1) through (2–4), each of the a_j's can be made arbitrarily large since $\mathscr{A}$ can be made arbitrarily large. Thus we can use an argument here very similar to that used in Section 1–5. We need make only one modification or extension. We have shown that $W(\mathbf{a})$ is a maximum when all the a_j's are equal, under the one constraint Eq. (2–1). We have now an additional constraint Eq. (2–2) on the a_j's. So instead of peaking at the point at which all the a_j's are equal, it will peak at some other set of a_j's but the spread, nevertheless, will be arbitrarily small. We shall determine this set of a_j's shortly. Let us denote this distribution by $\mathbf{a}^* = \{a_j^*\}$.

The spread of $W(\mathbf{a})$ about its maximum value can be made arbitrarily narrow by taking the a_j, that is, $\mathscr{A}$, to be arbitrarily large. Thus the $W(\mathbf{a})$ in Eq. (2–4) at any set of a_j's other than the set $\mathbf{a}^*$, which maximizes $W(\mathbf{a})$, are completely negligible. We can replace the summations in Eq. (2–4) over all distributions by just one term, evaluated at $\mathbf{a}^*$. Thus we can write

$$P_j = \frac{1}{\mathscr{A}} \frac{\sum_{\mathbf{a}} W(\mathbf{a}) a_j(\mathbf{a})}{\sum_{\mathbf{a}} W(\mathbf{a})} = \frac{1}{\mathscr{A}} \frac{W(\mathbf{a}^*) a_j^*}{W(\mathbf{a}^*)} = \frac{a_j^*}{\mathscr{A}} \qquad (\lim a_j \to \infty) \tag{2–6}$$

where a_j^* is the value of a_j in that distribution that maximizes $W(\mathbf{a})$, that is, the most probable distribution. The name of this section, the method of the most probable distribution, is derived from Eq. (2–6). Comparing Eqs. (2–6) with (2–4), we have

$$P_j = \frac{\bar{a}_j}{\mathscr{A}} = \frac{a_j^*}{\mathscr{A}} \tag{2–7}$$

Thus, to calculate the probabilities to be used in ensemble averages, we need determine only that distribution **a*** that maximizes $W(\mathbf{a})$ under the two constraints Eqs. (2–1) and (2–2). This is the problem to which we now turn.

As this is a problem of maximizing a function of many variables with given constraints on the variables, we have a direct application of Lagrange's method of undetermined multipliers. Following Section 1–5, the set of a_j's that maximizes $W(\mathbf{a})$, subject to Eqs. (2–1) and (2–2), is found from

$$\frac{\partial}{\partial a_j}\left\{\ln W(\mathbf{a}) - \alpha \sum_k a_k - \beta \sum_k a_k E_k\right\} = 0, \qquad j = 1, 2, \ldots \tag{2–8}$$

where α and β are the undetermined multipliers. Using Eq. (2–3) for $W(\mathbf{a})$ along with Stirling's approximation (which is exact here since each of the a_j's can be made arbitrarily large), one gets

$$-\ln a_j^* - \alpha - 1 - \beta E_j = 0 \qquad j = 1, 2, \ldots \tag{2–9}$$

or

$$a_j^* = e^{-\alpha'} e^{-\beta E_j} \qquad j = 1, 2, \ldots \tag{2–10}$$

where $\alpha' = \alpha + 1$. (See Problem 2–3.) This gives us the most probable distribution in terms of α and β. We now evaluate α' and β by using Eqs. (2–1) and (2–2) along with physical arguments.

2–3 THE EVALUATION OF THE UNDETERMINED MULTIPLIERS, α AND β

We can obtain an expression for α (or α') in terms of β by summing both sides of Eq. (2–10) over j and using Eq. (2–1) to get

$$e^{\alpha'} = \frac{1}{\mathscr{A}} \sum_j e^{-\beta E_j} \tag{2–11}$$

Equation (2–7) thus becomes

$$P_j = \frac{a_j^*}{\mathscr{A}} = \frac{e^{-\beta E_j(N,V)}}{\sum_j e^{-\beta E_j(N,V)}} \tag{2–12}$$

Substituting this into Eq. (2–5), with E_j taken to be the mechanical property, gives

$$\bar{E} = \bar{E}(N, V, \beta) = \frac{\sum_j E_j(N, V) e^{-\beta E_j(N,V)}}{\sum_j e^{-\beta E_j(N,V)}} \tag{2–13}$$

According to the postulate of the ensemble method of Gibbs, this average energy $\bar{E}(N, V, \beta)$ corresponds to the thermodynamic energy E.

The pressure is another important mechanical variable. When a system is in the state j, $dE_j = -p_j\, dV$ is the work done on the system when its volume is increased by dV (keeping the number of particles in the system fixed). Thus the pressure in the state j is given by

$$p_j = -\left(\frac{\partial E_j}{\partial V}\right)_N \tag{2–14}$$

The canonical ensemble average of p_j is

$$\bar{p} = \sum_j p_j P_j = -\frac{\sum_j \left(\frac{\partial E_j}{\partial V}\right) e^{-\beta E_j}}{\sum_j e^{-\beta E_j}} \tag{2-15}$$

We postulate that $\bar{p}$ corresponds to the thermodynamic pressure.

The sum in the denominator of Eqs. (2–13) and (2–15) occurs throughout the equations of the canonical ensemble. Let this be denoted by $Q(N, V, \beta)$:

$$Q(N, V, \beta) = \sum_j e^{-\beta E_j(N,V)} \tag{2-16}$$

We shall see that this function $Q(N, V, \beta)$ is the central function of the canonical ensemble.

We have made two connections with thermodynamics:

$$\begin{matrix} p \leftrightarrow \bar{p} \\ E \leftrightarrow \bar{E} \end{matrix} \qquad \text{(ensemble postulate of Gibbs)}$$

Equation (2–13) gives E as a function of β. In principle, one could solve this equation for β as a function of E, but in practice this is not feasible. Fortunately β turns out to be a more convenient quantity than E, so much so that it is preferable to have E as a function of β rather than the inverse. We shall now evaluate β in two different ways.

We differentiate Eq. (2–13) with respect to V, keeping N and β fixed:

$$\left(\frac{\partial \bar{E}}{\partial V}\right)_{N,\beta} = -\bar{p} + \beta\overline{Ep} - \beta\bar{E}\bar{p} \tag{2-17}$$

In this equation,

$$\overline{Ep} = \frac{\sum_j p_j E_j e^{-\beta E_j}}{Q} = -\frac{\sum_j \left(\frac{\partial E_j}{\partial V}\right) E_j e^{-\beta E_j}}{Q}$$

and

$$\bar{E}\bar{p} = \frac{\sum_j E_j e^{-\beta E_j}}{Q} \cdot \frac{\sum_j p_j e^{-\beta E_j}}{Q}$$

Similarly, we can differentiate Eq. (2–15) to get

$$\left(\frac{\partial \bar{p}}{\partial \beta}\right)_{N,V} = \bar{E}\bar{p} - \overline{Ep} \tag{2-18}$$

From Eqs. (2–17) and (2–18) we get

$$\left(\frac{\partial \bar{E}}{\partial V}\right)_{N,\beta} + \beta\left(\frac{\partial \bar{p}}{\partial \beta}\right)_{N,V} = -\bar{p} \tag{2-19}$$

Note that $\bar{E}$ is a function of N, V, and β, whereas the E_j's are functions of N and V only. This is an important distinction, that should be clearly and completely understood.

Let us now compare Eq. (2–19) with the purely thermodynamic equation. (See Problem 1–29.)

$$\left(\frac{\partial E}{\partial V}\right)_{T,N} - T\left(\frac{\partial p}{\partial T}\right)_{N,V} = -p \tag{2-20}$$

which we rewrite in terms of $1/T$ instead of T:

$$\left(\frac{\partial E}{\partial V}\right)_{N,1/T} + \frac{1}{T}\left(\frac{\partial p}{\partial 1/T}\right)_{N,V} = -p \tag{2-21}$$

A comparison of Eq. (2–19) with Eq. (2–21) allows us to deduce that $\beta = \text{const}/T$. It is customary to write $\beta = 1/kT$, where k is a constant, whose value could possibly vary from substance to substance. We shall show now, however, that k has the same value for all substances, that is, k is a universal constant.

Consider two closed systems A and B, each having its own kind of particles and energy states, but in thermal contact with each other and immersed in a heat bath of temperature T. We now construct a canonical ensemble of systems AB (as shown in Fig. 2–1) representative of a thermodynamic AB system at temperature T and apply the method of the most probable distribution to the AB system. Let the number of molecules and volume of the A and B systems be N_A, V_A, and N_B, V_B, respectively, and let their energy states be denoted by $\{E_{jA}\}$ and $\{E_{jB}\}$. If a_j denotes the number of A systems in state E_{jA}, and b_j denotes the number of B systems in state E_{jB}, then the number of states of the AB ensemble with compound distribution $\{a_j\}$ and $\{b_j\}$ is

$$W(\mathbf{a}, \mathbf{b}) = \frac{\mathscr{A}!}{\prod_j a_j!} \cdot \frac{\mathscr{B}!}{\prod_k b_k!} \tag{2-22}$$

where $\mathscr{A}$ and $\mathscr{B}$ ($\mathscr{A} = \mathscr{B}$) are the number of A and B system, respectively. Equation (2–22) turns out to be a product of the separate A and B factors, because we can arrange the A systems over their possible quantum states independently of the B systems, and vice versa. The a_j's and b_j's must satisfy the three relations:

$$\sum_j a_j = \mathscr{A}$$

$$\sum_j b_j = \mathscr{B} = \mathscr{A}$$

$$\sum_j (a_j E_{jA} + b_j E_{jB}) = \mathscr{E} \tag{2-23}$$

We now apply the method of the most probable distribution to Eqs. (2–22) and (2–23) to get Problem 2–9 for the simultaneous probability that the AB system has its A part in the ith quantum state and its B part in the jth quantum state:

$$P_{ij} = \frac{e^{-\beta E_{iA}}}{Q_A} \cdot \frac{e^{-\beta E_{jB}}}{Q_B} = P_{iA} P_{jB} \tag{2-24}$$

where

$$Q_A = \sum_k e^{-\beta E_{kA}} \quad \text{and} \quad Q_B = \sum_k e^{-\beta E_{kB}} \tag{2-25}$$

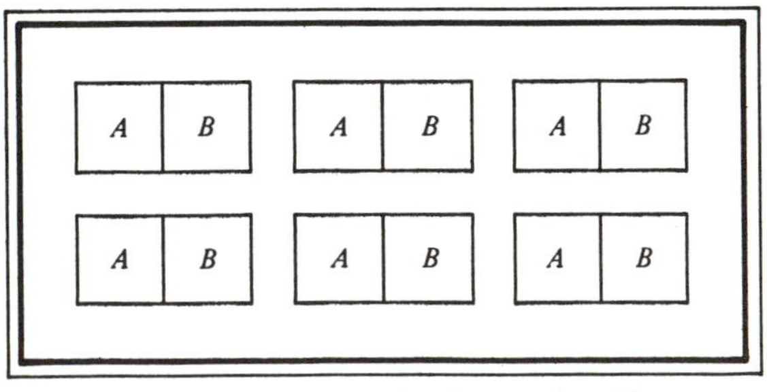

***Figure* 2–1. Canonical ensemble of composite *AB* systems.**

Thus we have shown that two arbitrary systems in thermal contact have the same β. But we have seen from Eqs. (2–19) and (2–21) that $\beta = 1/kT$, and so the two systems must have the same value of k. Since the nature of the two systems is completely arbitrary, k must have the same value for all systems. Thus k is a universal constant and can therefore be evaluated using any convenient system. The most convenient is an ideal gas, and one can determine from the equation of state of an ideal gas [*cf.* Eq. (5–18)] that $k = 1.3806 \times 10^{-16}$ erg-deg^{-1}, where the temperature is in units of degrees Kelvin.

There is an alternative way to determine β which utilizes the fact that $1/T$ is an integrating factor of dq_{rev}. We shall present this argument here, since it will bring the nonmechanical property of entropy into our formalism.

The argument based around Fig. 2–1 shows that if two systems are in thermal contact at equilibrium, they have the same value of β. Since the two systems can be quite arbitrary, this implies that β must be some function of the temperature. We shall now show that $\beta\, dq_{rev}$ is an exact differential.

Consider the function $f = \ln Q$. We regard f as a function of β and all the E_j's:

$$f(\beta, E_1, E_2, \ldots) = \ln \left\{ \sum_j e^{-\beta E_j} \right\} \tag{2–26}$$

The total derivative f is

$$df = \left(\frac{\partial f}{\partial \beta}\right)_{E_j\text{'s}} d\beta + \sum_k \left(\frac{\partial f}{\partial E_k}\right)_{\beta,\, E_i\text{'s}} dE_k \tag{2–27}$$

The partial derivatives occurring here are determined from Eq. (2–26) to be

$$\left(\frac{\partial f}{\partial \beta}\right)_{E_j\text{'s}} = \frac{-\sum_j E_j e^{-\beta E_j}}{Q} = -\bar{E}$$

$$\left(\frac{\partial f}{\partial E_k}\right)_{\beta,\, E_j\text{'s}} = \frac{-\beta e^{-\beta E_k}}{Q} = -\beta P_k$$

Thus Eq. (2–27) becomes

$$df = -\bar{E}\, d\beta - \beta \sum_j P_j\, dE_j$$

which can be written as

$$d(f + \beta\bar{E}) = \beta\left(d\bar{E} - \sum_j P_j\, dE_j\right) \tag{2–28}$$

We now subject the ensemble of systems to the following physical process. We change the volume of all the systems by dV, changing, of course, the E_j's for all of them alike in order to still have an ensemble of macroscopically identical systems. We also change the temperature of the ensemble by dT by coupling it with a large heat bath (of the same temperature), changing the temperature slightly and then isolating the ensemble from the heat bath.

If initially there were a_j systems of the ensemble in the energy state j with energy E_j, then $a_j\, dE_j$ is the work done on all these systems in changing the energy from E_j to $E_j + dE_j$. The total work done on the ensemble is $\sum a_j\, dE_j$ and $\sum_j P_j\, dE_j$ is the ensemble average reversible work that we do on the systems. And since $d\bar{E}$ is the average energy increase, the term enclosed in parentheses on the right-hand side of Eq. (2–28) is the average reversible heat supplied to a system. Thus Eq. (2–28) is

$$d(f + \beta\bar{E}) = \beta\, \delta q_{rev} \tag{2–29}$$

which says that $\beta\,\delta q_{rev}$ is the derivative of a state function, that is, that β is an integrating factor of δq_{rev}. One statement of the second law of thermodynamics says that β must be equal to constant/T, or $1/kT$.

The left-hand side of Eq. (2–29), therefore, must be dS/k, and so we can write that

$$S = \frac{\bar{E}}{T} + k \ln Q + \text{constant} \tag{2–30}$$

where the constant is independent of T and of the parameters (N, V, and so on) on which the E_j's depend. Since thermodynamics deals with ΔS only, the constant will always drop out of any calculations of entropy changes for chemical and/or physical changes. We shall, therefore, set this constant to zero and discuss the implications of this at the end of the chapter.

In the above argument that β was an integrating factor of δq_{rev}, we used the fact that the average work done on a system was $\sum_j P_j\,dE_j$. We do work, then, by changing the energies slightly, but keeping the population of these states fixed (the P_j's do not change). A molecular interpretation of thermodynamic work, then, is a change in the quantum mechanical energy states of the system, keeping the population over them fixed. That a molecular interpretation of the absorption of heat is the inverse of this can be seen from

$$\begin{aligned} d\bar{E} &= \sum_j E_j\,dP_j + \sum_j P_j\,dE_j \\ &= \delta q_{rev} - \delta \omega_{rev} \end{aligned}$$

Thus when a small quantity of heat is absorbed from the surroundings, the energy states of the system do not change (N and V are fixed), but the population of these states does.

2–4 THERMODYNAMIC CONNECTION

We now complete the connection between thermodynamics and the canonical ensemble. Equation (2–13) for $\bar{E}$ can be written as (see Problem 2–10):

$$\bar{E} = kT^2\left(\frac{\partial \ln Q}{\partial T}\right)_{N,V} \tag{2–31}$$

and we can also easily derive (see Problem 2–10)

$$\bar{p} = kT\left(\frac{\partial \ln Q}{\partial V}\right)_{N,T} \tag{2–32}$$

from Eq. (2–15). Equation (2–30) is an equation for the entropy S in terms of Q:

$$S = kT\left(\frac{\partial \ln Q}{\partial T}\right)_{N,V} + k \ln Q \tag{2–33}$$

We have E, p, and S now as functions of Q, and so it is possible to derive expressions for all the thermodynamic functions in terms of Q. The function Q is the central statistical thermodynamic function of the canonical ensemble (N, V, and T fixed) and is called the *canonical (ensemble) partition function*:

$$Q(N, V, T) = \sum_j e^{-E_j(N,V)/kT} \tag{2–34}$$

The partition function serves as a bridge between the quantum mechanical energy states of a macroscopic system and the thermodynamic properties of that system. If we can obtain Q as a function of N, V, and T, we can calculate thermodynamic properties in terms of quantum mechanical and molecular parameters. Although the E_j's are the energy states of an N-body system and consequently appear to be unobtainable in practice, we shall see that in a great many cases, we shall be able to reduce the N-body problem to a one-body, two-body, three-body problem, and so on, or approximate the system by classical mechanics. Both of these routes turn out to be very useful. For now, however, we need only assume that there is such a set of energies.

We can derive an equation for the Helmholtz free energy A in terms of Q by using Eqs. (2–31) and (2–33) along with the fact that $A = E - TS$. The result is

$$A(N, V, T) = -kT \ln Q(N, V, T) \tag{2–35}$$

Notice that of all the thermodynamic functions, it is A that is directly proportional to $\ln Q(N, V, T)$, and that A is the thermodynamic potential whose natural independent variables are those of the canonical ensemble. Equation (2–35) can be considered to be the most important connection between thermodynamics and the canonical partition function, since it is possible to derive many equations starting with its differential form (see Problem 2–11). Table 3–1 contains a summary of the formulas of the canonical ensemble.

In this chapter, we have developed the connection between thermodynamics and the quantum mechanical states available to a macroscopic system characterized by N, V, and T. This connection can be summarized by Eq. (2–35). Before concluding this chapter, we shall discuss the second and third laws of thermodynamics from a statistical thermodynamic point of view. A statement of the second law of thermodynamics for closed, isothermal systems is that $\Delta A < 0$ for a spontaneous process. We wish to derive this inequality starting with Eq. (2–35). To do this, we first write Eq. (2–34) in a slightly different form.

Consider Eq. (2–34) for $Q(N, V, T)$. The summation is over all the possible quantum states of the N-body system. In carrying out the summation, a particular value of $\exp(-E_j/kT)$ will occur $\Omega(E_j)$ times, where $\Omega(E_j)$ is the degeneracy. Instead of listing $\exp(-E_j/kT)$ $\Omega(E_j)$ times, we could simply write $\Omega(E_j)\exp(-E_j/kT)$, and then sum over different values of E. If we do this, Eq. (2–34) is

$$Q(N, V, T) = \sum_E \Omega(N, V, E)e^{-E(N,V)/kT} \tag{2–36}$$

where we have dropped the no longer necessary j subscript of E_j. In Eq. (2–34), we sum over the *states* of the system. In Eq. (2–36) we sum over *levels*. Equation (2–36) is a more useful form for discussing the second law of thermodynamics.

Consider a typical spontaneous processes, such as the expansion of a gas into a vacuum. Figure 2–2 shows the initial and final states of such a process. For simplicity, we consider the entire system to be isolated. Initially the gas might be confined to one half of the container. After removing the barrier, the gas occupies the entire container.

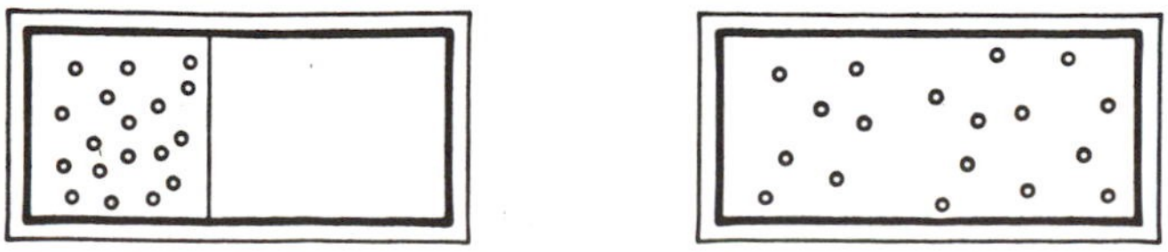

Figure **2–2. The initial and final states of the expansion of a gas into a vacuum.**

Equation (1–36) for $\Omega(N, V, E)$ of an ideal gas shows that the number of states is proportional to V^N. For the process illustrated in Fig. 2–2, the gas goes from a thermodynamic state of energy E, number of particles N, and volume $V/2$ to one with the same energy E (the system is isolated), the same number of particles N, but with volume V. Thus according to Eq. (1–36), the number of quantum states available or accessible to the system is increased.

Another example of a spontaneous process is the following. Initially we have an isolated system containing a mixture of hydrogen and oxygen gases. Although hydrogen and oxygen react to form water, in the absence of a catalyst the reaction is so slow that we can ignore it. Since the rate of this reaction (uncatalyzed) is very slow compared to any thermodynamic measurement, we can consider the mixture of hydrogen and oxygen to be simply a mixture of two gases in equilibrium. If we now add a small amount of catalyst to the system, the hydrogen and oxygen will readily form water, so that the system contains hydrogen, oxygen, and water. Thus the addition of a small amount of catalyst makes all the energy states associated with water molecules available or accessible to the system, and, hence, the system proceeds spontaneously to populate these states. Since the originally accessible states are still accessible (there is still some hydrogen and oxygen in the system), this spontaneous process is associated with an increase in the number of states accessible to the system by the removal of some constraint. In this case, the constraint was a high activation energy barrier, which was removed by the addition of the catalyst.

Both of the spontaneous processes that we have discussed occurred, because some restraint, inhibition, or barrier was removed which made additional quantum states accessible to the system. In general, any spontaneous process in an isolated system can be viewed in this manner. The removal of some constraint allows a greater number of quantum states to be accessible to the system, thus the "flow" of the system into these states is observed as a spontaneous process.

The above discussion is limited to isolated systems. In order to discuss the condition $\Delta A < 0$, we must now consider isothermal processes. When a system is in a heat bath rather than isolated, we must include all possible energy states or levels of the system. When a restraint is removed, the number of accessible quantum states of each and every energy E cannot decrease, and will usually increase, since the original states are still available. Thus we have that $\Omega_2(N, V, E) \geq \Omega_1(N, V, E)$ for all E, where the subscripts 1 and 2 denote the initial and final states, respectively. We now use this inequality along with Eq. (2–36) to show that $\Delta A < 0$. Since no term can be negative and many are positive, we have

$$Q_2 - Q_1 = \sum_E \{\Omega_2(N, V, E) - \Omega_1(N, V, E)\} e^{-E/kT} > 0 \tag{2–37}$$

In Eq. (2–37) we sum over all the levels available to the final state. It follows immediately from the inequality in Eq. (2–37) that

$$\Delta A = A_2 - A_1 = -kT \ln \frac{Q_2}{Q_1} < 0 \tag{2–38}$$

for a spontaneous isothermal process, and thus we have written the second law of thermodynamics in terms of Eq. (2–35).

Lastly, we consider the implications of putting the "constant" of Eq. (2–30) equal to zero. We shall see that this gives us a statistical thermodynamic version of the third law of thermodynamics. If we write Eq. (2–33) for S more explicitly, we get

$$S = k \ln \sum_j e^{-E_j/kT} + \frac{1}{T}\frac{\sum_j E_j e^{-E_j/kT}}{\sum_j e^{-E_j/kT}} \tag{2–39}$$

We wish to study the behavior of this equation as $T \to 0$. Assume for generality that the first n states have the same energy ($E_1 = E_2 = \cdots = E_n$) and that the next m states have the same energy ($E_{n+1} = E_{n+2} = \cdots = E_{n+m}$), and so on. Then in the limit of small T, Eq. (2–39) becomes (see Problem 2–19):

$$S = k \ln n + \frac{km}{n} e^{-(E_{n+1}-E_1)/kT} + \frac{m}{nT}(E_{n+1} - E_1)e^{-(E_{n+1}-E_1)/kT}$$

and so

$$\lim_{T\to 0} S = k \ln n \tag{2–40}$$

Thus as $T \to 0$, S is proportional to the logarithm of the degeneracy of the lowest level. Unless n is very large, Eq. (2–40) says that S is practically zero. For example, if the system were a gas of N-point particles, and the degeneracy of the lowest level were of the order of N, $k \ln N$ would be practically zero compared to a typical order of magnitude of the entropy, namely, Nk. Thus setting the "constant" of Eq. (2–30) equal to zero is equivalent to adopting the convention that the entropy of most systems is zero at the absolute zero of temperature [*cf.* Eq. (1–44)].

ADDITIONAL READING

General

Andrews, F. C. 1963. *Equilibrium statistical mechanics*. New York: Wiley. Chapters 9–11.

Hill, T. L. 1960. *Statistical thermodynamics*. Reading, Mass.: Addison-Wesley. Sections 1–1 through 1–5.

Kestin, J., and Dorfman, J. R. 1971. *A course in statistical thermodynamics*. New York: Academic. Sections 5–1 through 5–3 and 5–11 through 5–17.

Kittel, C. 1969. *Thermal physics*. New York: Wiley. Chapters 1–4.

Knuth, E. L. 1966. *Statistical thermodynamics*. New York: McGraw-Hill. Chapters 2, 4, and 5.

Mayer, J. E., and Mayer, M. G. 1940. *Statistical mechanics*. New York: Wiley. Chapters 3, 4, and 10.

Schrödinger, E. 1952. *Statistical thermodynamics*. Cambridge: Cambridge University Press. Chapters 1–3.

PROBLEMS

2–1. From statistical mechanics we have shown

$$\left(\frac{\partial \bar{E}}{\partial V}\right)_{N,\beta} + \beta\left(\frac{\partial \bar{p}}{\partial \beta}\right)_{N,V} = -\bar{p}$$

and from the thermodynamics we have

$$\left(\frac{\partial E}{\partial V}\right)_{N,T} - T\left(\frac{\partial p}{\partial T}\right)_{N,V} = -p$$

Why can't β be linearly proportional to the temperature? That is, $\beta = \text{constant} \times T$.

2–2. To investigate the replacement of $\bar{n}_j$ by n_j^*, that is, the replacement of the average number of systems in state j by the most probable number in state j, consider the simple example in which $\Omega(\mathbf{n})$ is just a binomial distribution

$$\Omega(\mathbf{n}) = \frac{n!}{n_1!\,(n-n_1)!}$$

and actually calculate n_1^* and $\bar{n}_1$. Hint: Recall that

$$(1+x)^n = \sum_{n_1=0}^{n} \frac{n!x^{n_1}}{n_1!(n-n_1)!}$$

2–3. Show that Eq. (2–9) follows from Eq. (2–8). Note that in deriving this result, we have written $\ln W(\mathbf{a})$ as $\mathscr{A} \ln \mathscr{A} - \mathscr{A} - \sum_j a_j \ln a_j + \mathscr{A}$ and have considered $\mathscr{A}$ to be a constant. Show that Eq. (2–10) is independent of this assumption, that is, derive Eq. (2–10) treating $\mathscr{A}$ as $\sum_j a_j$.

2–4. Starting with Eq. (2–31), prove that the Boltzmann constant k must be positive, using the fact that the heat capacity C_V is always positive.

2–5. Show that the entropy can be written as

$$S = -k \sum_j P_j \ln P_j$$

where P_j is given by Eq. (2–12).

2–6. Maximize the function defined as "information" in information theory.

$$I = \sum_j P_j \ln P_j$$

subject to the two constraints

$$\sum_j P_j = 1$$

and

$$\sum_j E_j P_j = E = \text{fixed}$$

Compare this result to that of Problem 1–51.

2–7. Obtain the most probable distribution of N molecules of an ideal gas contained in two equal and connected volumes at the same temperature by minimizing the Helmholtz free energy for the two systems.

2–8. Differentiate Eq. (2–16) with respect to β to derive Eq. (2–13).

2–9. Derive Eq. (2–24).

2–10. Derive Eqs. (2–31) and (2–32).

2–11. Derive Eqs. (2–31) through (2–33) by starting with $A = -kT \ln Q$.

2–12. We can derive Eq. (2–36) directly by the method of Lagrange multipliers. We label the *levels* rather than the states by a subscript l. The degeneracy of the lth level, whose energy is E_l, is Ω_l. The number of ways of distributing systems over levels, with degeneracy Ω_l, is

$$W(\mathbf{a}) = \frac{\mathscr{A}!}{\prod_l a_l!} \prod_l \Omega_l^{a_l}$$

where a_l is the number of systems in the lth level. Maximize this, subject to the constraints

$$\sum_l a_l = \mathscr{A}$$

$$\sum_l a_l E_l = \mathscr{E}$$

to get

$$a_l^* = \frac{\Omega_l e^{-E_l/kT}}{\sum_l \Omega_l e^{-E_l/kT}}$$

2–13. Show that for a particle confined to a cube of length a that

$$p_j = \frac{2}{3} \frac{E_j}{V}$$

By taking the ensemble average of both sides, we have

$$\bar{p} = \frac{2}{3}\frac{\bar{E}}{V}$$

If we use the fact that $\bar{E} = \frac{3}{2}NkT$ (to be proved in Chapter 5), we get the ideal gas equation of state.

2–14. We shall show in Chapter 5 that the partition function of a monatomic ideal gas is

$$Q(N, V, T) = \frac{1}{N!}\left(\frac{2\pi mkT}{h^2}\right)^{3N/2} V^N$$

Derive expressions for the pressure and the energy from this partition function. Also show that the ideal gas equation of state is obtained if Q is of the form $f(T)V^N$, where $f(T)$ is any function of temperature.

2–15. In Chapter 11 we shall approximate the partition function of a crystal by

$$Q = \left(\frac{e^{-h\nu/2kT}}{1 - e^{-h\nu/kT}}\right)^{3N} e^{U_0/kT}$$

where $h\nu/k \equiv \Theta_E$ is a constant characteristic of the crystal, and U_0 is the sublimation energy of the crystal. Calculate the heat capacity from this simple partition function and show that at high temperatures, one obtains the law of Dulong and Petit, namely, that $C_V \rightarrow 3Nk$ as $T \rightarrow \infty$.

2–16. In Chapter 13 of this author's textbook *Statistical Thermodynamics*, it is shown that the partition function of an ideal gas of diatomic molecules in an external electric field $\mathscr{E}$ is

$$Q(N, V, T, \mathscr{E}) = \frac{[q(V, T, \mathscr{E})]^N}{N!}$$

where

$$q(V, T, \mathscr{E}) = V\left(\frac{2\pi mkT}{h^2}\right)^{3/2}\left(\frac{8\pi^2 IkT}{h^2}\right)\frac{e^{-h\nu/2kT}}{(1 - e^{-h\nu/kT})}\left(\frac{kT}{\mu\mathscr{E}}\right)\sinh\left(\frac{\mu\mathscr{E}}{kT}\right)$$

Here I is the moment of inertia of the molecule; ν is its fundamental vibrational frequency; and μ is its dipole moment. Using this partition function along with the thermodynamic relation,

$$dA = -S\,dT - p\,dV - M\,d\mathscr{E}$$

where $M = N\bar{\mu}$, where $\bar{\mu}$ is the average dipole moment of a molecule in the direction of the external field $\mathscr{E}$, show that

$$\bar{\mu} = \mu\left[\coth\left(\frac{\mu\mathscr{E}}{kT}\right) - \frac{kT}{\mu\mathscr{E}}\right]$$

Sketch this result versus $\mathscr{E}$ from $\mathscr{E} = 0$ to $\mathscr{E} = \infty$ and interpret it.

2–17. In Chapter 14 we shall derive an *approximate* partition function for a dense gas, which is of the form

$$Q(N, V, T) = \frac{1}{N!}\left(\frac{2\pi mkT}{h^2}\right)^{3N/2}(V - Nb)^N e^{aN^2/VkT}$$

where a and b are constants that are given in terms of molecular parameters. Calculate the equation of state from this partition function. What equation of state is this? Calculate the thermodynamic energy and the heat capacity and compare it to Problem 1–30.

2–18. From electrostatics, the displacement vector D is given by $D = \mathscr{E} + 4\pi P$, where $\mathscr{E}$ is the electric field, and P is the polarization, i.e., the total dipole moment M per unit volume.

The dielectric constant ε is defined by $D=\varepsilon\mathscr{E}$. In the simple case of a parallel plate capacitor, D is the field produced by a set of charges on the plates, and so we can consider it to be external field; M is the total moment (both permanent and induced) of the substance between the plates; and $\mathscr{E}$ is the field between the plates, that is, the force that an infinitesimal change would feel. If only a vacuum existed between the plates, D and $\mathscr{E}$ would be the same. A real substance modifies D such that $D=\mathscr{E}+4\pi(M/V)$, where V is the volume. Since D, $\mathscr{E}$, and M are all in the same direction (at least for simple fluids), D must be $\geq\mathscr{E}$, which says that $\varepsilon\geq 1$.

When an external electric field D is present, the first law of thermodynamics becomes

$$dE=T\,dS-p\,dV-M\,dD+\mu\,dN$$

Problem: Describe *concisely* how one would calculate ε (at least in principle) from statistical mechanics.

2–19. Derive equation 2–40.

CHAPTER 3

OTHER ENSEMBLES AND FLUCTUATIONS

In Chapter 2 we considered an ensemble in which N, V, and T are held fixed for each system. This ensemble is one of many possible ensembles that can be constructed. For example, if we allow the walls of the containers to be permeable to molecular transport, N is no longer fixed for each system, and we no longer have a canonical ensemble. The ensemble in this case is called a grand canonical ensemble and is discussed in Section 3–1. In Section 3–2 we discuss two other ensembles that are often used in statistical thermodynamics: the microcanonical ensemble, in which N, V, and E are fixed, and the isothermal-isobaric ensemble, in which N, T, and p are fixed. The last section, Section 3–3, is devoted to an investigation of fluctuations in statistical thermodynamics.

One of our basic assumptions is that the ensemble average of a mechanical property can be equated to the corresponding thermodynamic function; hence it is important that we investigate the expected spread about the mean value. We show in Section 3–3 that for macroscopic systems the probability distribution of observing some mechanical property is a very narrow Gaussian distribution whose mean is the ensemble average. One important deduction from this result is that the various ensembles are essentially equivalent and that one can choose to work with a partition function on the basis of mathematical convenience rather than on the basis of which thermodynamic variables are used to specify the system of interest.

3–1 GRAND CANONICAL ENSEMBLE

In the previous chapter we treated the canonical ensemble, in which each system is enclosed in a container whose walls are heat conducting, but impermeable to the passage of molecules. The entire ensemble is placed in a heat bath at temperature T until equilibrium is reached, and then is isolated from its surroundings. Each system of the ensemble is specified by N, V, and T. In this section we shall treat a grand canonical ensemble. In a grand canonical ensemble, each system is enclosed in a container whose walls are both heat conducting and permeable to the passage of molecules. The number of molecules in a system, therefore, can range over all possible values, that is, each system is open with respect to the transport of matter. We construct a grand canonical ensemble by placing a collection of such systems in a large heat bath at temperature T and a large reservoir of molecules. After equilibrium is reached, the entire ensemble is isolated from its surroundings. Since the entire ensemble is at equilibrium with respect to the transport of heat and matter, each system is specified by V, T, and μ, where μ is the chemical potential. (If there is more than one component, the chemical potential of each component is the same from system to system.) Figure 3–1 shows a schematic picture of a grand canonical ensemble.

We proceed now in the same manner as in the treatment of the canonical ensemble. In this case, however, we must specify a system not only by which quantum state it is in but also by the number of molecules in the system. For each value of N, there is a set of energy states $\{E_{Nj}(V)\}$. We let a_{Nj} be the number of systems in the ensemble that contain N molecules and are in the state j. Each value of N has a particular set of levels associated with it, so we first specify N and then j. The set of occupation numbers $\{a_{Nj}\}$ is a distribution. By the postulate of equal a priori probabilities, we assume that all states associated with all possible distributions are to be given equal weight or equal probability of occurrence in the ensemble. Each possible distribution must satisfy the following three conditions:

$$\sum_N \sum_j a_{Nj} = \mathscr{A} \tag{3–1}$$

$$\sum_N \sum_j a_{Nj} E_{Nj} = \mathscr{E} \tag{3–2}$$

$$\sum_N \sum_j a_{Nj} N = \mathscr{N} \tag{3–3}$$

The three symbols $\mathscr{A}$, $\mathscr{E}$, and $\mathscr{N}$ denote the number of systems in the ensemble, the total energy of the ensemble (the ensemble is isolated), and the total number of molecules in the ensemble.

For any possible distribution, the number of states is given by

$$W(\{a_{Nj}\}) = \frac{\mathscr{A}!}{\prod_N \prod_j a_{Nj}!} \tag{3–4}$$

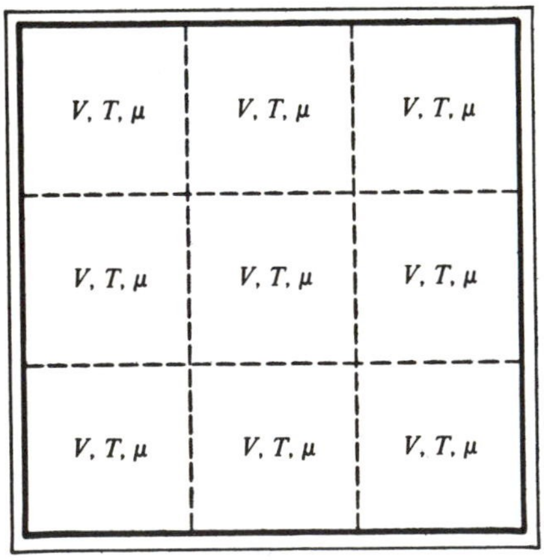

Figure 3–1. **A schematic picture of a grand canonical ensemble. Each system has a fixed volume and temperature, but is open with respect to molecular transport.**

As in the treatment of the canonical ensemble, the distribution that maximizes W subject to the appropriate constraints completely dominates all others. Thus we maximize Eq. (3–4) under the constraints of Eqs. (3–1) through (3–3), respectively, and we get (see Problem 3–1)

$$a_{Nj}^* = e^{-\alpha}e^{-\beta E_{Nj}(V)}e^{-\gamma N} \tag{3–5}$$

As before, the parameter α is easily determined in terms of the other parameter(s). We sum both sides of Eq. (3–5) over N and j and use Eq. (3–1) to get

$$P_{Nj}(V, \beta, \gamma) = \frac{a_{Nj}^*}{\mathscr{A}} = \frac{e^{-\beta E_{Nj}(V)}e^{-\gamma N}}{\sum_N \sum_j e^{-\beta E_{Nj}(V)}e^{-\gamma N}} \tag{3–6}$$

where $P_{Nj}(V, \beta, \gamma)$ is the probability that any randomly chosen system contains N molecules and be in the jth energy state, with energy $E_{Nj}(V)$.

The averages of the mechanical properties E, P, and N are

$$\bar{E}(V, \beta, \gamma) = \frac{1}{\Xi}\sum_N \sum_j E_{Nj}(V)e^{-\beta E_{Nj}(V)}e^{-\gamma N}$$

$$= -\left(\frac{\partial \ln \Xi}{\partial \beta}\right)_{V,\gamma} \tag{3–7}$$

$$\bar{p}(V, \beta, \gamma) = \frac{1}{\Xi}\sum_N \sum_j \left(-\frac{\partial E_{Nj}}{\partial V}\right)e^{-\beta E_{Nj}(V)}e^{-\gamma N}$$

$$= \frac{1}{\beta}\left(\frac{\partial \ln \Xi}{\partial V}\right)_{\beta,\gamma} \tag{3–8}$$

$$\bar{N}(V, \beta, \gamma) = \frac{1}{\Xi}\sum_N \sum_j N e^{-\beta E_{Nj}(V)}e^{-\gamma N}$$

$$= -\left(\frac{\partial \ln \Xi}{\partial \gamma}\right)_{V,\beta} \tag{3–9}$$

where

$$\Xi(V, \beta, \gamma) = \sum_N \sum_j e^{-\beta E_{Nj}(V)}e^{-\gamma N} \tag{3–10}$$

We now determine β and γ. In our treatment of the canonical ensemble, one of the methods used to determine β was to derive an equation that related $(\partial \bar{E}/\partial V)_{N,\beta}$ to $(\partial \bar{p}/\partial \beta)_{N,V}$ and to compare this with a purely thermodynamic equation relating $(\partial E/\partial V)_{N,T}$ to $(\partial p/\partial T)_{N,V}$ [*cf.* Eqs. (2–17) to (2–21)]. This comparison suggested that β was proportional to $1/T$. We then showed that any two systems at the same temperature have the same value of β, thus proving that $\beta = 1/kT$, where k is a universal constant. We can do the same thing here (Problem 3–2), but it is not necessary.

A grand canonical ensemble can be considered to be a collection of canonical ensembles in thermal equilibrium with each other but with all possible values of N. Each of the systems has the same value of β, regardless of the number of molecules it contains. That β has the same value as in the canonical ensemble can be seen by imagining that we suddenly make the walls of the containers impermeable to the molecules but still heat conducting. This gives us a collection of canonical ensembles with V, N, and T fixed, and the arguments of Chapter 2 can be used to show that $\beta = 1/kT$.

The value of γ can be found by using the same method that we used in Chapter 2 to show that β was an integrating factor of δq_{rev}. Consider the function

$$f(\beta, \gamma, \{E_{Nj}(V)\}) = \ln \Xi = \ln \sum_N \sum_j e^{-\beta E_{Nj}(V)} e^{-\gamma N}$$

As the notation indicates, we regard f to be a function of β, γ, and the E_{Nj}'s. The total derivative of f is

$$df = \left(\frac{\partial f}{\partial \beta}\right)_{\gamma, \{E_{Nj}\}} d\beta + \left(\frac{\partial f}{\partial \gamma}\right)_{\beta, \{E_{Nj}\}} d\gamma + \sum_N \sum_j \left(\frac{\partial f}{\partial E_{Nj}}\right)_{\beta, \gamma, E_{Nj}\text{'s}} dE_{Nj}$$

Using Eqs. (3–6) through (3–10), we have

$$df = -\bar{E}\, d\beta - \bar{N}\, d\gamma - \beta \sum_N \sum_j P_{Nj}\, dE_{Nj}$$

The last term here is the ensemble average reversible work done by the systems. For simplicity, we assume only $p-V$ work to get

$$df = -\bar{E}\, d\beta - \bar{N}\, d\gamma + \beta \bar{p}\, dV$$

Paralleling our development in Chapter 2, we add $d(\beta \bar{E}) + d(\gamma \bar{N})$ to both sides of this equation:

$$d(f + \beta \bar{E} + \gamma \bar{N}) = \beta\, d\bar{E} + \beta \bar{p}\, dV + \gamma\, d\bar{N}$$

If we compare this to the purely thermodynamic equation

$$T\, dS = dE + p\, dV - \mu\, dN$$

and use the fact that $\beta = 1/kT$, we can conclude that

$$\gamma = \frac{-\mu}{kT} \tag{3–11}$$

$$S = \frac{\bar{E}}{T} - \frac{\bar{N}\mu}{T} + k \ln \Xi \tag{3–12}$$

In Eq. (3–12), we have set the constant of integration equal to zero in accord with the third law of thermodynamics (see Problem 3–5).

We have now brought the entropy, a nonmechanical property, into our discussion. Equation (3–12), along with Eqs. (3–6) through (3–9), allows us to express any thermodynamic function of interest in a grand canonical ensemble in terms of $\Xi(V, T, \mu)$. This function is called the *grand (canonical ensemble) partition function*:

$$\Xi(V, T, \mu) = \sum_N \sum_j e^{-E_{Nj}(V)/kT} e^{\mu N/kT} \tag{3–13}$$

As the canonical partition function is the connection between thermodynamics and statistical thermodynamics for closed, isothermal systems (N, V, and T fixed), the grand partition function serves as the link for open, isothermal systems (V, T, and μ fixed). If we can determine Ξ for a system, we can calculate its thermodynamic properties.

By summing over j for fixed N in Eq. (3–13), we see that it is possible to write Ξ in the form

$$\Xi(V, T, \mu) = \sum_N Q(N, V, T) e^{\mu N/kT} \tag{3–14}$$

The term $e^{\mu/kT}$ is often denoted by λ. Since $\mu = kT \ln \lambda$, λ is an absolute activity, for the difference in chemical potentials between two states is given by $\Delta\mu = kT \ln(a_2/a_1)$, where a_1 and a_2 are activities.

Since we take the number of systems in an ensemble to be arbitrarily large, the number of particles in an ensemble becomes arbitrarily large, and hence the possible number of particles in any one system can approach infinity. Therefore the summation in Eq. (3–14) can be taken from 0 to ∞:

$$\Xi(V, T, \mu) = \sum_{N=0}^{\infty} Q(N, V, T)\lambda^N \tag{3–15}$$

Even though it may appear from Eq. (3–15) that Ξ would be more difficult to obtain than Q, it actually turns out in many problems that Ξ is easier to obtain, since the constraint of constant N is often mathematically awkward. This constraint can be avoided by using a grand partition function, that is, by summing over all values of N (see Section 4–2). Furthermore, there are many systems in which the many-body problem can be reduced to a one-body, two-body problem, and so on. In these cases, the grand partition function is particularly useful.

To complete our discussion of the grand canonical ensemble, we shall show that pV is the thermodynamic characteristic function of $\ln \Xi$. To see this, compare Eq. (3–12) with the thermodynamic equation

$$G = \mu N = E + pV - TS$$

Thus we have

$$pV = kT \ln \Xi(V, T, \mu) \tag{3–16}$$

Problem 1–37 shows that pV is the thermodynamic function whose natural variables are V, T, and μ. Equations (3–8), (3–9), and (3–12) can be derived from Eq. (3–16) and the thermodynamic equation $d(pV) = S\,dT + N\,d\mu + p\,dV$ (see Problem 3–6). Table 3–1 summarizes the formulas of the grand canonical ensemble.

3–2 OTHER ENSEMBLES

We could go on to consider other ensembles. For example, we could construct an ensemble of systems in which the containing walls of each system are heat conducting and flexible, so that each system of the ensemble is described by N, T, and p. The constraints would be on the total energy and total volume of the ensemble, and the partition function would turn out to be (see Problem 3–9):

$$\Delta(N, T, p) = \sum_E \sum_V \Omega(N, V, E) e^{-E/kT} e^{-pV/kT} \tag{3–17}$$

whose characteristic thermodynamic function is the Gibbs free energy, that is,

$$G = -kT \ln \Delta(N, T, p) \tag{3–18}$$

Equation (3–17) is called the isothermal-isobaric partition function. Notice that the natural variables of G are N, T, and p, the variables associated with this ensemble.

If we compare Eq. (3–17) with the two other partition functions that we have derived [Eqs. (2–34) and (3–13)], we see that all three can be obtained by starting with $\Omega(N, V, E)$, multiplying by some appropriate exponential, and summing over one or two of the variables N, V, and E. In a sense, $\Omega(N, V, E)$ is fundamental to all ensembles and, in fact, is itself the partition function for conceptually the most simple ensemble, the one representative of isolated systems. This is called the microcanonical ensemble.

We can apply the results of the previous section to a treatment of an isolated system. The grand canonical ensemble represents a collection of systems whose containing walls allow heat and molecules to pass freely from one system to another. From a physical point of view, the entire grand canonical ensemble is equivalent to one isolated system of volume $\mathscr{A}V$, containing $\mathscr{N}$ molecules and having energy $\mathscr{E}$. The partitions in Fig. 3–1 can be considered to be a conceptual division of one isolated system into $\mathscr{A}$ subsystems. The entropy of the entire ensemble S_e, considered as one isolated system, is $\mathscr{A}S$, where S is the entropy of each of the open, isothermal systems. This entropy is given by Eq. (3–12):

$$S = k(\beta\bar{E} + \gamma\bar{N} + \ln \Xi) \tag{3–19}$$

where we use the notation β and γ for convenience. We use Eqs. (3–7) and (3–9) for $\bar{E}$ and $\bar{N}$:

$$\begin{aligned} S &= k \ln \Xi + k\left(\sum_{N,j} \beta E_{Nj} \frac{e^{-\beta E_{Nj}} e^{-\gamma N}}{\Xi} + \sum_{N,j} \gamma \frac{N e^{-\beta E_{Nj}} e^{-\gamma N}}{\Xi}\right) \\ &= k \ln \Xi + k \sum_{N,j} (\beta E_{Nj} + \gamma N) \frac{e^{-\beta E_{Nj}} e^{-\gamma N}}{\Xi} \\ &= k \ln \Xi - k \sum_{N,j} (\ln a_{Nj}{}^* + \ln \Xi - \ln \mathscr{A}) \frac{a_{Nj}{}^*}{\mathscr{A}} \end{aligned} \tag{3–20}$$

where we have used Eq. (3–6) to write the last line. We can perform the summation over the second two terms in parentheses in Eq. (3–20):

$$S = k \ln \Xi - \frac{k}{\mathscr{A}} \sum_{N,j} a_{Nj}{}^* \ln a_{Nj}{}^* - k \ln \Xi + k \ln \mathscr{A}$$

or

$$\begin{aligned} S_e = \mathscr{A}S &= k\mathscr{A} \ln \mathscr{A} - k \sum_{N,j} a_{Nj}{}^* \ln a_{Nj}{}^* \\ &= k \ln W(\{a_{Nj}{}^*\}) \end{aligned} \tag{3–21}$$

We see that for an isolated system, the entropy is proportional to the logarithm of the number of states available to the system. In another notation, we can write

$$S = k \ln \Omega(N, V, E) \tag{3–22}$$

Equation (3–22) shows that the more states there are available to an isolated system, the higher is its entropy. This equation serves as the basis for qualitative statements concerning entropy and disorder, randomness, and so on. In practice, Eq. (3–22) is not used for the calculation of thermodynamic functions since N, V, and E are all mechanical variables.

The argument leading to Eq. (2–38) incidentally can be immediately applied to Eq. (3–22). For any spontaneous process in an isolated system,

$$\Delta S = k \ln \frac{\Omega_2}{\Omega_1} > 0$$

where 1 and 2 represent the initial and final states, respectively.

Equation (3–22) is due to Boltzmann and is possibly the best-known equation in statistical thermodynamics, mainly for historical reasons. Of course, Boltzmann (1844–1906) did not express his famous equation in terms of quantum states, but rather in a classical mechanical framework. We shall take up classical statistical mechanics in Chapter 7. Boltzmann, in fact, was a great contributor to both equilibrium and nonequilibrium statistical mechanics. He was one of the first to see clearly how probability ideas could be combined with mechanics. Equation (3–22) is carved on his tombstone in the Zentralfriedhoff in Vienna, although the equation is not often used today. However, his contribution to nonequilibrium statistical mechanics is such that to this day the so-called Boltzmann equation (which we shall study in Chapters 18 and 19, *Statistical Mechanics*) still is the fundamental equation describing the transport of dilute gases. It is interesting to note that Boltzmann, who contributed so much to understanding macroscopic phenomena in terms of molecular mechanics, lived at a time when the atomic theory was not so generally accepted as it is today, and his work was severely criticized by some of the leading physicists of the day. He committed suicide in 1906 (for reasons not entirely clear) and never lived to see the full acceptance of his work in statistical mechanics.

Although $\Omega(N, V, E)$ is not generally available, we have determined it for an ideal gas in Section 1–3 [*cf.* Eq. (1–37)]. If we calculate $k \ln \Omega$, neglecting terms of order less than $O(N^{-1})$, we get (see Problem 3–11):

$$S = Nk \ln\left[\left(\frac{2\pi mkT}{h^2}\right)^{3/2} \frac{Ve^{5/2}}{N}\right] \tag{3–23}$$

We shall see later that this equation gives excellent agreement with experiment, but now we simply show that if we use

$$dS = \frac{1}{T}dE + \frac{p}{T}dV - \frac{\mu}{T}dN \tag{3–24}$$

to get

$$\frac{p}{T} = \left(\frac{\partial S}{\partial V}\right)_{N,E}$$

and substitute Eq. (3–23) into this, we find

$$pV = NkT$$

which is the ideal gas equation of state. See Table 3–1 for a summary of the formulas related to the microcanonical ensemble.

It is possible to derive partition functions appropriate to other sets of independent variables, but the four that we have considered above are sufficient for most applications. We shall show that in the limit of large systems in equilibrium, one can choose an ensemble and its partition function on the basis of mathematical convenience rather than on the basis of which thermodynamic variables are used to describe the system. This result will come out of a study of fluctuations, which we turn to now.

3–3 FLUCTUATIONS

The methods that we have developed allow us to calculate ensemble averages of mechanical variables, which we then equate to thermodynamic functions. Equations such as Eq. (2–12) or (3–6) are the probability distributions over which these ensemble

Table 3–1. **A summary of formulas for several types of ensemble**

microcanonical ensemble, $\Omega(N, V, E)$

$$S = k \ln \Omega$$

$$dS = \frac{1}{T} dE + \frac{p}{T} dV - \frac{\mu}{T} dN$$

$$\frac{1}{kT} = \left(\frac{\partial \ln \Omega}{\partial E}\right)_{N,V} \tag{3–25}$$

$$\frac{p}{kT} = \left(\frac{\partial \ln \Omega}{\partial V}\right)_{N,E} \tag{3–26}$$

$$\frac{\mu}{kT} = -\left(\frac{\partial \ln \Omega}{\partial N}\right)_{V,E} \tag{3–27}$$

canonical ensemble, $Q(N, V, T)$

$$A = -kT \ln Q$$

$$dA = -S\,dT - p\,dV + \mu\,dN$$

$$S = k \ln Q + kT\left(\frac{\partial \ln Q}{\partial T}\right)_{N,V} \tag{3–28}$$

$$p = kT\left(\frac{\partial \ln Q}{\partial V}\right)_{N,T} \tag{3–29}$$

$$\mu = -kT\left(\frac{\partial \ln Q}{\partial N}\right)_{V,T} \tag{3–30}$$

$$E = kT^2\left(\frac{\partial \ln Q}{\partial T}\right)_{N,V} \tag{3–31}$$

grand canonical ensemble, $\Xi(V, T, \mu)$

$$pV = kT \ln \Xi$$

$$d(pV) = S\,dT + N\,d\mu + p\,dV$$

$$S = k \ln \Xi + kT\left(\frac{\partial \ln \Xi}{\partial T}\right)_{V,\mu} \tag{3–32}$$

$$N = kT\left(\frac{\partial \ln \Xi}{\partial \mu}\right)_{V,T} \tag{3–33}$$

$$p = kT\left(\frac{\partial \ln \Xi}{\partial V}\right)_{\mu,T} = kT\frac{\ln \Xi}{V} \tag{3–34}$$

isothermal-isobaric ensemble, $\Delta(N, T, p)$

$$G = -kT \ln \Delta$$

$$dG = -S\,dT + V\,dp + \mu\,dN$$

$$S = k \ln \Delta + kT\left(\frac{\partial \ln \Delta}{\partial T}\right)_{N,p} \tag{3–35}$$

$$V = -kT\left(\frac{\partial \ln \Delta}{\partial p}\right)_{N,T} \tag{3–36}$$

$$\mu = -kT\left(\frac{\partial \ln \Delta}{\partial N}\right)_{T,p} \tag{3–37}$$

averages are taken. In Section 1–5 we saw that the average is the first of a family of moments. Another important moment is the second central moment or the variance, $\overline{(x-\bar{x})^2}$, which is a measure of the spread of a probability distribution about the mean value. Furthermore, we saw toward the end of Section 1–5 that the most meaningful measure of the spread of a distribution is the square root of the variance, that is, the standard deviation, relative to the mean value. A standard deviation of 10^{10} may be large as an absolute number, but it is extremely small if the mean of the probability distribution is 10^{20}. In this section we shall calculate the variances of several mechanical variables and compare these to the mean values.

Any deviation of a mechanical variable from its mean value is called a fluctuation, and the investigation of the probability of such deviations is called fluctuation theory. Fluctuation theory is important in statistical mechanics for a number of reasons. The most obvious reason is to determine to what extent we expect to observe deviations from the mean values that we calculate. If the spread about these is large, then experimentally we would observe a range of values, whose mean or average is given by statistical thermodynamics. We shall see, however, that the probability of observing any value other than the mean value is extremely remote. As a corollary to this important result, we shall see that all of the ensembles that we have considered earlier are equivalent for all practical purposes. In addition, there are several statistical thermodynamical theories of solutions and light scattering based on fluctuation theory, and one formulation of the statistical mechanical theory of transport focuses on the rate of decay of spontaneous fluctuations.

Let us consider first fluctuations in a canonical ensemble. In a canonical ensemble, N, V, and T are held fixed, and we can investigate fluctuations in the energy, pressure, and related properties since these are the ones that vary from system to system. It is important to be aware of the properties that can vary and those properties that are fixed in each ensemble. We shall consider fluctuations in the energy. Thus we use Eq. (2–12) for the probability distribution of the energy and write for the variance

$$\begin{aligned}\sigma_E{}^2 &= \overline{(E-\bar{E})^2} = \overline{E^2} - \bar{E}^2 \\ &= \sum_j E_j{}^2 P_j - \bar{E}^2\end{aligned} \tag{3–38}$$

where

$$P_j = \frac{e^{-\beta E_j}}{Q(N, V, \beta)} \tag{3–39}$$

We can write Eq. (3–38) in a more convenient form by noting that

$$\begin{aligned}\sum_j E_j{}^2 P_j &= \frac{1}{Q}\sum_j E_j{}^2 e^{-\beta E_j} = -\frac{1}{Q}\frac{\partial}{\partial\beta}\sum_j E_j e^{-\beta E_j} \\ &= -\frac{1}{Q}\frac{\partial}{\partial\beta}(\bar{E}Q) = -\frac{\partial\bar{E}}{\partial\beta} - \bar{E}\frac{\partial \ln Q}{\partial\beta} \\ &= kT^2\frac{\partial\bar{E}}{\partial T} + \bar{E}^2\end{aligned} \tag{3–40}$$

Thus Eq. (3–38) becomes

$$\sigma_E{}^2 = kT^2\left(\frac{\partial\bar{E}}{\partial T}\right)_{N,V} \tag{3–41}$$

and if we associate $\bar{E}$ with the thermodynamic energy, we have

$$\sigma_E^2 = kT^2C_V \tag{3–42}$$

where C_V is the molar heat capacity.

To explore the *relative* magnitude of this spread, we look at

$$\frac{\sigma_E}{\bar{E}} = \frac{(kT^2C_V)^{1/2}}{\bar{E}} \tag{3–43}$$

To get an order-of-magnitude estimate of this ratio, we use the values of $\bar{E}$ and C_V for an ideal gas, namely, $O(NkT)$ and $O(Nk)$, respectively. If we use these values in Eq. (3–43), we find that $\sigma_E/\bar{E}$ is $O(N^{-1/2})$, showing that in a typical macroscopic system, the relative deviations from the mean are extremely small. The probability distribution of the energy may, therefore, be regarded as a Gaussian distribution which is practically a delta function.

We can derive a Gaussian distribution approximation to $P(E)$, the probability of observing a particular value of E in a canonical ensemble. According to Eq. (2–36), $P(E)$ is given by $C\Omega(E)e^{-E/kT}$, where C is a normalization factor which is independent of E. Since $\Omega(E)$ is an increasing function of E, and $e^{-E/kT}$ is a decreasing function of E, their product $P(E)$ peaks at some value of E, say E^*. But we have just seen above that the spread about the maximum value is extremely small, and so E^* and $\bar{E}$ are essentially the same point. The width of $P(E)$ is $O(N^{-1/2})$, and so E^* and $\bar{E}$ differ by $O(N^{-1/2})$.

Let us now expand $P(E)$ in a Taylor series about E^*, or $\bar{E}$. As in Section 1–5, it is more convenient to work with $\ln P(E)$. From the definition of $E^*(\approx \bar{E})$ as the value of E at the maximum in $P(E)$,

$$\left(\frac{\partial \ln P}{\partial E}\right)_{E=E^*=\bar{E}} = \left(\frac{\partial \ln \Omega}{\partial E}\right)_{E=E^*=\bar{E}} - \beta = 0 \tag{3–44}$$

Equation (3–44) determines $\bar{E}$ as a function of β. The second derivative of $\ln P(E)$ is

$$\left(\frac{\partial^2 \ln P}{\partial E^2}\right) = \left(\frac{\partial^2 \ln \Omega}{\partial E^2}\right)$$

which is to be evaluated at $E = E^* = \bar{E}$. Since

$$\left(\frac{\partial^2 \ln \Omega}{\partial E^2}\right)_{E=\bar{E}} = \frac{\partial^2 \ln \Omega(\bar{E})}{\partial \bar{E}^2} = \frac{\partial}{\partial \bar{E}}\left(\frac{\partial \ln \Omega(\bar{E})}{\partial \bar{E}}\right)$$
$$= \frac{\partial}{\partial \bar{E}}\left(\frac{\partial \ln \Omega}{\partial E}\right)_{E=\bar{E}} = \frac{\partial \beta}{\partial \bar{E}}$$

where the last term follows from Eq. (3–44), we have

$$\left(\frac{\partial^2 \ln P}{\partial E^2}\right)_{E=E^*=\bar{E}} = \frac{\partial \beta}{\partial \bar{E}} = -\frac{1}{kT^2}\frac{\partial T}{\partial \bar{E}} = -\frac{1}{kT^2C_V} \tag{3–45}$$

The Taylor expansion of $\ln P(E)$ through quadratic terms is

$$\ln P(E) = \ln P(\bar{E}) - \frac{(E-\bar{E})^2}{2kT^2C_V} + \cdots \tag{3–46}$$

or

$$P(E) = P(\bar{E})\exp\left\{-\frac{(E-\bar{E})^2}{2kT^2C_V}\right\} \tag{3–47}$$

Problem 3–16 involves showing that terms beyond the quadratic terms can be ignored in Eq. (3–46).

If we compare Eq. (3–47) to the standard form of a Gaussian distribution [Eq. (1–72)], we see that $\sigma_E^2 = kT^2C_V$ (in agreement with Eq. (3–42)] and that the normalization constant $P(\bar{E})$ is $(2\pi\sigma_E^2)^{-1/2}$. Equation (3–47) can be used to calculate the probability of observing a value of E that differs from $\bar{E}$. For example, the probability of observing an energy that differs by 0.1 percent from the average energy of 1 mole of an ideal gas $O(e^{-10^6})$, an extremely small number. (See Problem 3–12.)

Incidentally, the derivation of Eq. (3–47) is a case where we must be careful not to confuse the variable E in $P(E)$ with E^*, $\bar{E}$, or the thermodynamic quantity E, which unfortunately is also called "E." This is especially true of Eq. (3–44), where E is a variable, and $\bar{E}$ is that particular value of the variable for which the quantity $\partial \ln \Omega/\partial E$, a function of E, is equal to the preassigned value of β (N, V, T are given in a canonical ensemble).

We could also calculate the fluctuations in the pressure in a canonical ensemble, but this is left to Problem 3–18. Instead, we consider the fluctuations in a grand canonical ensemble. In a grand canonical ensemble, V, T, and μ are held fixed, while the energy and number of particles in each system are allowed to vary. We can calculate the fluctuation in the number of particles in the same manner as we treated the fluctuation in energy in a canonical ensemble. If σ_N^2 is the variance in the number of particles, then

$$\sigma_N^2 = \overline{N^2} - \bar{N}^2 = \sum_{N,j} N^2 P_{Nj} - \bar{N}^2 \tag{3–48}$$

where

$$P_{Nj} = \frac{e^{-\beta E_{Nj}} e^{-\gamma N}}{\Xi(V, \beta, \gamma)}$$

We treat $\overline{N^2}$ in analogy to Eq. (3–40):

$$\begin{aligned}\sum_{N,j} N^2 P_{Nj} &= \frac{1}{\Xi} \sum_{N,j} N^2 e^{-\beta E_{Nj}} e^{-\gamma N} = -\frac{1}{\Xi}\frac{\partial}{\partial \gamma} \sum_{N,j} N e^{-\beta E_{Nj}} e^{-\gamma N} \\ &= -\frac{1}{\Xi}\frac{\partial}{\partial \gamma}(\bar{N}\Xi) = -\frac{\partial \bar{N}}{\partial \gamma} - \bar{N}\frac{\partial \ln \Xi}{\partial \gamma} \\ &= kT\left(\frac{\partial \bar{N}}{\partial \mu}\right)_{V,T} + \bar{N}^2 \end{aligned}\tag{3–49}$$

Thus Eq. (3–48) becomes

$$\sigma_N^2 = kT\left(\frac{\partial \bar{N}}{\partial \mu}\right)_{V,T} \tag{3–50}$$

The right-hand side of this equation can be written in a more familiar form by thermodynamic manipulations. Problem 3–26 proves that

$$\left(\frac{\partial \mu}{\partial \bar{N}}\right)_{V,T} = -\frac{V^2}{N^2}\left(\frac{\partial p}{\partial V}\right)_{N,T}$$

and so

$$\sigma_N^2 = \frac{\bar{N}^2 kT\kappa}{V} \tag{3–51}$$

where κ is the isothermal compressibility

$$\kappa \equiv -\frac{1}{V}\left(\frac{\partial V}{\partial p}\right)_{N,T} \tag{3-52}$$

The value of σ_N relative to $\bar{N}$ is

$$\frac{\sigma_N}{\bar{N}} = \left(\frac{kT\kappa}{V}\right)^{1/2} \tag{3-53}$$

To get an order-of-magnitude estimate of this ratio, we use the fact that $\kappa = 1/p$ for an ideal gas to get $\sigma_N/\bar{N} = N^{-1/2}$. Again we find that relative deviations from the mean are very small. The result, $0(N^{-1/2})$, is typical of fluctuations in statistical thermodynamics.

Since V is fixed in the grand canonical ensemble, the fluctuation in the number of particles is proportional to the fluctuation in the density ρ, and so

$$\frac{\sigma_\rho}{\bar{\rho}} = \frac{\sigma_N}{\bar{N}} = \left(\frac{kT\kappa}{V}\right)^{1/2} \tag{3-54}$$

There is a condition under which the fluctuations in density are not negligible. At the critical point of a substance, $(\partial p/\partial V)_{N,T}$ is zero, and hence its isothermal compressibility is infinite. Thus there are large fluctuations in the density from point to point in a fluid at its critical point. This is observed macroscopically by the phenomenon of critical opalescence, in which a pure substance becomes turbid at its critical point.

We can also derive a Gaussian approximation to $P(N)$. Let $N^*(=\bar{N})$ be the value of N at the peak in $P(N)$. We have

$$P(N) = CQ(N, V, T)e^{\beta\mu N}$$

where C is a normalization constant. Then

$$\left(\frac{\partial \ln P}{\partial N}\right)_{N=N^*} = \left(\frac{\partial \ln Q}{\partial N}\right)_{N=\bar{N}} + \beta\mu = 0$$

This equation determines $\bar{N}$ as a function of $\beta\mu$. Also,

$$\left(\frac{\partial^2 \ln P}{\partial N^2}\right)_{N=\bar{N}} = \left(\frac{\partial^2 \ln Q}{\partial N^2}\right)_{N=\bar{N}} = \frac{\partial}{\partial \bar{N}}\frac{\partial \ln Q(\bar{N}, V, T)}{\partial \bar{N}} = -\frac{\partial \beta\mu}{\partial \bar{N}}$$

$$= -\frac{1}{kT(\partial \bar{N}/\partial\mu)_{V,T}}$$

Thus we find

$$P(N) = P(\bar{N})\exp\left[\frac{-(N-\bar{N})^2}{2kT(\partial\bar{N}/\partial\mu)_{V,T}}\right] \tag{3-55}$$

which gives the same expression for σ_N^2 as Eq. (3–50). Problems 3–19 through 3–20 involve the determination of fluctuations in the isothermal-isobaric ensemble.

An interesting application of the above fluctuation formulas is to the scattering of light by the atmosphere. It can be shown that if light of intensity I_0 is incident on a region of volume V with a dielectric constant ε, which differs from the average value of ε for the medium $\bar{\varepsilon}$, the intensity of light scattered at an angle θ at a distance R is

$$\frac{I(\theta)}{I_0} = \frac{\pi^2 V^2 \sigma_\varepsilon^2}{2\lambda^4}\frac{(1+\cos^2\theta)}{R^2} \tag{3-56}$$

where σ_ε^2 is the variance of ε, and λ is the wavelength of the incident light in vacuum.

This is called Rayleigh scattering. The dielectric constant ε is related to the density by the so-called Clausius-Mossotti equation

$$\frac{\varepsilon - 1}{\varepsilon + 2} = A\rho \tag{3-57}$$

which we shall derive in Chapter 13, but which is also in most physical chemistry texts. The quantity A is a constant, and ρ is the density. We can see from this equation that fluctuations in ρ lead to fluctuations in ε, and hence to Rayleigh scattering by Eq. (3–56). If we calculate σ_ε^2 in terms of σ_ρ^2 from Eq. (3–57), and use Eq. (3–54) for σ_ρ^2, we find (see Problem 3–21)

$$\frac{I(\theta)}{I_0} = \frac{\pi^2 kT}{18\lambda^4}\kappa(\varepsilon - 1)^2(\varepsilon + 2)^2 V \frac{(1 + \cos^2\theta)}{R^2}$$

where κ is the isothermal compressibility. By integrating this over the surface of a sphere of radius R, we obtain finally

$$\frac{I_{\text{scattered}}}{I_0} = \int \frac{I(\theta)}{I_0} R^2 \sin\theta \, d\theta \, d\phi$$

$$= \frac{8\pi^3}{27\lambda^4} kT\kappa(\varepsilon - 1)^2(\varepsilon + 2)^2 V$$

This equation shows that the blue color of the sky is due to fluctuations in the density of the atmosphere. The λ^4 in the denominator gives rise to a strong dependence on wavelength, so that the short wavelengths (blue) of the sun's light are scattered more than the red, and hence the sky appears blue. Similarly, red sunsets and sunrises are due to the fact that the long wavelengths (red) are not scattered as much as the blue.

There is one result of fluctuation theory which will be very useful to us. We have stated above that the various ensembles and their partition functions are essentially equivalent to each other, and that one can choose to work with a partition function on the basis of mathematical convenience. We now show why this is so.

Consider the canonical partition function:

$$Q(N, V, T) = \sum_E \Omega(N, V, E)e^{-E/kT} \tag{3-58}$$

We have seen in Eq. (3–47) that $P(E) = C\Omega(E)\exp(-E/kT)$ is an extremely narrow Gaussian function of E. In the limit of large N (and it is only in the limit of large N that classical thermodynamics is valid), only one value of E is important, namely, $E = E^* = \bar{E}$. Thus in the summation in Eq. (3–58), only the term with $E = \bar{E}$ contributes, and Eq. (3–58) becomes

$$Q(N, V, T) = \Omega(N, V, \bar{E})e^{-\bar{E}/kT} \tag{3-59}$$

Although the systems of a canonical ensemble can, in principle, assume any value of E (as long as it is an eigenvalue of the N-particle Schrödinger equation), it happens that the energy of the entire ensemble is distributed uniformly throughout the ensemble, and each system is almost certain to be found with the average energy $\bar{E}$. A canonical ensemble degenerates, in a sense, to a microcanonical ensemble.

If we take the logarithm of Eq. (3–59) and use Eq. (2–35), we find that

$$A = \bar{E} - kT \ln \Omega(N, V, \bar{E})$$

or that

$$S = k \ln \Omega(N, V, \bar{E})$$

This is an alternative derivation of the fundamental relation between the entropy and the number of states accessible to the system.

The general results we have obtained here are also obtained for other ensembles. For example, although the systems of a grand canonical ensemble can assume any value of N and E, in practice it turns out that the total energy and the total number of molecules of the entire ensemble are distributed uniformly throughout the ensemble, and each system has the average energy and contains the average number of molecules. This, of course, is exactly what one expects intuitively, as long as the systems are of macroscopic size and the density is not *extremely* low.

These results can be used to write down, by inspection, the characteristic thermodynamic function of any partition function. Equations (3–58) and (3–59) are a good example. Suppose we did not know that $A = -kT \ln Q$. We do know that $S = k \ln \Omega$, however, and so if we take the logarithm of Eq. (3–59), we get that $\ln Q = S/k - \beta E$, which shows that $A = -kT \ln Q$. Since partition functions, in general, are a sum of $\Omega(N, V, E)$ multiplied by exponential factors, this method can always be used to determine the thermodynamic characteristic function. (See Problem 3–15.)

ADDITIONAL READING

General

HILL, T. L. 1960. *Statistical thermodynamics.* Reading, Mass.: Addison-Wesley. Sections 1–5 through 1–7 and Chapter 2.

———. 1956. *Statistical mechanics.* New York: McGraw-Hill. Sections 14, 15, and Chapter 4.

KESTIN, J., and DORFMAN, J. R. 1971. *A course in statistical thermodynamics.* New York: Academic. Section 5–18.

KNUTH, E. 1966. *Statistical thermodynamics.* New York: McGraw-Hill. Chapters 3, 4, and 5.

KUBO, R. 1965. *Statistical mechanics.* Amsterdam: North-Holland Publishing Co. Sections 1–12 to 1–14.

REIF, F. 1965. *Statistical and thermal physics.* New York: McGraw-Hill. Chapter 3.

RUSHBROOKE, G. S. 1949. *Statistical mechanics.* London: Oxford University Press. Chapters 15 and 17.

Fluctuations

ANDREWS, F. C. 1963. *Equilibrium statistical mechanics.* New York: Wiley. Chapter 33.

DAVIDSON, N. 1962. *Statistical mechanics.* New York: McGraw-Hill. Chapter 14.

KESTIN, J., and DORFMAN, J. R. 1971. *A course in statistical thermodynamics.* New York: Academic. Chapter 14.

LANDAU, L. D., and LIFSHITZ, E. M. 1958. *Statistical physics.* London: Pergamon Press. Chapter 12.

MÜNSTER, A. 1969. *Statistical thermodynamics,* Vol. I. Berlin: Springer-Verlag. Chapter 3.

SCHRÖDINGER, E. 1952. *Statistical thermodynamics.* Cambridge: Cambridge University Press. Chapter 5.

TOLMAN, R. C. 1938. *Statistical mechanics.* London: Oxford University Press. Section 141.

PROBLEMS

3–1. Derive Eq. (3–5).

3–2. Using a grand canonical formalism, show that any two systems at the same temperature have the same value of β.

3–3. For a grand canonical ensemble show that

$$\left(\frac{\partial \bar{E}}{\partial V}\right)_{\gamma, \beta} + \beta\left(\frac{\partial \bar{p}}{\partial \beta}\right)_{\gamma, V} = -\bar{p}$$

Compare this to the thermodynamic equation (see Problem 1–31)

$$\left(\frac{\partial E}{\partial V}\right)_{\mu/T, 1/T} + \frac{1}{T}\left(\frac{\partial p}{\partial(1/T)}\right)_{\mu/T, V} = -p$$

to suggest that $\beta = \text{const}/T$ for a grand canonical ensemble.

3–4. State and use Euler's theorem to show

$$p = kT\left(\frac{\partial \ln \Xi}{\partial V}\right)_{\mu, T} = kT\frac{\ln \Xi}{V}$$

3–5. Show that the entropy given by Eq. (3–12) goes to zero as T goes to zero.

3–6. Derive the principal thermodynamic connection formulas of the grand canonical ensemble starting from

$$pV = kT \ln \Xi$$

and

$$d(pV) = S\, dT + N\, d\mu + p\, dV$$

3–7. Show that for a two-component system

$$\Xi(\mu_1, \mu_2, T, V) = \sum_{N_1} \sum_{N_2} Q(N_1, N_2, V, T)\lambda_1^{N_1}\lambda_2^{N_2}$$

where $\lambda_i = e^{\mu_i/kT}$ $(i = 1, 2)$. From this derive the corresponding thermodynamic connection formulas.

3–8. In the next chapter we shall see that the grand partition function of an ideal monatomic gas is

$$\Xi = e^{q\lambda}$$

where $q = (2\pi mkT/h^2)^{3/2}V$. Derive the thermodynamic properties of an ideal monatomic gas from Ξ.

3–9. Show that the partition function appropriate to an isothermal-isobaric ensemble is

$$\Delta(N, p, T) = \sum_{E} \sum_{V} \Omega(N, V, E)e^{-E/kT}\, e^{-pV/kT}$$

Derive the principal thermodynamic connection formulas for this ensemble.

3–10. In Problem 5–17 we shall show that the isothermal-isobaric partition function of an ideal monatomic gas is

$$\Delta = \left[\frac{(2\pi m)^{3/2}(kT)^{5/2}}{ph^3}\right]^N$$

Derive the thermodynamic properties of an ideal monatomic gas from Δ.

3–11. Derive Eq. (3–23) starting from Eq. (1–37).

3–12. Calculate the probability of observing an energy that differs by 10^{-4} percent from the average energy of 1 mole of an ideal gas.

3–13. Show that for macroscopic ideal systems, ones obtains the same result for the entropy whether one uses $S = k \ln \Phi(E)$, where Φ is the number of quantum states with energy $\leq E$ [Eq. (1–36)], or $S = k \ln \Omega(E, \Delta E)$, where $\Omega(E, \Delta E)$ is the number of quantum states within energy ΔE about E [Eq. (1–37)] as long as $\Delta E/E$ is small, but not zero. Show that S is insensitive to ΔE over a wide range of ΔE. The next problem discusses this remarkable result more generally.

3–14. Let $\Omega(E)\, dE$ be the number of quantum states between E and $E + dE$. In Chapter 1 we showed that $\Omega(E)$ is a monotonically increasing function of E (at least for an ideal gas). We can write two obvious inequalities for $\Omega(E)$:

$$\Phi(E) \equiv \int_0^E \Omega(E')\, dE' \geq \Omega(E)\, \Delta E$$

$$E\Omega(E) \geq \int_0^E \Omega(E')\, dE' = \Phi(E)$$

where ΔE is a small region surrounding E. By multiplying the second inequality by $\Delta E/E$, we get

$$\Omega(E)\,\Delta E \geq \frac{\Delta E}{E}\int_0^E \Omega(E')\,dE'$$

Combining this inequality with the first one above gives

$$\frac{\Delta E}{E}\int_0^E \Omega(E')\,dE' \leq \Omega(E)\,\Delta E \leq \int_0^E \Omega(E')\,dE'$$

Taking logarithms gives

$$\ln \Phi(E) - \ln\left(\frac{E}{\Delta E}\right) \leq \ln\,[\Omega(E)\,\Delta E] \leq \ln \Phi(E)$$

Now unless ΔE is extremely small, $\ln(E/\Delta E)$ is completely negligible compared to $\ln \Phi(E)$, since the total number of states with energies equal to or less than E is at least $0(e^N)$. Show that even if the energy could be measured to a millionth of a percent, $\ln(E/\Delta E) \approx 18$, which is completely negligible compared to N.

3–15. Fluctuation theory provides a simple method to determine the characteristic function associated with a particular partition function. Consider the canonical partition function

$$Q(N, V, T) = \sum_E \Omega(N, V, E)\,e^{-E/kT}$$

According to the theory of fluctuations, there is effectively only one term in this summation, and so we write

$$Q(N, V, T) = \Omega(N, V, \bar{E})\,e^{-\bar{E}/kT}$$

Remembering that $S = k \ln \Omega$, we have, upon taking logarithms, that

$$\ln Q = \frac{S}{k} - \frac{E}{kT}$$

or that

$$\ln Q = \frac{-A}{kT}$$

Proceeding in a like manner, determine the characteristic thermodynamic function of the following partition functions:

$$\Xi(V, T, \mu) = \sum_N Q(N, V, T)\,e^{\beta\mu N}$$

$$\Delta(p, T, N) = \sum_V Q(N, V, T)\,e^{-\beta p V}$$

$$\phi(V, E, \beta\mu) = \sum_N \Omega(N, V, E)\,e^{\beta\mu N}$$

$$\Psi(V, T, \mu_1, N_2) = \sum_{N_1} Q(N_1, N_2, T, V)\,e^{\beta\mu N_1}$$

$$W(p, \gamma, T, N) = \sum_V \sum_{\mathscr{A}} Q(N, V, \mathscr{A}, T)\,e^{-\beta p V}\,e^{\beta\gamma\mathscr{A}}$$

where $\mathscr{A}$ is surface area, and γ is the surface tension.

3–16. When we derived the Gaussian expression for $P(E)$ in a canonical ensemble, we expanded $\ln P(E)$ in a Taylor expansion about $E = E^* \approx \bar{E}$, dropping terms after the quadratic term. Show that these terms are negligible.

3–17. Show that

$$\overline{(E-\bar{E})^3}=k^2\left\{T^4\left(\frac{\partial C_V}{\partial T}\right)+2T^3C_V\right\}$$

and that

$$\frac{\overline{(E-\bar{E})^3}}{\bar{E}^3}=0(N^{-2})$$

for a canonical ensemble.

3–18. Derive an expression for the fluctuation in the pressure in a canonical ensemble.

3–19. Show that for an isothermal-isobaric ensemble

$$P(V)=P(V^*)\exp\left\{\frac{(V-\bar{V})^2}{2kT\left(\frac{\partial V}{\partial p}\right)_{N,T}}\right\}$$

3–20. Derive an equation for the fluctuation in the volume in an isothermal-isobaric ensemble. In other words, derive an equation for $\overline{V^2}-\bar{V}^2$. Express your answer in terms of the isothermal compressibility, defined by

$$\kappa=-\frac{1}{V}\left(\frac{\partial V}{\partial p}\right)_{N,T}$$

Show that $\sigma_v/\bar{V}$ is of the order of $N^{-1/2}$.

3–21. By calculating σ_ε^2 in terms of σ_ρ^2 from Eq. (3–57) and using Eq. (3–54) for σ_ρ^2, show that

$$\frac{I(\theta)}{I_0}=\frac{\pi^2kT}{18\lambda^4}\,\kappa(\varepsilon-1)^2(\varepsilon+2)^2V\frac{(1+\cos^2\theta)}{R^2}$$

3–22. Show that the fluctuation in energy in a grand canonical ensemble is

$$\sigma_E^2=(kT^2C_V)+\left(\frac{\partial\bar{E}}{\partial\bar{N}}\right)_{T,V}\sigma_N^2$$

3–23. Show that in a two-component open, isothermal ensemble that

$$\overline{N_1N_2}-\bar{N}_1\bar{N}_2=kT\left(\frac{\partial\bar{N}_1}{\partial\mu_2}\right)_{V,T,\mu_1}$$

$$=kT\left(\frac{\partial\bar{N}_2}{\partial\mu_1}\right)_{V,T,\mu_2}$$

3–24. Show that

$$\overline{H^2}-\bar{H}^2=kT^2C_p$$

in an N, p, T ensemble.

3–25. Use the formulas in Table 3–1 to derive expressions for any other thermodynamic functions for each of the four ensembles listed there.

3–26. Show that

$$\left(\frac{\partial\mu}{\partial N}\right)_{V,T}=-\frac{V^2}{N^2}\left(\frac{\partial p}{\partial V}\right)_{N,T}$$

CHAPTER 4

BOLTZMANN STATISTICS, FERMI-DIRAC STATISTICS, AND BOSE-EINSTEIN STATISTICS

The results that we have derived up to now are valid for macroscopic systems. In order to apply these equations, it is necessary to have the set of eigenvalues $\{E_j(N, V)\}$ of the N-body Schrödinger equation. In general, this is an impossible task. There are many important systems, however, in which the N-body Hamiltonian operator can be written as a sum of independent individual Hamiltonians. In such cases the total energy of the system can be written as a sum of individual energies. This leads to a great simplification of the partition function, and allows us to apply the results with relative ease.

We shall see that the final equations depend upon whether the individual particles of the system are fermions (that is, the N-body wave function is antisymmetric under the interchange of identical particles) or bosons (the N-body wave function is symmetric under the interchange of identical particles). These two types of particles obey different laws, called Fermi-Dirac or Bose-Einstein statistics. We shall show that under normal conditions (for example, sufficiently high temperatures), both of these distribution laws can be approximately reduced to an even simpler one, called Boltzmann statistics. The Boltzmann distribution law can also be derived from $Q(N, V, T)$ at high temperature without first deriving the Fermi-Dirac and Bose-Einstein distribution laws, and this is done in Section 4–1. We shall discuss in this section just what is meant by "normal" conditions or "sufficiently high" temperatures. Then in Section 4–2 we derive the two fundamental distribution laws, Fermi-Dirac and Bose-Einstein statistics, and show how both of them reduce to Boltzmann statistics in the appropriate limit.

4–1 THE SPECIAL CASE OF BOLTZMANN STATISTICS

In Section 1–3 it was shown that if the Hamiltonian of a many-body system can be written as a sum of one-body Hamiltonians, the energy of the system is the sum of individual energies, and the wave function is a product of the single-particle wave functions. In addition, the wave functions of a system of identical particles must satisfy certain symmetry requirements with respect to the interchange of the particles. All known particles fall into two classes: those whose wave function must be symmetric under the operation of the interchange of two identical particles, and those whose wave function must be antisymmetric under such an exchange. Particles belonging to the first class are called bosons, and the others are called fermions. There is no restriction of the distribution of bosons over their available energy states, but fermions have the very severe restriction that no two identical fermions can occupy the same single-particle energy state. This restriction follows immediately from the requirement that the wave function be antisymmetric (see Problem 1–26). These considerations become important in enumerating the many-body energy states available to the system.

There are many problems in which the Hamiltonian can be written as a sum of simpler Hamiltonians. The most obvious example perhaps is the case of a dilute gas, where the molecules are on the average far apart, and hence their intermolecular interactions can be neglected. Another example, which may be familiar from physical chemistry, is the decomposition of the Hamiltonian of a polyatomic molecule into its various degrees of freedom:

$$\mathscr{H} \approx \mathscr{H}_{\text{translational}} + \mathscr{H}_{\text{rotational}} + \mathscr{H}_{\text{vibrational}} + \mathscr{H}_{\text{electronic}} \tag{4–1}$$

Equation (4–1) is a good first approximation and can be systematically corrected by the introduction of small interaction terms.

There are many other problems in physics in which the Hamiltonian, by a proper and clever selection of variables, can be written as a sum of individual terms. Although these individual terms need not be Hamiltonians for actual individual molecules, they are nevertheless used to define the so-called quasi-particles, which mathematically behave like independent real particles. Some of the names of quasi-particles that are found in the literature are photons, phonons, plasmons, magnons, rotons, and other "ons." In spite of the apparent limitation of this requirement on the Hamiltonian, we shall see that it is very useful and can be used to study solids (Chapter 11) and liquids (Chapter 12), systems in which the decomposition of a many-body Hamiltonian into a sum of independent terms would hardly appear to be justified. First let us consider the canonical partition function for a system of distinguishable particles, in which the Hamiltonian can be written as a sum of individual terms. Denote the individual energy states by $\{\varepsilon_j^{\,a}\}$, where the superscript denotes the particle (they are distinguishable), and the subscript denotes the state. In this case, the canonical partition function becomes

$$\begin{aligned} Q(N, V, T) &= \sum_j e^{-E_j/kT} = \sum_{i,j,k,\ldots} e^{-(\varepsilon_i^a + \varepsilon_j^b + \varepsilon_k^c + \cdots)/kT} \\ &= \sum_i e^{-\varepsilon_i^a/kT} \sum_j e^{-\varepsilon_j^b/kT} \sum_k e^{-\varepsilon_k^c/kT} \cdots \\ &= q_a q_b q_c \cdots \end{aligned} \tag{4–2}$$

where

$$q(V, T) = \sum_i e^{-\varepsilon_i/kT} \tag{4–3}$$

Equation (4–2) is a very important result. It shows that if we can write the N-particle Hamiltonian as a sum of independent terms, and if the particles are *distinguishable*, then the calculation of $Q(N, V, T)$ reduces to a calculation of $q(V, T)$. Since $q(V, T)$

requires a knowledge only of the energy values of an individual particle or quasi-particle, its evaluation is quite feasible. In most cases $\{\varepsilon_i\}$ is a set of molecular energy states; thus $q(V, T)$ is called a molecular partition function.

If the energy states of all the particles are the same, then Eq. (4–2) becomes

$$Q(N, V, T) = [q(V, T)]^N \qquad \text{(distinguishable particles)} \tag{4–4}$$

Equation (4–4) shows that the original N-body problem (the evaluation of $Q(N, V, T)$) can be reduced to a one-body problem (the evaluation of $q(V, T)$) if the particles are independent and distinguishable. Although particles are certainly not distinguishable in general, there are many important cases where they can be treated as such. An excellent example of this is a perfect crystal. In a perfect crystal each atom is confined to one and only one lattice point, which we could, in principle, identify by a set of three numbers. Since each particle, then, is confined to a lattice point and the lattice points are distinguishable, the particles themselves are distinguishable. Furthermore, we shall see in Chapter 11 that although there are strong intermolecular interactions in crystals, we can treat the vibration of each particle about its lattice point as independent to a first approximation.

Another useful application of the separation indicated in Eq. (4–2) is to the molecular partition function itself. Equation (4–1) shows that the molecular Hamiltonian can be approximated by a sum of Hamiltonians for the various degrees of freedom of the molecule. Consequently we get the useful result that

$$q_{\text{molecule}} = q_{\text{translational}} q_{\text{rotational}} q_{\text{vibrational}} q_{\text{electronic}} \cdots \tag{4–5}$$

where, for example,

$$q_{\text{translational}} = \sum_i e^{-\varepsilon_i^{\text{trans}}/kT} \tag{4–6}$$

Thus not only can we reduce an N-body problem to a one-body problem, but it is possible to reduce it further into the individual degrees of freedom of the single particles.

Equation (4–4) is an attractive result, but atoms and molecules are, in general, not distinguishable; thus the utility of Eq. (4–4) is severely limited. The situation becomes more complicated when the inherent indistinguishability of atoms and molecules is considered. In this case, the N-body energy is

$$E_{ijkl\cdots} = \varepsilon_i + \varepsilon_j + \varepsilon_k + \varepsilon_l + \cdots \tag{4–7}$$

and the partition function is

$$Q(N, V, T) = \sum_{i, j, k, l \cdots} e^{-(\varepsilon_i + \varepsilon_j + \varepsilon_k + \varepsilon_l + \cdots)/kT} \tag{4–8}$$

Because the molecules are indistinguishable, one cannot sum over $i, j, k, l, \ldots$ separately as we did to get Eq. (4–2).

Consider, for example, the case of fermions. The antisymmetry of the wave function requires that no two identical fermions can occupy the same single-particle energy state. Thus in Eq. (4–8), terms in which two or more indices are the same cannot be included in the summation. The indices i, j, k, l, and so on, are not independent of one another, and a direct evaluation of $Q(N, V, T)$ for fermions by means of Eq. (4–8) is very difficult.

Bosons do not have the restriction that no two can occupy the same molecular state, but the summation in Eq. (4–8) is still complicated. Consider a term in Eq. (4–8) in which all of the indices are the same except one, that is, a term of the form $\varepsilon_i + \varepsilon_j + \varepsilon_j + \varepsilon_j + \cdots$ with $i \neq j$. Because the particles are indistinguishable, the position of ε_i is unimportant, and so this state is identical with $\varepsilon_j + \varepsilon_i + \varepsilon_j + \varepsilon_j + \cdots$ or $\varepsilon_j + \varepsilon_j + \varepsilon_i + \varepsilon_j + \varepsilon_j + \cdots$, and so on. Such a state should be included only once in Eq. (4–8), but an unrestricted summation over the indices in Eq. (4–8) would produce N terms of this type. Consider the other extreme in which all of the particles are in different molecular states, that is, the state with energy $\varepsilon_i + \varepsilon_j + \varepsilon_k + \cdots$ with $i \neq j \neq k \neq \cdots$. Because the particles are indistinguishable, the $N!$ states obtained by permuting the N different subscripts are identical and should occur only once in Eq. (4–8). Such terms will, of course, appear $N!$ times in an unrestricted summation. Consequently, a direct evaluation of Q for bosons by means of Eq. (4–8) also is difficult.

The terms that introduce complications are those in which two or more indices are the same. If it were not for this kind of term, one could carry out the summation in Eq. (4–8) in an unrestricted manner, and then correct the sum by dividing by $N!$ It turns out that this procedure yields an excellent approximation in many (most) cases for the following reason.

We showed in Section 1–3 that for a particle in a box, the number of molecular quantum states with energy $\leq \varepsilon$ is

$$\Phi(\varepsilon) = \frac{\pi}{6}\left(\frac{8ma^2\varepsilon}{h^2}\right)^{3/2} \tag{4–9}$$

For $m = 10^{-22}$ g, $a = 10$ cm, and $T = 300°$K, $\Phi(\varepsilon) = O(10^{30})$. Although this calculation is done for one particle in a cube (i.e., one molecule of an ideal gas), the order of magnitude of the result is general. Thus we see that the number of molecular quantum states available to a molecule at room temperature, say, is much greater than the number of molecules in the system for all but the most extreme densities. Since each particle has many individual states to choose from, it will be a rare event for two particles to be in the same molecular state. Therefore the vast majority of terms in Eq. (4–8) will have all different indices. This allows us to sum over all the indices unrestrictedly and divide by $N!$ to get

$$Q(N, V, T) = \frac{q^N}{N!} \qquad \text{(indistinguishable particles)} \tag{4–10}$$

with

$$q(V, T) = \sum_j e^{-\varepsilon_j/kT}$$

for a system of identical, indistinguishable particles satisfying the condition that the number of available molecular states is much greater than the number of particles.

Equation (4–10) is an extremely important result, since it reduces a many-body problem to a one-body problem. No longer is there a condition of distinguishability; the indistinguishability of the particles has been included by dividing by $N!$, a valid procedure for most systems under most conditions. We can investigate this condition in more detail using Eq. (4–9) for an ideal gas. Mathematically, we require that

$$\Phi(\varepsilon) \gg N$$

Using Eq. (4–9), we have the condition

$$\frac{\pi}{6}\left(\frac{12mkT}{h^2}\right)^{3/2} \gg \frac{N}{V} \tag{4–11}$$

where we have set $\varepsilon = 3kT/2$. Clearly this condition is favored by large mass, high temperature, and low density. Numerically it turns out that (4–11) is satisfied for all but the very lightest molecules at very low temperatures. Table 4–1 examines this condition for a number of systems. We see that the use of Eq. (4–10) is justified in most cases. We have examined (4–11) for only monatomic systems, but the results are valid for polyatomic molecules as well, since the translational energy states account for almost all of the energy states available to any molecule.

When Eq. (4–10) is valid, that is, when the number of available molecular states is much greater than the number of particles in the system, we say that the particles obey *Boltzmann statistics*. Boltzmann statistics is an approximation that becomes increasingly better at higher temperatures. We shall show in Chapter 7 that at high enough temperatures, one can describe the energy of a system by classical mechanics. Since the limiting case of Boltzmann statistics and the use of classical mechanics both require a high-temperature limit, Boltzmann statistics is also called the classical limit.

Let us examine Eq. (4–10). The total energy of the N-body system is

$$E = N\bar{\varepsilon} = kT^2\left(\frac{\partial \ln Q}{\partial T}\right)_{N,V} = N\sum_j \varepsilon_j \frac{e^{-\varepsilon_j/kT}}{q} \tag{4–12}$$

The first equality is valid, because the molecules are assumed to be independent, and hence their energies are additive. We see from Eq. (4–12) that the average energy of a particle is

$$\bar{\varepsilon} = \sum_j \varepsilon_j \frac{e^{-\varepsilon_j/kT}}{q} \tag{4–13}$$

We can conclude from this equation that the probability that a molecule is in the jth energy state

$$\pi_j = \frac{e^{-\varepsilon_j kT}}{\sum_j e^{-\varepsilon_j kT}} = \frac{e^{-\varepsilon_j kT}}{q} \tag{4–14}$$

It is interesting to note that the fluctuations in ε are of the same order as ε itself (see Problems 4–18 and 4–19), that is, the probability distribution for single molecules is not sharp. A sharp probability distribution is a many-body effect.

Table 4–1. **The quantity $(6N/\pi V)(h^2/12mkT)^{3/2}$ for a number of simple systems***

	T (°K)	$\frac{6N}{\pi V}\left(\frac{h^2}{12mkT}\right)^{3/2}$
liquid helium	4	1.6
gaseous helium	4	0.11
gaseous helium	20	2.0×10^{-3}
gaseous helium	100	3.5×10^{-5}
liquid neon	27	1.1×10^{-2}
gaseous neon	27	8.2×10^{-5}
gaseous neon	100	3.1×10^{-6}
liquid argon	86	5.1×10^{-4}
gaseous argon	86	1.6×10^{-6}
liquid krypton	127	5.4×10^{-5}
gaseous krypton	127	2.0×10^{-7}
electrons in metals (sodium)	300	1465

* This quantity must be much less than unity for Eq. (4–10) to be valid. The temperatures associated with the liquid states are the normal boiling points [*cf.* Eq. (4–11)].

The similarity between Eq. (4–14) for molecular states and Eq. (2–12) for states of the entire N-body system is not fortuitous. Equation (4–14) can be derived by the same *mathematical* formalism of Chapter 2. The ensemble is considered to be the N actual molecules in thermal contact with each other. The number of molecules n_j in the state with energy ε_j is found by maximizing a combinatorial factor similar to Eq. (1–77). This point of view was the one originally proposed by Boltzmann. It is valid only for systems in which the total energy is a sum of individual molecular energies, that is, only for dilute gases. The conceptual generalization of these ideas by Gibbs was a magnificent achievement, which allowed statistical thermodynamics to be applicable to all physical systems. Furthermore, the derivation given in Chapter 2 is rigorous, since macroscopic systems can be labeled, and the size of the ensemble can be increased arbitrarily. This is not so for the Boltzmann approach, since the molecules cannot be labeled, and the system is finite.

Equation (4–14) can be reduced further if we assume that the energy of the molecule can be written in the form [*cf.* Eq. (4–1)]

$$\varepsilon = \varepsilon_i^{\text{trans}} + \varepsilon_j^{\text{rot}} + \varepsilon_k^{\text{vib}} + \varepsilon_l^{\text{elec}} + \cdots$$

Then Eq. (4–14) and Eq. (4–5) can be combined to give, for example,

$$\pi_j^{\text{vib}} = \frac{e^{-\varepsilon_j^{\text{vib}}}}{q_{\text{vib}}} \tag{4–15}$$

for the probability that a molecule is in the jth vibrational state irrespective of the other degrees of freedom.

Although Eq. (4–10) is applicable to most systems, it is important to complete the development of systems of independent, indistinguishable particles by evaluating Eq. (4–8) for the general case. The exact evaluation of Eq. (4–8) is necessary for several systems that we shall study. We must return, then, to a consideration of the effect of the symmetry requirements of N-body wave functions on the sum over states in Eq. (4–8).

4–2 FERMI-DIRAC AND BOSE-EINSTEIN STATISTICS

There are two cases to consider in the evaluation of Eq. (4–8). The resultant distribution function in the case of fermions is called Fermi-Dirac statistics, and that in the case of bosons is called Bose-Einstein statistics. Since all known particles are either fermions or bosons, these two "statistics" are the only exact distributions. We shall see, however, that in the case of high temperature and/or low density, both of these distributions go over into the Boltzmann or classical distribution.

It is most convenient to treat the general case by means of the grand canonical ensemble for reasons that we shall see shortly. Let $E_j(N, V)$ be the energy states available to a system containing N molecules. Let ε_k be the molecular quantum states. Finally, let $n_k = n_k(E_j)$ be the number of molecules in the kth molecular state when the system itself is in the quantum state with energy E_j. A quantum state of the entire system is specified by the set $\{n_k\}$. The energy of the system is

$$E_j = \sum_k \varepsilon_k n_k \tag{4–16}$$

and, of course,

$$N = \sum_k n_k \tag{4–17}$$

We can write $Q(N, V, T)$ as

$$Q(N, V, T) = \sum_j e^{-\beta E_j} = \sum_{\{n_k\}}^{*} e^{-\beta \sum_i \varepsilon_i n_i} \tag{4–18}$$

where the asterisk in the summation signifies the restriction that

$$\sum n_k = N$$

This restriction turns out to be mathematically awkward. We can avoid this restriction by using the grand canonical partition function instead. This will be an excellent example where one partition function is much easier to evaluate than another. Since we have demonstrated the equivalence of ensembles, we are free to make the choice strictly on mathematical convenience. We then use

$$\Xi(V, T, \mu) = \sum_{N=0}^{\infty} e^{\beta\mu N} Q(N, V, T)$$

We use Eq. (4–18) for $Q(N, V, T)$ and the absolute activity $\lambda = e^{\beta\mu}$ to get

$$\begin{aligned}\Xi(V, T, \mu) &= \sum_{N=0}^{\infty} \lambda^N \sum_{\{n_k\}}^{*} e^{-\beta \sum_i \varepsilon_i n_i} \\ &= \sum_{N=0}^{\infty} \sum_{\{n_k\}}^{*} \lambda^{\sum n_i} e^{-\beta \sum_j \varepsilon_j n_j} \\ &= \sum_{N=0}^{\infty} \sum_{\{n_k\}}^{*} \prod_k (\lambda e^{-\beta \varepsilon_k})^{n_k} \end{aligned} \tag{4–19}$$

Now comes the crucial step (which requires some thought). Since we are summing over all values of N, each n_k ranges over all possible values, and Eq. (4–19) can be written as (see Problem 4–6)

$$\Xi(V, T, \mu) = \sum_{n_1=0}^{n_1^{\max}} \sum_{n_2=0}^{n_2^{\max}} \cdots \prod_k (\lambda e^{-\beta \varepsilon_k})^{n_k} \tag{4–20}$$

Equations (4–19) and (4–20) are completely equivalent. Equation (4–20) can be written in a more lucid form:

$$\Xi(V, T, \mu) = \sum_{n_1=0}^{n_1^{\max}} (\lambda e^{-\beta \varepsilon_1})^{n_1} \sum_{n_2=0}^{n_2^{\max}} (\lambda e^{-\beta \varepsilon_2})^{n_2} \cdots$$

or

$$= \prod_k \sum_{n_k=0}^{n_k^{\max}} (\lambda e^{-\beta \varepsilon_k})^{n_k} \tag{4–21}$$

Equation (4–21) is a simple product and is a general result. The crucial step in this series of equations is the step from Eq. (4–19) to Eq. (4–20), from which Eq. (4–21) follows immediately. The step from Eqs. (4–19) to (4–20) is possible only because we are summing over all values of N, or, in other words, since we are using the grand canonical partition function.

We now apply Eq. (4–21) to fermions and bosons. In *Fermi-Dirac* statistics, each of the n_k in Eq. (4–21) can be only either 0 or 1, since no two particles can be in the same quantum state. In this case $n_1^{\max} = 1$, and Eq. (4–21) is simply

$$\Xi_{\text{FD}} = \prod_k (1 + \lambda e^{-\beta \varepsilon_k}) \tag{4–22}$$

where FD, of course, signifies Fermi-Dirac.

In *Bose-Einstein* statistics, on the other hand, the n_k can be 0, 1, 2, ..., since there is no restriction on the occupancy of each state. Therefore, $n_k^{\max} = \infty$, and Eq. (4–21) becomes

$$\Xi_{BE} = \prod_k \sum_{n_k=0}^{\infty} (\lambda e^{-\beta\varepsilon_k})^{n_k} = \prod_k (1 - \lambda e^{-\beta\varepsilon_k})^{-1} \qquad \lambda e^{-\beta\varepsilon_k} < 1 \tag{4–23}$$

To get Eq. (4–23), we have used the fact that

$$\sum_{j=0}^{\infty} x^j = (1 - x)^{-1}$$

for $x < 1$.

Equations (4–22) and (4–23) are the two fundamental distributions of the statistical thermodynamics of systems of independent particles. We can combine these two equations into

$$\Xi_{\substack{FD\\BE}} = \prod_k (1 \pm \lambda e^{-\beta\varepsilon_k})^{\pm 1} \tag{4–24}$$

where as the notation indicates, the upper sign refers to Fermi-Dirac statistics, and the lower sign refers to Bose-Einstein statistics.

Using Eq. (3–33), we see that

$$\bar{N} = N = \sum_k \bar{n}_k = kT\left(\frac{\partial \ln \Xi}{\partial \mu}\right)_{V,T} = \lambda\left(\frac{\partial \ln \Xi}{\partial \lambda}\right)_{V,T} = \sum_k \frac{\lambda e^{-\beta\varepsilon_k}}{1 \pm \lambda e^{-\beta\varepsilon_k}} \tag{4–25}$$

The average number of particles in the kth quantum state is

$$\bar{n}_k = \frac{\lambda e^{-\beta\varepsilon_k}}{1 \pm \lambda e^{-\beta\varepsilon_k}} \tag{4–26}$$

Equation (4–26) is the quantum statistical counterpart of Eq. (4–14). We multiply Eq. (4–26) by ε_k and sum over k to get the quantum statistical version of Eq. (4–13).

$$\bar{E} = N\bar{\varepsilon} = \sum_k \bar{n}_k \varepsilon_k = \sum_k \frac{\lambda \varepsilon_k e^{-\beta\varepsilon_k}}{1 \pm \lambda e^{-\beta\varepsilon_k}} \tag{4–27}$$

Lastly, Eq. (3–16) gives

$$pV = \pm kT \sum_k \ln[1 \pm \lambda e^{-\beta\varepsilon_k}] \tag{4–28}$$

Equations (4–25) through (4–28) are the fundamental formulas of Fermi-Dirac (+) and Bose-Einstein (−) statistics. Note that the molecular partition function q is not a relevant quantity when we are dealing with quantum statistics, that is, Fermi-Dirac or Bose-Einstein statistics. In spite of the fact that we have neglected intermolecular forces, the individual particles of the system are not independent because of the symmetry requirements of the wave functions.

We noted above that both kinds of statistics should go over into Boltzmann or classical statistics in the limit of high temperature or low density, where the number of available molecular quantum states is much greater than the number of particles. This condition implies that the average number of molecules in any state is very small, since most states will be unoccupied and those few states that are occupied will most likely contain only one molecule. This means that $\bar{n}_k \to 0$ in Eq. (4–26). This is achieved by letting $\lambda \to 0$. Thermodynamically, this means the limit of $N/V \to 0$ for fixed T, or $T \to \infty$ for fixed N/V. (See Problem 4–3.) For small λ, Eq. (4–26) becomes

$$\overline{n_k} = \lambda e^{-\beta\varepsilon_k} \qquad (\lambda \text{ small})$$

If we sum both sides of this equation over k to eliminate λ, we have

$$\frac{\bar{n}_k}{N} = \frac{e^{-\beta\varepsilon_k}}{q} \tag{4–29}$$

where

$$q = \sum_j e^{-\beta\varepsilon_j} \tag{4–30}$$

Equation (4–26) then goes over to the Boltzmann or classical limit for both Fermi-Dirac and Bose-Einstein statistics.

Equations (4–27) and (4–28) also reduce to the formulas of Section 4–1 as $\lambda \to 0$. Equation (4–27) becomes

$$\bar{E} \to \sum_j \lambda\varepsilon_j e^{-\beta\varepsilon_j}$$

and since $n_j \to \lambda e^{-\beta\varepsilon_j}$, we have

$$\bar{\varepsilon} = \frac{\bar{E}}{\bar{N}} \to \frac{\sum_j \varepsilon_j e^{-\beta\varepsilon_j}}{\sum_j e^{-\beta\varepsilon_j}} \tag{4–31}$$

This is the same as Eq. (4–13). Similarly, for small λ we can expand the logarithm in Eq. (4–28) to get

$$pV \to (\pm kT)\left(\pm\lambda \sum_j e^{-\beta\varepsilon_j}\right) \tag{4–32}$$

We have used the fact that $\ln(1 + x) \approx x$ for small x. Using Eq. (4–30), this becomes

$$pV = \lambda kT \sum_j e^{-\beta\varepsilon_j} = \lambda kTq \tag{4–33}$$

or

$$\beta pV = \ln \Xi = \lambda q \tag{4–34}$$

Equation (3–33) can be used to show that $\lambda q = N$, and so Eq. (4–34) is the perfect gas law as expected. Thus the formulas of Fermi-Dirac and Bose-Einstein statistics reduce to those of Boltzmann statistics in the classical limits.

We can also derive Eq. (4–10) directly from Eq. (4–34) for Ξ:

$$\Xi = e^{\lambda q} = \sum_{N=0}^{\infty} \frac{(\lambda q)^N}{N!}$$

If we compare this to Eq. (3–15), see that

$$Q(N, V, T) = \frac{q^N}{N!}$$

We shall defer a discussion of the equations of Fermi-Dirac and Bose-Einstein statistics to Chapter 10. There are a few systems such as electrons in metals, liquid helium, electromagnetic radiation, for which one must use quantum statistics. For most systems that we shall study in this book, however, we shall be able to use Boltzmann or classical statistics. In the next chapter we shall apply the limit of Boltzmann statistics to the simplest system, namely, a monatomic ideal gas.

ADDITIONAL READING

General

ANDREWS, F. C. 1963. *Equilibrium statistical mechanics.* New York: Wiley. Chapter 17.
HILL, T. L. 1960. *Statistical thermodynamics.* Reading, Mass.: Addison-Wesley. Chapter 3.
KITTEL, C. 1969. *Thermal physics.* New York: Wiley. Chapter 9.
KNUTH, E. 1966. *Statistical thermodynamics.* New York: McGraw-Hill. Chapter 6.
KUBO, R. 1965. *Statistical mechanics.* Amsterdam: North-Holland Publishing Co. Section 1–15 and Example 1–12.
MANDL, F. 1971. *Statistical physics.* New York: Wiley. Chapter 11.
REIF, F. 1965. *Statistical and thermal physics.* New York: McGraw-Hill. Sections 9–1 through 9–8.
RUSHBROOKE, G. S. 1949. *Statistical mechanics.* London: Oxford University Press. Chapters 2 and 3.
TOLMAN, R. C. 1938. *Statistical mechanics.* London: Oxford University Press. Chapters 10 and 14.

PROBLEMS

4–1. Calculate the temperature below which each of the substances listed below cannot be treated classically at 1 atmosphere. Compare this with the normal boiling temperature for each substance.

He, Ne, Ar, Kr, CO_2, N_2, H_2, Cl_2, H_2O

4–2. Show that the quantity

$$\frac{6N}{\pi V}\left(\frac{h^2}{12mkT}\right)^{3/2}$$

given in Table 4–1 is indeed very large for electrons in metals at room temperature.

4–3. Show that the condition that $\lambda \to 0$ corresponds thermodynamically to the limit $N/V \to 0$ for fixed T, or $T \to \infty$ for fixed density. Remember that $\lambda = e^{\beta\mu}$.

4–4. In deriving the limiting case of Boltzmann statistics, we claimed that if the number of quantum states M far exceeds the number of particles N, then the terms in the product of the molecular partition functions in which each particle is in a different quantum state constitute the overwhelming number of terms. Show, in fact, that the ratio of this type of term to the total number of terms approaches unity as $N/M \to 0$, N and M both large. Hint: remember that

$$\lim_{x\to\infty}\left(1+\frac{a}{x}\right)^x = e^a$$

4–5. For an ideal gas, show that the relation

$$P = \frac{2}{3}\frac{E_{\text{kin}}}{V}$$

holds irrespective of its statistics, where E_{kin} is the total kinetic energy.

4–6. To convince yourself of the step leading from Eq. (4–19) to Eq. (4–20), consider the summation

$$S = \sum_{N=0}^{\infty} \sum_{\{n_j\}}{}^{*} x_1{}^{n_1} x_2{}^{n_2}$$

where n_1 and $n_2 = 0$, 1, and 2. Show by directly expanding S for this simple case that this is equivalent to [Eq. (4–21)]

$$S = \prod_{K=1}^{2} (1 + x_K + x_K{}^2)$$

4–7. Recall that the equation of state for an ideal quantum gas is

$$pV = kT \ln \Xi = \pm kT \sum_j \ln\,[1 \pm \lambda e^{-\epsilon_j/kT}]$$

where $\lambda = e^{\mu/kT}$. Using the fact that the summation over states can be replaced by an integration over energy levels

$$\omega(\varepsilon)\, d\varepsilon = 2\pi\left(\frac{2m}{h^2}\right)^{3/2} V\varepsilon^{1/2}\, d\varepsilon$$

derive the quantum virial expansion

$$\frac{p}{kT} = \mp\frac{1}{\Lambda^3}\sum_{j=1}^{\infty}\frac{(\mp 1)^j\lambda^j}{j^{5/2}}$$

where $\Lambda = (h^2/2\pi mkT)^{1/2}$.

4–8. Show that the entropy of an ideal quantum gas can be written as

$$S = -k\sum_j [\bar{n}_j \ln \bar{n}_j \pm (1 \mp \bar{n}_j)\ln(1 \mp \bar{n}_j)]$$

where the upper (lower) sign denotes Fermi-Dirac (Bose-Einstein) statistics.

4–9. Show that $pV \geq \langle N\rangle kT$ for fermions, and $pV \leq \langle N\rangle kT$ for bosons.

4–10. Consider a system of independent, distinguishable particles, each of which has only two accessible states; a ground state of energy 0 and an excited state of energy ε. If the system is in equilibrium with a heat bath of temperature T, calculate A, E, S, and C_v. Sketch C_v versus T. Does the choice of the ground-state energy $= 0$ affect P, C_v, or S? How would your results change if ε_0 were added to both energy values?

4–11. Generalize Eq. (4–10) to the case of a mixture of several different species of non-interacting particles.

4–12. Consider a system of N distinguishable independent particles, each of which can be in the state $+\varepsilon_0$ or $-\varepsilon_0$. Let the number of particles with energy $\pm\varepsilon_0$ be $N_\pm$, so that the energy is

$$E = N_+\varepsilon_0 - N_-\varepsilon_0 = 2N_+\varepsilon_0 - N\varepsilon_0$$

Evaluate the partition function Q by summing $\exp(-E/kT)$ over levels and compare your result to $Q = q^N$. Do not forget the degeneracy of the levels, which in this case is the number of ways that N_+ particles out of N can be in the $+$ state. Calculate and plot the heat capacity C_V for this system.

4–13. The vibrational energy levels of a diatomic molecule can be approximated by a quantum mechanical harmonic oscillator. The fundamental vibrational frequency ν is $0(10^{13}\ \text{sec}^{-1})$ for many diatomic molecules. Calculate the fraction of molecules in the first few vibrational levels in an ideal diatomic gas at 25°C. Derive a closed expression for the fraction of molecules in all excited states.

4–14. The rotational energy of diatomic molecules can be well approximated by a quantum mechanical rigid rotor. According to Eq. (1–32), the energy levels depend upon the moment of inertia, which for a diatomic molecule is $0(10^{-40}\ \text{g-cm}^2)$. Calculate and plot the population of rotational levels of a diatomic ideal gas at 25°C. Do not forget to include the degeneracy $2J + 1$.

4–15. Show that $Q(N, V, T) = [q(V, T)]^N/N!$ implies that $q(V, T) = f(T)V$. Do this in both the canonical ensemble and grand canonical ensemble formalisms.

4–16. Show that the most probable distribution of $2N$ molecules of an ideal gas contained in two equal and connected volumes at the same temperature is N molecules in each volume.

4–17. In Fermi-Dirac statistics, the maximum occupancy of any state is 1, while in Bose-Einstein statistics, it is ∞. All particles appear to obey one of these two statistics. In 1940, however, Gentile* investigated the implications of an intermediate statistics, in which the maximum occupancy is m. Derive the distribution law for this case.

* G. Gentile, *Nuovo Cimento* **17**, p. 493, 1940.

4–18. Derive the equation

$$C_V = \frac{N}{kT^2}[\overline{\varepsilon^2} - \bar{\varepsilon}^2]$$

for independent particles and show that fluctuations of molecular energies are not at all negligible.

4–19. Starting from Eqs. (4–13) and (4–14), show that the fluctuations in ε, the energy of a single particle, are not small, and in fact, are given by

$$\frac{\sigma_\varepsilon}{\bar{\varepsilon}} = N^{1/2}\frac{\sigma_E}{\bar{E}}$$

4–20. Consider a gas in equilibrium with the surface of a solid. Some of the molecules of the gas will be adsorbed onto the surface, and the number adsorbed will be a function of the pressure of the gas. A simple statistical mechanical model for this system is to picture the solid surface to be a two-dimensional lattice of M sites. Each of these sites can be either unoccupied, or occupied by at most one of the molecules of the gas. Let the partition function of an unoccupied site be 1 and that of an occupied site be $q(T)$. (We do not need to know $q(T)$ here.) Assuming that molecules adsorbed onto the lattice sites do not interact with each other, the partition function of N molecules adsorbed onto M sites is then

$$Q(N, M, T) = \frac{M!}{N!(M-N)!}[q(T)]^N$$

The binomial coefficient accounts for the number of ways of distributing the N molecules over the M sites. By using the fact the adsorbed molecules are in equilibrium with the gas phase molecules (considered to be an ideal gas), derive an expression for the fractional coverage, $\theta \equiv N/M$, as a function of the pressure of the gas. Such an expression, that is, $\theta(p)$, is called an adsorption isotherm, and this model gives the so-called Langmuir adsorption isotherm.

4–21. Consider a lattice of M equivalent noninteracting magnetic dipoles, μ (associated, say, with electron or nuclear spins). When placed in a magnetic field H, each dipole can orient itself either in the same direction, $\uparrow$, or opposed to, $\downarrow$, the field. The energy of a dipole is $-\mu H$ if oriented with the field, and $+\mu H$ if oriented against the field. Let N be the number of $\downarrow$ states and $M-N$ the number of $\uparrow$ states. For a given value of N, the total energy is

$$\mu HN - \mu H(M-N) = (2N-M)\mu H$$

The total magnetic moment I is

$$I = (M - 2\bar{N})\mu$$

where $\bar{N}$ is the average value of N for a given M, H, and T. The work necessary to increase H by dH is $-IdH$. Find the specific heat C and the total magnetic moment for this system, and sketch both I versus $\mu H/kT$, that is, the total magnetization versus the applied field, and C/Nk versus $kT/\mu H$.

4–22. (a) Consider a system of M independent and distinguishable macromolecules on which any number from 0 to m small molecules may bind. Let $q(j)$ be the macromolecular partition function when j molecules are bound. If there are N small molecules (or ions) and M macromolecules (say proteins), then

$$Q(N, M, T) = \sum_{\mathbf{a}}{}^{*}\frac{M!\,q(0)^{a_0}q(1)^{a_1}\cdots q(m)^{a_m}}{a_0!\,a_1!\cdots a_m!}$$

where the number of macromolecules having j bound molecules is a_j, and where the asterisk indicates the restrictions

$$\sum_{j=0}^{m} a_j = M \qquad \sum_{j=0}^{m} ja_j = N$$

Show that the grand partition function for this system can be written in the form

$$\Xi(M, T, \mu) = \xi(\mu, T)^M$$

where

$$\xi(\mu, T) = q(0) + q(1)\lambda + \cdots + q(m)\lambda^m$$

Interpret this result.

(b) Extend this result to the case in which the macromolecules are not distinguishable.

CHAPTER 5

IDEAL MONATOMIC GAS

In this chapter, we shall apply the general results of the preceding chapters to an ideal monatomic gas. By ideal, we mean a gas dilute enough that intermolecular interactions can be neglected. The results that we derive here will be applicable to real monatomic gases at pressures and temperatures for which the equation of state is well represented by $pV = NkT$, that is, pressures below 1 atmosphere and temperatures greater than room temperature.

We have shown in Section 4–1 that the number of available quantum states far exceeds the number of particles for an ideal gas. Thus we can write the partition function of the entire system in terms of the individual atomic partition functions:

$$Q(N, V, T) = \frac{[q(V, T)]^N}{N!} \tag{5–1}$$

A monatomic gas has translational, electronic, and nuclear degrees of freedom. The translational Hamiltonian is separable from the electronic and nuclear degrees of freedom, and the electronic and nuclear Hamiltonians are separable to a very good approximation. Thus we have

$$q(V, T) = q_{\text{trans}}\, q_{\text{elect}}\, q_{\text{nucl}} \tag{5–2}$$

We shall study each of these factors separately in the following sections of this chapter.

5–1 THE TRANSLATIONAL PARTITION FUNCTION

In this section we shall evaluate the translational partition function. The energy states are given by

$$\varepsilon_{n_x n_y n_z} = \frac{h^2}{8ma^2}(n_x^2 + n_y^2 + n_z^2) \qquad n_x, n_y, n_z = 1, 2, \ldots \tag{5–3}$$

We substitute this into q_{trans} to get

$$q_{\text{trans}} = \sum_{n_x, n_y, n_z=1}^{\infty} e^{-\beta\varepsilon_{n_x n_y n_z}}$$
$$= \sum_{n_x=1}^{\infty} \exp\left(-\frac{\beta h^2 n_x^2}{8ma^2}\right) \sum_{n_y=1}^{\infty} \exp\left(-\frac{\beta h^2 n_y^2}{8ma^2}\right) \sum_{n_z=1}^{\infty} \exp\left(-\frac{\beta h^2 n_z^2}{8ma^2}\right)$$
$$= \left(\sum_{n=1}^{\infty} \exp\left(-\frac{\beta h^2 n^2}{8ma^2}\right)\right)^3 \tag{5–4}$$

This summation cannot be evaluated in closed form, that is, it cannot be expressed in terms of any simple analytic function. This does not present any difficulty, however, for the following reason. The successive terms in these summations differ so little from each other that the terms vary essentially continuously, and so the summation can, for all practical purposes, be replaced by an integral. To prove this, we show that the argument of the exponential changes little in going from n_x to $n_x + 1$. This difference, Δ, is given by

$$\Delta = \frac{\beta h^2(n_x + 1)^2}{8ma^2} - \frac{\beta h^2 n_x^2}{8ma^2} = \frac{\beta h^2(2n_x + 1)}{8ma^2}$$

At room temperature, for $m = 10^{-22}$ g and $a = 10$ cm, this difference is

$$\Delta \approx (2n_x + 1) \times 10^{-20}$$

A typical value of n_x at room temperature is $O(10^{10})$ (see Problem 5–3), so Δ is indeed very small for all but very large values of n_x. A value of n_x for which Δ is as large as 10^{-5} would correspond to an energy of $(10^{10}kT)$, an extremely improbable energy. Thus we can replace the summation in Eq. (5–4) by an integration:

$$q_{\text{trans}}(V, T) = \left(\int_0^{\infty} e^{-\beta h^2 n^2/8ma^2}\, dn\right)^3 = \left(\frac{2\pi mkT}{h^2}\right)^{3/2} V \tag{5–5}$$

where we have written V for a^3.

It is instructive to evaluate q_{trans} in another way. Equation (5–4) is a sum over the states of the system. We could also write q_{trans} as a sum over levels. Recognizing that the levels are very densely distributed, we can write q_{trans} as an integral:

$$q_{\text{trans}} = \int_0^{\infty} \omega(\varepsilon) e^{-\beta\varepsilon}\, d\varepsilon \tag{5–6}$$

The function $\omega(\varepsilon)$ is the number of energy states between ε and $\varepsilon + d\varepsilon$, or, in other words, the effective degeneracy. Equation (5–6) is simply a continuous form of a sum over levels rather than a sum over states. We have already evaluated $\omega(\varepsilon)$ in Section 1–3. It is given by Eq. (1–35)

$$\omega(\varepsilon)\, d\varepsilon = \frac{\pi}{4}\left(\frac{8ma^2}{h^2}\right)^{3/2} \varepsilon^{1/2}\, d\varepsilon \tag{5–7}$$

If we substitute this into Eq. (5–6), we get

$$q_{\text{trans}} = \frac{\pi}{4}\left(\frac{8ma^2}{h^2}\right)^{3/2} \int_0^{\infty} \varepsilon^{1/2} e^{-\beta\varepsilon}\, d\varepsilon$$
$$= \left(\frac{2\pi mkT}{h^2}\right)^{3/2} V \tag{5–8}$$

Of course, we obtain the same result as Eq. (5–5).

The factor $(h^2/2\pi mkT)^{1/2}$ that occurs in the translational partition function has units of length and is usually denoted by Λ. In this notation, Eq. (5–5) or (5–8) read

$$q_{\text{trans}} = \frac{V}{\Lambda^3} \tag{5-9}$$

The quantity Λ can be given the following interpretation. The average translational or kinetic energy of an ideal gas molecule can be calculated immediately from Eq. (5–8) and Eq. (4–13), which in terms of q_{trans} is

$$\bar{\varepsilon}_{\text{trans}} = kT^2 \left(\frac{\partial \ln q_{\text{trans}}}{\partial T}\right)$$

We find that $\bar{\varepsilon}_{\text{trans}} = \frac{3}{2}kT$, and since $\varepsilon_{\text{trans}} = p^2/2m$, where p^2 is the momentum of the particle, we can say that the average momentum is essentially $(mkT)^{1/2}$. Thus Λ is essentially h/p, which is equal to the De Broglie wavelength of the particle. Consequently, Λ is called the thermal De Broglie wavelength. The condition for the applicability of classical or Boltzmann statistics is equivalent to the condition that $\Lambda^3/V \ll 1$, which physically says that the thermal De Broglie wavelength must be small compared to the dimensions of the container. Such a condition is similar to the condition that quantum effects decrease as the De Broglie wavelength becomes small (*cf.* Table 4–1).

5–2 THE ELECTRONIC AND NUCLEAR PARTITION FUNCTIONS

In this section we shall investigate the electronic and nuclear contributions to q.

It is more convenient to write the electronic partition function as a sum of levels rather than a sum over states. We have, then,

$$q_{\text{elect}} = \sum \omega_{ei} e^{-\beta\varepsilon_i} \tag{5-10}$$

where ω_{ei} is the degeneracy, and ε_i the energy of the ith electronic level. We first fix the arbitrary zero of energy such that $\varepsilon_1 = 0$, that is, we shall measure all of our electronic energies relative to the ground state. The electronic contribution to q can then be written as

$$q_{\text{elect}} = \omega_{e1} + \omega_{e2} e^{-\beta\Delta\varepsilon_{12}} + \cdots \tag{5-11}$$

where $\Delta\varepsilon_{1j}$ is the energy of the jth electronic level relative to the ground state. These $\Delta\varepsilon$'s are typically of the order of electron volts, and so $\beta\Delta\varepsilon$ is typically quite large at ordinary temperatures (see Problem 5–10). Therefore at ordinary temperatures, only the first term in the summation for q_{elect} is significantly different from zero. However, there are some cases, such as the halogen atoms, where the first excited state lies only a fraction of an electron volt above the ground state, so that several terms in q_{elect} are necessary. Even in these cases the sum converges extremely rapidly.

The electronic energies of atoms and ions are determined by atomic spectroscopy and are well tabulated. The standard reference is the tables of Moore* which list the energy levels and energies of many atoms and ions. Table 5–1 lists the first few levels for H, He, Li, O, and F. A look at this table will indicate that electronic states are labeled or characterized by a so-called term symbol, which is briefly explained in Section 5–4. (A knowledge of the meaning of atomic term symbols is not necessary for the calculation of q_{elect}, but they are explained in Section 5–4 for completeness.)

* See Table 5–1.

Table 5–1. **Atomic energy states**

atom	electron configuration	term symbol	degeneracy $g = 2J+1$	energy (cm^{-1})	energy (eV)
H	$1s$	$^2S_{1/2}$	2	0	0
	$2p$	$^2P_{1/2}$	2	82258.907	10.20
	$2s$	$^2S_{1/2}$	2	82258.942	
	$2p$	$^2P_{3/2}$	4	82259.272	
He	$1s^2$	$^1S_0$	1	0	
	$1s2s$	$^3S_1$	3	159850.318	19.82
		$^1S_0$	1	166271.70	
Li	$1s^22s$	$^2S_{1/2}$	2	0	
	$1s^22p$	$^2P_{1/2}$	2	14903.66	1.85
		$^2P_{3/2}$	4	14904.00	
	$1s^23s$	$^2S_{1/2}$	2	27206.12	
O	$1s^22s^22p^4$	$^3P_2$	5	0	
		$^3P_1$	3	158.5	0.02
		$^3P_0$	1	226.5	0.03
		$^1D_2$	5	15867.7	1.97
		$^1S_0$	1	33792.4	4.19
F	$1s^22s^22p^5$	$^2P_{3/2}$	4	0	
		$^2P_{1/2}$	2	404.0	0.05
	$1s^22s^22p^43s$	$^4P_{5/2}$	6	102406.50	12.70
		$^4P_{3/2}$	4	102681.24	
		$^4P_{1/2}$	2	102841.20	
		$^2P_{3/2}$	4	104731.86	
		$^2P_{1/2}$	2	105057.10	

Source: C. E. Moore, "Atomic Energy States," *Natl. Bur. Standards, Circ.*, **1**, p. 467, 1949.

Some general observations about Table 5–1 are: All the rare gases have a ground state $^1S_0$ (called a singlet S) with the first excited state O(10 eV) higher; the alkali metals have a $^2S_{1/2}$ (called a doublet S) ground state with the next state O(1 eV) higher; the halogen atoms have a $^2P_{3/2}$ (called a doublet P) ground state with the next one, a $^2P_{1/2}$ (also a doublet P) only O(0.1 eV) higher. Thus at ordinary temperatures the electronic partition function of the rare gases is essentially unity and that of the alkali metals is 2, while those for halogen atoms consist of two terms.

Using the data in Table 5–1, we can now calculate the fraction of He atoms in the lowest triplet state, $^3S_1$. This fraction is given by

$$f_2 = \frac{\omega_{e2}\, e^{-\beta\Delta\varepsilon_{12}}}{\omega_{e1} + \omega_{e2}\, e^{-\beta\Delta\varepsilon_{12}} + \omega_{e3}\, e^{-\beta\Delta\varepsilon_{13}} + \cdots}$$

$$= \frac{3e^{-\beta\Delta\varepsilon_{12}}}{1 + 3e^{-\beta\Delta\varepsilon_{12}} + \omega_{e3}\, e^{-\beta\Delta\varepsilon_{13}} + \cdots} \tag{5–12}$$

At 300°K, $\beta\,\Delta\varepsilon_{12} = 770$, and so $f_2 \approx 10^{-334}$. Even at 3000°K, $f_2 \approx 10^{-33}$. This is typical of the rare gases. The energy separation must be less than a few hundred cm^{-1} or so before any population of that level is significant. Incidentally, it is useful to know that Boltzmann's constant in units of cm^{-1}/deg-molecule is 0.695 (almost ln 2), and 1 eV = 8065.73 cm^{-1}.

Table 5–2 gives the fraction of fluorine atoms in the first excited electronic state as a function of temperature. It can be seen that fluorine is a case where it is necessary to use two terms in q_{elec}.

We shall write the electronic partition function as

$$q_{\text{elec}}(T) \approx \omega_{e1} + \omega_{e2}\, e^{-\beta\varepsilon_{12}} \tag{5–13}$$

Table 5–2. **The fraction of fluorine atoms in the first excited electronic state as a function of temperature**

T(°K)	f_2
200	0.027
400	0.105
600	0.160
800	0.195
1000	0.219
1200	0.236
2000	0.272

but at temperatures at which the second term is not negligible with respect to the first term, we must check the possible contribution of higher terms as well. This will rarely be necessary, however.

We now consider the nuclear partition function. The nuclear partition function has a form similar to that of the electronic partition function. Nuclear energy levels are separated by millions of electron volts, however, which means that it requires temperatures of the order of 10^{10}°K to produce excited nuclei. At terrestrial temperatures then, we need consider only the first term, that is, the degeneracy of the ground nuclear state ω_{n1}. We take our zero of nuclear energy states to be the ground state. Note that we have taken the overall atomic ground state to be the atom in its ground translational, electronic, and nuclear states. The nuclear partition function, $q_n = \omega_{n1}$, then contributes only a multiplicative constant to Q, and hence affects only the entropy and free energies by a constant additive factor. Since the nuclear state is rarely altered in any chemical process, it does not contribute to thermodynamic changes, and so we shall usually not include it in q. We cannot do this in the case of the electronic contribution since there are many chemical processes in which the electronic states change.

This completes the partition function of monatomic ideal gases. In summary then, we have

$$Q = \frac{(q_{\text{trans}} q_{\text{elec}} q_{\text{nucl}})^N}{N!} \tag{5–14}$$

where

$$\begin{aligned} q_{\text{trans}} &= \left(\frac{2\pi mkT}{h^2}\right)^{3/2} V = \frac{V}{\Lambda^3} \\ q_{\text{elect}} &= \omega_{e1} + \omega_{e2} e^{-\beta\,\Delta\varepsilon_{12}} + \cdots \\ q_{\text{nucl}} &= \omega_{n1} + \cdots \end{aligned} \tag{5–15}$$

The nuclear partition function, although not always equal to unity, is usually omitted. We can now calculate thermodynamic properties of a monatomic ideal gas.

5–3 THERMODYNAMIC FUNCTIONS

The Helmholtz free energy is given by

$$A(N, V, T) = -kT \ln Q = -NkT \ln\left[\left(\frac{2\pi mkT}{h^2}\right)^{3/2} \frac{Ve}{N}\right] - NkT \ln(\omega_{e1} + \omega_{e2} e^{-\beta\Delta\varepsilon_{12}}) \tag{5–16}$$

The argument of the first logarithm here is much larger than the argument of the second logarithm, and so the electronic contribution to A is quite small.

The thermodynamic energy is

$$E = kT^2\left(\frac{\partial \ln Q}{\partial T}\right)_{N,V} = \tfrac{3}{2}NkT + \frac{N\omega_{e2}\,\Delta\varepsilon_{12}\,e^{-\beta\Delta\varepsilon_{12}}}{q_{\text{elec}}} + \cdots \tag{5–17}$$

The contribution of the electronic degrees of freedom to the energy is small at ordinary temperatures (see Problem 5–12). Since we have neglected the contribution of the intermolecular potential to the total energy of the gas, the first term of Eq. (5–17) represents only kinetic energy. Furthermore, each atom has an average kinetic energy $3kT/2$, or $kT/2$ for each degree of translational freedom. We shall give an interesting interpretation of this result when we study classical statistical mechanics in Chapter 7. If we ignore the very small contribution from the electronic degrees of freedom, the molar heat capacity at constant volume is $3Nk/2$, a well-known experimental result for dilute gases.

The pressure is

$$p = kT\left(\frac{\partial \ln Q}{\partial V}\right)_{N,T} = \frac{NkT}{V} \tag{5–18}$$

Note that Eq. (5–18) results because $q(V, T)$ is of the form $f(T)V$, and the only contribution to the pressure is from the translational energy of the atoms. This is what one expects intuitively, since the pressure is due to bombardment of the walls of the container by the atoms and molecules of the gas.

The entropy is given by

$$S = \tfrac{3}{2}Nk + Nk\ln\left[\left(\frac{2\pi mkT}{h^2}\right)^{3/2}\frac{Ve}{N}\right] + Nk\ln(\omega_{e1} + \omega_{e2}\,e^{-\beta\Delta\varepsilon_{12}}) + \frac{Nk\omega_{e2}\,\beta\Delta\varepsilon_{12}\,e^{-\beta\Delta\varepsilon_{12}}}{q_{\text{elec}}} \tag{5–19}$$

$$= Nk\ln\left[\left(\frac{2\pi mkT}{h^2}\right)^{3/2}\frac{Ve^{5/2}}{N}\right] + S_{\text{elec}} \tag{5–20}$$

In Eq. (5–20), S_{elect} denotes the last two terms of Eq. (5–19). Equation (5–20) is called the Sackur-Tetrode equation. Table 5–3 compares the results of this equation with experimental values for several monatomic gases.

Table 5–3. **Comparison of experimental entropies at 1 atm and $T = 298°K$ to those calculated from the statistical thermodynamical equation for the entropy of an ideal monatomic gas***

	exp. (e.u.)	calc. (e.u.)
He	30.13	30.11
Ne	34.95	34.94
Ar	36.98	36.97
Kr	39.19	39.18
Xe	40.53	40.52
C	37.76	37.76
Na	36.72	36.70
Al	39.30	39.36
Ag	41.32	41.31
Hg	41.8	41.78

* The experimental values have been corrected for any nonideal gas behavior.

The chemical potential is

$$\mu(T, p) = -kT\left(\frac{\partial \ln Q}{\partial N}\right)_{V,T} = -kT \ln \frac{q}{N}$$

$$= -kT \ln\left[\left(\frac{2\pi mkT}{h^2}\right)^{3/2} \frac{V}{N}\right] - kT \ln q_e q_n$$

$$= -kT \ln\left[\left(\frac{2\pi mkT}{h^2}\right)^{3/2} \frac{kT}{p}\right] - kT \ln q_e q_n$$

$$= -kT \ln\left[\left(\frac{2\pi mkT}{h^2}\right)^{3/2} kT\right] - kT \ln q_e q_n + kT \ln p$$

$$= \mu_0(T) + kT \ln p \tag{5–21}$$

where the last line is the thermodynamic equation for $\mu(T, p)$ for an ideal gas [*cf.* Eq. (1–66)]. Thus statistical thermodynamics yields an expression for $\mu_0(T)$:

$$\mu_0(T) = -kT \ln\left[\left(\frac{2\pi mkT}{h^2}\right)^{3/2} kT\right] - kT \ln q_e q_n \tag{5–22}$$

The argument of the first logarithm here has units of pressure, but remember that there is a $kT \ln p$ term in Eq. (5–21) so that $\mu(T, p)$ itself has units of energy and is $O(kT)$.

5–4 A DIGRESSION ON ATOMIC TERM SYMBOLS

The electronic state of an atom is designated by a so-called atomic term symbol. Since one encounters atomic term symbols in the calculation of electronic partition functions, we discuss them in this section. (The quantum mechanical level of this section is above that of most of the book, and it is not necessary to read this section on first reading.)

In addition to the usual kinetic energy and electrostatic terms in the Hamiltonian of a many-electron atom, there are a number of magnetic and spin terms. The most important of these is the spin-orbit interaction term, which represents the interaction of the magnetic moment associated with the spin of the electron with the magnetic field generated by the electric current produced by its own orbital motion. There are other terms such as spin–spin and orbit–orbit interaction terms, but these are numerically much less important. The Hamiltonian can then be written as

$$H = -\frac{\hbar^2}{2m}\sum_j \nabla_j^2 - \sum_j \frac{Ze^2}{r_j} + \sum_{i<j} \frac{e^2}{r_{ij}} + \sum_j \xi(r_j)\mathbf{l}_j \cdot \mathbf{s}_j \tag{5–23}$$

where $\mathbf{l}_j$ and $\mathbf{s}_j$ are the individual electronic orbital and spin angular momenta, respectively, and $\xi(r_j)$ is a scalar function of r_j, whose form is not necessary here. We can abbreviate this equation by writing

$$H = H_0 + H_{ee} + H_{so} \tag{5–24}$$

where H_0 represents the first two terms (no interelectronic interactions), H_{ee} the third, and H_{so} the fourth.

For light atoms ($Z < 40$), H_{so} is small enough to be considered a small perturbation. If H_{so} is neglected altogether, it can be shown that the total orbital angular momentum **L** and the total spin angular momentum **S** are conserved (i.e., yield "good" quantum numbers, or commute with $H_0 + H_{ee}$). However, in this case, the *individual* orbital and spin angular momenta are not conserved; hence they are not useful concepts. The eigenvalues of the square of the total orbital angular momentum operator $\hat{L}^2$ and the square of the total spin angular momentum operator $\hat{S}^2$ are $L(L+1)\hbar^2$ and $S(S+1)\hbar^2$, respectively. One often interprets these eigenvalues by saying that the orbital angular momentum has the value L or that the total spin is S, but it should always be borne in mind that the orbital or spin angular momentum itself is not an eigenoperator in quantum mechanics.

The quantities $\mathbf{L}$ and $\mathbf{S}$ are the vector sums of the individual orbital $\mathbf{l}_j$ and spin $\mathbf{s}_j$, angular momenta. The possible ways of adding these $\mathbf{l}_j$ or $\mathbf{s}_j$ are governed by quantum mechanics with the result being that only certain values of the quantum numbers L and S are allowed. In the case of two electrons, L can take on only the values $l_1 + l_2, l_1 + l_2 - 1, \ldots |l_1 - l_2|$, with a similar result for S. What we really mean by this, of course, is that the only allowed values of the eigenvalues of L^2 are $(l_1 + l_2)(l_1 + l_2 + 1)\hbar^2$, $(l_1 + l_2 - 1)(l_1 + l_2)\hbar^2$, $\ldots$ $(|l_1 - l_2|)(|l_1 - l_2| + 1)\hbar^2$. The addition of electronic angular momenta to obtain $\mathbf{L}$ for cases involving more than two electrons can be accomplished using a scheme which is a straightforward but rather tedious electron-by-electron extension of the above two electron systems. Actually, rather specialized and advanced techniques have been developed to handle this problem, but we need not be concerned with them here.

The electronic energy states are designated by a term symbol, part of which is given by ^{2S+1}L. Terms with $L = 0, 1, 2, \ldots$ are denoted by $S, P, D, \ldots$.

When the spin-orbit term is taken into account, $\mathbf{L}$ and $\mathbf{S}$ are no longer conserved (that is, do not commute with the total H), and only the total angular momentum, $\mathbf{J} = \mathbf{L} + \mathbf{S}$, is conserved. The eigenvalues of $\hat{J}^2 = (\hat{L} + \hat{S})^2$ are $J(J+1)\hbar^2$, with a degeneracy $2J+1$, corresponding to the $2J+1$ eigenvalues of $\hat{J}_z$, namely, $J\hbar, (J-1)\hbar, \ldots, -J\hbar$. Just as in the addition of $\mathbf{l}_1$ and $\mathbf{l}_2$ above, the allowed values of J are $L+S, L+S-1, \ldots, |L-S|$. The spin–orbit term causes each of these values of J to have a slightly different energy, and so the value of J is included in the term symbol as a subscript to give $^{2S+1}L_J$ as a characterization of the electronic state of an atom.

Table 5–1 lists the first few electronic states for some of the first row atoms.

This light atom approximate coupling scheme, in which L and S are almost good quantum numbers (that is, not good quantum numbers because of the small spin–orbit perturbation term) and in which the total angular momentum $\mathbf{J}$ is found by adding $\mathbf{L}$ and $\mathbf{S}$, is called Russell-Saunders or L–S coupling. As the atomic number of the atom becomes larger, the spin–orbit term becomes larger than the interelectronic repulsion term, and H_{ee} can be considered to be a small perturbation on the others. In this case L and S are no longer useful, and the individual total angular momenta, $\mathbf{j}_i = \mathbf{l}_i + \mathbf{s}_i$, become the approximately conserved quantities. One then couples the $\mathbf{j}$'s to get the total angular momentum. This scheme is called j–j coupling and is applicable to heavier atoms. In spite of the deterioration of L–S coupling as Z increases, it is still approximately useful, and so the electronic states of even heavy atoms are designated by term symbols of the form $^{2S+1}L_J$.

We are ready to discuss an ideal gas of diatomic molecules. In addition to having translational and electronic degrees of freedom, diatomic molecules have rotational and vibrational degrees of freedom as well. It should be apparent at this point that the additional input into our statistical thermodynamical equations will be the rotation-vibration energy levels.

We could also leave the study of gases for a while and apply our general results to other systems such as solids and liquids. This would involve no more effort than continuing with ideal gases and perhaps would be a change of pace. We could go directly to Chapter 11, for example, but we shall finish gases before treating other systems.

ADDITIONAL READING

General

ANDREWS, F. C. 1963. *Equilibrium statistical mechanics.* New York: Wiley. Chapter 18.
HILL, T. L. 1960. *Statistical thermodynamics.* Reading, Mass.: Addison-Wesley. Chapter 4.
KESTIN, J., and DORFMAN, J. R. 1971. *A course in statistical thermodynamics.* New York: Academic. Sections 6–7 through 6–8.
KITTEL, C. 1969. *Thermal physics.* New York: Wiley. Chapters 11 and 12.
KNUTH, E. 1966. *Statistical thermodynamics.* New York: McGraw-Hill. Chapter 8.
MÜNSTER, A. 1969. *Statistical thermodynamics,* Vol. I. Berlin: Springer-Verlag. Sections 6–1 and 6–2.

Term symbols

STRAUSS, H. L. 1968. *Quantum mechanics.* Englewood Cliffs, N.J.: Prentice-Hall. Chapter 8.

PROBLEMS

5–1. Convert Boltzmann's constant $k = 1.3806 \times 10^{-16}$ ergs-molecule^{-1} deg^{-1} to cm^{-1}-molecule^{-1}-deg^{-1} and to eV-molecule^{-1}-deg^{-1}.

5–2. By considering the special case of an ideal gas, determine the order of magnitude of E, A, G, S, C_V, and μ. Express your answers in terms of N, kT, or Nk, whichever is appropriate.

5–3. Calculate the value of n_x, n_y, and n_z for the case $n_x = n_y = n_z$ for a hydrogen atom (atomic weight 1.00) in a box of dimensions 1 cc if the particle has a kinetic energy $3kT/2$, for $T = 27°$C. What significant fact does this calculation illustrate?

5–4. Calculate the entropy of Ne at 300°K and 1 atmosphere. Use the entropy, in turn, to estimate the "translational" degeneracy of the gas.

5–5. Calculate the entropy of 1 mole of argon at 298°K and 1 atm and compare this to Table 5–3.

5–6. The quantum mechanical energy of a particle confined to a rectangular parallelepiped of lengths a, b, and c is

$$\varepsilon_{n_x n_y n_z} = \frac{h^2}{8m}\left(\frac{n_x^2}{a^2} + \frac{n_y^2}{b^2} + \frac{n_z^2}{c^2}\right)$$

Show that the translational partition function for this geometry is the same as that of a cube of the same volume.

5–7. Given that the quantum mechanical energy levels of a particle in a two-dimensional box are

$$\varepsilon = \frac{h^2}{8ma^2}(s_x^2 + s_y^2) \qquad s_x, s_y = 1, 2, \ldots$$

First calculate the density of states $\omega(\varepsilon)\,d\varepsilon$, that is, the number of states between ε and $\varepsilon + d\varepsilon$, and use this to find the translational partition function of a two-dimensional ideal gas. Then find the partition function by another method. And finally find the equation of state, the thermodynamic energy E, the heat capacity C_A, and the entropy. This is a model for a gas adsorbed onto a surface or for long-chain fatty acids on the surface of water, say, as long as the number of molecules per unit area is small enough.

5–8. Calculate the entropy of a mixture of 50 percent neon and 50 percent argon at 500°K and 10 atm, assuming ideal behavior.

5–9. Calculate the De Broglie wavelength of an argon atom at 25°C and compare this with the average interatomic spacing at 1 atm.

5–10. Evaluate $\beta\Delta\varepsilon_{elec}$ at room temperature, given that electronic energy levels are usually separated by energies of the order of electron volts.

5–11. Using the data in Table 5–1, calculate the population of the first few electronic energy levels of an oxygen atom at room temperature.

5–12. Show that the contribution of the electronic degrees of freedom to the total energy is small at ordinary temperatures [*cf.* Eq. (5–17)].

5–13. Generalize the results of this chapter to an ideal binary mixture. In particular, show that

$$Q = \frac{q_1^{N_1} q_2^{N_2}}{N_1!\,N_2!}$$

$$E = \tfrac{3}{2}(N_1 + N_2)kT$$

and

$$S = N_1 k \ln\left(\frac{Ve^{5/2}}{\Lambda_1^3 N_1}\right) + N_2 k \ln\left(\frac{Ve^{5/2}}{\Lambda_2^3 N_2}\right)$$

if we ignore q_{elec} and q_{nucl}.

5–14. Derive the standard thermodynamic formula for the entropy of mixing by starting with Eq. (5–20).

5–15. Calculate A, E, μ, C_V, and S for 1 mole of Kr at 25°C and 1 atm (assuming ideal behavior).

5–16. Show that the most probable distribution of $2N$ molecules of an ideal gas contained in two equal and connected volumes at the same temperature is N molecules in each volume.

5–17. Evaluate the isothermal-isobaric partition function of a monatomic ideal gas by converting the summation over V in Eq. (3–17) to an integral. The result is

$$\Delta(N, p, T) = \left(\frac{kT}{p\Lambda^3}\right)^N$$

Using the fact that $G = -kT \ln \Delta$, derive expressions for S and V.

5–18. Consider a monatomic ideal gas of N particles in a volume V. Show that the number n of particles in some small subvolume v is given by the Poisson distribution

$$P_n = (\lambda q)^n \frac{e^{-\lambda q}}{n!}$$

$$= (\bar{n})^n \frac{e^{-\bar{n}}}{n!}$$

Hint: Use the grand canonical ensemble and particularly the result that $\Xi = \exp(\lambda q)$.

5–19. Calculate q_{elec} for a hydrogen atom. The energy levels are given by

$$E_n = -\frac{2\pi^2 me^4}{n^2 h^2} \qquad n = 1, 2, \ldots$$

and the degeneracy is $2n^2$. How is this seemingly paradoxical result explained? (See S. J. Strickler, *J. Chem. Educ.*, **43**, p. 364, 1966.)

CHAPTER 6

IDEAL DIATOMIC GAS

In this chapter we shall treat an ideal gas composed of diatomic molecules. In addition to translational and electronic degrees of freedom, diatomic molecules possess vibrational and rotational degrees of freedom as well. The general procedure would be to set up the Schrödinger equation for two nuclei and *n*-electrons and to solve this equation for the set of eigenvalues of the diatomic molecule. Such a general exact approach is very difficult and has been done only for H_2. Fortunately, a series of very good approximations can be used to reduce this complicated two-nuclei, *n*-electron problem to a set of simpler problems. The simplest of these approximations is the rigid rotor-harmonic oscillator approximation. In Section 6–1 we shall discuss this approximation, and then in Sections 6–2 and 6–3 we discuss the vibrational and rotational partition functions within this approximation. Section 6–4 is a discussion of the symmetry of the wave functions of homonuclear diatomic molecules under the interchange of the two nuclei, and Section 6–5 is an application of these results to the rotational partition function of homonuclear diatomic molecules. This section contains a detailed discussion of ortho- and para-hydrogen. Section 6–6 summarizes the thermodynamic functions under the rigid rotor–harmonic oscillator approximation.

6–1 THE RIGID ROTOR–HARMONIC OSCILLATOR APPROXIMATION

We first make the Born-Oppenheimer approximation. The physical basis of the Born-Oppenheimer approximation is that the nuclei are much more massive than the electrons, and thus move slowly relative to the electrons. Therefore the electrons can be considered to move in a field produced by the nuclei fixed at some internuclear separation. Mathematically, the Schrödinger equation approximately separates into two simpler equations. One equation describes the motion of the electrons in the field of the fixed nuclei. We denote the eigenvalues of this equation by $u_j(r)$, where r is the internuclear separation. The other equation describes the motion of the nuclei in the electronic potential $u_j(r)$, that is, the potential set up by the electrons in the electronic state j. Each electronic state of the molecule creates its own characteristic internuclear potential. As in the atomic case, the first excited electronic state usually lies several electron volts above the ground state, and so only the ground electronic state potential is necessary. The calculation of $u_j(r)$ for even the ground state is a difficult n-electron calculation, and so semiempirical approximations such as the Morse potential are often used. Figure 6–1 illustrates a typical internuclear potential. Given $u_0(r)$, we treat the motion of the two nuclei in this potential.

Problem 1–8 shows that the motion of two masses in a spherically symmetric potential can be rigorously separated into two separate problems by the introduction of center of mass and relative coordinates. The center-of-mass motion is that of a freely

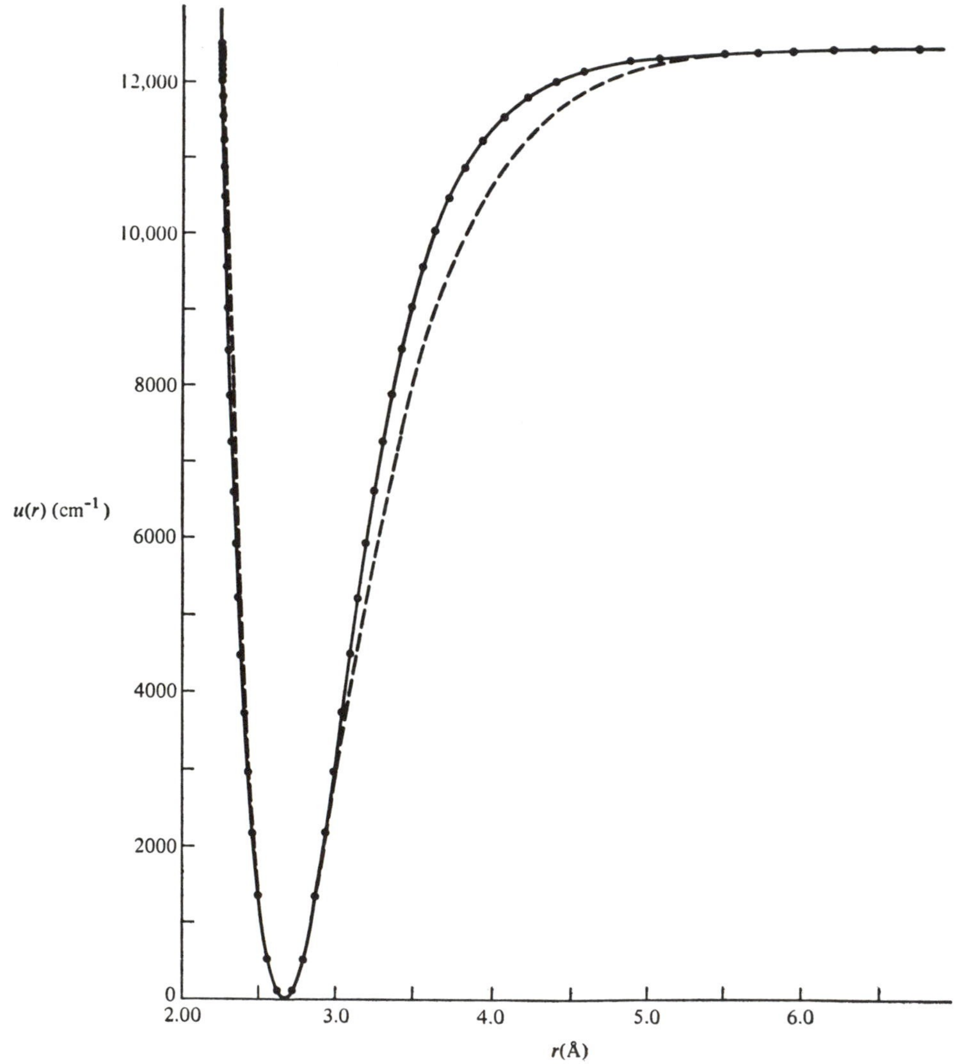

Figure 6–1. **The internuclear potential energy curve for the ground state of I_2 as computed from ultraviolet spectroscopy. The dashed curve is the Morse curve.** (From R. D. Verma, *J. Chem. Phys.*, **32,** 738, 1960.)

translating point of mass $m_1 + m_2$ situated at the center of mass. The other problem is that of the *relative* motion of the two bodies, which can be interpreted as one body of reduced mass $\mu = m_1 m_2/(m_1 + m_2)$ moving about the other one fixed at the origin.

The Hamiltonian can be written as

$$\mathscr{H} = \mathscr{H}_{\text{trans}} + \mathscr{H}_{\text{int}}$$

with eigenvalues

$$\varepsilon = \varepsilon_{\text{trans}} + \varepsilon_{\text{int}}$$

The partition function of the diatomic molecule, therefore, is

$$q = q_{\text{trans}} q_{\text{int}}$$

where

$$q_{\text{trans}} = \left[\frac{2\pi(m_1 + m_2)kT}{h^2}\right]^{3/2} V \tag{6–1}$$

The density of translational states alone is so great that we can write

$$Q(N, V, T) = \frac{q_{\text{trans}}^N q_{\text{int}}^N}{N!} \tag{6–2}$$

Thus we must investigate q_{int} to complete our treatment of diatomic molecules.

The relative motion of the two nuclei in the potential $u(r)$ consists of rotary motion about the center of mass and relative vibratory motion of the two nuclei. It turns out that the amplitude of the vibratory motion is very small, and so it is a good approximation to consider the angular motion to be that of a rigid dumbbell of fixed internuclear distance r_e. In addition, the internuclear potential $u(r)$ can be expanded about r_e:

$$\begin{aligned} u(r) &= u(r_e) + (r - r_e)\left(\frac{du}{dr}\right)_{r=r_e} + \tfrac{1}{2}(r - r_e)^2\left(\frac{d^2u}{dr^2}\right)_{r=r_e} + \cdots \\ &= u(r_e) + \tfrac{1}{2}k(r - r_e)^2 + \cdots \end{aligned} \tag{6–3}$$

The linear term vanishes because (du/dr) is zero at the minimum of $u(r)$. The parameter k is a measure of the curvature of the potential at the minimum and is called the force constant. A large value of k implies a stiff bond; a small value implies a loose bond.

The approximation introduced in the previous paragraph is called the rigid rotor–harmonic oscillator approximation. It allows the Hamiltonian of the relative motion of the nuclei to be written as

$$\mathscr{H}_{\text{rot, vib}} = \mathscr{H}_{\text{rot}} + \mathscr{H}_{\text{vib}} \qquad \text{(rigid rotor–harmonic oscillator approximation)} \tag{6–4}$$

and

$$\varepsilon_{\text{rot, vib}} = \varepsilon_{\text{rot}} + \varepsilon_{\text{vib}} \tag{6–5}$$

The partition function $q_{\text{rot, vib}}$, then, is

$$q_{\text{rot, vib}} = q_{\text{rot}} q_{\text{vib}} \qquad \text{(rigid rotor-harmonic oscillator approximation)} \tag{6–6}$$

The energy eigenvalues and the degeneracy of a rigid rotor are given in Eq. (1–32)

$$\varepsilon_J = \frac{\hbar^2 J(J+1)}{2I} \qquad J = 0, 1, 2, \ldots$$

$$\omega_J = 2J + 1 \tag{6–7}$$

where I is the moment of inertia, μr_e^2, of the molecule. The energy and degeneracy of an harmonic oscillator are [*cf.* Eq. (1–31)]

$$\varepsilon_{\text{vib}} = h\nu(n + \tfrac{1}{2}) \qquad n = 0, 1, 2, \ldots$$

$$\omega_n = 1 \quad \text{for all } n \tag{6–8}$$

where

$$\nu = \frac{1}{2\pi}\left(\frac{k}{\mu}\right)^{1/2} \tag{6–9}$$

Transitions from one rotational level to another can be induced by electromagnetic radiation. The selection rules for this are: (1) The molecule must have a permanent dipole moment, and (2) $\Delta J = \pm 1$. The frequency of radiation absorbed in the process of going from a level J to $J + 1$ is given by

$$\nu = \frac{\varepsilon_{J+1} - \varepsilon_J}{h} = \frac{h}{4\pi^2 I}(J+1) \qquad J = 0, 1, 2, \ldots \tag{6–10}$$

We thus expect absorption of radiation at frequencies given by multiples of $h/4\pi^2 I$ and should observe a set of equally spaced spectral lines, which for typical molecular values of μ and r_e^2 will be found in the microwave region. Experimentally one does see a series of almost equally spaced lines in the microwave spectra of linear molecules. The usual units of frequency in this region are wave numbers, or reciprocal wavelengths.

$$\bar{\omega}(\text{cm}^{-1}) = \frac{1}{\lambda} = \frac{\nu}{c} \tag{6–11}$$

Microwave spectroscopists define the rotational constant $\bar{B}$ by $h/8\pi^2 Ic$ (units of cm^{-1}), so that the energy of rigid rotor (in cm^{-1}) becomes

$$\bar{\varepsilon}_J = \bar{B}J(J+1) \tag{6–12}$$

Table 6–1 lists the values of $\bar{B}$ for several diatomic molecules.

For a molecule to change its vibrational state by absorbing radiation it must (1) change its dipole moment when vibrating and (2) obey the selection rule $\Delta n = \pm 1$. The frequency of absorption is, then, seen to be

$$\nu = \frac{\varepsilon_{n+1} - \varepsilon_n}{h} = \frac{1}{2\pi}\left(\frac{k}{\mu}\right)^{1/2} \tag{6–13}$$

Equation (6–13) predicts that the vibrational spectrum of a diatomic molecule will consist of just one line. This line occurs in the infrared, typically around 1000 cm^{-1}, giving force constants k of the order of 10^5 or 10^6 dynes/cm. (See Problem 6–5.) Table 6–1 gives the force constants of a number of diatomic molecules.

Table 6–1. **Molecular constants for several diatomic molecules***

molecule	electronic state	$\bar{\omega}$ (cm^{-1})	Θ_v (°K)	$\bar{B}$ (cm^{-1})	Θ_r (°K)	$k \times 10^{-5}$ (dynes/cm)	D_0 (kcal/mole)
H_2	$^1\Sigma_g^+$	4320	6215	59.3	85.3	5.5	103.2
D_2	$^1\Sigma_g^+$	3054	4394	29.9	42.7	5.5	104.6
Cl_2	$^1\Sigma_g^+$	561	808	0.244	0.351	3.2	57.1
Br_2	$^1\Sigma_g^+$	322	463	0.0809	0.116	2.4	45.4
I_2	$^1\Sigma_g^+$	214	308	0.0373	0.0537	1.7	35.6
O_2	$^3\Sigma_g^-$	1568	2256	1.437	2.07	11.6	118.0
N_2	$^1\Sigma_g^+$	2345	3374	2.001	2.88	22.6	225.1
CO	$^1\Sigma^+$	2157	3103	1.925	2.77	18.7	255.8
NO	$^2\Pi_{1/2}$	1890	2719	1.695	2.45	15.7	150.0
HCl	$^1\Sigma^+$	2938	4227	10.44	15.02	4.9	102.2
HBr	$^1\Sigma^+$	2640	3787	8.36	12.02	3.9	82.4
HI	$^1\Sigma^+$	2270	3266	6.46	9.06	3.0	70.5
Na_2	$^1\Sigma_g^+$	159	229	0.154	0.221	0.17	17.3
K_2	$^1\Sigma_g^+$	92.3	133	0.0561	0.081	0.10	15.8

* These parameters were obtained from a variety of sources and do not necessarily represent the most accurate values since they are obtained under the rigid rotor–harmonic oscillator approximation.

We furthermore assume that the electronic and nuclear degrees of freedom can be written separately, and thus we have

$$\mathscr{H} = \mathscr{H}_{\text{trans}} + \mathscr{H}_{\text{rot}} + \mathscr{H}_{\text{vib}} + \mathscr{H}_{\text{elec}} + \mathscr{H}_{\text{nucl}} \tag{6–14}$$

which implies that

$$\varepsilon = \varepsilon_{\text{trans}} + \varepsilon_{\text{rot}} + \varepsilon_{\text{vib}} + \varepsilon_{\text{elec}} + \varepsilon_{\text{nucl}} \tag{6–15}$$

and

$$q = q_{\text{trans}}\, q_{\text{rot}}\, q_{\text{vib}}\, q_{\text{elec}}\, q_{\text{nucl}} \tag{6–16}$$

The translational partition function is given by Eq. (6–1); the electronic partition function will be similar to Eq. (5–11) for a monatomic gas; and we shall usually adopt the convention that $q_{\text{nucl}} = 1$. Although Eq. (6–14) is not exact, it serves as a useful approximation. Within this approximation, the partition function of the gas itself is given by

$$Q(N, V, T) = \frac{(q_{\text{trans}}\, q_{\text{rot}}\, q_{\text{vib}}\, q_{\text{elec}}\, q_{\text{nucl}})^N}{N!} \tag{6–17}$$

We shall introduce several corrections to Eqs. (6–14) through (6–17) in Problems 6–23 and 6–24.

Only the vibrational and rotational contributions to the partition function are not known in Eq. (6–17), and we shall discuss these contributions in the next few sections. Before discussing these, however, we must choose a zero of energy for the rotational and vibrational states. The zero of rotational energy will usually be taken to be the $J = 0$ state. In the vibrational case we have two choices. One is to take the zero of vibrational energy to be that of the ground state, and the other is to take the zero to be the bottom of the internuclear potential well. In the first case, the energy of the ground vibrational state is zero, and in the second case it is $h\nu/2$. We shall choose the zero of

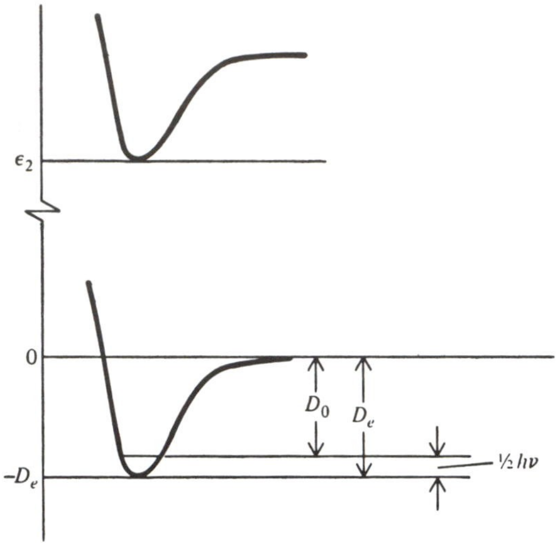

Figure 6–2. **The ground and first excited electronic states as a function of the internuclear separation r, illustrating the quantities D_0, D_e, and ε_2.**

vibrational energy to be the bottom of the internuclear potential well of the lowest electronic state. Lastly, we take the zero of the electronic energy to be the separated, electronically unexcited atoms at rest. If we denote the depth of the ground electronic state potential well by D_e, the energy of the ground electronic state is $-D_e$, and the electronic partition function is

$$q_{\text{elec}} = \omega_{e1}\, e^{D_e/kT} + \omega_{e2}\, e^{-\varepsilon_2/kT} + \cdots \tag{6–18}$$

where D_e and ε_2 are shown in Fig. 6–2. We also define a quantity D_0 by $D_e - \frac{1}{2}h\nu$. As Fig. 6–2 shows, D_0 is the energy difference between the lowest vibrational state and the dissociated molecule. The quantity D_0 can be measured spectroscopically (by predissociation spectra, for example) or calorimetrically from the heat of reaction at any one temperature and the heat capacities from 0°K to that temperature. The values of D_0 for a number of diatomic molecules are given in Table 6–1.

6–2 THE VIBRATIONAL PARTITION FUNCTION

Since we are measuring the vibrational energy levels relative to the bottom of the internuclear potential well, we have

$$\varepsilon_n = (n + \tfrac{1}{2})h\nu \qquad n = 0, 1, 2, \ldots \tag{6–19}$$

with $\nu = (k/\mu)^{1/2}/2\pi$, where k is the force constant of the molecule, and μ is its reduced mass [*cf.* Eq. (6–9)]. The vibrational partition function q_{vib}, then, becomes

$$\begin{aligned} q_{\text{vib}}(T) &= \sum_n e^{-\beta\varepsilon_n} \\ &= e^{-\beta h\nu/2} \sum_{n=0}^{\infty} e^{-\beta h\nu n} \\ &= \frac{e^{-\beta h\nu/2}}{1 - e^{-\beta h\nu}} \end{aligned} \tag{6–20}$$

where we have recognized the summation above as a geometric series. This is one of the rare cases in which q can be summed directly without having to approximate it by an integral, as we did in the translational case in Chapter 5 and shall do shortly in the rotational case. The quantity $\beta h\nu$ is ordinarily larger than 1, but if the temperature is high enough, $\beta h\nu \ll 1$, and we can replace the sum in Eq. (6–20) by an integral to get

$$q_{\text{vib}}(T) = e^{-\beta h\nu/2} \int_0^\infty e^{-\beta h\nu n}\, dn = \frac{kT}{h\nu} \qquad (kT \gg h\nu) \tag{6–21}$$

which we see is what results from Eq. (6–20) if $\beta h\nu \ll 1$. Although we shall rarely use this approximation since we have $q_{\text{vib}}(T)$ exactly, it will be interesting to compare this limit to some others which we shall derive later on. From $q_{\text{vib}}(T)$ we can calculate the vibrational contribution to the thermodynamic energy

$$E_v = NkT^2 \frac{d \ln q_v}{dT} = Nk\left(\frac{\Theta_v}{2} + \frac{\Theta_v}{e^{\Theta_v/T} - 1}\right) \tag{6–22}$$

where $\Theta_v \equiv h\nu/k$ and is called the vibrational temperature. Table 6–1 gives Θ_v for a number of diatomic molecules. The vibrational contribution to the heat capacity is

$$\left(\frac{\partial E_v}{\partial T}\right)_N = Nk\left(\frac{\Theta_v}{T}\right)^2 \frac{e^{\Theta_v/T}}{(e^{\Theta_v/T} - 1)^2} \tag{6–23}$$

Notice that as $T \rightarrow \infty$, $E_v \rightarrow NkT$ and $C_v \rightarrow Nk$, a result given in many physical chemistry courses and one whose significance we shall understand more fully when we discuss equipartition of energy.

Figure 6–3 shows the vibrational contribution of an ideal diatomic gas to the molar heat capacity as a function of temperature.

An interesting quantity to calculate is the fraction of molecules in excited vibrational states. The fraction of molecules in the vibrational state designated by n is

$$f_n = \frac{e^{-\beta h\nu(n+1/2)}}{q_{\text{vib}}} \tag{6–24}$$

This equation is shown in Fig. 6–4 for Br_2 at 300°K. Notice that most molecules are in the ground vibrational state and that the population of the higher vibrational states

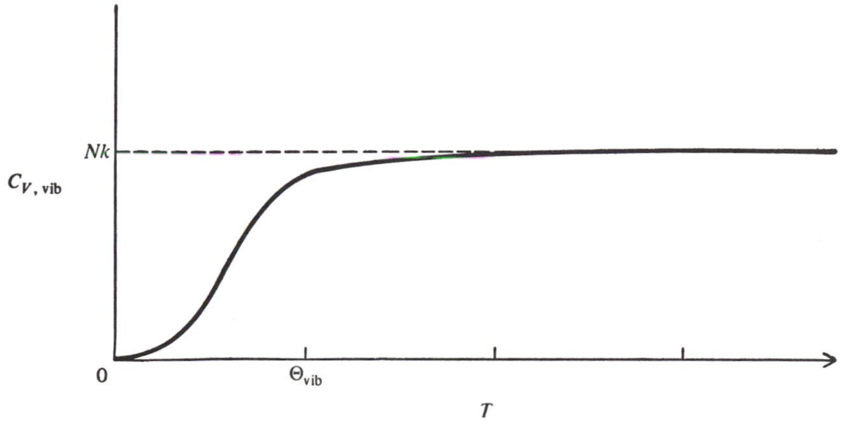

Figure 6–3. **The vibrational contribution of an ideal diatomic gas to the molar heat capacity as a function of temperature. Room temperature is typically $O(0.1\Theta_v)$(*Cf.* Table 6–1).**

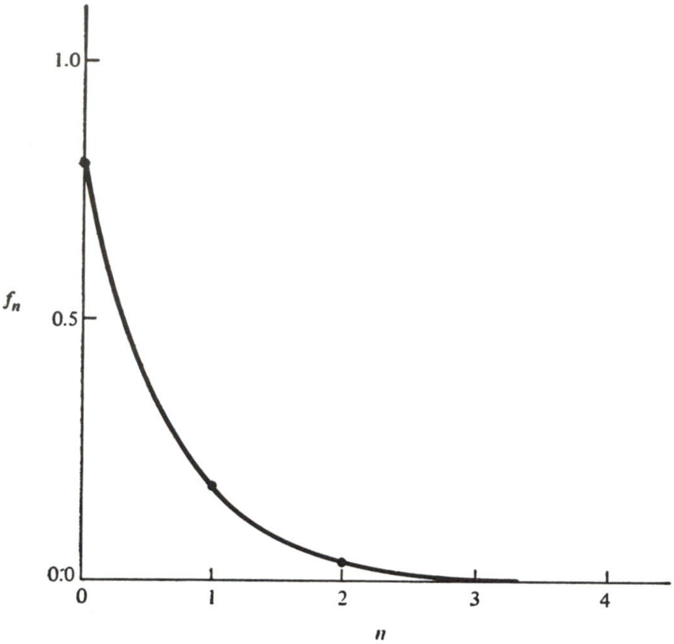

Figure 6–4. **The population of the vibrational levels of Br_2 at 300°K.**

decreases exponentially. Bromine has a force constant smaller than most molecules, however (*cf.* Table 6–1), and so the population of excited vibrational levels of Br_2 is greater than most other molecules. Table 6–2 gives the fraction of molecules in all excited states for a number of molecules. This fraction is given by

$$f_{n>0} = \sum_{n=1}^{\infty} \frac{e^{-\beta h\nu(n+1/2)}}{q_{vib}} = 1 - f_0 = e^{-\beta h\nu} = e^{-\Theta_v/T} \tag{6–25}$$

Table 6–2. **The fraction of molecules in excited vibrational states at 300°K and 1000°K**

		$e^{-\Theta_v/T}$	
gas	Θ_v, °K	300°K	1000°K
H_2	6215	1.04×10^{-9}	2.03×10^{-3}
HCl	4227	1.02×10^{-6}	1.59×10^{-2}
N_2	3374	1.51×10^{-5}	3.55×10^{-2}
CO	3100	3.71×10^{-5}	4.65×10^{-2}
Cl_2	810	6.72×10^{-2}	4.45×10^{-1}
I_2	310	3.56×10^{-1}	7.33×10^{-1}

6–3 THE ROTATIONAL PARTITION FUNCTION OF A HETERONUCLEAR DIATOMIC MOLECULE

For heteronuclear diatomic molecules, the calculation of the rotational partition function is straightforward. The rotational partition function is given by

$$q_{rot}(T) = \sum_{J=0}^{\infty} (2J+1)e^{-\beta BJ(J+1)} \tag{6–26}$$

In the nomenclature of Chapter 2, Eq. (6–26) is a summation over levels rather than over states.

The ratio $\bar{B}/k$ is denoted by Θ_r and is called the characteristic temperature of rotation. This is given in Table 6–1 for a number of molecules. Unlike the vibrational case, this sum cannot be written in closed form. However, because Θ_r/T is quite small at ordinary temperatures for most molecules, we can approximate this sum by an integral. (It is really $\Delta\varepsilon/kT = 2\Theta_r(J+1)/T$ that must be small compared to one, and this of course cannot be true as J increases. However, by the time J is large enough to contradict this, the terms are so small that it makes no difference.)

At high enough temperatures, then,

$$q_{\text{rot}}(T) = \int_0^\infty (2J+1)e^{-\Theta_r J(J+1)/T}\, dJ \tag{6–27}$$

$$= \int_0^\infty e^{-\Theta_r J(J+1)/T}\, d\{J(J+1)\} = \frac{T}{\Theta_r} \tag{6–28}$$

$$= \frac{8\pi^2 IkT}{h^2} \qquad \Theta_r \ll T \tag{6–29}$$

This result improves as the temperature increases and is called the high-temperature limit. For low temperatures or for molecules with large values of Θ_r, say HD with $\Theta_r = 42.7°\text{K}$, one can use the sum directly. For example,

$$q_{\text{rot}}(T) = 1 + 3e^{-2\Theta_r/T} + 5e^{-6\Theta_r/T} + 7e^{-12\Theta_r/T} \tag{6–30}$$

is sufficient to give the sum to within 0.1 percent for $\Theta_r > 0.7T$. For Θ_r less than $0.7T$ but not small enough for the integral to give a good approximation, we need some intermediate approximation.

The replacement of a sum by an integral can be viewed as the first of a sequence of approximations. The full scheme is a standard result of the field of the calculus of finite differences and is called the Euler-MacLaurin summation formula. It states that if $f(n)$ is a function defined on the integers and continuous in between, then

$$\sum_{n=a}^{b} f(n) = \int_a^b f(n)\, dn + \tfrac{1}{2}\{f(b) + f(a)\} + \sum_{j=1}^{\infty} (-)^j \frac{B_j}{(2j)!}\{f^{(2j-1)}(a) - f^{(2j-1)}(b)\} \tag{6–31}$$

where $f^{(k)}(a)$ is the kth derivative of f evaluated at a. The B_j's are the Bernoulli numbers, $B_1 = \frac{1}{6}$, $B_2 = \frac{1}{30}$, $B_3 = \frac{1}{42}$, ... Before applying this to $q_{\text{rot}}(T)$, let us apply it first to a case we can do exactly. Consider the sum [*cf.* Eq. (6–20)]

$$\sum_{j=0}^{\infty} e^{-\alpha j} = \frac{1}{1-e^{-\alpha}} \tag{6–32}$$

Applying the Euler-MacLaurin summation formula, we get

$$\sum_{j=0}^{\infty} e^{-\alpha j} = \frac{1}{\alpha} + \frac{1}{2} + \frac{\alpha}{12} - \frac{\alpha^3}{720} + \cdots \tag{6–33}$$

The expansion of $(1-e^{-\alpha})^{-1}$ is

$$\frac{1}{1-e^{-\alpha}} = \frac{1}{\alpha - \dfrac{\alpha^2}{2} + \dfrac{\alpha^3}{6} - \cdots} = \frac{1}{\alpha} + \frac{1}{2} + \frac{\alpha}{12} - \frac{\alpha^3}{720} + \cdots \tag{6–34}$$

We see that these two expansions are the same. If α is large, we can use the first few terms of Eq. (6–33); otherwise, we use the Euler-MacLaurin expansion in α.

Applying this formula to $q_{rot}(T)$ gives (see Problem 6–9):

$$q_{rot}(T) = \frac{T}{\Theta_r}\left\{1 + \frac{1}{3}\left(\frac{\Theta_r}{T}\right) + \frac{1}{15}\left(\frac{\Theta_r}{T}\right)^2 + \frac{4}{315}\left(\frac{\Theta_r}{T}\right)^3 + \cdots\right\} \quad (6\text{–}35)$$

which is good to within one percent for $\Theta_r < T$. For simplicity we shall use only the high-temperature limit in what we do here since Θ_r is $\ll T$ for most molecules at room temperature (*cf.* Table 6–1).

The rotational contribution to the thermodynamic energy is

$$E_{rot} = NkT^2\left(\frac{\partial \ln q_{rot}}{\partial T}\right) = NkT + \cdots \quad (6\text{–}36)$$

and the contribution to the heat capacity is

$$C_{V,rot} = Nk + \cdots \quad (6\text{–}37)$$

The fraction of molecules in the Jth rotational state is

$$\frac{N_J}{N} = \frac{(2J+1)e^{-\Theta_r J(J+1)/T}}{q_{rot}(T)} \quad (6\text{–}38)$$

Figure 6–5 shows this fraction for HCl at 300°K. Contrary to the vibrational case, most molecules are in excited rotational levels at ordinary temperatures. We can find the maximum of this curve by differentiating Eq. (6–38) with respect to J to get

$$J_{max} = \left(\frac{kT}{2\bar{B}}\right)^{1/2} - \frac{1}{2} \approx \left(\frac{T}{2\Theta_r}\right)^{1/2} = \left(\frac{kT}{2\bar{B}}\right)^{1/2}$$

We see then that J_{max} increases with T and is inversely related to $\bar{B}$, and so increases with the moment of inertia of the molecule since $\bar{B} \propto 1/I$.

The next two sections deal with the rotational partition function of homonuclear diatomic molecules. The wave function of a homonuclear diatomic molecule must possess a certain symmetry with respect to the interchange of the two identical nuclei in the molecule. In particular, if the two nuclei have integral spins, the wave function must be symmetric with respect to an interchange; if the nuclei have half odd integer

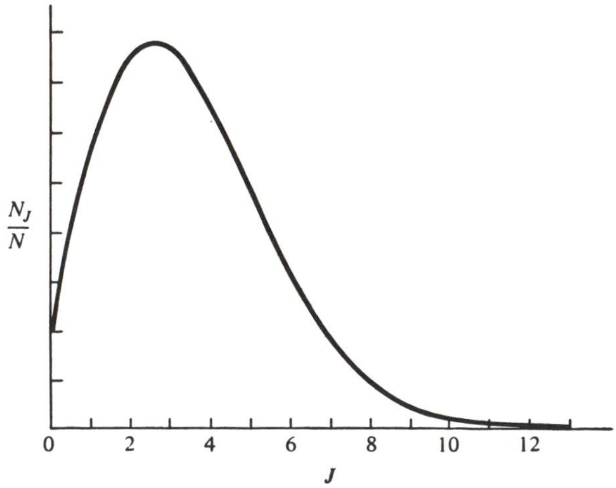

Figure 6–5. **The population of the rotational levels of hydrogen chloride at 300°K.**

spin, the wave function must be antisymmetric. This symmetry requirement has a profound effect on the rotational energy levels of a homonuclear diatomic molecule, which can be understood only by understanding the general symmetry properties of a homonuclear diatomic molecule. This is discussed in the next section. Then in Section 6–5 we apply these results to the rotational partition function. We shall see there that at low temperatures, the symmetry properties have an important effect on the thermodynamic properties of certain molecules, and in particular we shall discuss ortho- and para-hydrogen.

The discussion of the symmetry requirement is somewhat involved, however, and so we present here a summary of the high-temperature limit, which for most systems is completely adequate. At temperatures such that the summation in Eq. (6–26) can be approximated by an integral or even the Euler-MacLaurin expansion, the result for a homonuclear diatomic molecule is

$$q_{\text{rot}} = \frac{T}{2\Theta_r}\left\{1 + \frac{\Theta_r}{3T} + \frac{1}{15}\left(\frac{\Theta_r}{T}\right)^2 + \cdots\right\} \tag{6–39}$$

Note that this equation is the same as Eq. (6–35) except for the factor of 2 in the denominator. This factor is due to the symmetry of the homonuclear diatomic molecule and, in particular, is due to the fact that there are two indistinguishable orientations of a homonuclear diatomic molecule. There is a two-fold axis of symmetry perpendicular to the internuclear axis.

Equations (6–35) and (6–39) can be written as one equation by introducing a factor σ into the denominator of Eq. (6–35). If $\sigma = 1$, we have Eq. (6–35), and if $\sigma = 2$, we have Eq. (6–39). The factor σ is called the *symmetry number* of the molecule and represents the number of indistinguishable orientations that the molecule can have. The exact origin of σ can only be completely understood from the arguments presented in Sections 6–4 and 6–5, but on first reading it is possible to accept the factor of σ and proceed directly to Eq. (6–47) from here.

6–4 THE SYMMETRY REQUIREMENT OF THE TOTAL WAVE FUNCTION OF A HOMONUCLEAR DIATOMIC MOLECULE

The calculation of the rotational partition function is not quite so straightforward for homonuclear diatomic molecules. The total wave function of the molecule, that is, the electronic, vibrational, rotational, translational, and nuclear wave function, must be either symmetric or antisymmetric under the interchange of the two identical nuclei. It must be symmetric if the nuclei have integral spins (bosons), or antisymmetric if they have half-integral spins (fermions). This symmetry requirement has profound consequences on the thermodynamic properties of homonuclear diatomic molecules at low temperatures. We shall discuss the interchange of the two identical nuclei of a homonuclear diatomic molecule in this section, and then apply the results to the calculation of q_{rot} in the next section.

It is convenient to imagine this interchange as a result of (1) an inversion of all the particles, electrons and nuclei, through the origin, and then (2) an inversion of just the electrons back through the origin. This two-step process is equivalent to an exchange of the nuclei. Let us write ψ_{total} *exclusive* of the nuclear part as

$$\psi'_{\text{total}} = \psi_{\text{trans}}\psi_{\text{rot}}\psi_{\text{vib}}\psi_{\text{elec}}$$

where the prime on ψ_{total} indicates that we are ignoring the nuclear contribution for now. The translational wave function depends only upon the coordinates of the center of mass of the molecule, and so this factor is not affected by inversion. Furthermore, ψ_{vib} depends only upon the magnitude of $(r - r_e)$, and so this part of the total wave function is unaffected by any inversion operation. Therefore, we concentrate on ψ_{elec} and ψ_{rot}.

The property of ψ_{elec} under the inversions in both Steps (1) and (2) above depends upon the symmetry of the ground electronic state of the molecule. The ground electronic state of most molecules is symmetric under both of these operations. Such a state is designated by the term symbol Σ_g^+. Thus it is ψ_{rot} that controls the symmetry of ψ_{total}.

Only Step (1) above, the inversion of both electrons and nuclei through the origin, affects ψ_{rot}. The effect of this inversion is to change the coordinates (r, θ, ϕ) that describe the orientation of the diatomic molecule into $(r, \pi - \theta, \phi + \pi)$. One can see this either analytically from the eigenfunctions themselves or pictorially from the rotational wave functions shown in Fig. 6–6. Notice that the rigid rotor wave functions are the same functions as the angular functions of the hydrogen atom.

The net result then, when the ground electronic state is symmetric, that is, Σ_g^+ is that ψ_{total} remains unchanged for even J and changes sign for odd J. This result applies to the total wave function, exclusive of nuclear spin.

Now consider a molecule such as H_2, whose nuclei have a spin of $\frac{1}{2}$. Just as in the case of the two electrons in the helium atom, the two nuclei of spin $\frac{1}{2}$ have three symmetric spin functions $\alpha\alpha$, $\beta\beta$, and $2^{-1/2}(\alpha\beta + \beta\alpha)$, and one antisymmetric spin function $2^{-1/2}(\alpha\beta - \beta\alpha)$. Since nuclei with spin $\frac{1}{2}$ act as fermions, the total wave function must be antisymmetric in the exchange of these two nuclei. Now states with both even and odd values of J can be brought to the required antisymmetry by coupling them with the right spin functions. Since three symmetric nuclear spin functions can be combined with the odd J levels to achieve the correct overall antisymmetry for $^1\Sigma_g^+$ electronic states, we see that the odd J levels have a statistical weight of 3, compared to a weight of 1 for even J levels. This leads to the existence of ortho- (parallel nuclear spins) states and para- (opposed nuclear spins) states in H_2. This weighting of the rotational states will be seen shortly to have a profound effect on the low-temperature thermodynamics of H_2.

More generally, for nuclei of spin I, there are $2I + 1$ spin states for each nucleus. Let the eigenfunctions of these spin states be denoted by $\alpha_1, \alpha_2, \ldots, \alpha_{2I+1}$. There are $(2I + 1)^2$ nuclear wave functions to include in ψ_{total}. (In the case of H_2, $I = \frac{1}{2}$, there are two spin states α and β, and there are four nuclear spin functions, three of which are symmetric and one of which is antisymmetric.) The antisymmetric nuclear spin functions are of the form $\alpha_i(1)\alpha_j(2) - \alpha_i(2)\alpha_j(1)$, $1 \le i, j \le 2I + 1$. There are $(2I + 1)(2I)/2$ such combinations, and so this is the number of antisymmetric nuclear spin functions. (For H_2, we find that there is only one antisymmetric choice in agreement with the above paragraph.) All the remaining $(2I + 1)^2$ total nuclear spin functions are symmetric, and so their number is $(2I + 1)^2 - I(2I + 1) = (I + 1)(2I + 1)$. Thus we can write the following summary for Σ_g^+ states;

integral spin

$I(2I + 1)$ antisymmetric nuclear spin functions couple with odd J

$(I + 1)(2I + 1)$ symmetric nuclear spin functions couple with even J

half-integral spin

$I(2I + 1)$ antisymmetric nuclear spin functions couple with even J

$(I + 1)(2I + 1)$ symmetric nuclear spin functions couple with odd J

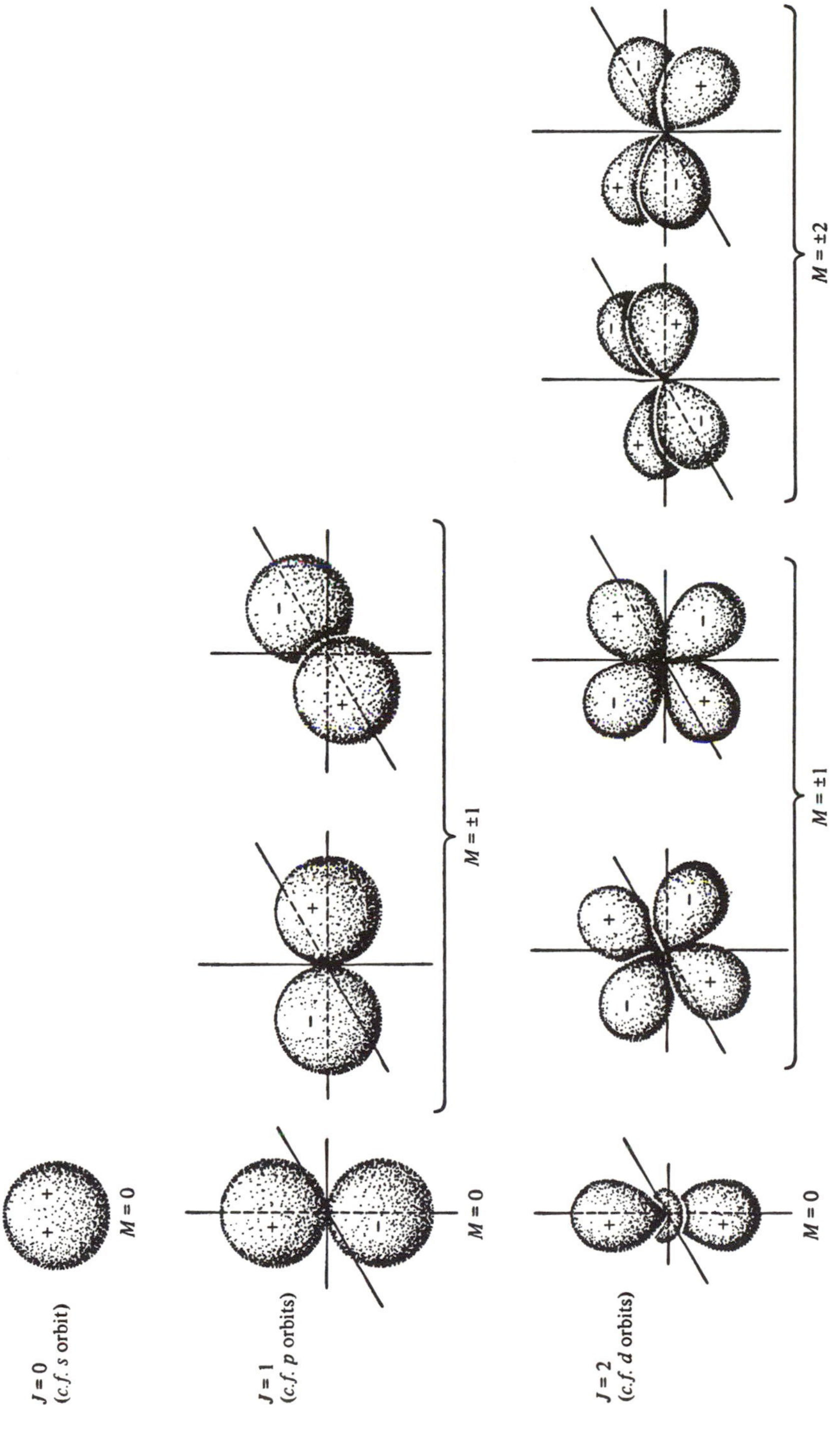

Figure 6–6. **Diagrammatic representations of the rotational wave functions for $J = 0$, 1, and 2.**

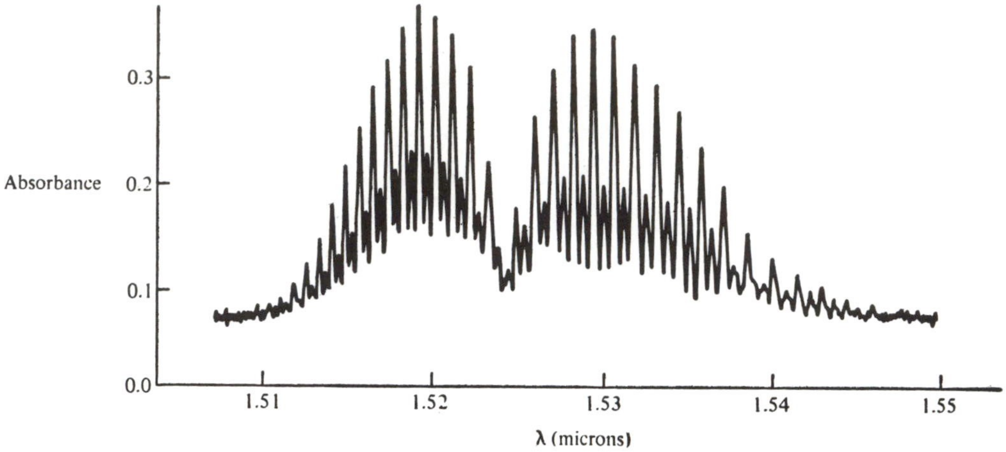

Figure 6–7. **The vibration–rotation spectrum of acetylene. This represents one vibrational line. The alternation in the intensity of the lines is due to the statistical weights of the rotational levels.** (From L. W. Richards, *J. Chem. Ed.*, **43**, p. 645, 1966.)

These combinations of nuclear and rotational wave functions produce the correct symmetry required of the total wave function under interchange of identical nuclei. Remember that all of these conclusions are for $\sum_g^+$ electronic states, the most commonly occurring ground state. (See Problem 6–26 for a discussion of O_2.)

Even though we have considered only diatomic molecules here, the results of this section apply also to linear polyatomic molecules such as CO_2, H_2C_2. For example, the molecules $HC^{12}C^{12}H$ and $DC^{12}C^{12}D$ have their rotational states weighted in a similar way as H_2 and D_2. Figure 6–7 shows the vibration-rotation spectrum of H_2C_2. The alternation in the intensity of these rotational lines due to the statistical weights is very apparent.

6–5 THE ROTATIONAL PARTITION FUNCTION OF A HOMONUCLEAR DIATOMIC MOLECULE

The results of the previous section show that for homonuclear diatomic molecules with nuclei having integral spin, rotational levels with odd values of J must be coupled with the $I(2I + 1)$ antisymmetric nuclear spin functions, and that rotational levels with even values of J must be coupled with the $(I + 1)(2I + 1)$ symmetric nuclear spin functions. Thus we write

$$q_{\text{rot, nucl}}(T) = (I + 1)(2I + 1) \sum_{J \text{ even}} (2J + 1)e^{-\Theta_r J(J+1)/T} + I(2I + 1) \sum_{J \text{ odd}} (2J + 1)e^{-\Theta_r J(J+1)/T} \tag{6–40}$$

Likewise, for molecules with nuclei with half-integer spins, we have

$$q_{\text{rot, nucl}}(T) = I(2I + 1) \sum_{J \text{ even}} (2J + 1)e^{-\Theta_r J(J+1)/T} + (I + 1)(2I + 1) \sum_{J \text{ odd}} (2J + 1)e^{-\Theta_r J(J+1)/T} \tag{6–41}$$

Notice that in this case the combined rotational and nuclear partition function *does not factor* into $q_{\text{rot}} q_{\text{nucl}}$. This is a situation in which we cannot ignore q_{nucl}. For most molecules at ordinary temperatures, $\Theta_r \ll T$, and we can replace the sum by an integral. We see then that

$$\sum_{J\,\text{even}} \approx \sum_{J\,\text{odd}} \approx \frac{1}{2}\sum_{\text{all}\,J} \approx \frac{1}{2}\int_0^\infty (2J+1)e^{-\Theta_r J(J+1)/T}\,dJ = \frac{T}{2\Theta_r} \tag{6–42}$$

and so both Eqs. (6–40) and (6–41) become

$$q_{\text{rot, nucl}}(T) = \frac{(2I+1)^2 T}{2\Theta_r} \tag{6–43}$$

which can be written as $q_{\text{rot}}(T)q_{\text{nucl}}(T)$, where

$$q_{\text{rot}}(T) = \frac{T}{2\Theta_r} \quad \text{and} \quad q_{\text{nucl}} = (2I+1)^2 \tag{6–44}$$

For Eq. (6–42) to be valid, Θ_r/T must be less than about 0.20.

This result is to be compared to the result for a heteronuclear diatomic molecule, namely, $q_{\text{rot}}(T) = T/\Theta_r$. The factor of 2 that appears above in the high-temperature limit takes into account that the molecule is homonuclear, and so its rotational partition function is given by Eq. (6–40) or (6–41) instead of (6–26). This factor of 2 is called the symmetry number and is denoted by σ. It legitimately appears only when Θ_r is less than approximately $0.2T$, since only then can we use Eq. (6–42). Understanding the origin of this fact then, we can write

$$q_{\text{rot}}(T) \approx \frac{8\pi^2 IkT}{\sigma h^2} \approx \frac{1}{\sigma}\sum_{J=0}^{\infty}(2J+1)e^{-\Theta_r J(J+1)/T} \qquad \Theta_r \ll T \tag{6–45}$$

where $\sigma = 1$ for heteronuclear molecules, and $\sigma = 2$ for homonuclear diatomic molecules. Remember that this is applicable only to the high-temperature limit or its Euler-MacLaurin correction. A similar factor will appear for polyatomic molecules also.

There are some interesting systems in which Θ_r/T is not small. Hydrogen is one of the most important such cases. Each nucleus in H_2 has nuclear spin $\frac{1}{2}$, and so

$$q_{\text{rot, nucl}} = \sum_{J\,\text{even}} (2J+1)e^{-\Theta_r J(J+1)/T} + 3\sum_{J\,\text{odd}} (2J+1)e^{-\Theta_r J(J+1)/T} \tag{6–46}$$

The hydrogen with only even rotational levels allowed (antisymmetric nuclear spin function or "opposite" nuclear spins) is called para-hydrogen; that with only odd rotational levels allowed (symmetric nuclear spin function or "parallel" nuclear spins) is called ortho-hydrogen. The ratio of the number of ortho-H_2 molecules to the number of para-H_2 molecules is

$$\frac{N_{\text{ortho}}}{N_{\text{para}}} = \frac{3\sum_{J\,\text{odd}}(2J+1)e^{-\Theta_r J(J+1)/T}}{\sum_{J\,\text{even}}(2J+1)e^{-\Theta_r J(J+1)/T}}$$

Figure 6–8 shows the percentage of p–H_2 versus temperature in an equilibrium mixture of ortho- and para-hydrogen. Note that the system is all para- at 0°K and 25 percent para- at high temperatures.

Figure 6–9 illustrates an interesting situation that occurs with low-temperature heat capacity measurements on H_2. Equation (6–46) can be used to calculate the heat capacity of H_2, and this is plotted in Fig. 6–9, along with the experimental results. It can be seen that the two curves are in great disagreement. These calculations and

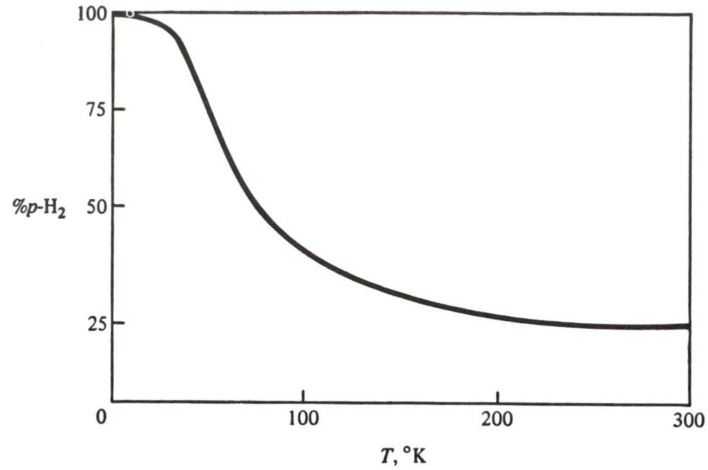

Figure 6–8. **The percentage of para-hydrogen in an equilibrium mixture as a function of temperature.**

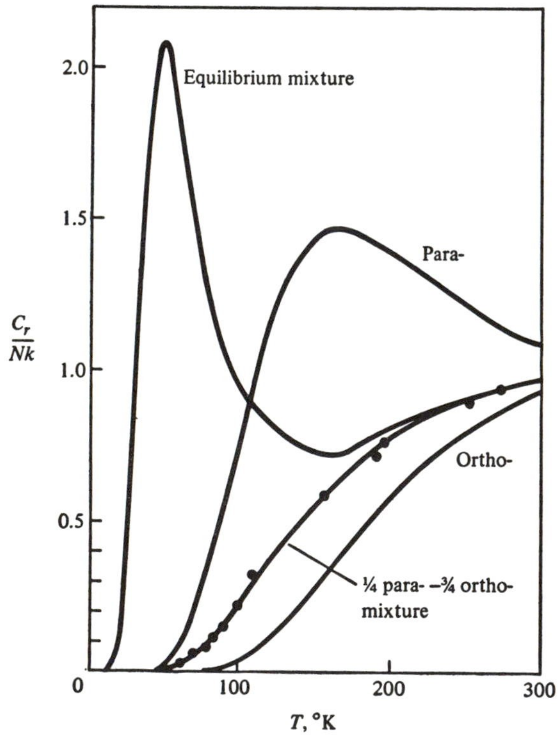

Figure 6–9. **The rotational and nuclear contribution to the molar heat capacity for ortho-hydrogen, para-hydrogen, an equilibrium mixture of ortho- and para-hydrogen, a metastable 75 percent ortho- and 25 percent para- mixture, and the experimental data. (From K. F. Bonhoeffer and P. Harteck, *Z. Physikal. Chem.*, 4B, p. 113, 1929.)**

measurements were made at a time when quantum mechanics was being developed, and was not accepted by all scientists. For a while, the disagreement illustrated in Fig. 6–9 was a blow to the proponents of the new quantum mechanics. It was Dennison* who finally realized that the conversion between ortho- and para-hydrogen is extremely slow in the absence of a catalyst, and so when hydrogen is prepared in the laboratory at room temperature and then cooled down for the low-temperature heat capacity measurements, the room-temperature composition persists instead of the equilibrium composition. Thus the experimental data illustrated in Fig. 6–9 are not for an equilibrium system of ortho- and para-hydrogen, but for a metastable system whose ortho-para composition is that of equilibrium room-temperature hydrogen, namely, 75 percent ortho- and 25 percent para-. If one calculates the heat capacity of such a system, according to

$$C_V = \tfrac{3}{4}C_V(\text{ortho-}) + \tfrac{1}{4}C_V(\text{para-})$$

where C_V(ortho-) is obtained from just the second term of Eq. (6–46), and C_V(para-) is obtained from the first term of Eq. (6–46), one obtains excellent agreement with the experimental curve. A clever confirmation of this explanation was shortly after obtained by Bonhoeffer and Harteck,† who performed heat capacity measurements on hydrogen in the presence of activated charcoal, a catalyst for the ortho–para conversion. This produces an equilibrium system at each temperature. The experimental data are in excellent agreement with the equilibrium calculation in Fig. 6–9.

The explanation of the heat capacity of H_2 was one of the great triumphs of post-quantum mechanical statistical mechanics. You should be able to go through a similar argument for D_2, sketching the equilibrium heat capacity, the pure ortho- and para-heat capacity, and finally what you should expect the experimental curve to be for D_2 prepared at room temperature and at some other temperature, say 20°K. (See Problem 6–17.)

In principle, such nuclear spin effects should be observable in other homonuclear molecules, but a glance at Table 6–1 shows that the characteristic rotational temperatures for all the other molecules are so small that these molecules reach the "high-temperature limit" while still in the solid state. Hydrogen is somewhat unusual in that its rotational constant is so much greater than its boiling point.

For most cases then, we can use Eq. (6–45) which, when we use the Euler-MacLaurin expansion, becomes

$$q_{\text{rot}}(T) = \frac{T}{\sigma\Theta_r}\left\{1 + \frac{\Theta_r}{3T} + \frac{1}{15}\left(\frac{\Theta_r}{T}\right)^2 + \frac{4}{315}\left(\frac{\Theta_r}{T}\right)^3 + \cdots\right\} \tag{6–47}$$

Usually only the first term of this is necessary. Some of the thermodynamic functions are

$$E_{\text{rot}} = NkT\left\{1 - \frac{\Theta_r}{3T} - \frac{1}{45}\left(\frac{\Theta_r}{T}\right)^2 + \cdots\right\} \tag{6–48}$$

$$C_{\text{rot}} = Nk\left\{1 + \frac{1}{45}\left(\frac{\Theta_r}{T}\right)^2 + \cdots\right\} \tag{6–49}$$

* D. M. Dennison, *Proc. Roy. Soc.* **A115**, 483, 1927.
† K. F. Bonhoeffer and P. Harteck, *Z. Phys. Chem.*, **4B**, 113, 1926.

$$S_{rot} = Nk\left\{1 - \ln\left(\frac{\sigma\Theta_r}{T}\right) - \frac{1}{90}\left(\frac{\Theta_r}{T}\right)^2 + \cdots\right\} \tag{6–50}$$

where all of these formulas are valid in the same region, in which σ itself is a meaningful concept, that is, $\Theta_r < 0.2T$. The terms in Θ_r/T and its higher powers are usually not necessary. Note that Eq. (6–47) is identical to Eq. (6–35) except for the occurrence of the symmetry number in Eq. (6–47).

6–6 THERMODYNAMIC FUNCTIONS

Having studied each contribution to the total partition function q in Eq. (6–17), we can write in the harmonic oscillator–rigid rotor approximation

$$q(V, T) = \left(\frac{2\pi mkT}{h^2}\right)^{3/2} V \frac{8\pi^2 IkT}{\sigma h^2} e^{-\beta h\nu/2}(1 - e^{-\beta h\nu})^{-1}\omega_{e_1} e^{D_e/kT} \tag{6–51}$$

Remember that this requires that $\Theta_r \ll T$, that only the ground electronic state is important, and that the zero of energy is taken to be the separated states at rest in their ground electronic states. Note that only q_{trans} is a function of V, and this is of the form $f(T)V$ which, we have seen before, is responsible for the ideal gas equation of state. The thermodynamic functions associated with Eq. (6–51) are

$$\frac{E}{NkT} = \frac{5}{2} + \frac{h\nu}{2kT} + \frac{h\nu/kT}{e^{h\nu/kT} - 1} - \frac{D_e}{kT} \tag{6–52}$$

$$\frac{C_V}{Nk} = \frac{5}{2} + \left(\frac{h\nu}{kT}\right)^2 \frac{e^{h\nu/kT}}{(e^{h\nu/kT} - 1)^2} \tag{6–53}$$

$$\frac{S}{Nk} = \ln\left[\frac{2\pi(m_1 + m_2)kT}{h^2}\right]^{3/2} \frac{Ve^{5/2}}{N} + \ln\frac{8\pi^2 IkTe}{\sigma h^2} + \frac{h\nu/kT}{e^{h\nu/kT} - 1} - \ln(1 - e^{-h\nu/kT}) + \ln\omega_{e1} \tag{6–54}$$

$$pV = NkT \tag{6–55}$$

$$\frac{\mu^0(T)}{kT} = -\ln\left[\frac{2\pi(m_1 + m_2)kT}{h^2}\right]^{3/2} kT - \ln\frac{8\pi^2 IkT}{\sigma h^2} + \frac{h\nu}{2kT} + \ln(1 - e^{-h\nu/kT}) - \frac{D_e}{kT} - \ln\omega_{e1} \tag{6–56}$$

Table 6–1 contains the characteristic rotational temperatures, the characteristic vibrational temperatures, and $D_0 = D_e - \frac{1}{2}h\nu$ for a number of diatomic molecules.

Table 6–3 presents a comparison of Eq. (6–54) with experimental data. It can be seen that the agreement is quite good and is typical of that found for the other thermodynamic functions. It is possible to improve the agreement considerably by including the first corrections to the rigid rotor–harmonic oscillator model. These include centrifugal distortion effects, anharmonic effects, and other extensions. The consideration of these effects introduces a new set of molecular constants, all of which are determined spectroscopically and are well tabulated. (See Problem 6–24.) The use of such

Table 6–3. **The entropies of some diatomic molecules calculated according to Eq. (6–54) compared to the experimental values at 1-atm pressure and 25°C***

	S(calc.) (e.u.)	S(exp.) (e.u.)
H_2	31.1	31.2
O_2	49.0	49.0
N_2	45.7	45.7
Cl_2	53.2	53.3
HCl	44.6	44.6
HBr	47.4	47.4
HI	49.4	49.3
CO	47.2	46.2

* The experimental values have been corrected for any nonideal gas behavior.

additional parameters from spectroscopic data can give calculated values of the entropy and heat capacity that are actually more accurate than experimental ones.

It should be pointed out, however, that extremely accurate calculations can require a sophisticated knowledge of molecular spectroscopy. For example, we said above that the electronic partition function was similar to that in the atomic case. This, however, is not entirely true. For molecules in states other than a $\sum$ state (which has zero total angular momentum), the total electronic angular momentum must be coupled with the overall rotational angular momentum, and this coupling must be treated in a detailed quantum mechanical way. This is too specialized to discuss here, but the result of this coupling is that the electronic and rotational partition functions do not separate. When $T \gg \Theta_r$, however, the molecules are in states with large enough rotational quantum numbers [*cf.* Eq. (6–39)] that the angular momentum coupling is no longer important, and the rotational-electronic partition function separates into a rotational part and an electronic part. Since we have chosen the zero of energy to be the separated electronically unexcited atoms at rest, the electronic partition function is

$$q_e(T) = \omega_{e1} e^{D_e/kT} + \omega_{e2} e^{-\varepsilon_2/kT} + \cdots$$

where the ω_{ej} are the degeneracies, and the ε_j's are measured relative to the ground electronic state of the molecule. Keep in mind, however, that for some molecules, such as NO, this equation is valid only at high temperatures, and that the low-temperature partition function requires a fairly specialized knowledge of the coupling of electronic and rotational angular momenta. See Herzberg under "Additional Reading" for a thorough discussion of this complication.

It seems logical at this time to go on to a discussion of polyatomic molecules in much the same manner as we have for diatomics. We would see very quickly, however, that unless the molecule possesses a certain degree of symmetry, it is impossible to write down any closed-form expression for its rotational energy levels. This means that a calculation of $q_{\text{rot}}(T)$ is at best a complicated numerical problem. This would appear to imply that we have come to the end of the line for statistical thermodynamical applications, and we have not even begun to consider *interactions* between molecules! Even just *two* simple molecules, to say nothing of N particles, interacting through any kind of realistic interatomic potential becomes an extremely complicated quantum mechanical problem. At this point we must go back and reconsider some of the results we have derived up to now.

ADDITIONAL READING

General

DAVIDSON, N. 1962. *Statistical mechanics.* New York: McGraw-Hill. Chapters 8 and 9.

FOWLER, R. H., and GUGGENHEIM, E. A. 1956. *Statistical thermodynamics.* Cambridge: Cambridge University Press. Sections 312 through 325.

HILL, T. L. 1960. *Statistical thermodynamics.* Reading, Mass.: Addison-Wesley. Chapter 8.

KESTIN, J., and DORFMAN, J. R. 1971. *A course in statistical thermodynamics.* New York: Academic. Sections 6–9, 6–10, and 6–13.

KNUTH, E. 1966. *Statistical and thermodynamics.* New York: McGraw-Hill. Chapter 13.

KUBO, R. 1965. *Statistical mechanics.* Amsterdam: North-Holland Publishing Co. Chapter 3 and Example 3–1.

MAYER, J. E., and MAYER, M. G. 1940. *Statistical mechanics.* New York: Wiley. Chapter 7.

MÜNSTER, A. 1969. *Statistical thermodynamics.* Vol. I. Berlin: Springer-Verlag. Sections 6–3 through 6–7.

RUSHBROOKE, G. S. 1949. *Statistical mechanics.* London: Oxford University Press. Chapters 6, 7, and 8.

Spectroscopy

BARROW, G. 1962. *Introduction to molecular spectroscopy.* New York: McGraw-Hill.

DUNFORD, H. B. 1968. *Elements of diatomic molecular spectra.* Reading, Mass.: Addison-Wesley.

HERZBERG, G. 1950. *Spectra of diatomic molecules,* 2nd ed. New York: Van Nostrand.

KING, G. W. 1964. *Spectroscopy and molecular structure.* New York: Holt, Rinehart and Winston.

PROBLEMS

6–1. The Morse potential is

$$U(r) = D_e(1 - e^{-\beta(r-r_e)})^2$$

Show that $\beta = \nu(2\pi^2\mu/D_e)^{1/2}$.

6–2. The dissociation energy D_0 of H_2 is 103.2 kcal/mole, and its fundamental vibrational frequency $\bar{\omega}$ is 4320 cm^{-1}. From this information, calculate D_0 and $\bar{\omega}$ for D_2, T_2, and HD, assuming the Born-Oppenheimer approximation.

6–3. Given that D_0 for H_2 is 103.2 kcal/mole and that Θ_v is 6215°K, calculate D_0 for both D_2 and T_2.

6–4. Show that the moment of inertia of a diatomic molecule is μr_e^2, where μ is the reduced mass, and r_e is the equilibrium separation.

6–5. Show that the force constants in Table 6–1 are consistent with the frequencies given there.

6–6. Using the data in Table 6–1, calculate the frequencies that are expected to be found in the rotational spectrum of HCl.

6–7. In the far infrared spectrum of HBr, there is a series of lines separated by 16.72 cm^{-1}. Calculate the moment of inertia and internuclear separation in HBr.

6–8. Show that the vibrational contribution to the heat capacity C_V of a diatomic molecule is Nk as $T \to \infty$.

6–9. Derive Eq. (6–35) from the Euler–MacLaurin summation formula.

6–10. Show that the rotational level that is most populated is given by $J_{max} = (kT/2\bar{B})^{1/2}$. Calculate J_{max} for CO_2 and H_2 at room temperature.

6–11. The rotational constant $\bar{B}$ for $HC^{12}N^{14}$ is 44,315.97 MHz (megahertz) and $DC^{12}N^{14}$ for 36,207.40 MHz. Deduce the moments of inertia for these molecules. Assuming that the bond lengths are independent of isotopic substitution, calculate the H–C and C–N bond length.

6–12. Given that the values of Θ_r and Θ_v for H_2 are 85.3°K and 6215°K, respectively, calculate these quantities for HD and D_2.

6–13. What is the most probable value of the rotational quantum number J of a gas phase N_2 molecule at 300°K? What is the most probable vibrational quantum number n for this same situation?

6–14. Using the Euler-MacLaurin expansion, derive the second- and third-order corrections to the (first-order) high-temperature limit of E_r and C_{Vr}. Express your result in terms of a power series of Θ_r/T.

6–15. Calculate the rotational contribution to the entropy of HD at 20°K, 100°K, and 300°K, using the formulas appropriate for each particular temperature, and estimate the error involved in each.

6–16. Discuss the statistical weights of a hypothetical diatomic molecule X_2 with a ground electronic state Σ_u^-, supposing the X nuclei have integral spin (bosons) and half-integral spin (fermions). Derive the rotational partition function and the rotational contribution to the heat capacity C_V for each case.

6–17. Calculate the percent of para-D_2 as a function of temperature (assuming equilibrium) and also calculate the heat capacity of the equilibrium mixture, para-D_2, ortho-D_2, and finally what you expect would be the experimental heat capacity.

6–18. Why does one not see discussions in the literature concerning the ortho-para forms of fluorine?

6–19. Show that the thermodynamic quantities p and C_V are independent of the choice of a zero of energy.

6–20. In the far infrared spectrum of HCl, there is a series of lines with an almost constant spacing of 20.7 cm^{-1}. In the near infrared spectrum, there is one intense band at 3.46 microns. Use these data to calculate the entropy of HCl at 300°K and 1 atm (assuming ideal behavior).

6–21. Molecular nitrogen is heated in an electric arc, and it is found spectroscopically that the relative populations of excited vibrational levels is

n	0	1	2	3	4	$\cdots$
$\frac{f_n}{f_0}$	1.000	0.200	0.040	0.008	0.002	$\cdots$

Is the nitrogen in thermodynamic equilibrium with respect to vibrational energy? What is the vibrational temperature of the gas? Is this necessarily the same as the translational temperature?

6–22. Without looking anything up, put in order of decreasing magnitudes the following "temperatures":

$$\Theta_v^{H_2},\ \Theta_r^{H_2},\ \Theta_v^{Cl_2},\ \Theta_r^{Cl_2},\ \Theta_v^{HCl},\ \Theta_r^{HCl}$$

6–23. A more accurate expression for the vibrational energy of a diatomic molecule is

$$\varepsilon_n = (n + \tfrac{1}{2})h\nu - x_e(n + \tfrac{1}{2})^2 h\nu$$

where x_e is called the anharmonicity constant. The additional term here represents the first deviations from strictly harmonic behavior. Treating x_e as a small parameter, calculate the anharmonic effect on the various thermodynamic functions at least to first order in x_e.

6–24. The model of a diatomic molecule presented in this chapter is called the rigid rotor–harmonic oscillator model. The rotational-vibrational energy in this approximation is

$$\varepsilon_{vr} = (n + \tfrac{1}{2})h\nu + \tilde{B}J(J + 1)$$

This expression can be improved in a number of ways. The harmonic oscillator approximation can be modified to include terms that reflect the deviations from harmonic behavior (anharmonicity) as the vibrational energy of the molecule increases. This is done by quantum mechanical perturbation theory, which gives

$$\varepsilon_v = (n + \tfrac{1}{2})h\nu - x_e(n + \tfrac{1}{2})^2 h\nu + \cdots$$

where x_e is a small constant called the anharmonicity constant. In addition to this, there is a correction due to the fact that the molecule is not a rigid rotor and, in fact, stretches some as the molecule rotates with greater energy. This is also handled by perturbation theory and gives

$$\varepsilon_r = \bar{B}J(J+1) - \bar{D}J^2(J+1)^2 + \cdots$$

where $\bar{D}$ is a small constant called the centrifugal distortion constant. Lastly, there exists a coupling between the rotational and vibrational modes of the molecule, since its moment of inertia changes as the molecule vibrates. Putting all this together gives

$$\varepsilon_{vr} = (n+\tfrac{1}{2})h\nu + \bar{B}J(J+1) - x_e(n+\tfrac{1}{2})^2 h\nu - \bar{D}J^2(J+1)^2 - \alpha(n+\tfrac{1}{2})J(J+1)$$

where α is the rotation-vibration coupling constant. These terms, which correct the rigid rotor–harmonic oscillator approximation, are usually quite small. Using this more rigorous expression for ε_{vr}, show that the molecular partition function can be written in the form

$$q(V, T) = q_{\text{rr-ho}}\, q_{\text{corr}}$$

where

$$q_{\text{corr}} = 1 + \frac{2kT}{\bar{B}}\left(\frac{\bar{D}}{\bar{B}}\right) + \frac{1}{e^{\beta h\nu} - 1}\left(\frac{\alpha}{\bar{B}}\right)$$

$$+ \frac{2\beta h\nu}{(e^{\beta h\nu} - 1)^2}\, x_e + \text{higher-order terms in } \bar{D},\ \alpha, \text{ and } x_e$$

Calculate the effect of q_{corr} on E and C_V for O_2 at 300°K, given the following values of the spectroscopic parameters: $x_e = 0.0076$, $\bar{D} = 4.8 \times 10^{-6}$ cm^{-1}, and $\alpha = 0.016$ cm^{-1}.

6–25. Consider a system of independent diatomic molecules constrained to move in a plane, that is, a two-dimensional ideal diatomic gas. How many degrees of freedom does a two-dimensional diatomic molecule have? Given that the energy eigenvalues of a two-dimensional rigid rotor are

$$\varepsilon_J = \frac{\hbar^2 J^2}{2I} \qquad J = 0, 1, 2, \ldots$$

with a degeneracy $\omega_J = 2$ for all J except $J = 0$, for which $\omega_J = 1$, calculate the rotational partition function. I is the moment of inertia of the molecule. The vibrational partition function is the same as for a three-dimensional diatomic gas. Write out

$$q(T) = q_{\text{trans}}(T)\, q_{\text{rot}}(T)\, q_{\text{vib}}(T)$$

and derive an expression for the average energy of this two-dimensional ideal diatomic gas.

6–26. Show that the molecule $O^{16}O^{16}$ has only odd rotational levels in its ground electronic state.

CHAPTER 7

CLASSICAL STATISTICAL MECHANICS

So far we have been able to derive translational, rotational, and vibrational partition functions for linear molecules. In each case we saw that if the temperature were high enough we could replace sums by integrals and obtain high-temperature limits. These prove to be numerically satisfactory for most gases at ordinary temperatures (with the exception perhaps of the vibrational case).

As the temperature increases, the average energy per molecule increases, and so in a quantum mechanical sense, the quantum numbers describing this motion (n_x for translational, J for rotational, etc.) also increase, meaning that the molecules are in the high quantum number limit. For example at room temperature translational quantum numbers are typically 10^8. (See Problem 5–3.) It is the recognition of this fact that will point the way to a solution to the problem discussed in the last paragraph of Chapter 6.

It is one of the fundamental principles of quantum mechanics that classical behavior is obtained in the limit of large quantum numbers. So we see that up to now our procedure has been to solve a particular quantum mechanical problem, use this result in the molecular partition function, use a high-temperature approximation, and then find that this high-temperature limit is satisfactory. In other words, we were starting with a quantum mechanical solution and then taking the classical limit at a later stage. It is natural to seek a procedure in which we can use classical mechanics throughout, and such an approach is developed in this chapter. For simplicity we shall first consider only molecular partition functions, although we shall generalize our results afterward.

7–1 THE CLASSICAL PARTITION FUNCTION

Consider the molecular partition function

$$q = \sum_j e^{-\beta \varepsilon_j} \tag{7–1}$$

This is of the form of a sum of $e^{-\beta(\text{energy})}$ over all possible quantum states. It is natural to assume that the corresponding classical expression is a similar sum, or since the energy in the classical sense is a continuous function of the momenta p_j and coordinates q_j, this sum would become an *integral* over all the possible classical "states" of the system. Since the classical energy is the Hamiltonian function $H(p, q)$, the molecular partition function $q(V, T)$ becomes

$$q_{\text{class}} \sim \int \cdots \int e^{-\beta H(p,q)}\, dp\, dq \tag{7–2}$$

In Eq. (7–2) the notation (p, q) denotes all the momenta and coordinates on which H depends; dp stands for $dp_1\, dp_2 \cdots dp_s$ and dq for $dq_1 \cdots dq_s$, where s is the number of momenta or coordinates necessary to completely specify the motion or position of the molecule. The quantity s represents the number of degrees of freedom of the molecule. The set of coordinates $\{q_j\}$ does not necessarily have to be a set of Cartesian coordinates, and more usually represents a set of generalized coordinates, that is, any set of coordinates that conveniently specifies the position of the molecule. For a mass point, for example, the generalized coordinates might be simply x, y, and z; for a rigid rotor, we might choose the two angles θ and ϕ needed to specify the orientation of the molecule. Usually the choice of generalized coordinates is obvious. The momenta $\{p_j\}$ in Eq. (7–2) are the generalized momenta conjugate to the $\{q_j\}$ [*cf.* Eq. (1–19)].

At this stage Eq. (7–2) is just a plausible conjecture. Let us now pursue this idea by considering a monatomic ideal gas once again. From Eq. (5–8), we have

$$q_{\text{trans}}(V, T) = \left(\frac{2\pi mkT}{h^2}\right)^{3/2} V$$

The classical Hamiltonian of one atom of a monatomic ideal gas is simply the kinetic energy:

$$H = \frac{1}{2m}(p_x{}^2 + p_y{}^2 + p_z{}^2)$$

According to Eq. (7–2), then,

$$q_{\text{class}} \sim \int \cdots \int \exp\left\{-\frac{\beta(p_x{}^2 + p_y{}^2 + p_z{}^2)}{2m}\right\} dp_x\, dp_y\, dp_z\, dx\, dy\, dz \tag{7–3}$$

Notice here that since it takes three coordinates to specify the position of a point particle, q_{class} is a six-fold integral. The integral over $dx\, dy\, dz$ simply yields the volume of the container V, and so we have

$$q_{\text{class}} \sim V\left\{\int_{-\infty}^{\infty} e^{-\beta p^2/2m}\, dp\right\}^3 = (2\pi mkT)^{3/2} V \tag{7–4}$$

We see that except for a factor of Planck's constant cubed, this is just the translational partition function that we obtained before. Of course, we cannot expect to derive a purely classical expression that contains h, and so although our conjecture may be incomplete, there seems to be some element of truth to it.

Let us see how this procedure works for the other partition functions that we have evaluated. For the rigid rotor, the Hamiltonian is

$$H = \frac{1}{2I}\left(p_\theta{}^2 + \frac{p_\phi{}^2}{\sin^2\theta}\right)$$

where I is the moment of inertia of the molecule. The generalized coordinates and momenta in this case are θ, ϕ, p_θ, and p_ϕ, and so Eq. (7–2) is

$$q_{\text{rot}} \sim \int_{-\infty}^{\infty} \int dp_\theta \, dp_\phi \int_0^{2\pi} d\phi \int_0^{\pi} d\theta e^{-\beta H} = (8\pi^2 IkT) \tag{7–5}$$

For the classical harmonic oscillator,

$$H = \frac{p^2}{2\mu} + \frac{k}{2} x^2 \tag{7–6}$$

and

$$q_{\text{vib}} \sim \int_{-\infty}^{\infty} dp \int_{-\infty}^{\infty} dx e^{-\beta H} = \frac{kT}{\nu} \tag{7–7}$$

where

$$\nu = \frac{1}{2\pi}\left(\frac{k}{\mu}\right)^{1/2}$$

We can see from these three examples that the translational partition function is incorrect by a factor of h^3; the rotational partition function is incorrect by a factor of h^2; and the vibrational partition function is incorrect by a factor of h. It appears that a factor of h results for each product $dp_j \, dq_j$ occurring in q_{class}. Since partition functions are dimensionless, and h has units of momentum times length, we see that this at least automatically satisfies a dimensional requirement. We shall therefore *assume* that

$$q = \sum_j e^{-\beta \varepsilon_j} \to \frac{1}{h^s} \int \cdots \int e^{-\beta H} \prod_{j=1}^{s} dp_j \, dq_j \tag{7–8}$$

We now shall extend this assumption to systems of molecules. Equation (4–10) says that at high enough temperatures, we can write for a system of N independent indistinguishable particles

$$\begin{aligned} Q &= \frac{q^N}{N!} \\ &= \frac{1}{N!} \prod_{j=1}^{N} \left\{ \frac{1}{h^s} \int \cdots \int e^{-\beta H_j} \prod_{i=1}^{s} dp_{ji} \, dq_{ji} \right\} \end{aligned}$$

where H_j is the Hamiltonian of the jth molecule and is a function of $p_{j1}, \ldots, p_{js}$, $q_{j1}, \ldots, q_{js}$. We now simply relabel the momenta and coordinates such that p_1 through p_s represent p_{11} through p_{1s}; p_{s+1} through p_{s+s} represent p_{21} through p_{2s}, and so on, and write

$$\begin{aligned} Q &= \frac{1}{N! \, h^{sN}} \int \cdots \int e^{-\beta \Sigma_j H_j} \prod_{i=1}^{sN} dp_i \, dq_i \\ &= \frac{1}{N! \, h^{sN}} \int \cdots \int e^{-\beta H} \prod_{i=1}^{sN} dp_i \, dq_i \end{aligned}$$

where H is the Hamiltonian of the N-body system. This form suggests the classical limit of Q for systems of interacting particles. We *conjecture* that

$$Q = \frac{1}{N! \, h^{sN}} \int \cdots \int e^{-\beta H(p,q)} \, dp \, dq \tag{7–9}$$

where $H(p, q)$ is the classical N-body Hamiltonian for *interacting* particles. The notation (p, q) represents the set of p_j's and q_j's that describes the entire system, and $dp\,dq$ represents

$$\prod_{j=1}^{sN} dp_j\, dq_j$$

We have assumed then that the classical limit of $Q(N, V, T)$ is given by

$$Q = \sum_j e^{-\beta E_j} \rightarrow \frac{1}{N!\,h^{sN}} \int \cdots \int e^{-\beta H(p, q)}\, dp\, dq \tag{7-10}$$

For a monatomic gas, for example,

$$H(p, q) = \frac{1}{2m} \sum_{j=1}^{N} (p_{xj}^2 + p_{yj}^2 + p_{zj}^2) + U(x_1, y_1, z_1, \ldots, x_N, y_N, z_N) \tag{7-11}$$

Equation (7–10) is, in fact, the correct classical limit of Q, although we have not proved it here. It is actually possible to start with the quantum mechanical sum in Eq. (7–10) and to derive the integral as the classical limit, that is, the limiting result as $h \rightarrow 0$ (*cf.* Section 10–7).

If we substitute Eq. (7–11) into Eq. (7–10), the momentum integrations can be done easily, and we get

$$Q_{\text{class}} = \frac{1}{N!} \left(\frac{2\pi mkT}{h^2}\right)^{3N/2} Z_N \tag{7-12}$$

where

$$Z_N = \int_V e^{-U(x_1, \ldots, z_N)/kT}\, dx_1 \cdots dz_N \tag{7-13}$$

In Eq. (7–13), Z_N is called the *classical configuration integral.* Since the intermolecular forces depend upon the relative distances between molecules, this integral is, in general, extremely difficult and is essentially responsible for the research in equilibrium statistical mechanics. In the absence of intermolecular forces, $U = 0$ and $Z_N = V^N$. Equations (7–12) and (7–13) are fundamental equations in the study of monatomic, classical, imperfect gases and liquids.

It often happens that not all of the degrees of freedom of a molecule can be treated classically. For example, we have seen that the spacing between translational and rotational levels is small enough that the sum over states or levels can be replaced by an integral, that is, these degrees of freedom can be treated classically. This is not the case, however, with the vibrational degrees of freedom, and these degrees of freedom must be treated quantum mechanically.

Suppose, then, that the Hamiltonian of a molecule can be written as

$$H = H_{\text{class}} + H_{\text{quant}} \tag{7-14}$$

where H_{class} refers to the s degrees of freedom that can be treated classically, and H_{quant} refers to the degrees of freedom that cannot be treated classically. Then

$$q = q_{\text{class}}\, q_{\text{quant}} \tag{7-15}$$

where

$$q_{\text{class}} = \frac{1}{h^s} \int e^{-H_{\text{class}}(p, q)/kT}\, dp_1\, dq_1 \cdots dp_s\, dq_s \tag{7-16}$$

Note that Eq. (6–51) is of the form of Eq. (7–15), where the translational and rotational degrees of freedom are treated classically, and the vibrational and electronic degrees of freedom are treated quantum mechanically.

Equations (7–14) to (7–16) are immediately generalizable to a system of interacting molecules. If the Hamiltonian of the entire system is separable into a classical part and a quantum part, then

$$H = H_{\text{class}} + H_{\text{quant}} \tag{7–17}$$

$$Q = Q_{\text{class}} Q_{\text{quant}} \tag{7–18}$$

$$= \frac{Q_{\text{quant}}}{N!\, h^{sN}} \int e^{-H_{\text{class}}/kT} \, dp_{\text{class}} \, dq_{\text{class}} \tag{7–19}$$

7-2 PHASE SPACE AND THE LIOUVILLE EQUATION

Until now our approach has been to go to the classical limit only when it was necessary. Historically, however, statistical mechanics was originally formulated by Boltzmann, Maxwell, and Gibbs in the nineteenth century before the evolution of quantum mechanics. Their formulation, therefore, was based on classical mechanics, and since this still is a most useful limit, we shall now discuss the classical mechanical formulation of statistical mechanics. This formalism forms the basis of most of the work involving interacting systems in equilibrium and nonequilibrium statistical mechanics that is done today.

Consider any classical system containing N (interacting) molecules. Let each molecule have s degrees of freedom, that is, each molecule requires s coordinates to completely describe its position. Let the number of coordinates necessary to describe the positions of all N molecules be $l = sN$. The l coordinates, $q_1, q_2, \ldots, q_l$, then completely describe the spatial orientation of the entire N-body system. To each of these l coordinates, there corresponds a conjugate momentum p_j, say, defined by Eq. (1–19). The l spatial coordinates $\{q_j\}$ and the l momenta $\{p_j\}$ completely specify the classical mechanical state of the N-body system. These $2l$ coordinates, along with the equations of motion of the system, completely determine the future and past course of the system.

We now construct a conceptual Euclidean space of $2l$ dimensions, with $2l$-rectangular axes, one for each of the spatial coordinates $q_1, \ldots, q_l$ and one for each of the momenta $p_1, \ldots, p_l$. Following Gibbs, we speak of such a conceptual space as a *phase space* for the system under consideration. The state of the classical N-body system at any time t is completely specified by the location of *one* point in phase space. Such a point is called a *phase point*. As the system evolves in time, its dynamics is completely described by the motion or trajectory of the phase point through phase space. The trajectory of the phase point is given by Hamilton's equations of motion:

$$\dot{q}_j = \frac{\partial H}{\partial p_j} \quad \text{and} \quad \dot{p}_j = -\frac{\partial H}{\partial q_j} \qquad j = 1, 2, \ldots, l = sN \tag{7–20}$$

In principle, these $2l$ equations can be integrated to give $\{q_j(t)\}$ and $\{p_j(t)\}$. For notational simplicity, we shall denote the set of l q's by $q(t)$, and the set of l p's by $p(t)$. The $2l$ constants of integration can be fixed by the location of the phase point at some initial time, say t_0. Of course, in practice, such an integration is not feasible.

We now introduce the concept of an ensemble of systems in phase space. For simplicity we shall consider a microcanonical ensemble, that is, an ensemble representative of an isolated system. Consider a large number $\mathscr{A}$ of isolated systems, each of which having the same values of macroscopic variables N, V, and E.

The detailed classical state of each system in the ensemble has a representative phase point in the *same* phase space. The entire ensemble then appears as a cloud of points in phase space. As time evolves, each point will trace out its *independent* trajectory. The trajectories are independent, since each one represents an isolated system and is, therefore, independent of all the others. The postulate of equal a priori probabilities requires that there is a representative phase point in phase space for each and every set of coordinates and momenta consistent with the few fixed macroscopic variables. In particular, the postulate of equal a priori probabilities states that for a microcanonical ensemble, the density is *uniform* over the constant energy "surface" in phase space, where the value of the energy on the surface is that of the isolated system. We consider all parts of phase space equally important, as long as the (p, q)'s are consistent with all that we know macroscopically about the system, that is, consistent with the values of N, V, and E for the system that the ensemble represents. Just as every quantum state was equally likely before, now we consider every classical state to be equally probable.

This cloud of points is very dense then, and we can define a number density $f(p, q, t)$, such that the number of systems in the ensemble that have phase points in $dp\,dq$ about the point p, q at time t is $f(p, q, t)\,dp\,dq$. Clearly we must have

$$\int \cdots \int f(p, q)\,dp\,dq = \mathscr{A} \tag{7-21}$$

The ensemble average of any function, say $\phi(p, q)$, of the momenta and coordinates of the system is defined as

$$\bar{\phi} = \frac{1}{\mathscr{A}} \int \cdots \int \phi(p, q) f(p, q, t)\,dp\,dq \tag{7-22}$$

It is Gibbs' postulate to equate this ensemble average to the corresponding thermodynamic function. Note the similarity between this equation and Eq. (2–5), its quantum mechanical analog.

Since the equations of motion determine the trajectory of each phase point, they must also determine the density $f(p, q, t)$ at any time if the dependence of f on p and q is known at some initial time t_0. The time dependence of f is thus controlled by the laws of mechanics and is not arbitrary. The time dependence of f is given by the Liouville equation, which we now derive.

Consider the small volume element $\delta p_1 \cdots \delta p_l\, \delta q_1 \cdots \delta q_l$ about the point $p_1, \cdots p_l, q_1, \cdots q_l$. The number of phase points inside this volume at any instant is

$$\delta N = f(p_1, \ldots, p_l, q_1, \ldots, q_l, t)\, \delta p_1 \cdots \delta p_l\, \delta q_1 \cdots \delta q_l$$

This number will, in general, change with time since the natural trajectories of phase points will take them into and out of this volume element, and the number passing through any one "face" will, in general, be different from the number passing through the opposite "face." Let us calculate the number entering one face and leaving through the opposite. Consider two faces perpendicular to the q_1-axis and located at q_1 and $q_1 + \delta q_1$. The number of phase points entering the first of these faces per unit time is

$$f\dot{q}_1\, \delta q_2 \cdots \delta q_l\, \delta p_1 \cdots \delta p_l \tag{7-23}$$

(See Problem 7–33.) The number passing through the other face per unit time is

$$f(q_1 + \delta q_1, q_2, \ldots, q_l, p_1, \ldots, p_l) \times \dot{q}_1(q_1 + \delta q_1, q_2, \ldots, q_l, p_1, \ldots, p_l)\, \delta q_2 \cdots \delta q_l\, \delta p_1 \cdots \delta p_l$$

which, if we expand f and $\dot{q}_1$ to linear terms in δq_1, gives

$$\left(f + \frac{\partial f}{\partial q_1}\delta q_1\right)\left(\dot{q}_1 + \frac{\partial \dot{q}_1}{\partial q_1}\delta q_1\right) \delta q_2 \cdots \delta q_l\, \delta p_1 \cdots \delta p_l + \cdots \tag{7–24}$$

Subtracting (7–24) from (7–23), we get the *net* flow of phase points in the q_1-direction into the volume element $\delta q_1 \cdots \delta q_l\, \delta p_1 \cdots \delta p_l$:

$$\text{net flow} = -\left(\frac{\partial f}{\partial q_1}\dot{q}_1 + f\frac{\partial \dot{q}_1}{\partial q_1}\right) \delta p_1 \cdots \delta p_l\, \delta q_1 \cdots \delta q_l$$

in the q_1-direction. In a similar manner, the net flow in the p_1-direction is (remember that momenta and spatial coordinates have equal status in phase space):

$$-\left(\frac{\partial f}{\partial p_1}\dot{p}_1 + f\frac{\partial \dot{p}_1}{\partial p_1}\right) \delta p_1 \cdots \delta p_l\, \delta q_1 \cdots \delta q_l$$

Thus the change in the number of phase points through all the faces is

$$-\sum_{j=1}^{l}\left(\frac{\partial f}{\partial q_j}\dot{q}_j + f\frac{\partial \dot{q}_j}{\partial q_j} + \frac{\partial f}{\partial p_j}\dot{p}_j + f\frac{\partial \dot{p}_j}{\partial p_j}\right) \delta p_1 \cdots \delta p_l\, \delta q_1 \cdots \delta q_l.$$

This must be equal to the change of δN with time, and so we have

$$\frac{d(\delta N)}{dt} = -\sum_{j=1}^{l}\left[f\left(\frac{\partial \dot{q}_j}{\partial q_j} + \frac{\partial \dot{p}_j}{\partial p_j}\right) + \left(\frac{\partial f}{\partial q_j}\dot{q}_j + \frac{\partial f}{\partial p_j}\dot{p}_j\right)\right] \delta p_1 \cdots \delta p_l\, \delta q_1 \cdots \delta q_l \tag{7–25}$$

This result can be immediately simplified. Since

$$\dot{q}_j = \frac{\partial H}{\partial p_j} \qquad \dot{p}_j = -\frac{\partial H}{\partial q_j} \tag{7–26}$$

the first term in parentheses in Eq. (7–25) is

$$\frac{\partial \dot{q}_j}{\partial q_j} + \frac{\partial \dot{p}_j}{\partial p_j} = 0 \tag{7–27}$$

Furthermore, we divide Eq. (7–25) by the volume element $\delta p_1 \cdots \delta p_l\, \delta q_1 \cdots \delta q_l$. This gives the rate of change in the density itself around the point $p_1, \ldots p_l, q_1, \ldots q_l$, so that we can write

$$\frac{\partial f}{\partial t} = -\sum_{j=1}^{l}\left(\frac{\partial f}{\partial q_j}\dot{q}_j + \frac{\partial f}{\partial p_j}\dot{p}_j\right) \tag{7–28}$$

where we have written $\partial f/\partial t$ to indicate that we have fixed our attention on a given stationary point in the phase space.

Equation (7–28) can be written in a more conventional form by using Eqs. (7–26) for $\dot{q}_j$ and $\dot{p}_j$. The result is

$$\frac{\partial f}{\partial t} = -\sum_{j=1}^{l}\left(\frac{\partial H}{\partial p_j}\frac{\partial f}{\partial q_j} - \frac{\partial H}{\partial q_j}\frac{\partial f}{\partial p_j}\right) \tag{7–29}$$

This is the Liouville equation, the most fundamental equation of classical statistical mechanics. In fact, it can be shown that the Liouville equation is equivalent to the $6N$ Hamiltonian equations of motion of the N-body system. See Mazo under "Additional Reading" for a proof of this. In Cartesian coordinates, the Liouville equation for N point masses is (see Problem 7–11):

$$\frac{\partial f}{\partial t} + \sum_{j=1}^{N} \frac{\mathbf{p}_j}{m_j} \cdot \nabla_{\mathbf{r}_j} f + \sum_{j=1}^{N} \mathbf{F}_j \cdot \nabla_{\mathbf{p}_j} f = 0 \tag{7–30}$$

In this equation $\nabla_{\mathbf{r}_j}$ denotes the gradient with respect to the spatial variables in f; $\nabla_{\mathbf{p}_j}$ denotes the gradient with respect to the momentum variables in f; and $\mathbf{F}_j$ is the total force on the jth particle. The Liouville equation forms the starting point of most theories of nonequilibrium statistical mechanics.

There are several interesting deductions from the Liouville equation which we now discuss. Consider Eq. (7–28)

$$\frac{\partial f}{\partial t} + \sum_{j=1}^{l} \left(\frac{\partial f}{\partial p_j}\right) \dot{p}_j + \sum_{j=1}^{l} \left(\frac{\partial f}{\partial q_j}\right) \dot{q}_j = 0 \tag{7–31}$$

Since $f = f(p, q, t)$, this equation is equivalent to

$$\frac{df}{dt} = 0 \tag{7–32}$$

Physically, this equation says that the density in the neighborhood of any selected moving phase point is a constant along the trajectory of that phase point. Thus the cloud of phase points behaves as an incompressable fluid. Gibbs called this the principle of the conservation of density in phase. An equivalent statement of this is that if p, q are the coordinates of a phase point at time t, which at time t_0 were (p_0, q_0), then Liouville's equation implies that (see Problem 7–12)

$$f(p, q; t) = f(p_0, q_0; t_0) \tag{7–33}$$

Because of the equations of motions, the point (p, q) should be considered a function of the initial point (p_0, q_0) and the elapsed time t. That is

$$p = p(p_0, q_0; t)$$
$$q = q(p_0, q_0; t)$$

Now let us select a small element of volume at (p_0, q_0) at time t_0. At a later time, $t_0 + t$, the phase points originally on the surface of this volume element will have formed a new surface enclosing a volume element of different shape at the phase point (p, q). The volume element at (p, q) must contain the same number of phase points as the original volume element at (p_0, q_0). This follows because a phase point outside or inside the volume element can never cross the surface as the element moves through phase space, for otherwise there would be two different trajectories through the same point in phase space. This is impossible, however, because of the uniqueness of the equations of motion of a phase point. Trajectories of phase points can never cross. Now since the density and number of phase points in the volume element are the same at p_0, q_0 and p, q, it follows that although the shape of this volume may change and contort itself as it moves through phase space, its volume remains constant. Gibbs called this result *conservation of extension in phase space*. This fact is expressed mathematically by writing

$$\delta p \, \delta q = \delta p_0 \, \delta q_0 \qquad \text{for all } t \tag{7–34}$$

Another way of expressing this is to say that the Jacobian of the set (p, q) to (p_0, q_0) is unity. This can be proved directly from the equations of motion of the system. See Mazo under "Additional Reading."

A corollary of this theorem, whose proof demands a more extensive knowledge of classical mechanics, is that if we are given two sets of coordinates and their conjugate momenta, say,

$$q_1, q_2, \ldots, q_{3n}, p_1, p_2, \ldots, p_{3n}$$
$$Q_1, Q_2, \ldots, Q_{3n}, P_1, P_2, \ldots, P_{3n}$$

which can describe a system in phase space equally well, then

$$dq_1\, dq_2 \cdots dq_{3n}\, dp_1 \cdots dp_{3n} = dQ_1 \cdots dQ_{3n}\, dP_1 \cdots dP_{3n}$$

For example, a single particle in three dimensions may be described by the coordinates (x, y, z) or the spherical coordinates (r, θ, ϕ). It is straightforward, albeit lengthy to show that

$$dp_x\, dp_y\, dp_z\, dx\, dy\, dz = dp_r\, dp_\theta\, dp_\phi\, dr\, d\theta\, d\phi \tag{7-35}$$

Notice that although the volume elements in ordinary coordinate space are $dx\,dy\,dz$ and $r^2 \sin\theta\, dr\, d\theta\, d\phi$, the $r^2 \sin\theta$ factor does not occur in the phase space transformation. These simple volume element transformations would not generally be true if we have chosen the generalized coordinates and velocities instead of momenta. This is one reason why momenta and not velocities are used to describe classical systems.

7-3 EQUIPARTITION OF ENERGY

We have seen that classical statistical mechanics is applicable when the temperature is high enough to replace the quantum statistical summation by an integral. Under these conditions, it is not necessary to know the eigenvalues of the quantum mechanical problem, only the classical Hamiltonian is required. There is an interesting theorem of classical statistical mechanics which can be used to understand more fully some of the results of the last two chapters.

Consider the expression for the average energy of a molecule in a system of independent molecules,

$$\bar{\varepsilon} = \frac{\iint He^{-\beta H}\, dp_1 \cdots dq_s}{\iint e^{-\beta H}\, dp_1 \cdots dq_s} \tag{7-36}$$

which can be evaluated in principle for any known dependence of H on the p's and the q's. Multiplying by the total number of molecules gives an expression for the total energy of the system, and by differentiating with respect to T, we obtain an expression for its heat capacity at constant volume.

If it so happens that the Hamiltonian is of the form

$$H(p_1, p_2, \ldots, q_s) = \sum_{j=1}^{m} a_j p_j^2 + \sum_{j=1}^{n} b_j q_j^2 + H(p_{m+1}, \ldots, p_s, q_{n+1}, \ldots, q_s) \tag{7-37}$$

where the a_j and b_j are constants, then it is easy to show that each of these quadratic terms will contribute $kT/2$ to the energy and $k/2$ to the heat capacity. (See Problem 7-29.) This result is called the principle of equipartition of energy. It should be emphasized that the principle is a consequence of the quadratic form of terms in the Hamiltonian, rather than a general consequence of classical statistical mechanics.

Let us apply this general theorem to some of the cases we have treated in Chapters 5 and 6. For instance, for a monatomic ideal gas, the Hamiltonian is

$$H = \frac{p_x^2 + p_y^2 + p_z^2}{2m} \tag{7–38}$$

Since there are three quadratic terms, each atom contributes $3kT/2$ to the total energy and so $3k/2$ to the constant volume heat capacity. This is exactly our result in Chapter 5. For the case of a rigid rotor, the Hamiltonian is

$$H = \frac{1}{2I}\left(p_\theta^2 + \frac{p_\phi^2}{\sin^2\theta}\right)$$

The $\sin^2\theta$ in the p_ϕ^2 term would seem to exclude the p_ϕ^2 term from the principle of equipartition, since Eq. (7–37) requires that the coefficients a_j be constants. There is a more general version, however, that allows the a_j and b_j to be functions of the momenta and coordinates not involved in the quadratic terms, that is, to be functions of $p_{m+1}, \ldots, p_s$ and $q_{n+1}, \ldots, q_s$ in Eq. (7–37). The proof of this is more difficult than the proof of the simpler version. (See either Problem 7–30 or Tolman under "Additional Reading.") Because of this, each quadratic term above still contributes its equipartition value, and so the rotational contribution of a rigid rotor to the energy is kT per molecule, just as we obtained in Chapter 6 [*cf.* Eq. (6–36)].

Note that equipartition is a classical concept, that is, the degree of freedom contributing must be such that $\Delta\varepsilon/kT$ is small in passing from one level to another. We have seen that this is true for translational and rotational degrees of freedom at ordinary temperatures, but not vibrational degrees of freedom. The heat capacity for an ideal diatomic gas in the rigid rotor–harmonic oscillator approximation is [*cf.* Eq. (6–53)]

$$C_V = \tfrac{5}{2}Nk + \frac{Nk(\Theta_v/T)^2 e^{\Theta_v/T}}{(e^{\Theta_v/T} - 1)^2} \tag{7–39}$$

where the $\frac{5}{2}Nk$ comes from the translational plus rotational degrees of freedom which, we have seen, are excited enough to be treated classically. The second term is the vibrational contribution, which reaches its expected classical limit of Nk, since the classical Hamiltonian for a harmonic oscillator is $(p^2/2m) + (k/2)x^2$), when Θ_v/T becomes small, which is far above room temperature for most molecules. A value of the vibrational contribution to C_V differing from Nk is thus a quantum mechanical result.

There are more general formulations of the principle of equipartition of energy than we have given here, but they are not necessary for most purposes. In fact, the principle itself is perhaps more of historical interest today than actual practical interest. It is interesting to note in this regard that when the electronic structure of atoms and metals evolved toward the end of the nineteenth century, it was of great concern to Gibbs that the electrons contributed only a very small fraction of their equipartition value to the heat capacities of metals. He did not live to see this anomalous result completely explained by quantum statistics. Since electrons have such a small mass, they behave not at all classically and should, therefore, not be governed by the equipartition of energy (*cf.* Section 10–2).

We have made this long detour through phase space for more than just historical reasons. As we said earlier, most of the systems of interest to chemists can be treated very satisfactorily by classical methods. In fact, the quantum statistical theories of systems of interacting particles are quite a demanding and specialized subject whose techniques are still being developed. Fortunately, being chemists, we are spared from having to master these techniques. Even today the classical Liouville equation forms the starting point for most of the rigorous approaches to nonequilibrium statistical mechanics. We shall now discuss the problem that sent us here in the first place, namely, the study of ideal polyatomic gases.

ADDITIONAL READING

General

EYRING, H., HENDERSON, D., STOVER, B. J., and EYRING, E. M. 1964. *Statistical mechanics and dynamics*. New York: Wiley. Chapter 7.
GIBBS, J. W. 1960. *Elementary principles in statistical mechanics*. New York: Dover.
HILL, T. L. 1956. *Statistical mechanics*. New York: McGraw-Hill. Chapter 1.
HUANG, K. 1963. *Statistical mechanics*. New York: Wiley. Chapter 7.
KHINCHIN, A. I. 1949. *Mathematical foundations of statistical mechanics*. New York: Dover.
KILPATRICK, J. E. 1967. In *Physical chemistry, an advanced treatise*, Vol. II, ed. by H. Eyring, D. Henderson, and W. Jost. New York: Academic.
KUBO, R. 1965. *Statistical mechanics*. Amsterdam: North-Holland Publishing Co. Sections 1–1 to 1–14.
MAZO, R. M. 1967. *Statistical mechanical theories of transport processes*. New York: Pergamon. Chapter 2.
MÜNSTER, A. 1969. *Statistical thermodynamics*, Vol. I. Berlin: Springer-Verlag. Chapter 1.
RUSHBROOKE, G. S. 1949. *Statistical mechanics*. London: Oxford University Press. Chapter 4.
TER HAAR, D. 1966. *Elements of thermostatistics*. London: Oxford University Press. Chapter 5.
———. 1955. *Rev. Mod. Phys.* **27**, p. 289, 1955.
TOLMAN, R. C. 1938. *Statistical mechanics*. London: Oxford University Press. Chapter 3.

PROBLEMS

7–1. Show that at room temperature the translational quantum numbers are typically around 10^8 or so.

7–2. What is the constant energy surface in phase space for a simple linear harmonic oscillator? What is it for a single-point mass? What is it for an ideal gas of N-point masses?

7–3. Convince yourself that trajectories in phase space can never cross, also that surfaces (really hypersurfaces) of constant energy can never intersect if the energies are different.

7–4. Consider a classical ideal gas enclosed in an infinitely tall cylinder in a gravitational field. Assuming that the temperature is uniform up the cylinder, derive the barometric formula

$$p(z) = p(0)\exp\left(\frac{-mgz}{kT}\right)$$

From this calculate the atmospheric pressure at the top of Mt. Everest.

7–5. An ideal gas consisting of N particles of mass m is enclosed in an infinitely tall cylindrical container placed in a uniform gravitational field, and is in thermal equilibrium. Calculate the classical partition function, Helmholtz free energy, mean energy, and heat capacity of this system.

7–6. Consider a perfect gas of molecules with permanent electric dipole moments $\boldsymbol{\mu}$ in an electric field $\mathscr{E}$. Neglecting the polarizability of the molecules, the potential energy is

$$U = -\mu\mathscr{E}\cos\theta$$

where θ is the angle between $\boldsymbol{\mu}$ and $\mathscr{E}$. Using classical mechanics, derive an expression for the additional effect of $\mathscr{E}$ on the energy E and heat capacity of the gas.

7–7. The potential energy of N molecules in a container V can often be fairly well approximated by a sum of pair-wise potentials:

$$U(\mathbf{r}_1, \ldots, \mathbf{r}_N) = \sum_{i<j} u(\mathbf{r}_i, \mathbf{r}_j)$$

In addition, the pair-wise potentials $u(\mathbf{r}_i, \mathbf{r}_j)$ are often assumed to depend only upon the distance $r_{ij} = |\mathbf{r}_i - \mathbf{r}_j|$ between the two molecules. Thus one often writes

$$U(\mathbf{r}_1, \ldots, \mathbf{r}_N) = \sum_{i<j} u(r_{ij})$$

Convince yourself that even these two simplifications of U do not help in trying to evaluate the configuration integral Z.

7–8. It is possible to determine the value of Boltzmann's constant by observing the distribution of suspended Brownian particles in a gravitational field as a function of their height z. Given that the particles have a mass of 1.0×10^{-14} g, that the temperature is 300°K, and the following data:

z(cm)	Number of particles
0.0000	100
0.0025	55
0.0050	31
0.0075	17
0.0100	9

calculate the value of the Boltzmann constant.

7–9. We can calculate the microcanonical ensemble partition function for a classical monatomic ideal gas in the following way. This partition function is given by

$$\Omega(E, \Delta E) = \frac{1}{N!h^{3N}} \int \cdots \int^* dp_1\, dp_2 \cdots dq_{3N}$$

where the asterisk indicates that one integrates over the region of phase space such that

$$E - \Delta E \leq \frac{1}{2m} \sum_{j=1}^{3N} p_j^2 \leq E$$

We have seen in the quantum mechanical case that the thermodynamic consequences of this equation are remarkably insensitive to the value of ΔE. (See Problem 3–14.) We can find Ω most readily by first evaluating

$$I(E) = \frac{1}{N!h^{3N}} \int \cdots \int^* dp_1\, dp_2 \cdots dq_{3N}$$

where now the asterisk signifies the constraint

$$0 \leq \frac{1}{2m} \sum_{j=1}^{3N} p_j^2 \leq E$$

Note that $\Omega(E, \Delta E)$ is given by $I(E) - I(E - \Delta E)$. The integration of $dq_1 \cdots dq_{3N}$ in $I(E)$ immediately gives V^N, and the remaining integration over the momenta is just the volume of a $3N$-dimensional sphere of radius $(2mE)^{1/2}$. The volume of a $3N$-dimensional sphere of radius R is (see Problem 1–24)

$$\frac{\pi^{3N/2}}{(3N/2)!} R^{3N}$$

(Note that this reduces correctly when $3N=2$ and 3.) Using this formula then, show that

$$I(E)=\frac{\pi^{3N/2}V^N(2mE)^{3N/2}}{N!\,h^{3N}(3N/2)!}$$

is in agreement with Eq. (1–36).

7–10. In Problem 3–14 we showed that the entropy could be calculated from $k\ln\Omega(E)\Delta E$ or $k\ln\Phi(E)$, where $\Omega(E)\Delta E$ is the number of states with energies between E and $E+\Delta E$, and $\Phi(E)$ is the total number of states with energies less than E. In addition to this, we showed that the result is remarkably insensitive to the choice of ΔE. We shall now discuss the classical analog of this. In particular, this problem involves showing that the volume of an N-dimensional sphere is essentially the same as the volume of the hypershell of thickness s. First write the volume of the hypersphere as

$$V_{\text{sphere}}(R)=\text{const}\times R^N$$

Now show that if N is large enough such that $sN\gg R$, then

$$\begin{aligned}V_{\text{shell}}&=V(R)-V(R-s)\\&=\text{const}\times R^N(1-e^{-sN/R})\\&\approx V_{\text{sphere}}\end{aligned}$$

7–11. Show that in Cartesian coordinates, the Liouville equation takes the form of Eq. (7–30).

7–12. Convince yourself that a corollary of Liouville's equation is

$$f(p,q;t)=f(p_0,q_0;t_0)$$

Although we did not discuss it explicitly, much of the kinetic theory of gases is contained in this chapter. Problems 7–13 through 7–25 develop some of the kinetic theory of gases.

7–13. Consider a system of N interacting molecules, whose vibrational degrees of freedom are treated quantum mechanically and whose translational and rotational degrees of freedom are treated classically with Hamiltonian

$$H_{\text{class}}=K_{\text{trans}}+K_{\text{rot}}+U$$

where K represents kinetic energy, and U represents potential energy. Substitute this into Eq. (7–19); integrate over all the coordinates except the $3N$ translational momentum coordinates; and derive

$$\text{prob}\{K_{\text{trans}}\}=\frac{e^{-K_{\text{trans}}/kT}\,dp_{\text{trans}}}{\int e^{-K_{\text{trans}}/kT}\,dp_{\text{trans}}}$$

Now realize that

$$K_{\text{trans}}=\sum_{j=1}^{N}\frac{1}{2m}(p_{xj}^2+p_{yj}^2+p_{zj}^2)$$

and derive the normalized Maxwell-Boltzmann distribution, namely,

$$f(p_x,p_y,p_z)\,dp_x\,dp_y\,dp_z=(2\pi mkT)^{-3/2}\,e^{-(p_x^2+p_y^2+p_z^2)/2mkT}\,dp_x\,dp_y\,dp_z \tag{7–40}$$

One can derive all of the usual expressions of the kinetic theory of gases from this.

7–14. An integral that appears often in statistical mechanics and particularly in the kinetic theory of gases is

$$I_n=\int_0^\infty x^n\,e^{-ax^2}\,dx$$

This integral can be readily generated from two basic integrals. For even values of n, we first consider

$$I_0 = \int_0^\infty e^{-ax^2}\, dx$$

The standard trick to evaluate this integral is to square it, and then transform the variables into polar coordinates:

$$\begin{aligned} I_0^2 &= \int_0^\infty \int_0^\infty e^{-ax^2} e^{-ay^2}\, dx\, dy \\ &= \int_0^\infty \int_0^{\pi/2} e^{-ar^2} r\, dr\, d\theta \\ &= \frac{\pi}{4a} \\ I_0 &= \frac{1}{2}\left(\frac{\pi}{a}\right)^{1/2} \end{aligned}$$

Using this result, show that for even n

$$I_n = \frac{1 \cdot 3 \cdot 5 \cdots (n-1)}{2(2a)^{n/2}} \left(\frac{\pi}{a}\right)^{1/2} \qquad n \text{ even}$$

For odd values of n, the basic integral I_1 is easy. Using I_1, show that

$$I_n = \frac{\Gamma\left(\dfrac{n+1}{2}\right)}{2a^{(n+1)/2}} \qquad n \text{ odd}$$

7–15. Convert Eq. (7–40) (see Problem 7–13) from a Cartesian coordinate to a spherical coordinate representation by writing

$$\begin{aligned} &p^2 = p_x^2 + p_y^2 + p_z^2 \\ &p_z = p \cos\theta \\ &p_x = p \sin\theta \cos\phi \\ &p_y = p \sin\theta \sin\phi \\ &dp_x\, dp_y\, dp_z \to p^2 \sin\theta\, dp\, d\theta\, d\phi \end{aligned}$$

and integrating over θ and ϕ to get

$$f(p)\, dp = 4\pi(2\pi mkT)^{-3/2} p^2\, e^{-p^2/2mkT}\, dp$$

for the fraction of molecules with momentum between p and $p + dp$. By substituting $p = mv$, we get the fraction of molecules with *speeds* between v and $v + dv$:

$$f(v)\, dv = 4\pi \left(\frac{m}{2\pi kT}\right)^{3/2} v^2 e^{-mv^2/2kT}\, dv$$

7–16. Prove that the most probable molecular speed is $v_{mp} = (2kT/m)^{1/2}$, that the mean speed is $\langle v \rangle = (8kT/\pi m)^{1/2}$, and that the root-mean-square speed is $\langle v^2 \rangle^{1/2} = (3kT/m)^{1/2}$. Evaluate these for H_2 and N_2 at 25°C.

7–17. Show that the mean-square fluctuation of the velocity of the Maxwell-Boltzmann distribution is

$$\overline{v^2} - \bar{v}^2 = \frac{kT}{m}\left(3 - \frac{8}{\pi}\right)$$

7–18. Show that the average velocity in any direction (say x, y, or z) vanishes. What does this mean?

7–19. Derive an expression for the fraction of molecules with translational energy between ε and $\varepsilon + d\varepsilon$ from both Eqs. (7–40) and (5–7).

7–20. According to Problem 1–35, the speed of sound in an ideal gas is given by

$$c_0 = \left(\gamma \frac{RT}{M}\right)^{1/2}$$

where M is the molecular weight of the gas, and $\gamma = C_p/C_V$. Show that $c_0 = 0.81\bar{v}$ for an ideal monatomic gas.

7–21. Calculate the probability that two molecules will have a total kinetic energy between ε and $\varepsilon + d\varepsilon$.

7–22. Calculate the fraction of molecules with x-component of velocity between $\pm n(2kT/m)^{1/2}$, where $n = 1$, 2, and 3. Remember that the integral of $\exp(-x^2)$ with finite limits cannot be evaluated in closed form and is expressed in terms of the error function erf (x) by

$$\text{erf}\,(x) = \frac{2}{\pi^{1/2}} \int_0^x e^{-t^2}\, dt$$

7–23. What is the average kinetic energy $\bar{\varepsilon}$ and the most probable kinetic energy ε_{mp} of a gas molecule?

7–24. Show that the number of molecules striking a unit area per unit time is $\rho\bar{v}/4$, where $\rho = N/V$.

7–25. How would you interpret the velocity distribution

$$\phi(\mathbf{v}) = \left(\frac{m}{2\pi kT}\right)^{3/2} \exp\left\{-\frac{m}{2kT}\,[(v_x - a)^2 + (v_y - b)^2 + (v_z - c)^2]\right\}$$

in which a, b, and c are constants?

7–26. The relativistic dependence of the kinetic energy on momentum is

$$\varepsilon = c(p_x^2 + p_y^2 + p_z^2 + m_0^2c^2)^{1/2}$$

where m_0 is the rest mass of the particle, and c is the speed of light. Determine the thermodynamic properties of an ideal gas in the extreme relativistic limit, where $p \gg m_0 c$.

7–27. If an atom is radiating light of wavelength λ_0, the wavelength measured by an observer will be

$$\lambda = \lambda_0\left(1 + \frac{v_z}{c}\right)$$

if moving away from or toward the observer with velocity v_z. In this equation c is the speed of light. This is known as the Doppler effect. If one observes the radiation emitted from a gas at temperature T, it is found that the line at λ_0 will be spread out by the Maxwellian distribution of velocities v_z of the molecules emitting the radiation. Show that $I(\lambda)\, d\lambda$, the intensity of radiation observed between wavelengths λ and $\lambda + d\lambda$, is

$$I(\lambda) \propto \exp\left\{-\frac{mc^2(\lambda - \lambda_0)^2}{2\lambda_0^2 kT}\right\}$$

This spreading about the line at λ_0 is known as Doppler broadening. Estimate the Doppler line width for HCl radiating microwave radiation at room temperature.

7–28. Plot C_V in Eq. (7–39) versus temperature and see that the vibrational contribution does not contribute until the temperature approaches Θ_V.

7–29. Prove that if the Hamiltonian is given by Eq. (7–37), then each of the quadratic terms will contribute $kT/2$ to the average molecular energy and $k/2$ to the molecular heat capacity.

7–30. Prove that even if the a_j and b_j in the Hamiltonian of Eq. (7–37) are functions of the momenta and coordinates not involved in the quadratic terms, the law of equipartition still applies. In particular, show how this more general version of the law of equipartition applies to the rigid rotor Hamiltonian.

7–31. Let $H(p, q)$ be the classical Hamiltonian for a classical system of N interacting particles. Let x_j be one of the $3N$ momentum components or one of the $3N$ spatial coordinates. Prove the generalized equipartition theorem, namely, that

$$\left\langle x_i \frac{\partial H}{\partial x_j} \right\rangle = kT\delta_{ij}$$

and from this derive the principle of equipartition of energy that we discussed earlier. Hint: Realize that the potential $U \rightarrow \infty$ at the walls of the container.

7–32. Consider a two-dimensional harmonic oscillator with Hamiltonian

$$H = \frac{1}{2m}(p_x^2 + p_y^2) + \frac{k}{2}(x^2 + y^2)$$

According to the principle of equipartition of energy, the average energy will be $2kT$. Now transform this Hamiltonian to plane polar coordinates to get

$$H = \frac{1}{2m}\left(p_r^2 + \frac{p_\theta^2}{r^2}\right) + \frac{k}{2}r^2$$

What would you predict for the average energy now? Show by direct integration in plane polar coordinates that $\bar{\varepsilon} = 2kT$. Is anything wrong here? Why not?

7–33. Convince yourself that the number of phase points passing through a face perpendicular to q_1 per unit time is

$$f\dot{q}_1\, \delta q_2\, \delta q_3 \cdots \delta p_l$$

CHAPTER 8

IDEAL POLYATOMIC GAS

In this chapter we make a direct extension to polyatomic molecules of the methods and approximations used in the previous chapters for diatomic molecules and classical statistical mechanics. The introduction of the concept of normal coordinates will allow us to treat the vibrational problem of polyatomic molecules as a simple extension of the vibration of diatomic molecules (Section 8–1). We shall see, however, that the rotational problem must be treated by the method of classical statistical mechanics since, for most polyatomic molecules, the quantum mechanical rotational energy levels cannot be written in any convenient closed form, and most temperatures are such that classical statistical mechanics is applicable anyway (Section 8–2). One new feature that arises with polyatomic molecules is hindered internal rotation (in molecules such as ethane). This type of consideration is important in organic chemistry, where the results can be directly applied to a determination of the various conformers that exist in certain molecules. Hindered rotation is treated in Section 8–4.

The discussion in Section 6–1 for diatomic molecules applies equally well to polyatomic molecules. After making the Born-Oppenheimer approximation, we transform the Schrödinger equation that describes the motion of the nuclei in the Born-Oppenheimer potential into center of mass and relative coordinates. This allows us to write

$$\mathscr{H} = \mathscr{H}_{\text{trans}} + \mathscr{H}_{\text{int}}$$

$$\varepsilon = \varepsilon_{\text{trans}} + \varepsilon_{\text{int}}$$

$$q = q_{\text{trans}} q_{\text{int}}$$

where

$$q_{\text{trans}} = \left[\frac{2\pi MkT}{h^2}\right]^{3/2} V \qquad (8\text{–}1)$$

and M is the total mass of the molecule. As before, the density of translational energy states alone is sufficient to guarantee that the number of energy states available to any molecule is much greater than the number of molecules in the system, and so

$$Q(N, V, T) = \frac{q_{\text{trans}}^N q_{\text{int}}^N}{N!}$$

The Born-Oppenheimer potential, that is, the potential set up by the electrons moving in the field of the fixed nuclei, depends upon a number of internuclear distances. Each of the n atoms in a polyatomic molecule requires three coordinates to locate it. Thus it takes $3n$ coordinates to specify the polyatomic molecule itself. Of these $3n$ coordinates, 3 are needed to specify the center of mass of the molecule. In addition, it requires two coordinates (θ and ϕ) to specify its orientation if it is linear, and three coordinates if it is nonlinear. The remaining $3n - 5$ or $3n - 6$ coordinates are internal coordinates that are required to specify the *relative* location of the n nuclei. Thus the Born-Oppenheimer potential of a polyatomic molecule depends upon $3n - 5$ (linear) or $3n - 6$ (nonlinear) coordinates. Since these coordinates specify the relative locations of the n nuclei, they are referred to as vibrational degrees of freedom. Similarly, since the two (linear) or three (nonlinear) orientation angles are used to specify the orientation of the molecule about its center of mass, they represent rotational degrees of freedom. The three coordinates used to locate the center of mass of the molecule are translational degrees of freedom.

As in the case of diatomic molecules, we use a rigid rotor-harmonic oscillator approximation. This allows us to separate the rotational motion from the vibrational motion of the molecule, and we can treat each one separately. Both problems are somewhat more complicated for polyatomic molecules than for diatomic molecules. Nevertheless, we can write the polyatomic analog of Eqs. (6–14) to (6–17):

$$Q(N, V, T) = \frac{(q_{\text{trans}} q_{\text{rot}} q_{\text{vib}} q_{\text{elec}} q_{\text{nucl}})^N}{N!} \tag{8–2}$$

We choose as the zero of energy all n atoms completely separated in their ground electronic states. Thus the energy of the ground electronic state is $-D_e$. As before then, the electronic partition function is

$$q_{\text{elect}} = \omega_{e1} e^{D_e/kT} + \cdots \tag{8–3}$$

and we set $q_{\text{nucl}} = 1$.

To calculate $Q(N, V, T)$ then, we must investigate q_{vib} and q_{rot}, and this is done in the next two sections.

8–1 THE VIBRATIONAL PARTITION FUNCTION

For a polyatomic molecule, the potential in which the nuclei vibrate is a function of $3n - 5$ or $3n - 6$ relative coordinates. This potential is a generalization of Fig. 6–1 to $3n - 5$ or $3n - 6$ dimensions, and thus is a complicated energy surface. As in the case of diatomic molecules, however, the amplitude of the nuclear vibrations is very small, and we can expand the potential function about its stable or equilibrium configuration. Let the Cartesian coordinates of each nucleus in the molecule be $x_1, y_1, z_1, x_2, \ldots, x_n, y_n, z_n$. For small vibrations about the equilibrium configuration, we have a generalization of Eq. (6–3) to $3n - 5$ or $3n - 6$ relative coordinates, and the potential energy will be a quadratic function of $3n - 5$ or $3n - 6$ relative coordinates such as $x_2 - x_1, x_3 - x_1, y_2 - y_1$, and so on. When terms such as $(x_2 - x_1)^2$ are multiplied out, there result cross terms of the form x_1x_2, x_1x_3, y_1y_2, and so on. The presence of such cross terms makes the potential energy and hence the Hamiltonian a complicated mixture of $x_1, x_2, \ldots$. By the introduction of certain linear combinations of the Cartesian coordinates $x_1, x_2, \ldots$, it is possible to eliminate the cross terms in the potential and to write it as a sum of squares of these new coordinates, say $Q_1, Q_2, \ldots$. In other words, it is possible to transform a Hamiltonian which contains cross terms in terms of $x_1, x_2, \ldots, y_1, y_2, \ldots$, to a Hamiltonian, which when written in terms of $Q_1, Q_2, \ldots$, becomes

$$\mathscr{H} = -\sum_{j=1}^{\alpha} \frac{\hbar^2}{2\mu_j} \frac{\partial^2}{\partial Q_j^2} + \sum_{j=1}^{\alpha} \frac{k_j}{2} Q_j^2 \tag{8–4}$$

In this equation α is the number of vibrational degrees of freedom, that is, $3n - 5$ for a linear molecule and $3n - 6$ for a nonlinear molecule, and μ_j and k_j are effective reduced masses and force constants.

Equation (8–4) is the Hamiltonian of a sum of *independent* harmonic oscillators, and so the total energy is of the form

$$\varepsilon = \sum_{j=1}^{\alpha} (n_j + \tfrac{1}{2})h\nu_j \qquad n_j = 0, 1, 2, \ldots \tag{8–5}$$

where

$$\nu_j = \frac{1}{2\pi}\left(\frac{k_j}{\mu_j}\right)^{1/2}$$

There are straightforward methods to determine the Q_j's that allow the Hamiltonian to be written as a sum of independent terms as in Eq. (8–4). The Q_j's are called normal coordinates, and their determination for any particular molecule is called a normal coordinate analysis. The α fundamental frequencies ν_j are obtained automatically in a normal coordinate analysis, but in practice they are usually determined spectroscopically. The normal coordinates for a linear triatomic molecule (CO_2) are as shown in the following figure.

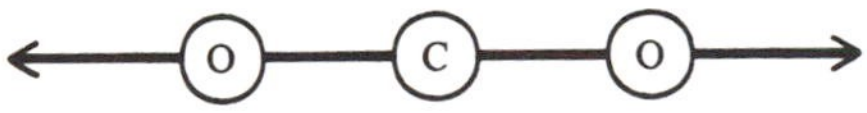

(a)

(b)

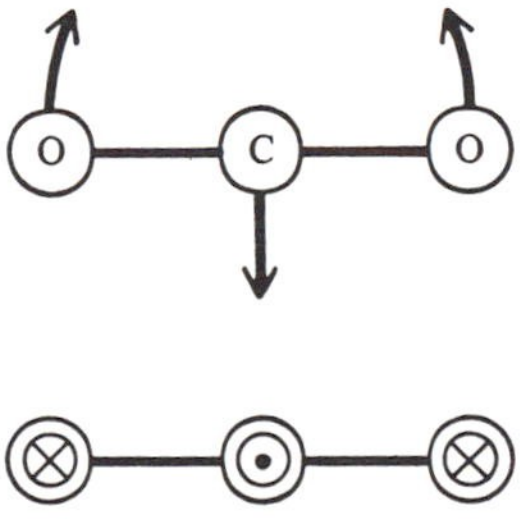

(c)

The mode labeled (a) is a symmetric stretch; the one labeled (b) is an asymmetric stretch; and the one labeled (c) is a bending mode. This mode is doubly degenerate, with one of the modes being in the plane of the page and the other being perpendicular to the page.

For a nonlinear triatomic molecule such as water, we have the following three modes.

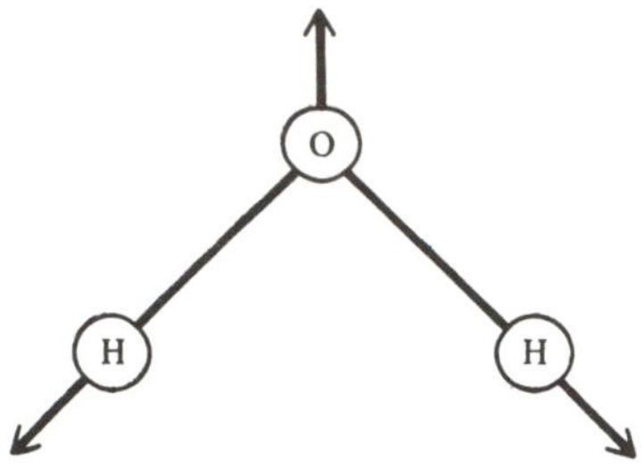

(a)

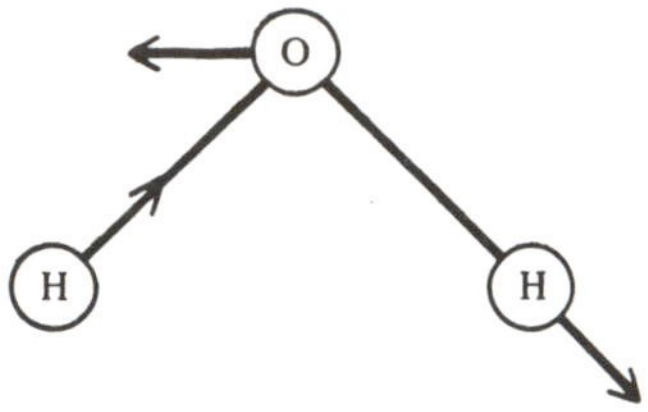

(b)

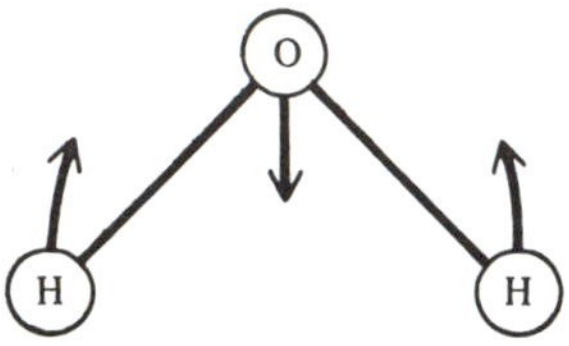

(c)

Because of Eq. (8–4), each normal mode of vibration makes an independent contribution to thermodynamic functions such as E, C_V, S, and so on. If the normal frequencies $\nu_1, \nu_2, \ldots, \nu_\alpha$, where $\alpha = 3n - 5$ or $3n - 6$, are known, we have immediately [see Eqs. (6–22) and (6–23)] equations such as

$$q_{\text{vib}} = \prod_{j=1}^{\alpha} \frac{e^{-\Theta_{vj}/2T}}{(1 - e^{-\Theta_{vj}/T})} \tag{8–6}$$

$$E_{\text{vib}} = Nk \sum_{j=1}^{\alpha} \left(\frac{\Theta_{vj}}{2} + \frac{\Theta_{vj} e^{-\Theta_{vj}/T}}{1 - e^{-\Theta_{vj}/T}} \right) \tag{8–7}$$

$$C_{V,\text{vib}} = Nk \sum_{j=1}^{\alpha} \left[\left(\frac{\Theta_{vj}}{T} \right)^2 \frac{e^{-\Theta_{vj}/T}}{(1 - e^{-\Theta_{vj}/T})^2} \right] \tag{8–8}$$

where

$$\Theta_{vj} = \frac{h\nu_j}{k}$$

Table 8–1 contains values of Θ_{vj} for a number of molecules.

Table 8–1. **Values of the characteristic rotational temperatures, the characteristic vibrational temperatures, and D_0 for polyatomic molecules***

molecule	Θ_{rot}(°K)			Θ_{vib}(°K)	D_0(kcal/mole)
CO_2		0.561		3360, 954(2), 1890	381.5
H_2O	40.1	20.9	13.4	5360, 5160, 2290	219.3
NH_3	13.6	13.6	8.92	4800, 1360, 4880(2), 2330(2)	276.8
ClO_2	2.50	0.478	0.400	1360, 640, 1600	90.4
SO_2	2.92	0.495	0.422	1660, 750, 1960	254.0
N_2O		0.603		3200, 850(2), 1840	263.8
NO_2	11.5	0.624	0.590	1900, 1980, 2330	221.8
CH_4	7.54	7.54	7.54	4170, 2180(2), 4320(3), 1870(3)	392.1
CH_3Cl	7.32	0.637	0.637	4270, 1950, 1050, 4380(2), 2140(2), 1460(2)	370.7
CCl_4	0.0823	0.0823	0.0823	660, 310(2), 1120(3), 450(3)	308.8

* These parameters were obtained from a variety of sources and do not necessarily represent the most accurate values since they are obtained under the rigid rotor–harmonic oscillator approximation.

8–2 THE ROTATIONAL PARTITION FUNCTION

The rotational properties of a polyatomic molecule depend upon the general shape of the molecule. If the molecule is linear, such as CO_2 and C_2H_2, the problem is exactly the same as for a diatomic molecule. The energy levels are given by

$$\varepsilon_J = \frac{J(J+1)h^2}{8\pi^2 I} \qquad J = 0, 1, 2, \ldots$$

$$\omega_J = 2J + 1 \tag{8–9}$$

where, in this case, the moment of inertia I is

$$I = \sum_{j=1}^{n} m_j d_j^2 \tag{8–10}$$

where d_j is the distance of the jth nucleus from the center of mass of the molecule. Recall that the coordinates of the center of mass of a molecule are given by

$$x_{cm} = \frac{1}{M}\sum_{j=1}^{n} m_j x_j$$

$$y_{cm} = \frac{1}{M}\sum_{j=1}^{n} m_j y_j$$

$$z_{cm} = \frac{1}{M}\sum_{j=1}^{n} m_j z_j \tag{8–11}$$

where x_j, y_j, and z_j are the Cartesian coordinates of the jth nucleus in an arbitrary coordinate system, and $M = m_1 + m_2 + \cdots + m_n$. (See Problem 8–1.)

The rotational partition function of a linear polyatomic molecule is

$$q_{\text{rot}} = \frac{8\pi^2 IkT}{\sigma h^2} = \frac{T}{\sigma\Theta_r} \tag{8–12}$$

As before, we have introduced a symmetry number, which is unity for unsymmetrical molecules such as N_2O and COS and equal to two for symmetrical molecules such as CO_2 and C_2H_2. The symmetry number is the number of different ways the molecule can be rotated into a configuration indistinguishable from the original. Classically, it is a factor introduced to avoid over counting indistinguishable configurations in phase space. Table 8–1 gives Θ_r for several linear polyatomic molecules.

The energy calculated from Eq. (8–12) is kT, in accord with equipartition of energy, since there are two degrees of rotational freedom of a linear molecule.

The moment of inertia is a fundamental property of rigid bodies. The rotational properties of a rigid body are characterized by the *principal moments* of the body, which are defined in the following way. Choose any set of Cartesian axes with origin at the center of mass of the body. The moments of inertia about these three axes are

$$I_{xx} = \sum_{j=1}^{n} m_j[(y_j - y_{cm})^2 + (z_j - z_{cm})^2]$$

$$I_{yy} = \sum_{j=1}^{n} m_j[(x_j - x_{cm})^2 + (z_y - z_{cm})^2]$$

$$I_{zz} = \sum_{j=1}^{n} m_j[(x_j - x_{cm})^2 + (y_j - y_{cm})^2]$$

In addition to these, there are also products of inertia, such as

$$I_{xy} = \sum_{j=1}^{n} m_j(x_j - x_{cm})(y_j - y_{cm}) \cdots$$

Now there is a theorem of rigid body motion that says that there always exists a particular set of Cartesian coordinates X, Y, Z, called the principal axes, passing through the center of mass of the body such that all the products of inertia vanish. The moments of inertia about these axes I_{XX}, I_{YY}, and I_{ZZ} are called the principal moments of inertia. The principal moments of inertia are customarily denoted by I_A, I_B, and I_C.

If the molecule possesses any degree of symmetry, the principal axes are simple to find. For example, if the molecule is planar, one of the principal axes will be perpendicular to the plane. Usually an axis of symmetry of the molecule will be a principal axis. The C–H bond of $CHCl_3$ is a three-fold axis of symmetry and also a principal axis. In general, however, it is not often necessary to calculate the principal moments of inertia, since there are extensive tables in the literature. They are usually given in terms of rotational constants in units of cm^{-1}, defined by

$$\bar{A} = \frac{h}{8\pi I_A c}$$

$$\bar{B} = \frac{h}{8\pi I_B c}$$

$$\bar{C} = \frac{h}{8\pi I_C e} \tag{8–13}$$

from which it is an easy matter to calculate the corresponding rotational temperatures Θ_A, Θ_B, and Θ_C. In these quantities, c is the speed of light. Table 8–1 contains values of Θ_A, Θ_B, and Θ_C for a number of polyatomic molecules.

The relative magnitudes of the three principal moments of inertia are used to characterize the rigid body. If all three are equal, the body is called a spherical top; if only two are equal, the body is called a symmetric top; and if all three are different, the body is called an asymmetric top. Table 8–1 shows that CH_4 and CCl_4 are spherical tops; CH_3Cl and NH_3 are symmetric tops; and H_2O and NO_2 are asymmetric tops. The quantum mechanical problem of thet otation of spherical tops and symmetric tops can be readily solved, but the rotation of an asymmetric top is fairly involved.

We shall start with the easiest example, namely, that of spherical top $I_A = I_B = I_C$. The quantum mechanical problem of a spherical top is readily solvable, having energy levels ε_J and degeneracy ω_J given by

$$\varepsilon_J = \frac{J(J+1)\hbar^2}{2I} \qquad J = 0, 1, 2, \ldots$$
$$\omega_J = (2J+1)^2 \tag{8–14}$$

The high-temperature limit of the partition function is

$$q_{\text{rot}} = \frac{1}{\sigma}\int_0^\infty (2J+1)^2 e^{-J(J+1)\hbar^2/2IkT}\, dJ \tag{8–15}$$

Here again we have introduced a symmetry number σ. The symmetry number for a polyatomic molecule is simply the number of ways that the molecule can be rotated "into itself." For example, $\sigma = 2$ for H_2O and $\sigma = 3$ for NH_3. For methane, $\sigma = 12$ since there is three-fold symmetry about each of the four carbon–hydrogen bonds. Similarly, $\sigma = 4$ for ethylene and 12 for benzene. Classically, the symmetry number avoids overcounting indistinguishable configurations in phase space. For those readers who know some group theory, σ is the number of pure rotational elements (including the identity) in the point group of a nonlinear molecule. Since high temperature means that high values of J are important, we may neglect 1 compared to J in Eq. (8–15) and write

$$q_{\text{rot}} = \frac{1}{\sigma}\int_0^\infty 4J^2 e^{-J^2\hbar^2/2IkT}\,dJ = \frac{\pi^{1/2}}{\sigma}\left(\frac{8\pi^2 IkT}{h^2}\right)^{3/2} \tag{8–16}$$

Problem 8–2 involves the derivation of Eq. (8–16) by evaluating the integral in Eq. (8–15) without neglecting 1 compared to J.

The quantum mechanical problem of a symmetric top ($I_A = I_B \neq I_C$) is also solvable in closed form. In this case the energy levels depend upon two quantum numbers, one of which is a measure of the total rotational angular momentum of the molecule J, and the other a measure of the component of the rotational angular momentum along the unique axis of the symmetric top K, that is, the axis having the unique moment of inertia (customarily denoted by I_C). It might be pointed out here that any molecule with an n-fold axis of symmetry, with $n \geq 3$, is at least a symmetric top. The expression for the energy levels is

$$\varepsilon_{JK} = \frac{\hbar^2}{2}\left\{\frac{J(J+1)}{I_A} + K^2\left(\frac{1}{I_C} - \frac{1}{I_A}\right)\right\}$$

where $J = 0, 1, 2, \ldots$; $K = J, J-1, \ldots, -J$; and the degeneracy is

$$\omega_{JK} = (2J+1)$$

The partition function is, then,

$$q_{\text{rot}} = \frac{1}{\sigma}\sum_{J=0}^{\infty}(2J+1)e^{-\alpha_A J(J+1)}\sum_{K=-J}^{+J} e^{-(\alpha_C - \alpha_A)K^2}$$

where

$$\alpha_j = \frac{\hbar^2}{2I_j kT} \qquad j = A \text{ or } C$$

Problem 8–3 converts this to a double integral over J and K, which results in

$$q_{\text{rot}} = \frac{\pi^{1/2}}{\sigma}\left(\frac{8\pi^2 I_A kT}{h^2}\right)\left(\frac{8\pi^2 I_C kT}{h^2}\right)^{1/2} \tag{8–17}$$

Notice that this reduces to the rotational partition of a spherical top [Eq. (8–16)] when $I_A = I_C$.

The next case is that of an asymmetric top $I_A \neq I_B \neq I_C$. This is the most commonly occurring type of molecule. The quantum mechanical problem of the rotational levels of an asymmetric top is a fairly involved problem and must be solved numerically. Consequently a quantum-statistical treatment is awkward, and it is desirable to use classical mechanics. Even the classical Hamiltonian of a rigid asymmetrical rotor is quite complicated. We confine ourselves, therefore, to the statement that insertion of the classical Hamiltonian into the classical phase integral leads, after a rather long but straightforward integration, to (see Problem 8–16)

$$q_{\text{rot}} = \frac{\pi^{1/2}}{\sigma}\left(\frac{8\pi^2 I_A kT}{h^2}\right)^{1/2}\left(\frac{8\pi^2 I_B kT}{h^2}\right)^{1/2}\left(\frac{8\pi^2 I_C kT}{h^2}\right)^{1/2} \tag{8–18}$$

Notice that this is a generalization of Eqs. (8–16) and (8–17).

If we introduce the characteristic rotational temperatures into Eq. (8–18), we have

$$q_{\text{rot}} = \frac{\pi^{1/2}}{\sigma}\left(\frac{T^3}{\Theta_A\Theta_B\Theta_C}\right)^{1/2} \tag{8–19}$$

Table 8–1 contains Θ_A, Θ_B, and Θ_C for several molecules. The rotational contributions to some thermodynamic functions are

$$E_{\text{rot}} = \tfrac{3}{2}NkT \tag{8–20}$$

$$C_{V,\,\text{rot}} = \tfrac{3}{2}Nk \tag{8–21}$$

$$S_{\text{rot}} = Nk \ln\left[\frac{\pi^{1/2}}{\sigma}\left(\frac{T^3e^3}{\Theta_A\Theta_B\Theta_C}\right)^{1/2}\right] \tag{8–22}$$

Note that since there are three degrees of rotational freedom, the rotational kinetic energy here is $3NkT/2$, in accord with equipartition of energy.

8–3 THERMODYNAMIC FUNCTIONS

We can now use the results of Sections 8–1 and 8–2 to construct $q(V, T)$. We get for linear polyatomic molecules

$$q = \left(\frac{2\pi MkT}{h^2}\right)^{3/2} V \cdot \frac{T}{\sigma\Theta_r} \cdot \left\{\prod_{j=1}^{3n-5} \frac{e^{-\Theta_{vj}/2T}}{(1 - e^{-\Theta_{vj}/T})}\right\}\omega_{e1}e^{D_e/kT} \tag{8–23}$$

$$-\frac{A}{NkT} = \ln\left[\left(\frac{2\pi MkT}{h^2}\right)^{3/2}\frac{Ve}{N}\right] + \ln\left(\frac{T}{\sigma\Theta_r}\right) - \sum_{j=1}^{3n-5}\left[\frac{\Theta_{vj}}{2T} + \ln(1 - e^{-\Theta_{vj}/T})\right] + \frac{D_e}{kT} + \ln\omega_{e1} \tag{8–24}$$

$$\frac{E}{NkT} = \frac{3}{2} + \frac{2}{2} + \sum_{j=1}^{3n-5}\left[\left(\frac{\Theta_{vj}}{2T}\right) + \frac{\Theta_{vj}/T}{(e^{\Theta_{vj}/T} - 1)}\right] - \frac{D_e}{kT} \tag{8–25}$$

$$\frac{C_V}{Nk} = \frac{3}{2} + \frac{2}{2} + \sum_{j=1}^{3n-5}\left(\frac{\Theta_{vj}}{T}\right)^2 \frac{e^{\Theta_{vj}/T}}{(e^{\Theta_{vj}/T} - 1)^2} \tag{8–26}$$

$$\frac{S}{Nk} = \ln\left[\left(\frac{2\pi MkT}{h^2}\right)^{3/2}\frac{Ve^{5/2}}{N}\right] + \ln\left(\frac{Te}{\sigma\Theta_r}\right) + \sum_{j=1}^{3n-5}\left[\frac{\Theta_{vj}/T}{e^{\Theta_{vj}/T} - 1} - \ln(1 - e^{-\Theta_{vj}/T})\right] + \ln\omega_{e1} \tag{8–27}$$

$$pV = NkT \tag{8–28}$$

and for nonlinear polyatomic molecules:

$$q = \left(\frac{2\pi MkT}{h^2}\right)^{3/2} V \cdot \frac{\pi^{1/2}}{\sigma}\left(\frac{T^3}{\Theta_A \Theta_B \Theta_C}\right)^{1/2} \cdot \left\{\prod_{j=1}^{3n-6} \frac{e^{-\Theta_{vj}/2T}}{(1-e^{-\Theta_{vj}/T})}\right\}\omega_{e1}e^{D_e/kT} \tag{8-29}$$

$$-\frac{A}{NkT} = \ln\left[\frac{2\pi MkT}{h^2}\right]^{3/2}\frac{Ve}{N} + \ln\frac{\pi^{1/2}}{\sigma}\left(\frac{T^3}{\Theta_A\Theta_B\Theta_C}\right)^{1/2} - \sum_{j=1}^{3n-6}\left[\frac{\Theta_{vj}}{2T} + \ln(1-e^{-\Theta_{vj}/T})\right] + \frac{D_e}{kT} + \ln \omega_{e1} \tag{8-30}$$

$$\frac{E}{NkT} = \tfrac{3}{2} + \tfrac{3}{2} + \sum_{j=1}^{3n-6}\left(\frac{\Theta_{vj}}{2T} + \frac{\Theta_{vj}/T}{e^{\Theta_{vj}/T}-1}\right) - \frac{D_e}{kT} \tag{8-31}$$

$$\frac{C_V}{Nk} = \tfrac{3}{2} + \tfrac{3}{2} + \sum_{j=1}^{3n-6}\left(\frac{\Theta_{vj}}{T}\right)^2 \frac{e^{\Theta_{vj}/T}}{(e^{\Theta_{vj}/T}-1)^2} \tag{8-32}$$

$$\frac{S}{Nk} = \ln\left[\frac{2\pi MkT}{h^2}\right]^{3/2}\frac{Ve^{5/2}}{N} + \ln\frac{\pi^{1/2}e^{3/2}}{\sigma}\left(\frac{T^3}{\Theta_A\Theta_B\Theta_C}\right)^{1/2} + \sum_{j=1}^{3n-6}\left[\frac{\Theta_{vj}/T}{e^{\Theta_{vj}/T}-1} - \ln(1-e^{-\Theta_{vj}/T})\right] + \ln\omega_{e1} \tag{8-33}$$

$$pV = NkT \tag{8-34}$$

Table 8–1 contains the characteristic rotational temperatures, the characteristic vibrational temperatures, and

$$D_0 = D_e - \sum_j \tfrac{1}{2}h\nu_j$$

for a number of polyatomic molecules. See Herzberg under "Additional Reading" for an excellent chapter dealing with practical statistical thermodynamical calculations for polyatomic molecules. He includes a number of corrections to the above formulas and discusses them numerically.

Table 8–2 gives the vibrational contribution to the heat capacity for a variety of molecules of different shapes. It can be seen that the vibrational contributions are far

Table 8–2. **Vibrational contribution to heat capacities of some polyatomic molecules at 300°K**

molecule						C^{vib}/Nk	total C_V/Nk (calc.)
CO_2, linear	Θ_v(°K)	1890	3360	954			
	degeneracy	1	1	2			
	contribution to C_V/Nk	0.073	0.000	0.458		0.99	3.49
N_2O, linear	Θ_v(°K)	1840	3200	850			
	degeneracy	1	1	2			
	contribution to C_V/Nk	0.082	0.003	0.533		1.15	2.65
NH_3, pyramidal	Θ_v(°K)	4800	1360	4880	2330		
	degeneracy	1	1	2	2		
	contribution to C_V/Nk	0.000	0.226	0.000	0.026	0.28	3.28
CH_4, tetrahedron	Θ_v(°K)	4170	2180	4320	1870		
	degeneracy	1	2	3	3		
	contribution to C_V/Nk	0.000	0.037	0.000	0.077	0.30	3.30
H_2O, isosceles triangle	Θ_v(°K)	2290	5160	5360			
	degeneracy	1	1	1			
	contribution to C_V/Nk	0.028	0.000	0.000		0.03	3.03

from their equipartition values at the temperatures listed and that the agreement between the calculated and experimental values of C_V/Nk is excellent. A calculation for more complicated molecules would show similar agreement between the calculated values and the experimental data.

Table 8–3 compares calculated values of the entropy to those measured calorimetrically. It can be seen again that the agreement with experiment is quite good. In fact, calculated values of the entropy are often more accurate than measured values, provided sophisticated enough spectroscopic models are used.

Table 8–3. **The entropy of several polyatomic gases at 25°C and 1-atm pressure***

	S(calc.) (e.u.)	*S*(exp.) (e.u.)
CO_2	51.1	51.0
NH_3	46.1	46.0
NO_2	57.5	57.5
ClO_2	59.4	59.6
CH_4	44.5	44.5
CH_3Cl	55.8	56.0
CCl_4	74.0	73.9
C_6H_6	64.5	64.4

* The experimental values have been corrected for nonideal gas behavior.

There is, however, a class of molecules for which the type of agreement in Table 8–3 is not found. For example, it is found that for carbon monoxide, $S_{\text{calc}} = 47.3$ e.u. and $S_{\text{exp}} = 46.2$, for a discrepancy of 1.1 e.u. Other such discrepancies are found, and in all cases $S_{\text{calc}} > S_{\text{exp}}$. This difference is often referred to as *residual* entropy. The explanation of this is the following. Carbon monoxide has a very small dipole moment, and so when carbon monoxide is crystallized, the molecules do not have a strong tendency to line up in an energetically favorable way. The resultant crystal, then, is a random mixture of the two possible orientations CO and OC. As the crystal is cooled down toward 0°K, each molecule gets locked into its orientation and cannot realize the state of lowest energy with $\Omega = 1$, that is, all the molecules oriented in the same direction. Instead, the number of configurations Ω of the crystal is 2^N, since each of the N molecules exists equally likely (almost equally likely since the dipole moment is so small) in two states. Thus the entropy of the crystal at 0°K is $S = k \ln \Omega = Nk \ln 2$ instead of zero. If $Nk \ln 2 = 1.4$, entropy units are added to the experimental entropy, the agreement in the case of carbon monoxide becomes satisfactory. If it were possible to obtain carbon monoxide in its true equilibrium state at $T = 0$, this discrepancy would not occur. A similar situation occurs with nitrous oxide. For H_3CD, the residual entropy is 2.8 esu, and this is explained by realizing that each molecule of monodeuterated methane can assume four different orientations in the low-temperature crystal, and so $S_{\text{residual}} = Nk \ln 4 = 2.7$ esu, in very close agreement with the experimental value.

8–4 HINDERED ROTATION

There is one extension or modification of the partition function of polyatomic molecules that we shall discuss in this section. In molecules such as ethane, one of the most important internal degrees of freedom is a rotation about the single carbon–carbon bond. Because of the interactions between the hydrogen atoms on each carbon, this rotation is not free, but is said to be restricted or hindered. As the two methyl

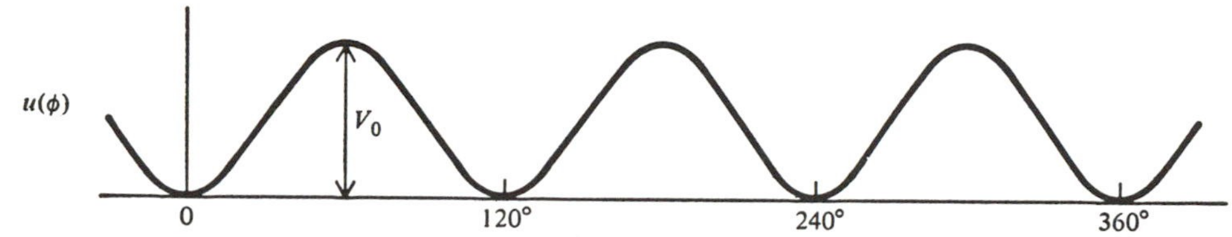

Figure 8–1. **Potential energy of internal rotation in ethane. This curve can be represented approximately by $\frac{1}{2}V_0(1 - \cos 3\phi)$.**

groups rotate about the carbon–carbon bond, the hydrogen atoms become alternately eclipsed (directly opposite each other) and staggered. The potential energy associated with this rotation is shown in Fig. 8–1. The maxima correspond to the configuration where the hydrogen atoms are eclipsed, and the minima correspond to the configuration in which they are staggered. At temperatures such that $kT \gg V_0$, the internal rotation is essentially free and can be treated by methods similar to the rigid rotor. At temperatures such that $kT \ll V_0$, the molecule is trapped at the bottom of the wells in Fig. 8–1, and the motion is that of a simple torsional vibration, which can be treated by a method similar to that used for the simple harmonic oscillator. Typical values of V_0 are such, however, that at ordinary temperatures, the motion is intermediate between that of free rotation and torsional vibration.

It is necessary to solve the Schrödinger equation for the potential shown in Fig. 8–1. This potential can be approximately represented by $\frac{1}{2}V_0(1 - \cos 3\phi)$, for which the Schrödinger equation is

$$-\frac{h^2}{8\pi^2 I_r}\frac{\partial^2\psi}{\partial\phi^2} + \tfrac{1}{2}V_0(1 - \cos 3\phi)\psi = \varepsilon\psi$$

where I_r is an effective moment of inertia whose precise form we shall not need. This differential equation is difficult to solve analytically, but the eigenvalues have been tabulated numerically as a function of V_0. These can be used to compute a partition function for the restricted rotation, which can then be used to compute thermodynamic properties. There are extensive tables of the various thermodynamic functions as a function of V_0/kT.

Figure 8–2 shows a sketch of the contribution of internal rotation to the heat capacity in an ethanelike molecule. One can use curves such as these to fit heat capacity

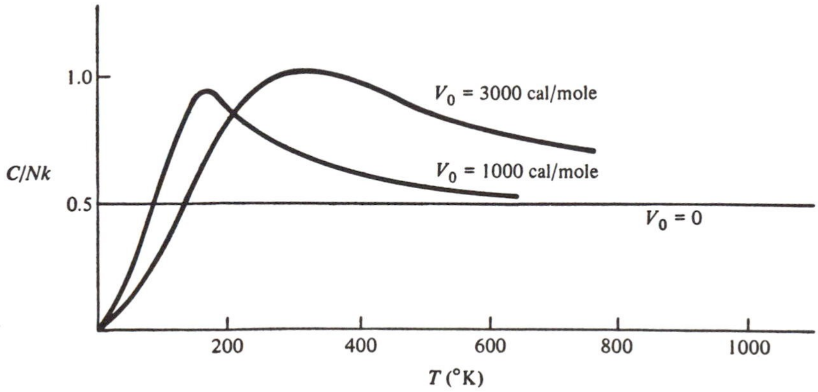

Figure 8–2. **The contribution of internal rotation to the heat capacity as a function of barrier height V_0.**

Table 8–4. **Potential barriers of some molecules**

molecule	V_0(kcal/mole)
CH_3CH_3	2.7–3.0
CH_3CCl_3	2.7
CH_3CH_2Cl	2.7–4.7
CH_3OH	1.1–1.6
CH_3SH	1.3–1.5
CH_3NH_2	1.9
$CH_3CH_2CH_3$	3.3
CH_3CHCCH_2	1.59–1.65

data and thus obtain a thermodynamic estimate of V_0. One can also generate curves for the entropy as a function of V_0 and hence determine V_0 by comparing the results to experimental values of the entropy. The values of V_0 determined from heat capacity data and entropy data are in fair agreement. There are a number of other ways of determining V_0 (see e.g. Section 9–2(F)) and the reader is referred to a review article by Wilson* for a general discussion of the problem of barriers to internal rotation in molecules and a comparison of the various methods for determining V_0 experimentally. Table 8–4 lists potential barriers for several molecules. The range of values given for each molecule is an indication of the agreement between different experimental methods of determining V_0.

ADDITIONAL READING

General

DAVIDSON, N. 1962. *Statistical mechanics.* New York: McGraw-Hill. Chapter 11.

FOWLER, R. H., and GUGGENHEIM, E. A. 1956. *Statistical thermodynamics.* Cambridge: Cambridge University Press. Sections 326 through 331.

HILL, T. L. 1952. *Statistical thermodynamics.* Cambridge: Cambridge University Press. Chapter 9.

KUBO, R. 1965. *Statistical mechanics.* Amsterdam: North-Holland Publishing Co. Sections 3–1 to 3–3. Example 3–2.

MAYER, J. E., and MAYER, M. G. 1940. *Statistical mechanics.* New York: Wiley. Chapter 8.

MÜNSTER, A. 1969. *Statistical thermodynamics,* Vol. I. Berlin: Springer-Verlag. Sections 6–8 through 6–11.

RUSHBROOKE, G. S. 1949. *Statistical mechanics.* London: Oxford University Press. Chapter 9.

Spectroscopy

BARROW, G. M. 1962. *Introduction to molecular spectroscopy.* New York: McGraw-Hill.

COSTAIN, C. C. 1970. In *Physical chemistry: an advanced treatise,* Vol. IV, ed. by H. Eyring, D. Henderson, and W. Jost. New York: Academic.

HALL, J. R. 1970. In *Physical chemistry, an advanced treatise,* Vol. IV, ed. by H. Eyring, D. Henderson, and W. Jost. New York: Academic.

HERZBERG, G. 1945. *Infrared and Raman spectra of polyatomic molecules.* New York: Van Nostrand.

KING, G. W. 1970. In *Physical chemistry, an advanced treatise,* Vol. IV, ed. by H. Eyring, D. Henderson, and W. Jost. New York: Academic.

———. 1964. *Spectroscopy and molecular structure.* New York: Holt, Rinehart & Winston.

PROBLEMS

8–1. The HOH bond angle in water is 104°, and the OH bond length is 0.96Å. Calculate the center of mass and the three moments of inertia of water. From this verify the results for Θ_{rot} for H_2O given in Table 8–1.

8–2. Evaluate Eq. (8–16) without neglecting 1 compared to J in Eq. (8–15).

8–3. Derive Eq. (8–17) from its corresponding summation by converting the sum to an integral.

* E. B. Wilson, Jr., *Adv. Chem. Phys.*, **2**, p. 367, 1959.

8–4. Verify the calculated entries in Table 8–3 for CH_4, using the data in Table 8–1.

8–5. Use the data in Table 8–1 to calculate the entropy of CO_2 at 25°C and 1 atm. Compare your result to that in Table 8–3.

8–6. The same as Problem 8–5, but for H_2O.

8–7. The same as Problem 8–5, but for CH_4.

8–8. What molar heat capacities would you expect under classical conditions for the following gases: (a) Ne, (b) O_2, (c) H_2O, (d) CO_2, and (e) $CHCl_3$?

8–9. Verify the results for methane in Table 8–3.

8–10. Verify the results for ammonia and water in Table 8–2.

8–11. Calculate the entropy of ClO_2 at 298°K and compare this to the experimental value of 61 e.u.

8–12. Verify from group theoretic character tables that the symmetry number is equal to the number of pure rotational elements (including the identity) in the point group of the molecule.

8–13. Show that C_V for NH_3 at 300°K is $3.3Nk$.

8–14. In Problem 1–36, heat-capacity data were listed for a calculation of the third-law entropy of nitromethane. From the following molecular data, calculate the statistical entropy S^0_{289}. Bond distances (Å): N–O 1.21; C–N 1.46; C–H 1.09. Bond angles: O–N–O 127°; H–C–N $109\frac{1}{2}$°. From these distances, calculate the principal moments of inertia, $I = 67.2$, 76.0, 137.9×10^{-40} g-cm^2. The fundamental vibration frequencies in cm^{-1} are 476, 599, 647, 921, 1097, 1153, 1384, 1413, 1449, 1488, 1582, 2905, 3048(2). One of the torsional vibrations has become a free rotation around the C–N bond with $I = 4.86 \times 10^{-40}$.

8–15. The classical rotational kinetic energy of a symmetric top molecule is

$$K = \frac{p_\theta^2}{2I_A} + \frac{(p_\phi - p_\psi \cos\theta)^2}{2I_A \sin^2\theta} + \frac{p_\psi^2}{2I_C}$$

where I_A, I_A, and I_C are the principal moments of inertia, and θ, ϕ, and ψ are the three Euler angles. Derive the classical limit of the rotational partition function for a symmetric top molecule. Hint: Recall that the Euler angles have the ranges:

$$0 \le \theta \le \pi$$
$$0 \le \phi \le 2\pi$$
$$0 \le \psi \le 2\pi$$

8–16. The classical Hamiltonian for an asymmetric top molecule with principal moments of inertia I_A, I_B, and I_C is given by

$$H = \frac{1}{2I_A \sin^2\theta}\{(p_\phi - p_\psi \cos\theta)\cos\psi - p_\theta \sin\theta \sin\psi\}^2$$

$$+ \frac{1}{2I_B \sin^2\theta}\{(p_\phi - p_\psi \cos\theta)\sin\psi + p_\theta \sin\theta \cos\psi\}^2 + \frac{1}{2I_C} p_\psi^2$$

Derive the classical limit of the rotational partition function for an asymmetric top molecule. Hint: It may help to rearrange the Hamiltonian and integrate over p_θ, p_ϕ, p_ψ in that order.

CHAPTER 9

CHEMICAL EQUILIBRIUM

One of the most important chemical applications of statistical thermodynamics is the calculation of equilibrium constants in terms of molecular parameters. Often the results of such calculations are more accurate than the experimental values. In Section 9–1 we derive the basic equations, giving the equilibrium constant in terms of partition functions. This is an easy derivation, since Eq. (1–68) gives the thermodynamic equilibrium condition in terms of chemical potentials, and in Chapters 5, 6, and 8 we derived thermodynamic functions such as the chemical potential in terms of partition functions. Thus there are no new principles introduced in this chapter. In a sense, it is simply a numerical application of the results of the previous chapters. The bulk of this chapter is a numerical discussion of the few equations of Section 9–1. In Section 9–2 we discuss six types of reactions: (1) the association of atoms or molecules in a vapor, (2) a simple isotopic exchange reaction, (3) a more complicated isotopic exchange reaction, (4) a chemical reaction involving only diatomic molecules, (5) a chemical reaction involving polyatomic molecules, and (6) a chemical reaction involving a molecule with restricted internal rotation. Then in Section 9–3 we discuss the use of thermodynamic tables to calculate equilibrium constants. We shall see that thermodynamic tables are equivalent to a tabulation of the partition function at various temperatures.

9–1 THE EQUILIBRIUM CONSTANT IN TERMS OF PARTITION FUNCTIONS

We shall consider the general homogeneous gas phase chemical reaction

$$\nu_A A + \nu_B B \rightleftharpoons \nu_C C + \nu_D D \qquad (9\text{–}1)$$

at equilibrium in a closed thermostated vessel. The ν's are stoichiometric coefficients and A, B, and so on, represent the reactants and products. The thermodynamic condition for chemical equilibrium is derived as follows.

We first write Eq. (9–1) algebraically as

$$\nu_C C + \nu_D D - \nu_A A - \nu_B B = 0 \tag{9–2}$$

We then define a variable λ such that $dN_j = \nu_j \, d\lambda$, where $j = A, B, C,$ or D and where ν_j is taken to be positive for products and negative for reactants. The Helmholtz free energy of the system is

$$dA = -S\,dT - p\,dV + \sum_j \mu_j \, dN_j$$

For a reaction vessel at fixed volume and temperature,

$$dA = \sum_j \mu_j \, dN_j = \left(\sum_j \nu_j \mu_j\right) d\lambda \qquad \text{(constant } T \text{ and } V\text{)} \tag{9–3}$$

For a system at equilibrium, the free energy must be a minimum with respect to all possible changes $d\lambda$, and so $(\partial A/\partial \lambda)_{T,V} = 0$. From Eq. (9–3) then, we have the condition for chemical equilibrium:

$$\sum_j \nu_j \mu_j = \nu_C \mu_C + \nu_D \mu_D - \nu_A \mu_A - \nu_B \mu_B = 0 \tag{9–4}$$

This is the general thermodynamic equation of chemical equilibrium. We shall now introduce statistical thermodynamics through the relation between chemical potential and partition functions. In a mixture of ideal gases, the species are independent and distinguishable, and so the partition function of the mixture is a product of the partition functions of the individual components. Thus

$$\begin{aligned} Q(N_A, N_B, N_C, N_D, V, T) &= Q(N_A, V, T)Q(N_B, V, T)Q(N_C, V, T)Q(N_D, V, T) \\ &= \frac{q_A(V, T)^{N_A}}{N_A!} \frac{q_B(V, T)^{N_B}}{N_B!} \frac{q_C(V, T)^{N_C}}{N_C!} \frac{q_D(V, T)^{N_D}}{N_D!} \end{aligned} \tag{9–5}$$

The chemical potential of each species is given by an equation such as

$$\mu_A = -kT \left(\frac{\partial \ln Q}{\partial N_A}\right)_{N_j, V, T} = -kT \ln \frac{q_A(V, T)}{N_A} \tag{9–6}$$

where Stirling's approximation has been used for $N_A!$. The N_j subscript on the partial derivative indicates that the numbers of particles of the other species are held fixed. Equation (9–6) simply says that the chemical potential of one species of an ideal gas mixture is calculated as if the other species were not present. This, of course, is obvious for an ideal gas mixture.

If we substitute Eq. (9–6) into Eq. (9–4), we get

$$\frac{N_C^{\nu_C} N_D^{\nu_D}}{N_A^{\nu_A} N_B^{\nu_B}} = \frac{q_C^{\nu_C} q_D^{\nu_D}}{q_A^{\nu_A} q_B^{\nu_B}} \tag{9–7}$$

For an ideal gas, the molecular partition function is of the form $f(T)V$ (see Problem 4–15), so that q/V is a function of temperature only. This allows us to write

$$K_c(T) = \frac{\rho_C^{\nu_C} \rho_D^{\nu_D}}{\rho_A^{\nu_A} \rho_B^{\nu_B}} = \frac{(q_C/V)^{\nu_C}(q_D/V)^{\nu_D}}{(q_A/V)^{\nu_A}(q_B/V)^{\nu_B}} \tag{9–8}$$

where $K_c(T)$ is the equilibrium constant of the reaction. For an ideal system, K_c is a function of temperature only.

Another commonly used equilibrium constant is $K_p(T)$, which is expressed in terms of partial pressures rather than concentrations. We can derive the equation for $K_p(T)$ by substituting $p_j = \rho_j kT$ into Eq. (9–8):

$$K_p(T) = \frac{p_C^{\nu_C} p_D^{\nu_D}}{p_A^{\nu_A} p_B^{\nu_B}} = (kT)^{\nu_C+\nu_D-\nu_A-\nu_B} K_c(T) \tag{9–9}$$

By means of Eq. (9–8) or (9–9), along with the results of Chapters 5 through 8, it is a simple matter to calculate equilibrium constants in terms of molecular parameters. This is best illustrated by means of examples.

9–2 EXAMPLES OF THE CALCULATION OF EQUILIBRIUM CONSTANTS

A. THE ASSOCIATION OF ALKALI METAL VAPORS

We use, as an example, the reaction

$$2Na \rightleftharpoons Na_2$$

The equilibrium constant for this reaction can be written as

$$K_p(T) = \frac{p_{\text{dimer}}}{p^2_{\text{monomer}}} \tag{9–10}$$

$$= (kT)^{-1} \frac{(q_{Na_2}/V)}{(q_{Na}/V)^2} \tag{9–11}$$

The equation for the partition function of a monatomic ideal gas [Eq. (5–15)] and a diatomic molecule [Eq. (6–51)] are

$$q_{Na}(T, V) = \left(\frac{2\pi m_{Na} kT}{h^2}\right)^{3/2} V q_{elec}(T) \tag{9–12}$$

$$q_{Na_2}(T, V) = \left(\frac{2\pi m_{Na_2} kT}{h^2}\right)^{3/2} V \frac{8\pi^2 IkT}{2h^2} \frac{e^{-\beta h\nu/2}}{(1 - e^{-\beta h\nu})} \omega_{1e} e^{D_e/kT}$$

$$= \left(\frac{2\pi m_{Na_2} kT}{h^2}\right)^{3/2} V \left(\frac{T}{2\Theta_r}\right)(1 - e^{-\Theta_v/T})^{-1} e^{D_0/kT} \tag{9–13}$$

Note that we have introduced $D_0 = D_e - \frac{1}{2}h\nu$ into Eq. (9–13). We shall usually do this since there are extensive tables of D_0. Tables 6–1 and 8–1 contain values of D_0 for diatomic and polyatomic molecules. In addition to values of D_0, Table 6–1 contains the other parameters needed in Eq. (9–13). From Table 6–1, $\Theta_v = 229°K$, $\Theta_r = 0.221°K$, and $D_0 = 17.3$ Kcal/mole. In addition, we need to know that the ground electronic state of a sodium atom is $^2S_{1/2}$, and that the next electronic state lies approximately 16,000 cm^{-1} above the ground state. At 1000°K, then

$$\frac{q_{Na}}{V} = \left(\frac{2\pi \times 23 \times 1.66 \times 10^{-24} \times 1.38 \times 10^{-16} \times 10^3}{6.626 \times 6.626 \times 10^{-54}}\right)^{3/2} q_{elec}$$

$$= (6.54 \times 10^{26}) \times 2 = 1.31 \times 10^{27} \tag{9–14}$$

$$\frac{q_{Na_2}}{V} = (1.85 \times 10^{27}) \times (2.26 \times 10^3)(4.88)(5.96 \times 10^3)$$

$$= (1.22 \times 10^{35}) \tag{9–15}$$

and the equilibrium constant is

$$K_p(T) = \frac{1.22 \times 10^{35}}{(1.38 \times 10^{-16})(1000)(1.72 \times 10^{54})}$$
$$= 0.50 \times 10^{-6}(\text{dyne/cm}^2)^{-1} = 0.50 \text{ atm}^{-1} \qquad (9\text{–}16)$$

The experimental value is 0.475 atm^{-1}. We have used the fact that 1 atm = 1.01×10^6 dynes/cm^2. This can easily be calculated from the gas constant or the Boltzmann constant in ergs/deg-mole and liter-atm/deg-mole. Note that K_p has units of 1/(pressure), even though it is a function only of the temperature for an ideal system. Table 9–1 gives $K_p(T)$ at a number of other temperatures. The agreement is seen to be good.

Table 9–1. **A comparison of the experimental values of the equilibrium constant for the reaction $2Na \rightleftharpoons Na_2$ with the values calculated from Eq. (9–11)**

T(°K)	K_p(calc.)$(\text{atm})^{-1}$	K_p(exp.)*$(\text{atm})^{-1}$
900	1.44	1.32
1000	0.50	0.47
1100	0.22	0.21
1200	0.11	0.10

* See C. T. Ewing, *et al.*, *J. Chem. Phys.*, **71**, 473, 1967.

A general principle of dissociation or association reactions of this type is that the dimer is favored by the energetics of the reaction but the monomers are favored by the entropy. Problems 9–2 and 9–3 involve similar calculations for the association of potassium vapors and the dissociation of I_2.

B. AN ISOTOPIC EXCHANGE REACTION

We consider the isotopic exchange reaction

$$H_2 + D_2 \rightleftharpoons 2HD$$

We can use this simple reaction to illustrate the consequences of the Born-Oppenheimer approximation in isotopic exchange reactions. The Born-Oppenheimer approximation is based upon the approximation that the nuclei are so much more massive than the electrons that it is legitimate to calculate the electronic state of a molecule in a field of fixed nuclei. Thus H_2, D_2, and HD all have the same internuclear potential function, and therefore have the same force constant k, the same depth of the potential D_e, and the same internuclear separation. This leads to a great deal of canceling between the numerator and denominator in the ratio of partition functions in the equilibrium constant.

The equilibrium constant is

$$K(T) = K_p(T) = K_c(T) = \frac{\rho_{HD}^{\,2}}{\rho_{H_2}\rho_{D_2}} = \frac{p_{HD}^{\,2}}{p_{H_2}p_{D_2}} = \frac{q_{HD}^{\,2}}{q_{H_2}q_{D_2}}$$

$$= \frac{\left(\dfrac{2\pi m_{HD}kT}{h^2}\right)^3\left(\dfrac{T}{\Theta_{r,HD}}\right)^2\left(\dfrac{e^{-\Theta_{v,HD}/2T}}{1-e^{-\Theta_{v,HD}/T}}\right)^2 e^{2D_e/kT}}{\left(\dfrac{2\pi m_{H_2}kT}{h^2}\right)^{3/2}\left(\dfrac{2\pi m_{D_2}kT}{h^2}\right)^{3/2}\left(\dfrac{T^2}{4\Theta_{r,H_2}\Theta_{r,D_2}}\right)\left(\dfrac{e^{-\Theta_{v,H_2}/2T}}{1-e^{-\Theta_{v,H_2}/T}}\right)\left(\dfrac{e^{-\Theta_{v,D_2}/2T}}{1-e^{-\Theta_{v,D_2}/T}}\right)e^{2D_e/kT}}$$

$$= \frac{m_{HD}^{\,3}}{(m_{H_2}m_{D_2})^{3/2}}\,\frac{4\Theta_{r,H_2}\Theta_{r,D_2}}{\Theta_{r,HD}^2}\,\frac{(1-e^{-\Theta_{v,H_2}/T})(1-e^{-\Theta_{v,D_2}/T})}{(1-e^{-\Theta_{v,HD}/T})^2}\,e^{-(2\Theta_{v,HD}-\Theta_{v,H_2}-\Theta_{v,D_2})/2T} \qquad (9\text{–}17)$$

From Table 6–1, $\Theta_{r,\,H_2} = 85.3°K$, $\Theta_{r,\,D_2} = 42.7°K$, $\Theta_{v,\,H_2} = 6215°K$, and $\Theta_{v,\,D_2} = 4394°K$. The values for HD are easy to calculate from these, since $\Theta_v = h\nu/k$, and $\nu = (k/\mu)^{1/2}/2\pi$, where μ is the reduced mass of the molecule. Since the force constants of all three molecules are equal under the Born-Oppenheimer approximation, we have

$$\frac{\nu_{HD}}{\nu_{H_2}} = \left(\frac{\mu_{H_2}}{\mu_{HD}}\right)^{1/2} \quad \text{or} \quad \frac{\Theta_{v,\,HD}}{\Theta_{v,\,H_2}} = \left(\frac{\mu_{H_2}}{\mu_{HD}}\right)^{1/2} \tag{9–18}$$

which gives

$$\Theta_{v,\,HD} = (\tfrac{3}{4})^{1/2}\Theta_{v,\,H_2} \tag{9–19}$$

Of course, we also have the relations

$$\Theta_{v,\,HD} = (\tfrac{3}{2})^{1/2}\Theta_{v,\,D_2} \quad \text{and} \quad \Theta_{v,\,D_2} = (\tfrac{1}{2})^{1/2}\Theta_{v,\,H_2} \tag{9–20}$$

Similarly, since $\Theta_r = h^2/8\pi^2 Ik$ and $I = \mu r_e^{\,2}$, where r_e is the internuclear separation, we can write

$$\frac{\Theta_{r,\,HD}}{\Theta_{r,\,H_2}} = \frac{\mu_{H_2}}{\mu_{HD}} = \frac{3}{4} \tag{9–21}$$

or

$$\frac{\Theta_{r,\,HD}}{\Theta_{r,\,D_2}} = \frac{3}{2} \tag{9–22}$$

Note that $\Theta_{r,\,H_2} = 2\Theta_{r,\,D_2}$. We shall calculate $K(T)$ between 195°K and 741°K. At these temperatures, the $(1 - e^{-\Theta_v/T})$ factors in Eq. (9–17) are approximately unity.

Substituting all of these reduction formulas into Eq. (9–17), we get

$$\begin{aligned} K(T) &= 4(\tfrac{9}{8})^{3/2}(\tfrac{8}{9})e^{-[3^{1/2}-1-(1/2)^{1/2}]\Theta_{v,H_2}/2T} \\ &= 4(1.06)\exp\left\{\frac{-77.7}{T}\right\} \end{aligned} \tag{9–23}$$

Table 9–2 compares Eq. (9–23) with the experimental data of Urey and Rittenberg, obtained during Urey's early investigations on heavy hydrogen. Equation (9–23) shows that the product or the symmetry numbers of the H_2 and D_2 predominates in the calculation of the equilibrium constant. All of the other factors are close to unity.

The hydrogen–deuterium exchange reaction, in fact, shows an unusually large deviation from $K = 4$, since the percentage mass difference between hydrogen and

Table 9–2. **A comparison of the experimental values of the equilibrium constant of the reaction $H_2 + D_2 \rightleftharpoons 2HD$ with the values calculated from Eq. (9–23)**

T(°K)	K(calc.)	K(exp.)
195	2.84*	2.92
273	3.18	3.24
298	3.26	3.28
383	3.46	3.50
543	3.67	3.85
670	3.77	3.8
741	3.81	3.75

* Quantum effects begin to be important at these low temperatures. If these effects are included, a value of 2.87 is obtained.

Source: D. Rittenberg, W. Bleakney, and H. C. Urey, *J. Chem. Phys.*, **2**, p. 362, 1934, and A. J. Gould, W Bleakney, and H. S. Taylor, *J. Chem. Phys.*, **2**, p. 362, 1934.

deuterium is large. For other diatomic isotopic exchange reactions such as $N_2^{14} + N_2^{15} \rightleftharpoons 2N^{14}N^{15}$, the equilibrium constant differs very little from the value 4. It is easy to derive a general expression for the equilibrium constant of such reactions by expanding the complete expression for $K(T)$ in a power series in Δ/M, where Δ is the difference in atomic mass between the two isotopes, and M is the atomic mass of the heavier isotope. To order $(\Delta/M)^2$, the result is (see Problem 9–6)

$$K(T) = 4\left(1 + \frac{\Delta^2}{8M^2}\right)e^{-\Delta^2\Theta_{M,\text{vib}}/32M^2T} \tag{9–24}$$

where $\Theta_{M,\text{vib}}$ is the characteristic vibrational temperature of the diatomic molecule containing two heavier nuclei. For the $N_2^{14} + N_2^{15} \rightleftharpoons 2N^{14}N^{15}$ reaction, $\Delta/M = 1/15$ and $\Theta_{M,\text{vib}} = 0.97$ times the value in Table 6–1, or $\Theta_{M,\text{vib}} = 3260°K$. Equation (9–24) becomes

$$K(T) = 4(1.0005)e^{-0.44/T} \tag{9–25}$$

which is essentially equal to 4 for all temperatures greater than 100°K, say. Thus the equilibrium constant is completely determined by the symmetry of the molecules involved in the reaction. This is not necessarily so for more complicated molecules, however, as the next section shows.

C. A MORE COMPLICATED ISOTOPIC EXCHANGE REACTION: AN ILLUSTRATION OF THE TELLER-REDLICH PRODUCT RULE

Consider the exchange reaction

$$CH_4 + DBr \rightleftharpoons CH_3D + HBr$$

Because of the Born-Oppenheimer approximation, the internuclear potential surfaces of CH_4 and CH_3D are the same, and the internuclear potential energy curves of HBr and DBr are the same. As before, this leads to a great deal of simplification in the ratio of partition functions. The equilibrium constant for this reaction is

$$K(T) = \frac{\rho(CH_3D)\rho(HBr)}{\rho(CH_4)\rho(Br)} = \frac{q(CH_3D)q(HBr)}{q(CH_4)q(DBr)}$$

$$= \frac{\sigma_{CH_4}\sigma_{DBr}}{\sigma_{CH_3D}\sigma_{HBr}}\left(\frac{M_{CH_3D}M_{HBr}}{M_{CH_4}M_{DBr}}\right)^{3/2}\frac{I_{HBr}}{I_{DBr}}\frac{(I_AI_BI_C)^{1/2}_{CH_3D}}{(I_AI_BI_C)^{1/2}_{CH_4}}\frac{q_{\text{vib},CH_3D}\,q_{\text{vib},HBr}}{q_{\text{vib},CH_4}\,q_{\text{vib},DBr}} \tag{9–26}$$

The fundamental vibrational frequencies of CH_4, CH_3D, HBr, and DBr are given in Table 9–3. All of these frequencies are high enough that the factors $(1 - e^{-\Theta_v/T})$ in

Table 9–3. **Fundamental vibrational frequencies of CH_4, CH_3D, HBr, and DBr**

CH_4		CH_3D	
frequency	degeneracy	frequency	degeneracy
2917 cm^{-1}	1	2200 cm^{-1}	1
1534	2	2945	1
3019	3	1310	1
1306	3	1471	2
		3021	2
		1155	2

HBr		DBr	
frequency	degeneracy	frequency	degeneracy
2650 cm^{-1}	1	1880 cm^{-1}	1

the vibrational partition functions are essentially unity at room temperature. The ratio of the vibrational partition functions in Eq. (9–26) includes then only the zero-point vibrational energy terms and is

$$\frac{q_{\text{vib},CH_3D}\,q_{\text{vib},HBr}}{q_{\text{vib},CH_4}\,q_{\text{vib},DBr}} \approx \exp\left\{-\sum_j \frac{(\Theta_{vj}^{CH_3D} + \Theta_{vj}^{HBr} - \Theta_{vj}^{CH_4} - \Theta_{vj}^{DBr})}{2T}\right\}$$

$$= \exp\left\{\frac{220\text{ cm}^{-1}}{kT}\right\} \tag{9–27}$$

using the frequencies given in Table 9–3.

The ratio of the symmetry numbers in Eq. (9–26) is $12 \times 1/(3 \times 1) = 4$. To calculate the remainder of Eq. (9–26), we could look up all the moments of inertia we need in the literature, but this is not necessary. We can avoid it by using the *Teller-Redlich product rule* for isotopically substituted compounds, which says that if molecules A and A' differ only by isotopic substitution, then

$$\left(\frac{M'}{M}\right)^{3/2} \frac{I'}{I} = \prod_{i=1}^{n} \left(\frac{m_i'}{m_i}\right)^{3/2} \prod_{j=1}^{3n-5} \frac{\nu_j'}{\nu_j} \quad \text{(linear)} \tag{9–28}$$

for linear molecules and

$$\left(\frac{M'}{M}\right)^{3/2} \frac{(I_A'I_B'I_C')^{1/2}}{(I_A I_B I_C)^{1/2}} = \prod_{i=1}^{n} \left(\frac{m'_i}{m_i}\right)^{3/2} \prod_{j=1}^{3n-6} \frac{\nu_j'}{\nu_j} \quad \text{(nonlinear)} \tag{9–29}$$

for nonlinear molecules. In these two equations, M denotes the total mass of molecule A; m_j is the mass of the jth atom; and ν_j is the j fundamental vibrational frequency. The proof of the Teller-Redlich product involves a detailed consideration of the normal coordinates of molecules, and so we shall not prove it here. It is easy to show, however, that it is valid for a diatomic molecule under the Born-Oppenheimer approximation (see Problem 9–7).

The Teller-Redlich product rule is useful to us, since we already have the ν_j in Table 9–3. Furthermore, when this is used in Eq. (9–26), the products of the atomic masses cancel, and we are left with the simple result:

$$K(T) = 4e^{317/T} \frac{\nu_{HBr}}{\nu_{DBr}} \prod_{j=1}^{9} \left(\frac{\nu_{j\,CH_3D}}{\nu_{j\,CH_4}}\right) \tag{9–30}$$

$$= 2.99e^{317/T}$$

$$= 8.65 \qquad \text{at } 298°\text{K} \tag{9–31}$$

D. A CHEMICAL REACTION INVOLVING DIATOMIC MOLECULES

We shall calculate the equilibrium constant for the reaction

$$H_2 + I_2 \rightleftharpoons 2HI$$

from 500°K to 1000°K. The equilibrium constant is given by

$$K(T) = \frac{(q_{HI}/V)^2}{(q_{H_2}/V)(q_{I_2}/V)} = \frac{q_{HI}^2}{q_{H_2}\,q_{I_2}}$$

$$= \left(\frac{m_{HI}^2}{m_{H_2}\,m_{I_2}}\right)^{3/2} \left(\frac{4\Theta_r^{H_2}\Theta_r^{I_2}}{(\Theta_r^{HI})^2}\right) \frac{(1 - e^{-\Theta_v^{H_2}/T})(1 - e^{-\Theta_v^{I_2}/T})}{(1 - e^{-\Theta_v^{HI}/T})^2}$$

$$\times \exp \frac{(2D_0^{HI} - D_0^{H_2} - D_0^{I_2})}{RT} \tag{9–32}$$

All of the necessary parameters are given in Table 6–1. Figure 9–1 shows ln K plotted

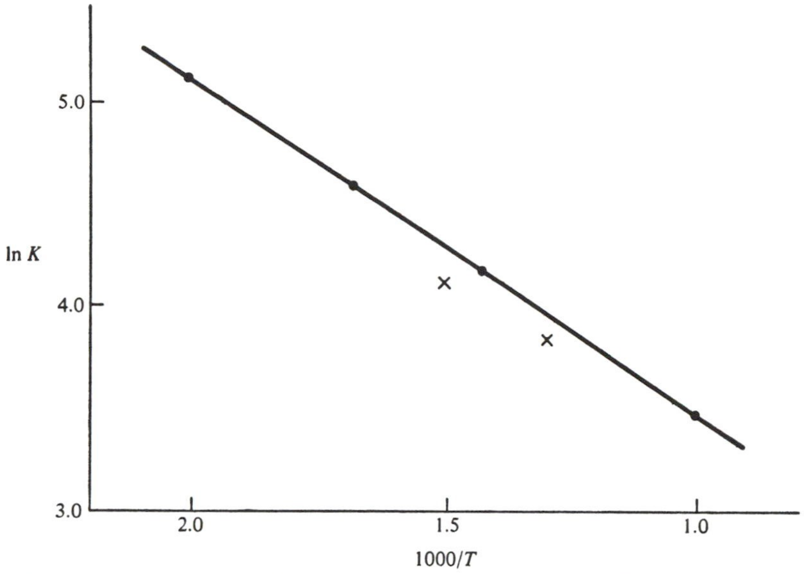

Figure 9–1. **The logarithm of the equilibrium constant versus $1/T$ for the reaction $H_2 + I_2 \rightleftharpoons 2HI$. The points are calculated from Eq. (9–32) and the crosses are the experimental values.** (From A. H. Taylor and R. H. Crist, *J. Am. Chem, Soc.*, **63**, 1377, 1941.)

versus $1/T$. The thermodynamic equation for the variation of the equilibrium constant with temperature is

$$d(\ln K) = -\frac{\Delta H}{R}\, d\left(\frac{1}{T}\right) \tag{9–33}$$

where ΔH is the heat of reaction. From Fig. 9–1 we get $\Delta H = -3100$ cal/mole, compared to the experimental value of -2950 cal/mole.

Problems 9–8 and 9–9 involve the calculation of the equilibrium constant for some other reactions involving only diatomic molecules.

E. A REACTION INVOLVING POLYATOMIC MOLECULES

As an example of a reaction involving a polyatomic molecule, consider the reaction

$$H_2 + \tfrac{1}{2}O_2 \rightleftharpoons H_2O$$

whose equilibrium constant is

$$K_p(T) = \frac{(q_{H_2O}/V)}{(kT)^{1/2}(q_{H_2}/V)(q_{O_2}/V)^{1/2}} \tag{9–34}$$

It is almost as convenient to calculate each partition function separately as to substitute them into K_p first. Furthermore, we shall use the values in the next section.

The necessary parameters are given in Tables 6–1 and 8–1. At 1500°K, the three partition functions are

$$\begin{aligned}\frac{q_{H_2}}{V} &= \left(\frac{2\pi m_{H_2} kT}{h^2}\right)^{3/2}\left(\frac{T}{2\Theta_r^{H_2}}\right)(1 - e^{-\Theta_v^{H_2}/T})^{-1} e^{D_0^{H_2}/RT} \\ &= 2.80 \times 10^{26} e^{D_0^{H_2}/RT}\end{aligned} \tag{9–35}$$

$$\begin{aligned}\frac{q_{O_2}}{V} &= \left(\frac{2\pi m_{O_2} kT}{h^2}\right)^{3/2}\left(\frac{T}{2\Theta_r^{O_2}}\right)(1 - e^{-\Theta_v^{O_2}/T})^{-1} 3 e^{D_0^{O_2}/RT} \\ &= 2.79 \times 10^{30} e^{D_0^{O_2}/RT}\end{aligned} \tag{9–36}$$

and

$$\frac{q_{H_2O}}{V} = \left(\frac{2\pi m_{H_2O} kT}{h^2}\right)^{3/2} \frac{\pi^{1/2}}{\sigma} \left(\frac{T^3}{\Theta_A{}^{H_2O}\Theta_B{}^{H_2O}\Theta_C{}^{H_2O}}\right)^{1/2} \prod_{j=1}^{3} (1 - e^{-\Theta_{vj}{}^{H_2O}/T})^{-1} e^{D_0{}^{H_2O}/RT}$$

$$= 5.33 \times 10^{29} e^{D_0{}^{H_2O}/RT} \tag{9–37}$$

The factor of 3 occurs in q_{O_2}/V because the ground state of O_2 is $^3\Sigma_g^-$. (See Table 6–1.) At 1500°K, the equilibrium constant is

$$K_p = 4.77 \times 10^5 (\text{atm})^{-1/2} \tag{9–38}$$

Table 9–4 compares the calculated values of log K_p with the experimental data of

Table 9–4. **The logarithm of the equilibrium constant for the reaction $H_2 + \frac{1}{2}O_2 \rightleftharpoons H_2O$**

T(°K)	log K(calc.)	log K(exp.)*
1000	10.2	10.1
1500	5.7	5.7
2000	3.7	3.5

* See H. Zeiss, *Z. Electrochem.*, **43**, p. 706, 1937.

Zeiss. The agreement can be considerably improved by using more sophisticated spectroscopic models. At high temperatures, the rotational energies of the molecules are high enough to warrant centrifugal distortion effects and other extensions of the simple rigid rotor–harmonic oscillator approximation. Problems 9–10 through 9–12 involve similar calculations.

F. A CHEMICAL REACTION INVOLVING A MOLECULE WITH RESTRICTED INTERNAL ROTATION, ETHYLENE–ETHANE EQUILIBRIUM

As a final example, we discuss the equilibrium constant for the reaction

$$C_2H_4 + H_2 \rightleftharpoons C_2H_6$$

This calculation is more complicated than the previous ones because of the hindered rotation in ethane. The equilibrium constant is calculated for various values of the barrier height V_0, and V_0 is chosen to give the best agreement with experiment. Table 9–5 gives some values for the equilibrium constant K_p for this reaction at various

Table 9–5. **The equilibrium constant for the reaction $C_2H_4 + H_2 \rightleftharpoons C_2H_6$ calculated with $V_0 = 3100$ kcal/mole**

	K_p(atm^{-1})	
T(°K)	calc.	obs.
673	1.1×10^4	1.2×10^4
773	4.3×10^2	4.2×10^2
873	33.5	32.0
973	4.5	5.0

Source: E. A. Guggenheim, *Trans. Faraday Soc.*, **37**, p. 97, 1941.

temperatures as calculated by Guggenheim.* The value of V_0 that seemed to give the best agreement was 3.1 kcal/mole. This value agrees fairly well with other determinations of V_0 for ethane. This is another method for determining barrier heights and some values determined by this method are included in Table 8–4.

* E. A. Guggenheim, *Trans. Faraday Soc.*, **37**, 97, 1941.

9–3 THERMODYNAMIC TABLES

In the previous section we have seen that the rigid rotor-harmonic oscillator approximation can be used to calculate equilibrium constants in reasonably good agreement with experiment, and because of the simplicity of the model, the calculations involved are not extensive. If greater accuracy is desired, however, one must include corrections to the rigid rotor–harmonic oscillator model, and the calculations become increasingly more laborious. It is natural, then, that a number of numerical tables of partition functions has evolved, and in this section we shall discuss the use of these tables. These tables are actually much more extensive than a compilation of partition functions. They include many experimentally determined values of thermodynamic properties, often complemented by theoretical calculations. The thermodynamic tables we are about to discuss in this section, then, represent a collection of the thermodynamic and/or statistical thermodynamic properties of many substances. In order to fully appreciate the use of these tables, it is necessary to express the equilibrium constant in terms of the standard chemical potential $\mu_0(T)$ and then to discuss at length the problem, or rather the convention, of the choice of the zero of energy.

First we shall derive the relation between the equilibrium constant and the standard chemical potential [*cf.* Eq. (1–68)]. Equation (9–4) gives the thermodynamic condition for chemical equilibrium:

$$\nu_C \mu_C + \nu_D \mu_D - \nu_A \mu_A - \nu_B \mu_B = 0$$

For an ideal gas, the chemical potential is of the form [*cf.* Eq. (1–66) or Eq. (5–21)]

$$\mu(T, p) = \mu_0(T) + kT \ln p \tag{9–39}$$

If we substitute this into Eq. (9–4), we get

$$\ln K_p = -\frac{\Delta\mu_0}{kT} \tag{9–40}$$

where

$$\Delta\mu_0 = \nu_C \mu_{0C} + \nu_D \mu_{0D} - \nu_A \mu_{0A} - \nu_B \mu_{0B} \tag{9–41}$$

The μ_0 are related to the molecular partition function in the following way. For an ideal gas, $Q = q^N/N!$. The chemical potential is given by $-kT(\partial \ln Q/\partial N)_{V,T}$, so

$$\mu = -kT \ln\left(\frac{q}{N}\right) \tag{9–42}$$

For an ideal gas, the molecular partition function is a function of temperature times the volume, and so q/V is a function of temperature only. We can rewrite Eq. (9–42) as

$$\mu = -kT \ln\left\{\left(\frac{q}{V}\right)\frac{V}{N}\right\}$$

and use the ideal gas equation of state for V/N to get

$$\mu = -kT \ln\left\{\left(\frac{q}{V}\right) kT\right\} + kT \ln p \tag{9–43}$$

This equation is of the form of Eq. (9–39) with $\mu_0(T)$ given by

$$\mu_0(T) = -kT \ln\left\{\left(\frac{q}{V}\right) kT\right\} \tag{9–44}$$

The argument of the logarithm here has units of pressure, but of course the units cancel when this term is combined with $kT \ln p$. Nevertheless, the numerical value of $\mu_0(T)$ does depend upon the units of pressure. In anticipation of the convention used in the thermodynamic tables, we shall express the pressure in atmospheres. If we use the fact that 1 atm = 1.01×10^6 dynes/cm^2, we can write Eq. (9–43) in the form

$$\mu = -kT \ln\left\{\left(\frac{q}{V}\right)\frac{kT}{1.01 \times 10^6}\right\} + kT \ln p(\text{atm}) \tag{9–45}$$

where (q/V) and kT in the $\mu_0(T)$ term are expressed in cgs units. The equilibrium constant calculated with these values of $\mu_0(T)$, however, will be in atmospheres, since we have explicitly included the conversion factor of 1.01×10^6. We shall see below that the thermodynamic tables are given in pressure units of atmospheres.

In principle, then, we could calculate $\mu_0(T)$ over a range of temperature and tabulate the results for future use in Eq. (9–40). This is almost the content of thermodynamic tables, but there is one last convention we must discuss, namely, the convention of the zero of energy. Throughout this book we have calculated partition functions in terms of energies relative to a zero of energy being taken as the infinitely separated atoms in their lowest energy states. From a molecular point of view, this is a satisfactory choice and is conceptually the most convenient choice. It is experimentally difficult to determine the energy required to separate a molecule into its constituent atoms, and so the thermodynamic tables utilize another convention which we describe below. Of course, it makes no difference what we choose as the zero of energy as long as we are consistent and use the same convention on both sides of a chemical equation. Let us go back now and reexamine at what point this arbitrariness in the choice of zero energy enters the calculation of partition functions.

We have seen that to a good approximation, the molecular partition function can be written as

$$q(V, T) = q_{\text{trans}}(V, T)q_{\text{rot}}(T)q_{\text{vib}}(T)q_{\text{elec}}(T) \tag{9–46}$$

where, of course, q_{rot} and q_{vib} do not appear in the case of an atom. If we recall the discussion centering around Fig. 6–2, it is in the calculation of $q_{\text{elec}}(T)$ that the zero-of-energy convention was used. The electronic partition function is

$$q_{\text{elec}}(T) = \omega_{e1}e^{-\varepsilon_{e1}/kT} + \omega_{e2}\,e^{-\varepsilon_{e2}/kT} + \cdots \tag{9–47}$$

If one chooses the separated atoms in their ground states as the zero of energy, then (*cf.* Fig. 6–2 or Fig. 9–2)

$$\begin{aligned}
\varepsilon_{e1} &= -D_e \\
\varepsilon_{e2} &= \varepsilon_{e1} + (\varepsilon_{e2} - \varepsilon_{e1}) = \varepsilon_{e1} + \Delta\varepsilon_{12} \\
&= -D_e + \Delta\varepsilon_{12} \\
&\cdots
\end{aligned} \tag{9–48}$$

and

$$\begin{aligned}
q_{\text{elec}}(T) &= \omega_{e1}e^{D_e/kT} + \omega_{e2}\,e^{D_e/kT}e^{-\Delta\varepsilon_{12}/kT} + \cdots \\
&= e^{D_e/kT}(\omega_{e1} + \omega_{e2}\,e^{-\Delta\varepsilon_{12}/kT} + \cdots) \\
&= e^{D_e/kT}q^0{}_{\text{elec}}(T)
\end{aligned} \tag{9–49}$$

The superscript on the $q^0{}_{elec}(T)$ in the last line indicates that this is the electronic partition function of a molecule with its ground electronic state taken to be zero.

If we substitute Eq. (9–49) into Eq. (9–46), we get

$$q(V, T) = q_{trans}(V, T)q_{rot}(T)q_{vib}(T)q^0{}_{elec}(T)e^{D_e/kT} \tag{9–50}$$

The vibrational partition function is calculated using energies whose zero is taken as the bottom of the potential well, so that there is an energy of $\frac{1}{2}h\nu(\sum_j \frac{1}{2}h\nu_j$ in a polyatomic molecule) in the ground vibrational state. It is this zero-point vibrational energy that leads to the numerator in the vibrational partition functions of Chapters 6 and 8, that is, to the $e^{-\Theta_{vj}/2T}$ in

$$q_{vib}(T) = \prod_j \frac{e^{-\Theta_{vj}/2T}}{(1 - e^{-\Theta_{vj}/T})} \tag{9–51}$$

If we substitute this into Eq. (9–50), we have

$$q(V, T) = q_{trans}(V, T)q_{rot}(T)\left\{\prod_j (1 - e^{-\Theta_{vj}/T})^{-1}\right\}q^0{}_{elec}(T)e^{(D_e-(1/2)\Sigma_j h\nu_j/kT)} \tag{9–52}$$

$$= q_{trans}(V, T)q_{rot}(T)q^0{}_{vib}(T)q^0{}_{elec}(T)e^{D_0/kT} \tag{9–53}$$

where we have written D_0 for $D_e - \frac{1}{2}\sum_j h\nu_j$ (*cf.* Fig. 9–2) and have superscripted q_{vib} with a ^{0} since the term in brackets in Eq. (9–52) is a vibrational partition function whose lowest vibrational state is taken to be zero. In addition, we could have superscripted q_{trans} and q_{rot} since these partition functions are calculated on the basis of the lowest translational and rotational states being zero. Therefore we can write the molecular partition function as

$$q(V, T) = q^0(V, T)e^{D_0/kT} \tag{9–54}$$

where $q^0(V, T)$ indicates that the partition function is calculated on the basis that the ground state of the entire molecule is taken to be zero. The factor of $\exp(D_0/kT)$ "scales" the partition function to the zero of energy being the separated atoms in their ground states.

Equation (9–54) can be interpreted as saying that the partition function can be written as the product of an internal part $q^0(V, T)$, and a scaling factor that accounts for the arbitrary zero of energy. Figure 9–2(a) shows the energy D_0, and Fig. 9–2(b)

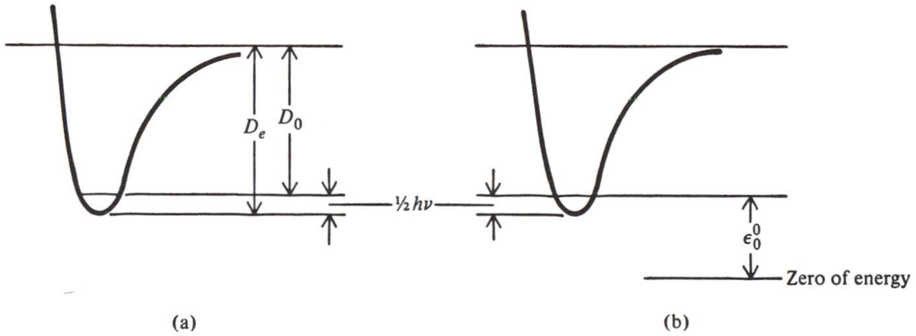

Figure 9–2. **An illustration of the arbitrary zero of energies used in Eqs. (9–54) and (9–55). (a) Shows the zero of energy to be the separated atoms in their ground states and (b) shows a "more arbitrary" zero in which the lowest vibrational state lies $\varepsilon_0{}^0$ above the zero.**

shows a completely arbitrary zero of energy. In terms of this other zero of energy, the partition function is

$$q(V, T) = q^0(V, T)e^{-\varepsilon_0{}^0/kT} \tag{9–55}$$

where $q^0(V, T)$ is the same partition function that occurs in Eq. (9–54). Note that if $\varepsilon_0{}^0 = -D_0$, Eq. (9–55) reduces to Eq. (9–54). In a sense, Eq. (9–55) is more arbitrary than Eq. (9–54) since the zero of energy has no physical basis and is completely arbitrary. If we substitute this into Eq. (9–45), we get

$$\mu - \varepsilon_0{}^0 = -kT \ln\left\{\left(\frac{q^0}{V}\right)\frac{kT}{1.01 \times 10^6}\right\} + kT \ln p(\text{atm}) \tag{9–56}$$

Equation (9–56) clearly displays the fact that the chemical potential is calculated relative to some zero of energy. As $T \to 0$, the molecule is found in its lowest energy state, which according to Fig. 9–2(b) is $\varepsilon_0{}^0$. This is the significance of the zero subscript on $\varepsilon_0{}^0$. It is easy to show mathematically that at 1 atm, $\mu \to \varepsilon_0{}^0$ as $T \to 0$ since $T \ln T \to 0$ as $T \to 0$.

We are now ready to introduce the notation and the conventions used in most thermodynamic tables. We shall adopt the convention that the energy of an element is zero at 0°K if the element is in the physical state (i.e., gas, liquid, or solid) characteristic of 25°C and 1 atm pressure. All energies, therefore, are calculated relative to this convention. We shall, in addition, multiply Eq. (9–56) by Avogadro's number and recognize that $N\mu$ is the Gibbs free energy per mole. We shall write Eq. (9–56) as

$$G^0 - E_0{}^0 = -RT \ln\left\{\left(\frac{q^0}{V}\right)\frac{kT}{1.01 \times 10^6}\right\} \tag{9–57}$$

The significance of the ^{0} superscript on G^0 indicates that this is the Gibbs free energy per mole at 1 atm pressure relative to the zero-of-energy convention adopted above. It is clear from Eq. (9–57) that $G^0 \to E_0{}^0$ as $T \to 0$, and so $E_0{}^0$ is also the standard free energy at 0°K. Thus we could write $G^0 - G_0{}^0$ instead of $G^0 - E_0{}^0$. This notation is often found in thermodynamic tables. Furthermore, since the enthalpy $H = E + pV$, and $p \to 0$ as $T \to 0$, we also have $H_0{}^0 = E_0{}^0$, and one also finds the notation $G^0 - H_0{}^0$. Since $G_0{}^0 = H_0{}^0 = E_0{}^0$, the quantities $G^0 - G_0{}^0$, $G^0 - E_0{}^0$, and $G^0 - H_0{}^0$ are equivalent. The important point is that $G^0 - E_0{}^0$ can be calculated without assuming any convention concerning the zero of energy, since this represents the standard Gibbs free energy relative to arbitrary zero, $E_0{}^0$. The right-hand side of Eq. (9–57) is based upon the energy of a molecule in its lowest state, that is, at $T = 0$°K, being zero, and hence can be calculated for any molecule independent of the choice of $E_0{}^0$.

Table 9–6 gives $-(G^0 - E_0{}^0)/T$ for a number of substances. The ratio $(G^0 - E_0{}^0)/T$ rather than $(G^0 - E_0{}^0)$ is given since $(G^0 - E_0{}^0)/T$ varies more slowly with temperature, and hence the tables are easier to interpolate. In addition to $-(G^0 - E_0{}^0)/T$, it is necessary to tabulate $E_0{}^0$ according to our chosen convention that the energy of an element is zero at 0°K if the element is in the physical state characteristic of 25°C and 1 atm. For a molecule, $E_0{}^0$ represents the energy of a molecule at 0°K relative to the elements, and therefore it also represents the heat of formation at 0°K. Consequently,

Table 9–6. **Thermodynamic functions of some selected substances**

	$-(G^0 - E_0{}^0)/T$ cal/deg-mole						
	298.15°K	500°K	1000°K	1500°K	2000°K	$H^0_{298} - E_0{}^0$	$E_0{}^0$(kcal/mole)
Ne	29.98	32.56	36.00	38.00	39.44	1.481	0
Ar	32.01	34.59	38.04	40.04	41.48	1.481	0
Kr	34.22	36.80	40.24	42.25	43.69	1.481	0
H(g)	22.42	24.99	28.44	30.45	31.88	1.481	51.62
Cl(g)	34.43	37.06	40.69	42.83	44.34	1.499	28.54
Br(g)	36.84	39.41	42.85	44.89	46.36	1.481	26.90
I(g)	38.22	40.78	44.23	46.24	47.68	1.481	25.61
O(g)	33.08	35.84	39.46	41.54	43.00	1.607	58.98
N(g)	31.65	34.22	37.66	39.67	41.10	1.481	112.54
H_2(g)	24.42	27.95	32.74	35.59	37.67	2.024	0
Cl_2(g)	45.93	49.85	55.43	58.85	61.34	2.194	0
Br_2(g)	50.85	54.99	60.80	64.31	66.83	2.325	8.37
I_2(g)	54.18	58.46	64.40	67.96	70.52	2.148	15.66
O_2(g)	42.06	45.68	50.70	53.81	56.10	2.07	0
N_2(g)	38.82	42.42	47.31	50.28	52.48	2.072	0
CO(g)	40.25	43.86	48.77	51.78	53.99	2.073	−27.202
NO(g)	42.98	46.76	51.86	54.96	57.24	2.194	21.48
HCl(g)	37.72	41.31	46.16	49.08	51.23	2.065	−22.019
HBr(g)	40.53	44.12	48.99	51.95	54.13	2.067	−8.1
HI(g)	42.40	45.99	50.90	53.90	56.11	2.069	6.7
CO_2(g)	43.56	47.67	54.11	58.48	61.85	2.238	−93.969
H_2O(g)	37.17	41.29	47.01	50.60	53.32	2.368	−57.107
NH_3(g)	37.99	42.28	48.63	53.03	56.56	2.37	−9.37
Cl_2O(g)	54.52	59.49	67.04	71.91	—	2.719	18.61
ClO_2(g)	52.79	57.48	64.65	69.33	—	2.577	25.59
N_2O(g)	44.89	49.11	55.76	60.27	—	2.291	20.31
NO_2(g)	49.19	53.60	60.23	64.58	67.88	2.465	8.68
CH_4(g)	36.46	40.75	47.65	52.84	57.1	2.397	−15.99
CH_3Cl(g)	47.45	52.06	59.78	65.54	—	2.489	−17.7
CH_2Cl_2(g)	55.08	60.36	69.58	76.17	—	2.834	−19
$CHCl_3$(g)	59.29	65.81	76.78	84.63	—	3.390	−23
CCl_4(g)	60.15	68.12	81.41	89.96	—	4.111	−25
C_2H_6(g)	45.27	50.77	61.11	69.46	—	2.856	−16.52
C_2H_4(g)	43.98	48.74	57.29	63.94	69.46	2.525	14.52
CH_2O(g)	44.25	48.54	55.11	59.81	63.58	2.393	−26.8
Na_2(g)	46.65	51.04	57.14	60.77	—	2.484	—
K_2(g)	51.06	55.57	61.76	65.47	—	2.566	—

Source: G. N. Lewis and M. Randall, *Thermodynamics*, revised by K. S. Pitzer and L. Brewer (New York: McGraw-Hill, 1961).

one often finds the notation $\Delta H_0{}^0$ instead of $E_0{}^0$. Thus we can calculate equilibrium constants from the tables since Eq. (9–40), is, in the notation introduced above,

$$-R \ln K_p = \frac{\Delta E_0{}^0}{T} + \Delta\left(\frac{G^0 - E_0{}^0}{T}\right) \qquad (9\text{–}58)$$

where Δ has the same significance here as in Eq. (9–41).

We shall illustrate the use of these tables by means of examples. First let us show how the tables are consistent with our previous calculations. Consider the reaction $H_2 + I_2 \rightleftharpoons 2HI$. For hydrogen,

$$q^0(V, T) = \left(\frac{2\pi mkT}{h^2}\right)^{3/2} V\left(\frac{T}{2\Theta_r}\right)(1 - e^{-\Theta_v/T})^{-1}$$

$$= 9.95 \times 10^{25}\, V \qquad \text{at } 1000°\text{K}$$

According to Eq. (9–57),

$$-\left(\frac{G^0 - E_0{}^0}{T}\right) = R \ln\left\{\frac{9.95 \times 10^{25} \times 1.38 \times 10^{-13}}{1.01 \times 10^6}\right\}$$
$$= 32.87 \qquad \text{at } 1000°\text{K}$$

in good agreement with Table 9–6. Of course, the value given in Table 9–6 must be considered to be more accurate than the value we have calculated here. Since hydrogen is a gas at 25°C and 1 atm, we find from Table 9–6 that $E_0{}^0 = 0$. For I_2, the parameters in Table 6–1 give $-(G^0 - E_0{}^0)/T = 64.40$ given in Table 9–6. Since I_2 is a solid at 25°C and 1 atm, $E_0{}^0$ is not zero, but is 15.66 kcal/mole, the heat of sublimation of I_2 at 0°K. If we substitute the values from Table 9–6 into Eq. (9–58), we get $K_p = 32.5$, compared to 30.9 from the simpler equations of Chapter 6.

For the reaction $H_2 + \frac{1}{2}O_2 \rightleftharpoons H_2O$ at 1000°K, the tables give $\log K_p = 10.00$ compared to the experimental value of 10.06 and the less accurate value 10.2 calculated in the previous section on the basis of the rigid rotor–harmonic oscillator model. Note that K_p is expressed in atmospheres, since the tables are based on the standard pressure being 1 atm.

Table 9–6 also gives $H^0_{298} - E_0{}^0$, which is the enthalpy at 298°K and 1 atm pressure relative to the zero of energy already explained. One often finds $H^0_{298} - H_0{}^0$, which, of course, is equivalent to $H^0_{298} - E_0{}^0$. For 1 mole of an ideal gas,

$$H = E + pV = E + RT$$
$$= RT^2\left(\frac{\partial \ln Q}{\partial T}\right)_{N,V} + RT \tag{9–59}$$

and so at temperatures high enough that the rotation is classical, we have

$$H_T{}^0 - E_0{}^0 = \tfrac{3}{2}RT + \tfrac{2}{2}RT + \frac{R\Theta_v}{e^{\Theta_v/T} - 1} + RT \qquad \text{(diatomic molecule)} \tag{9–60}$$

$$= \tfrac{3}{2}RT + \tfrac{2}{2}RT + \sum_j \frac{R\Theta_{vj}}{e^{\Theta_{vj}/T} - 1} + RT \qquad \text{(linear polyatomic molecule)} \tag{9–61}$$

$$= \tfrac{3}{2}RT + \tfrac{3}{2}RT + \sum_j \frac{R\Theta_{vj}}{e^{\Theta_{vj}/T} - 1} + RT \qquad \text{(nonlinear polyatomic molecule)} \tag{9–62}$$

Note that there is no contribution from the zero-point vibrational energy here as there is in Eq. (6–22), since we have taken the energy of the ground vibrational state to be zero. A value of 2.07 kcal/mole for a diatomic molecule comes from the $7RT/2$. Values higher than this are due to the vibrational contributions; a value lower than this (such as H_2) indicates that the rotation is not completely classical, and Eq. (6–35) should be used.

Although entropies are not given in Table 9–6, they can be easily obtained from the data given there since

$$S^0_{298} = \left(\frac{H_{298} - E_0{}^0}{T}\right) - \left(\frac{G^0_{298} - E_0{}^0}{T}\right) \tag{9–63}$$

The values of the entropy obtained in this way are in excellent agreement with those given in Tables 6–3 and 8–3.

Table 9–6 can also be used to calculate values of D_0 for various molecules. For example, consider D_0 for H_2. Table 9–6 shows that the energy of a hydrogen atom relative to a hydrogen molecule (both at 0°K) is 51.62 kcal/mole. The value given in Table 6–1 is twice this.

For H_2O, we have

$$H_2 + \tfrac{1}{2}O_2 \longrightarrow H_2O + 57.107 \text{ kcal} \qquad \text{at } 0°\text{K}$$

We add to this the following two reactions

$$\left.\begin{array}{lll} 2H & \longrightarrow & H_2 + 103.24 \text{ kcal} \\ O & \longrightarrow & \tfrac{1}{2}O_2 + 58.98 \text{ kcal} \end{array}\right\} \quad \text{at } 0°\text{K}$$

to get

$$2H + O \longrightarrow H_2O + 219.3 \text{ kcal} \qquad \text{at } 0°\text{K}$$

At 0°K, both reactants and products are in their ground states, and so the 219.3 kcal represents D_0 for H_2O, in agreement with Table 8–1.

As a last example, we calculate D_0 for HI. This case is slightly different since iodine is a solid at 25°C and 1 atm. Table 9–6 gives that

$$\tfrac{1}{2}H_2(g) + \tfrac{1}{2}I_2(s) \longrightarrow HI - 6.7 \text{ kcal} \qquad \text{at } 0°\text{K}$$

We add the following equations to this:

$$H \longrightarrow \tfrac{1}{2}H_2(g) + 51.62 \text{ kcal} \qquad \text{at } 0°\text{K}$$

$$I(g) \longrightarrow \tfrac{1}{2}I_2(s) + 25.61 \text{ kcal} \qquad \text{at } 0°\text{K}$$

The result is

$$H + I(g) \longrightarrow HI + 70.53 \text{ kcal} \qquad \text{at } 0°\text{K}$$

and hence D_0 for HI is 70.53 kcal/mole.

Thermodynamic tables contain a great deal of thermodynamic and/or statistical thermodynamic data. Their use requires some amount of practice, but it is well worth the effort. Problems 9–17 through 9–26 are meant to supply this practice.

ADDITIONAL READING

General

FOWLER, R. H., and GUGGENHEIM, E. A. 1956. *Statistical thermodynamics.* Cambridge: Cambridge University Press. Chapter 5.
KITTEL, C. 1969. *Thermal physics.* New York: Wiley. Chapter 21.
MAYER, J. E., and MAYER, M. G. 1940. *Statistical mechanics.* New York: Wiley. Chapter 9.
MÜNSTER, A. 1969. *Statistical thermodynamics*, Vol. I. Berlin: Springer-Verlag. Chapter 7.
RUSHBROOKE, G. S. 1949. *Statistical mechanics.* London: Oxford University Press. Chapters 11 and 12.
WIBERG, K. B. 1964. *Physical organic chemistry.* New York: Wiley. Part 2.
WILSON, Jr., E. B. 1940. *Chem. Rev.*, **27**, 17.

Thermodynamic Tables

LEWIS, G. N., and RANDALL, M. 1961. *Thermodynamics*, rev. ed. by K. S. Pitzer and L. Brewer. New York: McGraw-Hill.

PROBLEMS

9–1. Consider the reaction $A \rightleftharpoons 2B$. The canonical ensemble partition function for an ideal binary mixture is

$$Q(N_A, N_B, V, T) = \frac{q_A^{N_A} q_B^{N_B}}{N_A!N_B!}$$

Minimize the Helmholtz free energy with the stoichiometric constraint $2N_A + N_B =$ constant to show that

$$\frac{N_B^{*2}}{N_A^*} = \frac{q_B^2}{q_A}$$

where N_A^* and N_B^* are the equilibrium numbers of A and B. Can you generalize this approach to the reaction $\nu_A A + \nu_B B \rightleftharpoons \nu_C C + \nu_D D$ and derive Eq. (9–7)?

9–2. Using the data in Table 6–1 (plus atomic spectroscopic tables), calculate the equilibrium constant for the reaction $2K \rightleftharpoons K_2$ at 800°K and 1000°K. The experimental values* at these temperatures are 0.673 and 0.123, respectively.

9–3. Calculate the equilibrium constant K_p for the reaction $I_2 \rightleftharpoons 2I$, using the data in Table 6–1 and the fact that the ground electronic state of the iodine atom is $^2P_{3/2}$ and that the first excited electronic state lies 0.94 eV higher. The experimental values† are

T(°K)	800	900	1000	1100	1200
K_p	1.14×10^{-2}	4.74×10^{-2}	0.165	0.492	1.23

9–4. Show that the equilibrium constant for a reaction such as

$$HCl + DBr \rightleftharpoons DCl + HBr$$

approaches unity at sufficiently high temperatures.

9–5. Calculate the equilibrium constant for the reaction $N_2^{14} + N_2^{15} \rightleftharpoons 2N^{14}N^{15}$ and compare your results to the experimental values of Joris and Taylor.‡

9–6. Derive Eq. (9–24).

9–7. Show that the Teller-Redlich product rule is valid for a diatomic molecule under the Born-Oppenheimer approximation.

9–8. Calculate the enthalpy of reaction for $H_2 + I_2 \rightleftharpoons 2HI$ around 300°K.

9–9. Calculate the equilibrium constant for the reaction $\frac{1}{2}N_2 + \frac{1}{2}O_2 \rightleftharpoons NO$, using the data in Table 6–1. The observed values are

T(°K)	1500	2000	2500
K_p	2.4×10^{-3}	1.5×10^{-2}	4.5×10^{-2}

Why is the agreement not as good as you obtained in the other problems in this chapter?

9–10. Calculate the equilibrium constant for the reaction $CO_2 \rightleftharpoons CO + \frac{1}{2}O_2$ at 3000°K. The experimental value is 0.378 $(\text{atm})^{1/2}$.

9–11. Using the data in Tables 6–1 and 8–1, calculate the equilibrium constant for the water gas reaction $CO_2 + H_2 \rightleftharpoons CO + H_2O$ at 900°K and 1200°K. The experimental values at these two temperatures are 0.46 and 1.37, respectively.

9–12. Using the data in Tables 6–1 and 8–1, calculate the equilibrium constant of the reaction $3H_2 + N_2 \rightleftharpoons 2NH_3$ at 400°C. The accepted value is 3.3×10^{-4} $(\text{atm})^{-2}$.

9–13. Calculate the temperature at which molecular nitrogen is 99 percent dissociated at 100 atm, 1 atm, and 0.1 atm.

9–14. For the two ionization processes

$$H \rightleftharpoons H^+ + e^-$$
$$Cs \rightleftharpoons Cs^+ + e^-$$

derive an expression for the ratio of the equilibrium constants K_H/K_{Cs} where

$$K_H = \frac{p_{H+}\,p_{e-}}{p_H} \quad \text{and} \quad K_{Cs} = \frac{p_{Cs+}\,p_{e-}}{p_{Cs}}$$

* See C. T. Ewing, *et al.*, *J. Chem. Phys.*, **71**, p. 473, 1967.
† See Perlman and Rollefson, *J. Chem. Phys.*, **9**, p. 362, 1941.
‡ See G. C. Joris and H. S. Taylor, *J. Chem. Phys.*, **7**, p. 893, 1939.

in terms of the respective ionization potentials I_H and I_{Cs}. What is the high-temperature limit of K_H/K_{Cs}? At room temperature do you expect K_H/K_{Cs} to be greater than, less than, or approximately equal to one?

9–15. Calculate the equilibrium constant for the first ionization of argon at 10,000°K. Use the data in A. B. Cambel, *Plasma Physics and Magnetofluid Dynamics* (New York: McGraw-Hill, 1963, p. 119).

9–16. Estimate the temperature at which gaseous atomic hydrogen would be at least 99 percent dissociated at 10^{-3} torr pressure into electrons and protons. Assume classical statistics, but remember that the electron has a spin degeneracy of 2.

9–17. Calculate the equilibrium constants of Problem 9–2, using the data in Table 9–6 and the *Handbook of Chemistry and Physics.*

9–18. Calculate the equilibrium constants of Problem 9–8, using the data in Table 9–6.

9–19. Calculate the equilibrium constants of Problem 9–9, using the data in Table 9–6.

9–20. Calculate the equilibrium constants of Problem 9–10, using the data in Table 9–6.

9–21. Calculate the equilibrium constants of Problem 9–11, using the data in Table 9–6.

9–22. Calculate the equilibrium constants of Problem 9–12, using the data in Table 9–6.

9–23. Verify the results in Table 9–6 for Ne, using the results of Chapter 5.

9–24. Verify the results in Table 9–6 for HBr, using the results of Chapter 6.

9–25. Verify the results of Table 9–6 for CH_3Cl, using the results of Chapter 8.

9–26. Verify the results of Table 9–6 for NO_2, using the results of Chapter 8.

CHAPTER 10

QUANTUM STATISTICS

In Chapter 4 we derived the two fundamental distribution laws of statistical mechanics. One, the Fermi-Dirac distribution, applies to systems whose N-body wave function is antisymmetric with respect to an interchange of any two identical particles; the other, Bose-Einstein statistics, applies to systems whose N-body wave function is symmetric under such an interchange. All elementary particles that have a half-odd-integral spin, such as the electron and the proton, obey Fermi-Dirac statistics and are called fermions; elementary particles that have an integral spin, such as the deuteron and the photon, obey Bose-Einstein statistics and are called bosons. The classification of compound particles can become a delicate problem at times, but as long as the binding energy of the compound particle is large compared to all other energies in the problem (which will be so for all cases that we shall discuss), one can say that a compound particle containing an odd number of fermions, such as He-3, will obey Fermi-Dirac statistics, and one with an even number, such as He-4, will obey Bose-Einstein statistics. Incidentally, it should be pointed out that it is not obvious that all particles occurring in nature are necessarily fermions or bosons, but there appear to be no known exceptions.

The basic equations associated with the two fundamental distribution laws are [Eqs. (4–24) through (4–28)]

$$\Xi(V, T, \lambda) = \prod_k (1 \pm \lambda e^{-\beta\varepsilon_k})^{\pm 1} \tag{10–1}$$

$$N = \sum_k \frac{\lambda e^{-\beta\varepsilon_k}}{1 \pm \lambda e^{-\beta\varepsilon_k}} \tag{10–2}$$

$$\bar{n}_k = \frac{\lambda e^{-\beta\varepsilon_k}}{1 \pm \lambda e^{-\beta\varepsilon_k}} \tag{10–3}$$

$$E = \sum_k \frac{\lambda \varepsilon_k e^{-\beta\varepsilon_k}}{1 \pm \lambda e^{-\beta\varepsilon_k}} \tag{10–4}$$

and

$$pV = \pm kT \sum_k \ln(1 \pm \lambda e^{-\beta\varepsilon_k}) \tag{10-5}$$

where $\lambda = \exp(\mu/kT)$. In these equations the upper sign (+) corresponds to Fermi-Dirac statistics, and the lower sign (−) corresponds to Bose-Einstein statistics. In order to discuss the thermodynamic properties given by these equations, it is necessary to solve Eq. (10–2) for λ in terms of N and the $\beta\varepsilon$'s, and since the ε_k's are functions of V, this procedure gives λ as a function of N, V, and T. This solution for λ is then substituted into Eqs. (10–4) and (10–5) to give E and p, and hence other thermodynamic functions, in terms of N, V, and T. (See Problem 10–1.)

The difficulty is that it is not possible to solve Eq. (10–2) analytically for all values of λ. In Chapter 4 we showed that if λ is small, both Fermi-Dirac and Bose-Einstein reduce to Boltzmann, or classical, statistics. In this case $\lambda = N/q$ [*cf.* Eq. (10–2)], and all the mathematical manipulations are easy to perform. Thus the magnitude of λ can be regarded as a measure of the degree of quantum behavior of the system. Small values of λ correspond to classical or near-classical behavior, and large values of λ apparently correspond to quantum statistical behavior. Table 4–1 and the discussion surrounding it show that quantum effects become important for low temperatures and high densities. Thus we may expect λ to be small for high temperatures and low densities and large for low temperatures and high densities. We shall verify this later in the chapter.

Fortunately most systems of interest can be described by Boltzmann statistics and hence are characterized by small values of λ. This led us to the detailed discussion of classical ideal gases in Chapters 5 to 9.

There are several important and interesting systems, however (see Table 4–1), which cannot be described by classical statistics, and hence for which λ is not small. In this chapter we shall discuss these applications.

In Section 10–1 we shall study an ideal Fermi-Dirac gas for values of λ such that a series expansion of Eqs. (10–2) through (10–5) is useful. The first terms of these expansions represent the Boltzmann, or classical, limit, and so these expansions are useful in a temperature and density region where there are only small deviations from classical behavior, that is, high temperature and low density. We shall find the interesting result that even though we neglect the intermolecular forces in an ideal gas model, the equation of state will no longer be $p = \rho kT$, but will be a virial expansion, that is, a power series in the density. In Section 10–2 we shall examine the case for large values of λ (low values of the temperature or high values of the density). In this region we expect large deviations from classical behavior. Such systems are said to have large quantum effects or to be strongly *degenerate*. We shall see that although the results of these two sections are derived for an ideal gas, they serve as an interesting model for the electrons in a metal, and, in fact, Section 10–2 considers this model specifically.

Then in Section 10–3 we shall treat Bose-Einstein statistics in the range of λ, where a series expansion of Eqs. (10–2) through (10–5) is useful. The results are very similar to those of Section 10–1. In Section 10–4 we examine the low-temperature or high-density limit of Eqs. (10–2) through (10–5). This section, although more mathematically involved than the others, is very interesting, since we find that a system of ideal bosons undergoes a kind of phase transition as the temperature is lowered. We shall discuss the implications of this result with respect to the well-known λ-transition that occurs in He-4 at 2.18°K. In Section 10–5 we shall treat an ideal gas of photons or, in reality, electromagnetic radiation in thermal equilibrium. In this section we shall derive the fundamental blackbody radiation distribution law first derived by Planck and which led to the development of quantum mechanics.

Section 10–6 is devoted to an alternative formulation of quantum statistics, namely, through the density matrix, which is the quantum statistical analog of the density of phase points in classical phase space. Lastly, Section 10–7 is a rigorous discussion of the classical limit of Q.

10–1 A WEAKLY DEGENERATE IDEAL FERMI-DIRAC GAS

As the title of this section indicates, we shall consider an ideal gas of fermions in a region where λ is small enough that we may represent the deviations from classical behavior by a series expansion in λ. Consider Eqs. (10–2) and (10–4) with the upper signs:

$$N = \sum_k \frac{\lambda e^{-\beta\varepsilon_k}}{1 + \lambda e^{-\beta\varepsilon_k}} \tag{10-6}$$

$$pV = kT \sum_k \ln(1 + \lambda e^{-\beta\varepsilon_k}) \tag{10-7}$$

The ε_k's in these equations are the eigenvalues of a particle in a cube [*cf.* Eq. (5–3)], and the one index k really stands for n_x, n_y, and n_z:

$$\varepsilon_{n_x n_y n_z} = \frac{h^2}{8mV^{2/3}} (n_x{}^2 + n_y{}^2 + n_z{}^2) \qquad n_x, n_y, n_z = 1, 2, \ldots \tag{10-8}$$

where we have replaced a^2 by $V^{2/3}$. In Chapter 5 we converted sums over the energy states to integrals over energy levels. We used the argument that at room temperature, successive values of ε/kT differed so little from each other that ε_k ($\varepsilon_{n_x n_y n_z}$) was essentially a continuous function. We derived an expression for the density of states $\omega(\varepsilon)$ between ε and $\varepsilon + d\varepsilon$ on the same basis. In this chapter we do not wish to restrict our discussion to high temperatures, but the very same type of argument applies if we consider the thermodynamic limit of a large volume [*cf.* Eq. (10–8)]. Thus we may consider the energy states in Eq. (10–8) to be continuous and write

$$N = 2\pi \left(\frac{2m}{h^2}\right)^{3/2} V \int_0^\infty \frac{\lambda \varepsilon^{1/2} e^{-\beta\varepsilon}\, d\varepsilon}{1 + \lambda e^{-\beta\varepsilon}} \tag{10-9}$$

$$pV = 2\pi kT \left(\frac{2m}{h^2}\right)^{3/2} V \int_0^\infty \varepsilon^{1/2} \ln(1 + \lambda e^{-\beta\varepsilon})\, d\varepsilon \tag{10-10}$$

This procedure is valid only because the summands in Eqs. (10–6) and (10–7) are finite continuous functions of λ for all the ε_k. Equations (10–6) and (10–7) indicate that the range of λ is $0 \le \lambda < \infty$. Because of the minus sign in the denominator of the expressions for Bose-Einstein statistics, such a conversion of a summation into an integration must be done with care.

The integrals occurring in Eqs. (10–9) and (10–10) cannot be evaluated in closed form, but they can be written as power series in λ by expanding the denominator of Eq. (10–9) and the logarithm in Eq. (10–10) in a power series in λ and integrating term by term (see Problem 10–4):

$$\rho = \frac{1}{\Lambda^3} \sum_{l=1}^{\infty} \frac{(-1)^{l+1} \lambda^l}{l^{3/2}} \tag{10-11}$$

and

$$\frac{p}{kT} = \frac{1}{\Lambda^3} \sum_{l=1}^{\infty} \frac{(-1)^{l+1} \lambda^l}{l^{5/2}} \tag{10–12}$$

These series are valid only for small values of λ. Equation (10–11) gives ρ as a power series in λ, and Eq. (10–12) gives p/kT as a power series in λ. We wish to solve Eq. (10–11) for λ as a function of ρ, and then substitute this into Eq. (10–12) to give p/kT as a function of ρ.

The problem of solving an equation like Eq. (10–11) is called *reversion* of a series and can be done in general. The general solution is involved, however, and since we are interested only in small deviations from classical behavior in this section, we can use a straightforward algebraic method that readily gives the first few terms. We assume that λ is a power series in ρ and write

$$\lambda = a_0 + a_1 \rho + a_2 \rho^2 + \cdots$$

We then substitute this into Eq. (10–11) and equate coefficients of like powers of ρ on both sides of the equation to get

$$\begin{aligned} a_0 &= 0 \\ a_1 &= \Lambda^3 \\ a_2 - \frac{a_1^2}{2^{3/2}} &= 0 \\ a_3 - \frac{a_1 a_2}{2^{1/2}} + \frac{a_1^3}{3^{3/2}} &= 0 \\ &\cdots \end{aligned} \tag{10–13}$$

Thus λ is

$$\lambda = \rho\Lambda^3 + \frac{1}{2^{3/2}} (\rho\Lambda^3)^2 + \left(\frac{1}{4} - \frac{1}{3^{3/2}}\right)(\rho\Lambda^3)^3 + \cdots \tag{10–14}$$

We now substitute this into Eq. (10–12) and get

$$\frac{p}{kT} = \rho + \frac{\Lambda^3}{2^{5/2}} \rho^2 + \left(\frac{1}{8} - \frac{2}{3^{5/2}}\right)\Lambda^6 \rho^3 + \cdots \tag{10–15}$$

This equation is in the form of a virial expansion for the pressure, which is customarily written in the form

$$\frac{p}{kT} = \rho + B_2(T)\rho^2 + B_3(T)\rho^3 + \cdots \tag{10–16}$$

where $B_j(T)$ is a function of only the temperature and is called the jth virial coefficient. Virial coefficients reflect deviations from ideality, or intermolecular interactions. Although there are no intermolecular forces in this case, the particles, nevertheless, experience an effective interaction through the symmetry requirement of the N-body wave function. In the case of fermions, this interaction may be said to be repulsive since the first correction to ideal behavior, $B_2(T)$, is positive and hence increases the pressure above that of a classical ideal gas under the same conditions. Note also that $B_2(T)$ is $0(\Lambda^3)$. But Λ is just the thermal De Broglie wavelength, and so we see that quantum statistical effects decrease as the thermal De Broglie wavelength decreases In fact, if Eq. (10–15) is written in a dimensionless form, we see once again that it is the dimensionless ratio Λ^3/V that is a measure of the quantum effects.

The Fermi-Dirac equation for $\bar{E}$, Eq. (10–4) with the positive sign, also can be converted to an integral and then written as a series expansion in λ:

$$E = \tfrac{3}{2}VkT\frac{1}{\Lambda^3}\sum_{l=1}^{\infty}\frac{(-1)^{l+1}\lambda^l}{l^{5/2}} \tag{10–17}$$

If Eq. (10–14) for λ is substituted into this, we get

$$E = \tfrac{3}{2}NkT\left(1 + \frac{\Lambda^3}{2^{5/2}}\rho + \cdots\right) \tag{10–18}$$

All other thermodynamic functions can be obtained from Eqs. (10–14), (10–15), and (10–16). For example, the chemical potential μ follows immediately from Eq. (10–14) since $\lambda = \exp(\mu/kT)$. The entropy can be obtained from $G = \mu N = E - TS + pV$.

These series expansions for the thermodynamic functions are valid only for small values of λ, or for temperatures and densities such that quantum effects are a small correction to the classical limit. The leading terms in all these expansions are the classical limits obtained in Chapter 5. Now we shall discuss the case for large values of λ.

10–2 A STRONGLY DEGENERATE IDEAL FERMI-DIRAC GAS

In this section we shall treat the case of an ideal Fermi-Dirac gas at low temperature and/or high density. For concreteness we shall develop the results in terms of the free electron model of metals, where the valence electrons of the atoms of the metal are represented by an ideal gas of electrons. It is possible to understand a number of important physical properties of some metals, in particular the simple monovalent metals, in terms of the free electron model. There are several reasons why such a simple model can be used. One is that although the electrons do indeed interact with each other and with the atomic cores through a Coulombic potential, this potential is so long range $(1/r)$ that the total electronic potential that any one electron "sees" is almost constant from point to point in the metallic crystal. Furthermore, in general many of the physically observable properties of a metallic crystal are due more to quantum statistical effects than to the details of the electron–electron and electron–ionic core interactions.

Consider Eq. (10–3) (with the positive sign in the denominator), which gives the number of particles in the molecular energy state k:

$$\bar{n}_k = \frac{\lambda e^{-\beta\varepsilon_k}}{1 + \lambda e^{-\beta\varepsilon_k}} = \frac{1}{1 + e^{\beta(\varepsilon_k - \mu)}} \tag{10–19}$$

As in the previous section, ε_k is essentially a continuous parameter, and we can write

$$f(\varepsilon) = \frac{1}{1 + e^{\beta(\varepsilon - \mu)}} \tag{10–20}$$

where $f(\varepsilon)$ is the probability that a given state is occupied. This equation is plotted in Fig. 10–1, where, in particular, $f(\varepsilon)$ is plotted versus ε/μ for fixed values of $\beta\mu$. In the extreme limit of low temperatures, that is, $T = 0$, the distribution is unity for energies

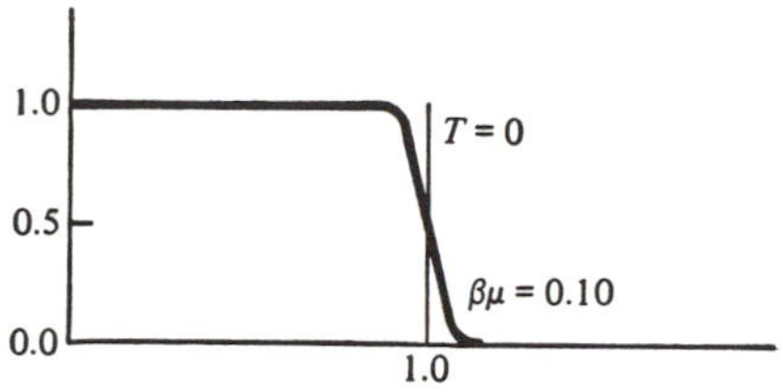

Figure **10–1. The Fermi-Dirac distribution as a function of ε (in units of μ) for $T=0$ and $\beta\mu = 0.10$.**

less than μ, and zero for energies greater than μ. The quantity μ is, in general, a function of temperature, and so we indicate the value of μ at $T = 0$ by μ_0.

The zero-temperature limit of $f(\varepsilon)$ is very simple, and it is instructive to understand its behavior. At the absolute zero of temperature, all the states with energy less than μ_0 are occupied and those with energy greater than μ_0 unoccupied. Thus μ_0 has the property of being a cutoff energy. According to Eq. (1–35), the number of states with energy between ε and $\varepsilon + d\varepsilon$ is

$$\omega(\varepsilon)\, d\varepsilon = 4\pi\left(\frac{2m}{h^2}\right)^{3/2} V\varepsilon^{1/2}\, d\varepsilon \tag{10–21}$$

where we have introduced a factor of 2, since an electron has two spin states ($\pm\frac{1}{2}$) associated with each translational state. We can find μ_0 immediately from the fact that all the states below $\varepsilon = \mu_0$ are occupied and all these above are unoccupied. Thus if N is the number of valence electrons,

$$\begin{aligned} N &= 4\pi\left(\frac{2m}{h^2}\right)^{3/2} V \int_0^{\mu_0} \varepsilon^{1/2}\, d\varepsilon \\ &= \frac{8\pi}{3}\left(\frac{2m}{h^2}\right)^{3/2} V(\mu_0)^{3/2} \end{aligned} \tag{10–22}$$

from which we write

$$\mu_0 = \frac{h^2}{2m}\left(\frac{3}{8\pi}\right)^{2/3}\left(\frac{N}{V}\right)^{2/3} \tag{10–23}$$

Using the fact that the molar volume of Na, say, is 23.7 cm^3/mole, and that a sodium atom has one valence electron, the quantity μ_0 is 3.1 eV. We see, then, that at $T = 0°\text{K}$, the conduction electrons in sodium metal fill all the energy states up to 3.1 eV according to the Pauli exclusion principle of allowing only two electrons to occupy each state. The quantity μ_0 is called the Fermi energy of a metal and is typically of the order of 1–5 eV (see Problem 10–6).

This is a very significant result since a plot as in Fig. 10–1 shows that at room temperature, where $\beta\mu_0$ is of order of 10^2, the distribution $f(\varepsilon)$ is still essentially a step function like at $T = 0°\text{K}$. Compared to a characteristic temperature μ_0/k, room temperature may be considered to be zero, and it is an excellent first approximation to use the distribution

$$\begin{aligned} f(\varepsilon) &= 1 \qquad \varepsilon < \mu_0 \\ &= 0 \qquad \varepsilon > \mu_0 \end{aligned} \tag{10–24}$$

at room temperature. The quantity μ_0/k is called the Fermi temperature and is denoted by T_F. Fermi temperatures are typically of the order of thousands of degrees Kelvin. In this approximation, the thermodynamic energy is [*cf.* Eq. (10–4)]

$$E_0 = 4\pi\left(\frac{2m}{h^2}\right)^{3/2} V\int_0^{\mu_0} \varepsilon^{3/2}\, d\varepsilon$$
$$= \tfrac{3}{5}N\mu_0 \qquad (10\text{–}25)$$

where we have written E_0 to emphasize that this is a $T = 0°\text{K}$ result. The energy E_0 is the zero-point energy of a Fermi-Dirac gas. Equation (10–25) implies that the contribution of the conduction electrons to the heat capacity is zero, in sharp contrast to the equipartition value of $3k/2$ for each electron. The physical explanation of this is that in order to contribute to the heat capacity, the electrons must be excited to higher quantum states, but since μ_0 is so large compared to kT, only a small fraction of all the particles will be within kT from the top of the distribution (near μ_0), where there are vacant states lying above. Thus a very small fraction of all the electrons can contribute to the heat capacity, and so the experimental heat capacity is almost zero.

The pressure is given by [*cf.* Eq. (10–10) with the additional factor of 2 due to the two allowed spin orientations of an electron]:

$$p = 4\pi kT\left(\frac{2m}{h^2}\right)^{3/2} \int_0^{\mu_0} \varepsilon^{1/2} \ln(1 + e^{\beta(\mu_0-\varepsilon)})\, d\varepsilon$$

In order to evaluate this integral, one can neglect unity compared to $\varepsilon^{\beta(\mu_0-\varepsilon)}$ since $\beta(\mu_0 - \varepsilon)$ is much larger than one over most of the range of integration. Therefore

$$p_0 = 4\pi\left(\frac{2m}{h^2}\right)^{3/2} \int_0^{\mu_0} \varepsilon^{1/2}(\mu_0 - \varepsilon)\, d\varepsilon$$
$$= \tfrac{2}{5}N\mu_0/V \qquad (10\text{–}26)$$

This "zero-point pressure" is $0(10^6)$ atm. The occurrence of h in this expression shows that the zero-point pressure is a quantum effect. It follows from $G = N\mu = E - TS + pV$ and Eqs. (10–23), (10–25), and (10–26) that $S_0 = 0$. This is to be expected, since there is only one way to put the N indistinguishable particles into the lowest possible quantum states.

Equations (10–22) through (10–26) are based on the zero-temperature distribution function, Eq. (10–24). It is not difficult to calculate corrections to these zero-temperature results as an expansion in powers of a parameter $\eta = kT/\mu_0$. At room temperature, $\eta = 0(10^{-2})$, so that such an expansion converges quickly. To determine this expansion, first notice that all the thermodynamic quantities $N, E, p, \ldots$ can be written as

$$I = \int_0^{\infty} f(\varepsilon)h(\varepsilon)\, d\varepsilon \qquad (10\text{–}27)$$

where the following table gives examples of I and $h(\varepsilon)$:

I	$h(\varepsilon)$
N	$4\pi\left(\frac{2m}{h^2}\right)^{3/2} V\varepsilon^{1/2}$
E	$4\pi\left(\frac{2m}{h^2}\right)^{3/2} V\varepsilon^{3/2}$

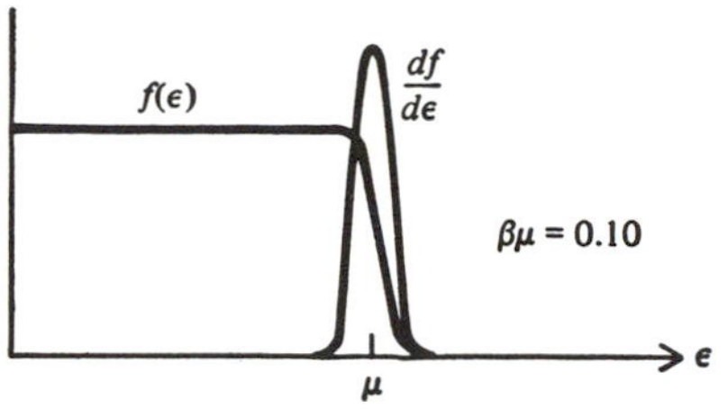

Figure 10–2. **The Fermi-Dirac distribution function and its derivative for a value of $\beta\mu = 0.10$.**

Figure 10–2 shows $f(\varepsilon)$ and the derivative of $f(\varepsilon)$, $f'(\varepsilon)$, versus ε for $\beta\mu = 0.10$. This shows that even at $\beta\mu = 0.10$ (which, we shall show below, corresponds to a temperature at which most metals are liquid), $f(\varepsilon)$ is a step function with rounded edges, and its derivative therefore is a function that is zero everywhere except around the region in which $\varepsilon = \mu$. Thus it is convenient to express the integral I in Eq. (10–27) in terms of $f'(\varepsilon)$. This is done by an integration by parts:

$$I = -\int_0^\infty f'(\varepsilon)H(\varepsilon)\,d\varepsilon \tag{10–28}$$

where

$$H(\varepsilon) = \int_0^\varepsilon h(x)\,dx \tag{10–29}$$

and we have used the fact that $h(\varepsilon) = 0$ at $\varepsilon = 0$. Now since $f'(\varepsilon)$ is nonzero only for some small region around $\varepsilon = \mu$, the only important values of $H(\varepsilon)$ will be for values of ε around $\varepsilon = \mu$. Thus we expand $H(\varepsilon)$ in a Taylor series about $\varepsilon = \mu$:

$$H(\varepsilon) = H(\mu) + (\varepsilon - \mu)\left(\frac{dH}{d\varepsilon}\right)_{\varepsilon=\mu} + \tfrac{1}{2}(\varepsilon - \mu)^2\left(\frac{d^2H}{d\varepsilon^2}\right)_{\varepsilon=\mu} + \cdots \tag{10–30}$$

If we substitute this into Eq. (10–28), we get

$$I = H(\mu)L_0 + \left(\frac{dH}{d\varepsilon}\right)_{\varepsilon=\mu} L_1 + \frac{1}{2}\left(\frac{d^2H}{d\varepsilon^2}\right)_{\varepsilon=\mu} L_2 + \cdots \tag{10–31}$$

where

$$L_j \equiv -\int_0^\infty (\varepsilon - \mu)^j f'(\varepsilon)\,d\varepsilon \tag{10–32}$$

The first of these integrals, L_0, is simply unity since it is equal to $f(0) - f(\infty)$. In the others, such as L_1 and L_2, we may replace the lower limit by $-\infty$ since $f'(\varepsilon)$ is so small from $-\infty$ to 0 that there is no contribution to the integral. If we let $x = \beta(\varepsilon - \mu)$, we have

$$L_j = \frac{1}{\beta^j}\int_{-\infty}^\infty \frac{x^j e^x}{(1 + e^x)^2}\,dx \qquad j = 0, 1, 2, \ldots . \tag{10–33}$$

All of these integrals with odd values of j vanish, since except for the x^j, the integrands are even functions of x. (See Problem 10–10.) The integral for L_2 is standard and may be found in tables:

$$\int_{-\infty}^\infty \frac{x^2 e^x\,dx}{(1 + e^x)^2} = \frac{\pi^2}{3}$$

This gives $L_2 = \pi^2(kT)^2/3$ and, by Eq. (10–31),

$$I = H(\mu) + \frac{\pi^2}{6}(kT)^2 H''(\mu) + \cdots \tag{10–34}$$

As an example of the use of this equation, we calculate N. In this case $h(\varepsilon) = 4\pi(2m/h^2)^{3/2} V \varepsilon^{1/2}$, and so

$$N = \frac{8\pi}{3}\left(\frac{2m}{h^2}\right)^{3/2} V \mu^{3/2}\left[1 + \frac{\pi^2}{8}(\beta\mu)^{-2} + \cdots\right] \tag{10–35}$$

Using Eq. (10–23) for μ_0, this becomes

$$\mu_0 = \mu\left[1 + \frac{\pi^2}{8}(\beta\mu)^{-2} + \cdots\right]^{2/3}$$

$$= \mu\left[1 + \frac{\pi^2}{12}(\beta\mu)^{-2} + \cdots\right]$$

This gives μ_0/μ as a power series in $(\beta\mu)^{-2}$. We can get μ/μ_0 as a power series in $\eta = (\beta\mu_0)^{-1}$ by taking the reciprocal of this equation

$$\frac{\mu}{\mu_0} = 1 - \frac{\pi^2}{12}(\beta\mu)^{-2} + \cdots$$

and then substituting $\mu = \mu_0(1 - \pi^2/12(\beta\mu)^2 + \cdots)$ into the right-hand side to get

$$\mu = \mu_0\left[1 - \frac{\pi^2}{12}\eta^2 + \cdots\right] \tag{10–36}$$

This equation shows that μ changes slowly with temperature and is approximately μ_0 throughout the entire solid-state range of a metal.

We can use Eq. (10–34) to calculate other thermodynamic properties such as E. This gives

$$E = \frac{8\pi}{5}\left(\frac{2m}{h^2}\right)^{3/2} V \mu^{5/2}[1 + \tfrac{5}{8}\pi^2(\beta\mu)^{-2} + \cdots]$$

$$= E_0\left(\frac{\mu}{\mu_0}\right)^{5/2}[1 + \tfrac{5}{8}\pi^2(\beta\mu)^{-2} + \cdots]$$

We can use Eq. (10–36) now to write E as a power series in η:

$$E = E_0\left[1 + \frac{5\pi^2}{12}\eta^2 + \cdots\right] \tag{10–37}$$

The contribution of the conductance electrons to the heat capacity is

$$C_V = \frac{\pi^2 NkT}{2(\mu_0/k)} = \frac{\pi^2}{2} Nk\left(\frac{T}{T_F}\right) \tag{10–38}$$

where we have introduced the Fermi temperature T_F. This equation predicts the molar electronic heat capacity of metals to be of the order of $10^{-4}T$ cal/deg-mole, and this is observed for many, but not all, metals to which the free electron model is expected to be applicable.

Fermi-Dirac statistics have been applied to other physical systems besides the electrons in metals. Two such applications are the theory of white dwarf stars and "nuclear" gases.

10-3 A WEAKLY DEGENERATE IDEAL BOSE-EINSTEIN GAS

In this section we shall treat an ideal Bose-Einstein gas in a region where λ is small enough that we may use series expansions in λ as we did in Section 10–1 for an ideal gas of fermions. We shall see that the results of this section are very similar to those of Section 10–1. In this case we use Eqs. (10–2) and (10–4) with the lower signs:

$$N = \sum_k \frac{\lambda e^{-\beta\varepsilon_k}}{(1 - \lambda e^{-\beta\varepsilon_k})} \tag{10–39}$$

$$pV = -kT \sum_k \ln(1 - \lambda e^{-\beta\varepsilon_k}) \tag{10–40}$$

The energy values here are also given by Eq. (10–8).

In Section 10–1 we converted these summations to integrals, since for large V the ε_k's vary essentially continuously. The discussion following Eq. (10–10) states that such a procedure is valid only because the summands are finite continuous functions of λ for all ε_k. If we let ε_0 be the ground translational energy state, we see from Eq. (10–39) that λ is restricted to the values $0 \leq \lambda < e^{\beta\varepsilon_0}$. Thus if we wish our results to be valid for all values of λ, we must take care in converting the summations in Eqs. (10–39) and (10–40) to integrals. In particular, we must single out the ground state and write

$$N = \frac{\lambda e^{-\beta\varepsilon_0}}{1 - \lambda e^{-\beta\varepsilon_0}} + \sum_{k \neq 0} \frac{\lambda e^{-\beta\varepsilon_k}}{1 - \lambda e^{-\beta\varepsilon_k}} \tag{10–41}$$

for Eq. (10–39), where the $k \neq 0$ means a summation over all states other than the ground state. Now since $0 \leq \lambda < e^{\beta\varepsilon_0}$, and $\varepsilon_k > \varepsilon_0$ for all $k \neq 0$, the summation here can be converted to an integral by the same method that we used in Section 10–1. Equation (10–41) becomes

$$N = \frac{\lambda e^{-\beta\varepsilon_0}}{1 - \lambda e^{-\beta\varepsilon_0}} + 2\pi\left(\frac{2m}{h^2}\right)^{3/2} V \int_{\varepsilon > \varepsilon_0}^{\infty} \frac{\lambda \varepsilon^{1/2} e^{-\beta\varepsilon}\, d\varepsilon}{1 - \lambda e^{-\beta\varepsilon}} \tag{10–42}$$

According to Eq. (10–8), the ground state $\varepsilon_0 = 3h^2/8mV^{2/3}$, but we can always redefine the energy such that $\varepsilon_0 = 0$. Physical results must be independent of where one sets the zero of energy of a system, and the following equations are simpler if we choose the energy such that $\varepsilon_0 = 0$. Equation (10–42) becomes

$$\rho = \frac{N}{V} = 2\pi\left(\frac{2m}{h^2}\right)^{3/2} \int_{\varepsilon > 0}^{\infty} \frac{\lambda \varepsilon^{1/2} e^{-\beta\varepsilon}\, d\varepsilon}{1 - \lambda e^{-\beta\varepsilon}} + \frac{\lambda}{V(1 - \lambda)} \tag{10–43}$$

Similarly, Eq. (10–40) is

$$\frac{p}{kT} = -2\pi\left(\frac{2m}{h^2}\right)^{3/2} \int_{\varepsilon > 0}^{\infty} \varepsilon^{1/2} \ln(1 - \lambda e^{-\beta\varepsilon})\, d\varepsilon - \frac{1}{V} \ln(1 - \lambda) \tag{10–44}$$

where in both Eqs. (10–43) and (10–44), $0 \leq \lambda < 1$.

The second terms of both equations contain a factor of $1/V$. Ordinarily it is legitimate to ignore such terms, since we are always interested only in the thermodynamic limit, where $V \to \infty$. We must be careful here, however, since if $\lambda \to 1$, the term $\lambda/(1-\lambda)$ becomes very large, and $\lambda/(1-\lambda)$ divided by V is not necessarily negligible compared to the first term in Eq. (10–43). In this section we shall consider only the case where λ is close to zero, and so the $1/V$ terms in Eqs. (10–43) and (10–44) are not important. But in the next section we shall consider the strongly degenerate case in which $\lambda \to 1$, and we shall see that in this case the $1/V$ terms contribute in an important way to the thermodynamic functions.

The integrals occurring in Eqs. (10–43) and (10–44) can be evaluated as power series in λ in much the same way as Eqs. (10–9) and (10–10). The result is

$$\rho = \frac{1}{\Lambda^3} g_{3/2}(\lambda) \tag{10–45}$$

and

$$\frac{p}{kT} = \frac{1}{\Lambda^3} g_{5/2}(\lambda) \tag{10–46}$$

where

$$g_n(\lambda) = \sum_{l=1}^{\infty} \frac{\lambda^l}{l^n} \tag{10–47}$$

Equation (10–45) gives ρ as a power series in λ, and Eq. (10–46) gives p/kT as a power series in λ. We can derive an expression for p/kT as an expansion in the density by *reverting* Eq. (10–45) and substituting this result into Eq. (10–46):

$$\frac{p}{\rho kT} = 1 - \frac{\Lambda^3}{2^{5/2}} \rho + \cdots \tag{10–48}$$

The second virial coefficient in this case is negative, implying that the effective interaction between ideal bosons is attractive, in contrast to the case for fermions.

The Bose-Einstein equation for E, Eq. (10–4) with the negative sign, can be converted to an integral and then written as a series expansion in λ:

$$E = \tfrac{3}{2} VkT \frac{1}{\Lambda^3} g_{5/2}(\lambda)$$

In terms of ρ, this becomes

$$E = \tfrac{3}{2} NkT \left(1 - \frac{\Lambda^3}{2^{5/2}} \rho + \cdots \right) \tag{10–49}$$

All other thermodynamic functions for a weakly degenerate ideal gas of bosons follow in a similar way. Just as in Section 10–1, these virial expansion expressions for the thermodynamic functions are useful only for small values of λ or ρ and represent small quantum corrections to the limiting classical results. In the next section we shall discuss the region of strong degeneracy, where λ approaches its largest allowed value ($\lambda \to 1$), and the $1/V$ terms in Eqs. (10–43) and (10–44) cannot be ignored.

10–4 A STRONGLY DEGENERATE IDEAL BOSE-EINSTEIN GAS

We now consider the situation when λ is not necessarily small. Let us return to Eqs. (10–43) and (10–44) for the density and pressure of an ideal Bose-Einstein gas

$$\rho = 2\pi\left(\frac{2m}{h^2}\right)^{3/2}\int_0^\infty \frac{\lambda\varepsilon^{1/2}e^{-\varepsilon/kT}}{1-\lambda e^{-\varepsilon/kT}}\,d\varepsilon + \frac{\lambda}{V(1-\lambda)}$$

$$\frac{p}{kT} = -2\pi\left(\frac{2m}{h^2}\right)^{3/2}\int_0^\infty \varepsilon^{1/2}\ln(1-\lambda e^{-\varepsilon/kT})\,d\varepsilon - \frac{1}{V}\ln(1-\lambda)$$

or

$$\rho = \frac{1}{\Lambda^3}g_{3/2}(\lambda) + \frac{\lambda}{V(1-\lambda)} \tag{10–50}$$

$$\frac{p}{kT} = \frac{1}{\Lambda^3}g_{5/2}(\lambda) - \frac{1}{V}\ln(1-\lambda) \tag{10–51}$$

where $g_{3/2}(\lambda)$ and $g_{5/2}(\lambda)$ are defined by Eq. (10–47). Since λ can approach unity, we cannot ignore the $1/V$ terms in these equations.

Equation (10–3) shows that the average number of particles in their ground state is

$$\bar{n}_0 = \frac{\lambda}{1-\lambda}$$

and so it is clear that $0 \le \lambda < 1$. Note that there is no such restriction on the range of λ in the Fermi-Dirac case. There $0 \le \lambda < \infty$. The function $g_{3/2}(\lambda)$ in Eq. (10–50) is a bounded, positive, and monotonically increasing function of λ in the range $0 \le \lambda < 1$. At $\lambda = 1$ the first derivative diverges, but $g_{3/2}$ itself is finite since

$$g_{3/2}(1) = \sum_{l=1}^{\infty}\frac{1}{l^{3/2}} = \zeta(\tfrac{3}{2}) = 2.612\ldots \tag{10–52}$$

where $\zeta(n)$ is the Riemann zeta function, defined by (see Problem 1–63)

$$\zeta(n) = \sum_{l=1}^{\infty}\frac{1}{l^n} \tag{10–53}$$

The function $g_{3/2}(\lambda)$ is plotted in Fig. 10–3.

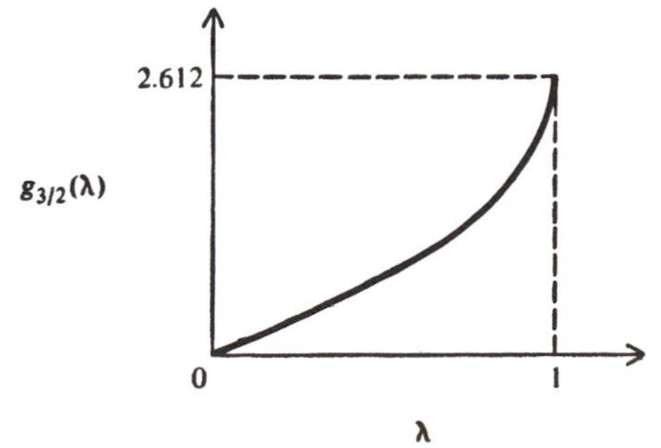

Figure 10–3. **The function $g_{3/2}(\lambda)$ versus λ.**

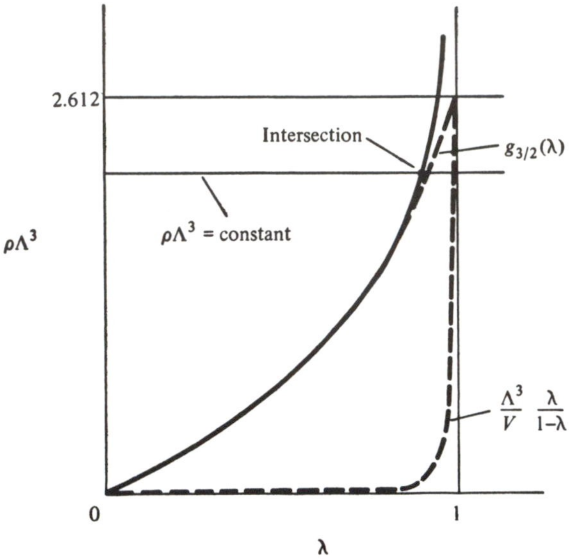

Figure 10–4. **The graphical solution of Eq. (10–50) for λ as a function of ρ and T. The dotted lines are $g_{3/2}(\lambda)$ and $\Lambda^3\lambda/V(1-\lambda)$. The solid lines are $\rho\Lambda^3$ = constant and $g_{3/2}(\lambda) + \Lambda^3\lambda/V(1-\lambda)$.**

In order to determine the equation of state, we must determine λ as a function of ρ and T by solving Eq. (10–50) for λ, and then substituting this into Eq. (10–51) for p/kT. Equation (10–50) cannot be solved analytically, but can be solved graphically by plotting both sides of the equation on the same graph and picking off the intersection. In particular, we shall plot $\rho\Lambda^3$ and $g_{3/2}(\lambda) + \Lambda^3\lambda/V(1-\lambda)$ versus λ. (See Fig. 10–4.) For fixed values of ρ and T, the function $\rho\Lambda^3$ is just a constant and so appears as a horizontal line. The function $g_{3/2}(\lambda) + \Lambda^3\lambda/V(1-\lambda)$, on the other hand, deserves some thought. Since V is large, the term $\Lambda^3\lambda/V(1-\lambda)$ is small for all values of λ except those where λ is very close to unity. In fact, this term is negligible compared to $g_{3/2}(\lambda)$ *except* when $\lambda = 1 - 0(1/V)$, say $1 - a/V$, where a is some positive number (remember that λ must be less than 1). Thus a graph of $g_{3/2}(\lambda) + \Lambda^3\lambda/V(1-\lambda)$ is indistinguishable from a graph of $g_{3/2}(\lambda)$ everywhere except where $\lambda \approx 1$. This sum $g_{3/2}(\lambda) + \Lambda^3\lambda/V(1-\lambda)$, along with its two contributions, is plotted in Fig. 10–4. The dashes are the two separate terms, and the solid line is their sum. Note that the value of λ given by the intersection of the curves $\rho\Lambda^3$ = constant and $g_{3/2}(\lambda) + \Lambda^3\lambda/V(1-\lambda)$, and the value of λ given by the intersection of $\rho\Lambda^3$ = constant and $g_{3/2}(\lambda)$ differ by only $0(1/V)$. This is so since $g_{3/2}(\lambda)$ and $g_{3/2}(\lambda) + \Lambda^3\lambda/V(1-\lambda)$ differ only when $\lambda = 1 - 0(1/V)$ in the first place.

Figure 10–5 shows the set of intersections obtained by varying $\rho\Lambda^3$, which may be done by varying the density at constant temperature or by varying the temperature at constant density. Note that above $\rho\Lambda^3 = 2.612$, λ is essentially equal to unity, being different from unity by $0(1/V)$. This may be immediately seen by noting that the solid curve in Fig. 10–4 lies very close to the line $\lambda = 1$ for values of $\rho\Lambda^3$ greater than 2.612. The point $\rho\Lambda^3 = 2.612$ represents some sort of a critical point as we shall see below.

We can summarize the results of Figs. 10–4 and 10–5 mathematically in the following way. We wish to solve Eq. (10–50) for λ as a function of ρ and T, that is, we wish to solve

$$\rho\Lambda^3 = g_{3/2}(\lambda) + \frac{\Lambda^3}{V}\frac{\lambda}{1-\lambda} \tag{10–54}$$

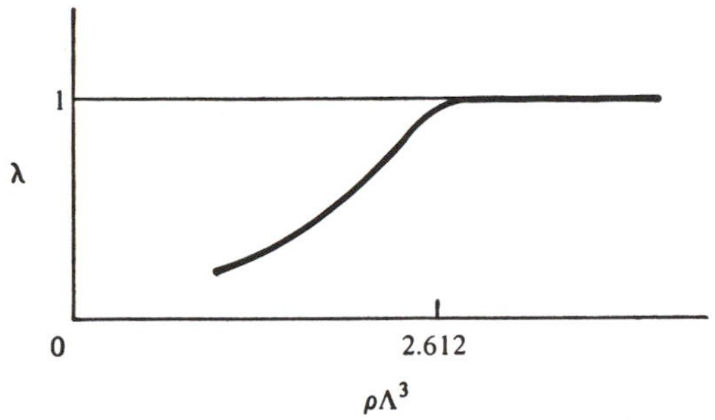

Figure 10–5. **A plot of λ versus $\Lambda\rho^3$, obtained from the locus of intercepts of plots such as the one in Figure 10–4.**

for λ. Figures 10–4 and 10–5 show that for values of $\rho\Lambda^3$ less than 2.612, λ is given to within $0(1/V)$ by the root of $g_{3/2}(\lambda) = \rho\Lambda^3 = \text{constant}$. For values of $\rho\Lambda^3$ greater than 2.612, on the other hand, $\lambda = 1 - 0(1/V) = 1 - a/V$, where a is some positive number. We can determine a by substituting $\lambda = 1 - a/V$ into Eq. (10–54) and solving for a:

$$a = \frac{\Lambda^3}{\rho\Lambda^3 - g_{3/2}(1)} \qquad (\rho\Lambda^3 > 2.612) \tag{10–55}$$

where we have written $g_{3/2}(1)$ for $g_{3/2}(1 - 0(1/V))$ since $g_{3/2}(\lambda)$ is a continuous function of λ. Thus we have

$$\begin{aligned} \lambda &= 1 - \frac{a}{V} \qquad \text{if } \rho\Lambda^3 > g_{3/2}(1) \\ &= \text{the root of } g_{3/2}(\lambda) = \rho\Lambda^3 \qquad \text{if } \rho\Lambda^3 < g_{3/2}(1) \end{aligned} \tag{10–56}$$

where a is given by Eq. (10–55). In the thermodynamic limit $V \to \infty$,

$$\begin{aligned} \lambda &= 1 \qquad \text{if } \rho\Lambda^3 > g_{3/2}(1) \\ &= \text{the root of } g_{3/2}(\lambda) = \rho\Lambda^3 \qquad \text{if } \rho\Lambda^3 < g_{3/2}(1) \end{aligned} \tag{10–57}$$

Clearly the point $\rho\Lambda^3 = g_{3/2}(1) = 2.612$ is a special point. To explore its physical significance, consider $\rho\Lambda^3 = \rho(h^2/2\pi mkT)^{3/2}$ to be a function of temperature for a fixed density.

At high temperatures such that $\rho\Lambda^3 < 2.612$, λ must be determined numerically from the equation $g_{3/2}(\lambda) = \rho\Lambda^3$. But at low temperatures such that $\rho\Lambda^3 > 2.612$, $\lambda = 1 - a/V$ where a is given by Eq. (10–55). The quantity $\lambda/(1 - \lambda)$ is the average number of particles in their ground state, and for temperatures such that $\rho\Lambda^3 > 2.612$, we have

$$\bar{n}_0 = \frac{\lambda}{1 - \lambda} = \frac{V}{a} = \frac{V}{\Lambda^3}(\rho\Lambda^3 - g_{3/2}(1)) \qquad (\rho\Lambda^3 > 2.612) \tag{10–58}$$

We can write this in a more instructive form by defining a temperature T_0 by

$$\rho\Lambda_0{}^3 = \rho\left(\frac{h^2}{2\pi mkT_0}\right)^{3/2} = g_{3/2}(1) \tag{10–59}$$

In terms of this T_0 then, Eq. (10–58) becomes

$$\frac{\bar{n}_0}{N} = 1 - \left(\frac{T}{T_0}\right)^{3/2} \qquad T < T_0 \tag{10–60}$$

For temperatures greater than T_0, the value of λ determined from $g_{3/2}(\lambda) = \rho\Lambda^3$ will not be in the neighborhood of $\lambda = 1$ (unless $T = T_0 + 0(1/V)$, which is, of course, of no practical interest). Let this value of λ be denoted by λ_0. Then $\bar{n}_0 = \lambda_0/(1 - \lambda_0)$ and so $\bar{n}_0/N$ would be vanishingly small. Thus, at temperatures above T_0,

$$\frac{\bar{n}_0}{N} = 0 \qquad T > T_0 \tag{10–61}$$

Equations (10–60) and (10–61) for $\bar{n}_0/N$ are plotted in Fig. 10–6.

It can be seen that when $T > T_0$, the fraction of molecules in their ground state is essentially zero. This is the normal situation, where the molecules are distributed smoothly over the many molecular quantum states available to each one. However, as the temperature is lowered past T_0, suddenly the ground state begins to be appreciably populated, and the population increases until at $T = 0$ all the molecules are in their ground state. The fact that one state (the ground state) out of the many available to each molecule starts to become greatly preferred abruptly at $T = T_0$ is analogous to an ordinary phase transition. This "condensation" of the molecules into their ground states is called *Bose-Einstein condensation.*

A similar result is found if the temperature is held fixed and the density is allowed to vary. There is a critical density, above which the ground states of the molecules are not preferred, and below which the molecules tend to populate their ground state. The analog of Eqs. (10–60) and (10–61) is (see Problem 10–36)

$$\begin{aligned} \frac{\bar{n}_0}{N} &= 1 - \frac{\rho_0}{\rho} \qquad && \rho > \rho_0 \\ &= 0 && \rho < \rho_0 \end{aligned} \tag{10–62}$$

To determine the properties of a Bose-Einstein condensation, consider the equation of state. From Eq. (10–51), the pressure is given by

$$\frac{p}{kT} = \frac{1}{\Lambda^3} g_{5/2}(\lambda) - \frac{1}{V} \ln(1 - \lambda) \tag{10–63}$$

We may neglect the logarithm term here, since its greatest value is achieved when $1 - \lambda = a/V$. In this case the second term is $(1/V)\ln(1/V)$, and this vanishes as $V \to \infty$ since $x \ln x \to 0$ as $x \to 0$. Thus Eq. (10–63) is simply

$$\frac{p}{kT} = \frac{1}{\Lambda^3} g_{5/2}(\lambda) \tag{10–64}$$

where λ is determined in terms of ρ and T by the arguments surrounding Figs. 10–4 and 10–5, which are summarized in Eq. (10–56).

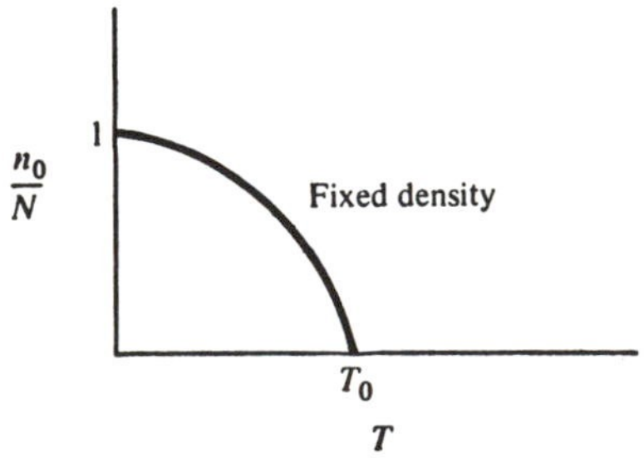

Figure **10–6. The fraction of particles in their ground state as a function of temperature.**

If we consider p/kT to be a function of the density for fixed temperature, we can write

$$\begin{aligned} \frac{p}{kT} &= \frac{1}{\Lambda^3} g_{5/2}(\lambda) \qquad \rho < \rho_0 \\ &= \frac{1}{\Lambda^3} g_{5/2}(1) \qquad \rho > \rho_0 \end{aligned} \tag{10-65}$$

where $g_{5/2}(1) = \zeta(\frac{5}{2}) = 1.342 \ldots$. The crucial point is that for $\rho > \rho_0$, p/kT is independent of the density and hence appears as a horizontal line when p/kT is plotted versus ρ. For $\rho < \rho_0$, on the other hand, p/kT is a function of ρ that can be determined numerically.

It is customary to plot isotherms of p versus v rather than p versus ρ, and these are shown in Fig. 10–7. Note that these isotherms are very similar to the isotherms observed for real gases.

The horizontal lines represent that region in which the system is a mixture of two phases. The points A and B correspond to the two phases in equilibrium: the condensed phase (A) and the dilute phase (B). The dilute phase has a specific volume v_0, and the condensed phase has a specific volume 0. The two phases have the same pressure, namely, the vapor pressure, which is given by

$$p_0(T) = \frac{kT}{\Lambda^3} g_{5/2}(1) \tag{10-66}$$

If we differentiate this and compare it to the Clapeyron equation,

$$\frac{dp}{dT} = \frac{\Delta H_{\text{cond}}}{T\,\Delta V} \tag{10-67}$$

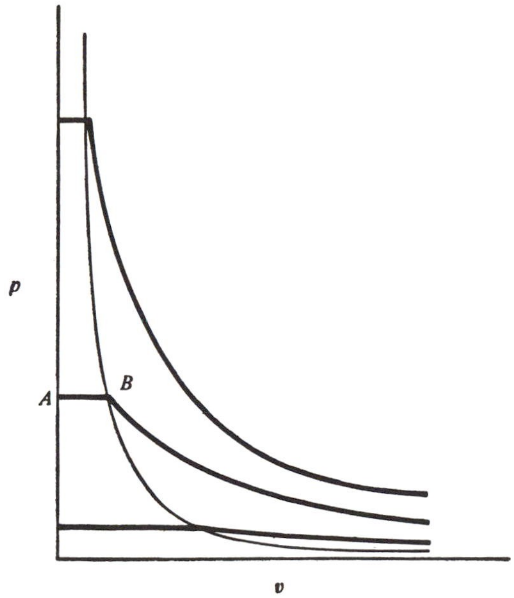

Figure 10–7. **The pressure-volume isotherms for an ideal Bose-Einstein gas. The points A and B correspond to the two phases in equilibrium.**

we see that there is a heat of transition associated with this process that is given by (see Problem 10–20)

$$\Delta H_{\text{cond}} = \tfrac{5}{2}kT \frac{g_{5/2}(1)}{g_{3/2}(1)} \tag{10–68}$$

Therefore the Bose-Einstein condensation is a first-order process. This is a very unusual first-order transition, however, since the condensed phase has no volume, and the system therefore has a uniform macroscopic density rather than the two different densities that are usually associated with first-order phase transitions. This is often interpreted by saying that the condensation occurs in momentum space rather than coordinate space, particularly since from a classical point of view, the particles in the condensed phase are found in the same region of momentum space, namely, zero momentum.

One observes similar behavior in the other thermodynamic functions. The heat capacity has a particularly interesting property. The thermodynamic energy is given by

$$E = \frac{3}{2}\frac{kTV}{\Lambda^3} g_{5/2}(\lambda) \tag{10–69}$$

which becomes

$$\begin{aligned} \frac{E}{N} &= \frac{3}{2}\frac{kTv}{\Lambda^3} g_{5/2}(\lambda) \qquad T > T_0 \\ &= \frac{3}{2}\frac{kTv}{\Lambda^3} g_{5/2}(1) \qquad T < T_0 \end{aligned} \tag{10–70}$$

We differentiate these with respect to T at constant N and V, we get (see Problem 10–19)

$$\begin{aligned} \frac{C_V}{Nk} &= \frac{15}{4}\frac{v}{\Lambda^3} g_{5/2}(\lambda) - \frac{9}{4}\frac{g_{3/2}(\lambda)}{g_{1/2}(\lambda)} \qquad T > T_0 \\ &= \frac{15}{4}\frac{v}{\Lambda^3} g_{5/2}(1) \qquad T < T_0 \end{aligned} \tag{10–71}$$

Again, λ must be determined numerically from Eq. (10–50), and the result of this is shown in Fig. 10–8(a), where C_V is plotted against T. There is no discontinuity in C_V at T_0, but there is a discontinuity in the slope there.

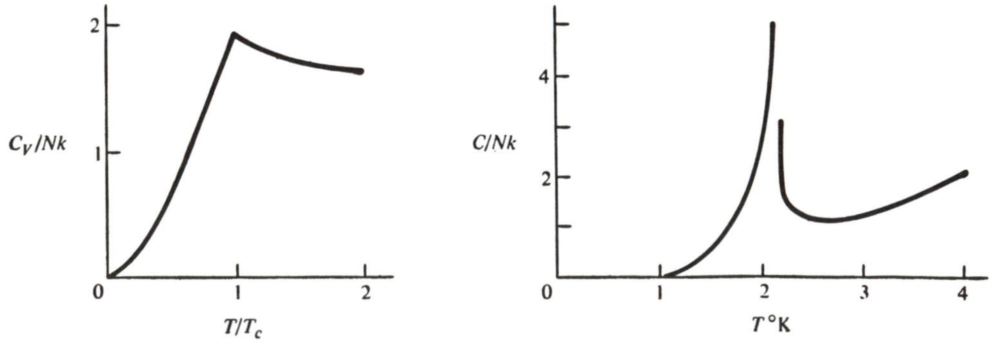

Figure **10–8. (a) The heat capacity of an ideal Bose-Einstein gas and (b) the experimental heat capacity of liquid helium under its saturated vapor.**

Bose-Einstein condensation is an interesting phenomenon since it occurs even though the particles do not interact with each other through an intermolecular potential. Indeed, however, there is an effective interaction through the symmetry requirement of the N-body wave function of the system, and it is this effective interaction that leads to the condensation. Even though the results derived here are valid only for an ideal gas, there is a real system to which they are approximately applicable. Helium exists in the form of two isotopes: He-3 and He-4. The thermodynamic properties of the pure isotopes have been extensively studied, and it turns out that He-4, which has a spin of zero and therefore must obey Bose-Einstein statistics, exhibits many remarkable properties, one of which being the heat capacity curve shown in Fig. 10–8(b). The similarity between this experimental curve and the Greek letter λ has led the transition to be referred to as a "lambda transition." Although the heat capacity appears to diverge logarithmically at $T = 2.18°K$, the similarity between it and the heat capacity curve of an ideal Bose-Einstein gas shown in Fig. 10–8(a) is striking. One cannot expect complete agreement since liquid He-4 is an inextricable combination of quantum statistics *and* intermolecular interactions, but it appears that the experimental heat capacity is due in part to the quantum statistics of the Bose-Einstein He-4 system. Furthermore, liquid He-3, which obeys Fermi-Dirac statistics, does not have any unusual behavior in its heat capacity curve, just as the heat capacity curve of an ideal Fermi-Dirac gas is "normal." Most intriguing, however, is that if we calculate the value of T_0 from Eq. (10–59) (using the density of liquid helium = 0.145 g/cm^3), we find that $T_0 = 3.14°K$, which is the right order of magnitude. One important difference between the λ-transition in He-4 and the Bose-Einstein condensation is that the λ-transition is not a first-order transition. Nevertheless, Bose-Einstein statistics seem to play an important role in the λ-transition in liquid He-4.

10–5 AN IDEAL GAS OF PHOTONS (BLACKBODY RADIATION)

In this section we shall apply statistical thermodynamics to electromagnetic radiation enclosed in a fixed volume V and at a fixed temperature T. The experimental system is obtained by making a cavity in any material, evacuating the cavity, and then heating the material to the temperature T. The atoms of the walls of the cavity constantly emit and absorb radiation, and so at equilibrium we will have a cavity filled with electromagnetic radiation. Such a cavity is called a blackbody cavity, and the radiation within the cavity is called blackbody radiation.

The quantum mechanical theory of electromagnetic radiation tells us that an electromagnetic wave may be regarded as a massless particle of spin angular momentum $\hbar = h/2\pi$ and with a momentum and energy that are functions of the wavelength. These massless particles are called photons. Since photons have a spin of 1 (in units of $\hbar$), they form an ideal Bose-Einstein gas. A new feature here is that since the walls of the blackbody cavity are constantly emitting and absorbing photons, the number of photons is not fixed at any instant, and thus N is not an independent thermodynamic variable. Thermodynamically, the blackbody cavity is described by V and T.

Let us briefly review some of the results of vibratory motion. In what follows, we shall consider only the "electro" part of an electromagnetic wave, since this is several orders of magnitude more important for our purposes. Consider a harmonic electromagnetic wave of unit amplitude traveling with velocity c in the positive x-direction. Mathematically, this is described by

$$E(x, t) = \sin\left[\frac{2\pi}{\lambda}(x - ct)\right] \tag{10–72}$$

and is called a traveling wave. The symbol λ is the wavelength of the wave. Physicists customarily write this equation in the form

$$E(x, t) = \sin(kx - \omega t) \tag{10–73}$$

where $k = 2\pi/\lambda$ and is called the wave vector, and $\omega = 2\pi\nu$. In some sense, this wave also describes a photon with energy and momentum given by

$$\varepsilon = h\nu = \hbar\omega = \hbar ck$$

$$\text{momentum} = \frac{h}{\lambda} = \hbar k \tag{10–74}$$

We shall consider blackbody radiation at equilibrium to be a system of standing waves set up in the cavity. It is not necessary to do this, since as the volume becomes large, the boundary conditions chosen have no effect on the thermodynamic properties, but imposing a boundary condition of standing waves is as convenient as any. Consider the superposition of two traveling harmonic waves of unit amplitude and traveling in opposite directions. This is given by

$$\begin{aligned}\phi(x, t) &= \sin(kx - \omega t) + \sin(kx + \omega t) \\ &= 2 \sin kx \cos \omega t\end{aligned} \tag{10–75}$$

This new wave does not move either backward or forward. It vanishes at the values of x for which $\sin kx = 0$ for all values of t. These points are called nodes. They occur at the values of x given by $kx = n\pi$, where $n = 1, 2, \ldots$. In between the nodes, the disturbance vibrates harmonically with time. Figure 10–9 shows the first few standing waves that can be set up between 0 and L. The appropriate boundary condition is that the standing wave be fixed at the end points 0 and L, which implies that the wave vector k be given by

$$k = \frac{n\pi}{L} \qquad n = 1, 2, \ldots \tag{10–76}$$

Note that this is equivalent to saying that there must be an integral number of half wavelengths between 0 and L, that is, $n(\lambda/2) = L$.

In order to discuss three-dimensional waves, it is convenient to use an exponential representation rather than a sine or a cosine. In one dimension, a wave of unit amplitude can be described by

$$E(x, t) = e^{i(kx - \omega t)} \tag{10–77}$$

Equation (10–73) is obtained by taking the imaginary part of this. This is also a harmonic wave traveling in the positive x-direction. A standing wave in this representation is given by the imaginary part of

$$\begin{aligned}E(x, t) &= e^{i(kx - \omega t)} + e^{i(kx + \omega t)} \\ &= 2e^{ikx} \cos \omega t\end{aligned} \tag{10–78}$$

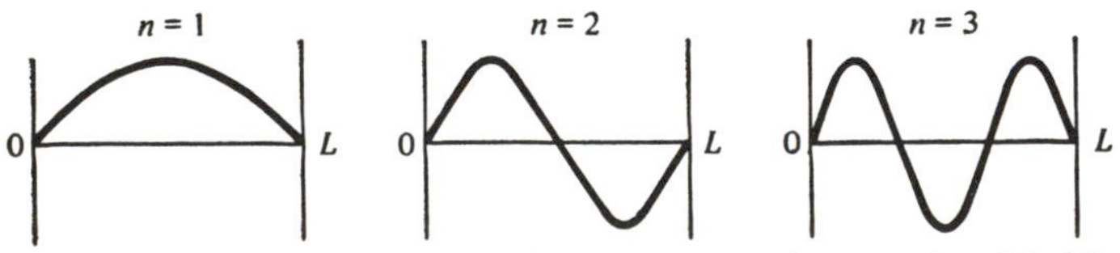

Figure 10–9. **The first three standing waves set up between 0 and L. The positions of the nodes do not move with time, but the wave vibrates harmonically up and down between the nodes.**

The boundary condition given in Eq. (10–76) is obtained from this by requiring that the imaginary part of Eq. (10–78) vanish at the endpoints 0 and L. Remember that the imaginary part of $e^{i\theta}$ vanishes whenever θ is an integral multiple of π.

In three dimensions, an electromagnetic wave is described by

$$\mathbf{E}(\mathbf{r}, t) = \boldsymbol{\sigma} e^{i(\mathbf{k}\cdot\mathbf{r} - \omega t)} \tag{10–79}$$

This represents a traveling wave propagated in the direction of the wave vector $\mathbf{k}$. The wavelength λ is given by $|\mathbf{k}| = 2\pi/\lambda$. The wave propagates in the direction of $\mathbf{k}$, but the direction of $\mathbf{E}$ itself is in the direction of $\boldsymbol{\sigma}$. The vector $\boldsymbol{\sigma}$ is called the polarization vector and is perpendicular to $\mathbf{k}$. Thus $\mathbf{E}$ represents a *transverse* wave, that is, one in which the disturbance vibrates in a direction perpendicular to the direction of propagation. In fact, $\mathbf{E}$ vibrates in the plane perpendicular to $\mathbf{k}$, and hence there are two and only two independent polarization vectors.

A standing wave can be formed from Eq. (10–79) (where it is understood that we take the imaginary part):

$$\mathbf{E}(\mathbf{r}, t) = 2\boldsymbol{\sigma} e^{i\mathbf{k}\cdot\mathbf{r}} \cos \omega t \tag{10–80}$$

If we let the components of the wave vector $\mathbf{k}$ be k_x, k_y, and k_z, the imaginary part of this vanishes when $k_x L = n_x \pi$, $k_y L = n_y \pi$, and $k_z L = n_z \pi$, or in vector notation, when

$$\mathbf{k} = \frac{\pi}{L}\mathbf{n} \qquad n_x, n_y, n_z = 1, 2, \ldots \tag{10–81}$$

We have assumed for simplicity here that the volume is a cube of length L. The energy and momentum are given by

$$\begin{aligned} \varepsilon &= \hbar c|\mathbf{k}| = \hbar ck \\ \text{momentum} &= \hbar|\mathbf{k}| = \hbar k \end{aligned} \tag{10–82}$$

Note that the energy depends only upon $|\mathbf{k}|$. We are going to need an expression for the number of standing waves with energy between ε and $\varepsilon + d\varepsilon$. This is found by the same method that we used to find the number of translational energy states between ε and $\varepsilon + d\varepsilon$ in Section 1–3. The square of $\mathbf{k}$ is

$$k^2 = \frac{\pi^2}{L^2}(n_x^2 + n_y^2 + n_z^2) \qquad n_x, n_y, n_z = 1, 2, \ldots \tag{10–83}$$

Thus the number of standing waves with the magnitude of the vector less than k is

$$\Phi(k) = \frac{\pi}{6}\left(\frac{Lk}{\pi}\right)^3 = \frac{L^3k^3}{6\pi^2} = \frac{Vk^3}{6\pi^2} \tag{10–84}$$

and the number between k and k and $k + dk$ is

$$\omega(k)\, dk = \frac{d\Phi}{dk}\, dk = \frac{Vk^2\, dk}{2\pi^2} \tag{10–85}$$

According to Eq. (10–82), the energy $\varepsilon = \hbar ck$. Furthermore, there are two polarizations with the energy ε, and so we can write

$$\omega(\varepsilon)\, d\varepsilon = \frac{V\varepsilon^2\, d\varepsilon}{\pi^2 c^3 \hbar^3} \tag{10–86}$$

for tne number of standing waves with energy between ε and $\varepsilon + d\varepsilon$.

The total energy of the system is given by

$$E(\{n_k\}) = \sum_k \varepsilon_k n_k \tag{10-87}$$

where the number of "particles" with energy ε_k (actually $\hbar ck$) is $n_k = 0, 1, 2, \ldots$ since photons are bosons. The partition function is

$$Q(V, T) = \sum_{\{n_k\}} e^{-\beta E(\{n_k\})} = \sum_{\{n_k\}} e^{-\beta \sum_k \varepsilon_k n_k} \tag{10-88}$$

This summation is similar to the one in Eq. (4–18), but in this case there is no restriction on the set $\{n_k\}$ since the number of photons is not conserved. This is why Q in Eq. (10–88) is a function of only V and T rather than the usual N, V, and T. The quantity N is not a thermodynamic variable in this case, since it is not fixed.

Without the restriction, a summation like the one in Eq. (10–88) is easy since

$$\begin{aligned} Q(V, T) &= \prod_k \left(\sum_{n=0}^{\infty} e^{-\beta \varepsilon_k n} \right) \\ &= \prod_k \frac{1}{1 - e^{-\beta \varepsilon_k}} \end{aligned} \tag{10-89}$$

All thermodynamic functions are expressed in terms of ln Q, which is

$$\ln Q = -\sum_k \ln(1 - e^{-\beta \varepsilon_k}) = -\sum_\varepsilon \ln(1 - e^{-\beta \varepsilon}) \tag{10-90}$$

As usual, this summation can be converted to an integral by introducing the density of states and treating ε_k to be a continuous variable. Thus

$$\ln Q = -\frac{V}{\pi^2 c^3 \hbar^3} \int_0^\infty \varepsilon^2 \ln(1 - e^{-\beta\varepsilon})\, d\varepsilon \tag{10-91}$$

This integral may be readily evaluated by expanding the logarithm and integrating term by term:

$$\begin{aligned} \ln Q &= \frac{V}{\pi^2 c^3 \hbar^3} \sum_{n=1}^{\infty} \frac{1}{n} \int_0^\infty \varepsilon^2 e^{-n\beta\varepsilon}\, d\varepsilon \\ &= \frac{V}{\pi^2 c^3 \hbar^3} \frac{2}{\beta^3} \sum_{n=1}^{\infty} \frac{1}{n^4} = \frac{2V}{\pi^2 (c\hbar\beta)^3} \zeta(4) \end{aligned} \tag{10-92}$$

where $\zeta(4)$ is the summation, which is a Riemann zeta function [*cf.* Eq. (10–52)] and is equal to $\pi^4/90$. (See Problem 1–63.)

The thermodynamic energy is given by

$$E = kT^2 \left(\frac{\partial \ln Q}{\partial T} \right)_V = \frac{\pi^2 V (kT)^4}{15 (\hbar c)^3} \tag{10-93}$$

This result can be used to derive the Stefan-Boltzmann law. In Problem 7–24 it is shown that the number of gas molecules striking a surface per unit area per unit time is $\rho\bar{v}/4$, where ρ is the number density, and $\bar{v}$ is the average velocity. By analogy, then, $cE(T)/4V$ is the energy incident per unit area per unit time on the wall of the enclosure containing the radiation. Thus if one cuts a small hole of unit area in the wall, the energy radiated per unit time

$$R = \frac{cE}{4V} = \frac{\pi^2 (kT)^4}{60 \hbar^3 c^2} \equiv \sigma T^4 \tag{10-94}$$

This is known as the Stefan-Boltzmann law, and σ is known as the Stefan-Boltzmann constant.

The pressure of the blackbody radiation is

$$p = kT\left(\frac{\partial \ln Q}{\partial V}\right)_T = \frac{2(kT)^4}{\pi^2(\hbar c)^3}\zeta(4) = \frac{\pi^2(kT)^4}{45(\hbar c)^3} \tag{10–95}$$

The pressure due to the radiation is negligible except for the highest temperatures.

It is instructive to go on and calculate the entropy and the chemical potential. The entropy is given by [*cf.* Eq. (2–33)]

$$\begin{aligned} S &= k \ln Q + kT\left(\frac{\partial \ln Q}{\partial T}\right)_V \\ &= \frac{4\pi^2 Vk(kT)^3}{45(\hbar c)^3} \end{aligned} \tag{10–96}$$

If we calculate $G = \overline{N}\mu$ from the equation $\overline{N}\mu = E - TS + pV$, we find that

$$\overline{N}\mu = 0$$

and since $\overline{N}$ is not zero, this implies that the chemical potential equals zero for an ideal gas of photons. We can prove on thermodynamic grounds that this is true for any system in which the number of particles is not conserved. In such a system we can write the chemical reaction $mA \rightleftharpoons nA$, where m and n are arbitrary integers. If we apply the criterion that the change in chemical potential $\Delta\mu = 0$ for a system in equilibrium, we find that $\Delta\mu = m\mu - n\mu = (m - n)\mu = 0$. But since $m - n$ does not equal zero, this implies that $\mu = 0$. Since $\lambda = \exp(\beta\mu)$, this also says that we could have derived all of the above results by setting $\lambda = 1$ in the Bose-Einstein formulas (see Problem 10–33).

Before leaving this section, we wish to calculate the energy density at each frequency. We can do this by noting that Eq. (10–90) is equivalent to

$$\ln Q = -\sum_{\omega} \ln(1 - e^{-\beta\hbar\omega})$$

where $\omega = \varepsilon/\hbar = ck$. If we use this directly in Eq. (10–93), we find that

$$E = \sum_{\omega} \frac{\hbar\omega e^{-\beta\hbar\omega}}{1 - e^{-\beta\hbar\omega}} \tag{10–97}$$

This can be converted to an integral over ω by introducing the number of states between ω and $\omega + d\omega$ from Eq. (10–86) and $\omega = \varepsilon/\hbar$. (It is unfortunate that the same notation is utilized for the density of states and the frequency, but since the density of states always occurs with an argument, such as $\omega(\varepsilon)$ or $\omega(k)$, there should be no confusion. We hesitate writing $\omega(\omega)$, however.) If we introduce the density of frequencies, Eq. (10–97) becomes

$$E = \frac{V\hbar}{\pi^2 c^3}\int_0^\infty \frac{\omega^3}{e^{\beta\hbar\omega} - 1}\, d\omega \tag{10–98}$$

The thermodynamic energy per unit volume can be written in terms of the energy density of each frequency by

$$\frac{E}{V} = \int_0^\infty \rho(\omega, T)\, d\omega$$

where

$$\rho(\omega, T)\, d\omega = \frac{\hbar}{\pi^2 c^3} \frac{\omega^3}{e^{\beta\hbar\omega} - 1}\, d\omega \tag{10–99}$$

This is the famous blackbody distribution law first derived by Planck in 1901. Clearly he did not derive this formula in the way that we have since his work preceded, and, in fact, led to the development of quantum mechanics. Problems 10–25 through 10–27 are involved with some well-known results of blackbody radiation.

10–6 THE DENSITY MATRIX

In the previous sections of this chapter, we have treated ideal Fermi-Dirac and Bose-Einstein gases by applying the results of Chapter 4. The basic equations there are derived on the basis of independent particles and so are not applicable when intermolecular forces are present. In this section we introduce a formalism of quantum statistical mechanics, which can be generalized to study nonideal quantum systems.

Let us return now to the completely general equations

$$Q = \sum_j e^{-\beta E_j} \tag{10–100}$$

$$\overline{M} = \frac{1}{Q} \sum_j M_j e^{-\beta E_j} \tag{10–101}$$

where $\overline{M}$ is the ensemble average of the mechanical property, and M_j is the quantum mechanical expectation value of the operator $\hat{M}$ in the jth quantum state.

We wish to express Q and $\overline{M}$ in terms of the quantum mechanical operators $\mathscr{H}$ and $\hat{M}$. Let $\{\psi_j\}$ be the set of normalized eigenfunctions of $\mathscr{H}$ and $\{E_j\}$ be the corresponding eigenvalues, that is,

$$\mathscr{H}\psi_j = E_j \psi_j \tag{10–102}$$

Since ψ_j is an eigenfunction of $\mathscr{H}$, we can also write

$$\mathscr{H}^n \psi_j = E_j^n \psi_j \tag{10–103}$$

for integral n. This allows us to define the result of an analytic function of $\mathscr{H}$ acting on ψ_j. In particular, we have

$$\begin{aligned} e^{-\beta\mathscr{H}}\psi_j &= \left(\sum_{n=0}^{\infty} \frac{(-\beta)^n}{n!} \mathscr{H}^n \right) \psi_j = \sum_{n=0}^{\infty} \frac{(-\beta)^n}{n!} \mathscr{H}^n \psi_j \\ &= \sum_{n=0}^{\infty} \frac{(-\beta)^n}{n!} E_j^n \psi_j = e^{-\beta E_j}\psi_j \end{aligned} \tag{10–104}$$

(In general, functions of operators are defined through their MacLaurin expansions.) Thus we have

$$e^{-\beta\mathscr{H}}\psi_j = e^{-\beta E_j}\psi_j \tag{10–105}$$

We multiply both sides of this equation by ψ_j^* and integrate over all the coordinates involved to get

$$e^{-\beta E_j} = \int \psi_j^* e^{-\beta\mathscr{H}} \psi_j \, d\tau \tag{10–106}$$

where we have used the fact that $\exp(-\beta E_j)$ is just a number and that ψ_j is normalized. The symbol $d\tau$ represents an integration over all the coordinates on which ψ_j depends. Using Eq. (10–106) then, Q becomes

$$Q = \sum_j e^{-\beta E_j} = \sum_j \int \psi_j{}^* e^{-\beta \mathscr{H}} \psi_j \, d\tau \tag{10–107}$$

Integrals of quantum mechanical operators, such as $\exp(-\beta\mathscr{H})$ above, between wave functions occur naturally in the matrix formulation of quantum mechanics and are called matrix elements. They can be represented by the notation

$$\int \psi_i{}^* e^{-\beta \mathscr{H}} \psi_j \, d\tau = (e^{-\beta \mathscr{H}})_{ij} \tag{10–108}$$

In this notation, the canonical partition function is

$$Q = \sum_j (e^{-\beta \mathscr{H}})_{jj} = \mathrm{Tr}(e^{-\beta \mathscr{H}}) \tag{10–109}$$

where we have employed the notation of representing a summation over the diagonal elements of a matrix by Tr, which stands for "trace."

Now there is a standard theorem of matrix algebra which says that the trace of a matrix is independent of the particular function ψ_i in Eq. (10–108) used to calculate the matrix elements. This theorem is easy to prove. Let $\{\psi_j\}$ be the orthonormal set of eigenfunctions of $\mathscr{H}$ and let $\{\phi_n\}$ be any orthonormal set of functions that can be expanded in terms of the ψ_j as

$$\phi_j = \sum_n a_{jn} \psi_n \tag{10–110}$$

where the a_{jn}'s are constant. We can calculate the a_{jn}'s by multiplying both sides of eq. (10–110) by $\psi_k{}^*$, integrating over all values of the coordinates, and using the fact that the ψ_j's are orthonormal:

$$a_{jn} = \int \psi_n{}^* \phi_j \, d\tau \tag{10–111}$$

Furthermore, since the ϕ_j's are normalized, we have that the a_{jn} must satisfy the conditions (see Problem 10–34)

$$\sum_n a_{jn}{}^* a_{jn} = 1 \tag{10–112}$$

We shall also need the coefficients for the expansion of ψ_j in terms of the ϕ's. Write

$$\psi_s = \sum_t b_{st} \phi_t \tag{10–113}$$

The b_{st} can be related to the a_{jn} by multiplying both sides of this equation by $\phi_t{}^*$ and using the fact that the ϕ_t's are orthonormal:

$$b_{st} = \int \phi_t{}^* \psi_s \, d\tau = a_{ts}{}^* \tag{10–114}$$

where we have used Eq. (10–111) to write the last equality. Thus b_{ij}'s are obtained from the a_{ij}'s by reversing the subscripts and taking the complex conjugate. Since the ψ's are normalized, we have

$$\sum_n b_{jn}{}^* b_{jn} = \sum_n a_{nj}{}^* a_{nj} = 1 \tag{10–115}$$

Note that this equation is similar to Eq. (10–112). Here we sum over the first subscript, and there we sum over the second. We shall need this relation to prove the theorem referred to above, namely, that

$$Q = \sum_j \int \psi_j^* e^{-\beta\mathscr{H}} \psi_j \, d\tau = \sum_j \int \phi_j^* e^{-\beta\mathscr{H}} \phi_j \, d\tau \tag{10–116}$$

To prove this, we substitute Eq. (10–110) into Eq. (10–116) to get

$$\int \phi_j^* e^{-\beta\mathscr{H}} \phi_j \, d\tau = \sum_{m,n} a_{jm}^* a_{jn} e^{-\beta E_n} \int \psi_m^* \psi_n \, d\tau \tag{10–117}$$

But the integral is just a Kroenecker delta δ_{mn} since the set $\{\psi_j\}$ is orthonormal. Thus the double summation over m and n becomes a single summation, and we have

$$\int \phi_j^* e^{-\beta\mathscr{H}} \phi_j \, d\tau = \sum_n a_{jn}^* a_{jn} e^{-\beta E_n} \tag{10–118}$$

We now sum both sides over j and use Eq. (10–115)

$$\sum_j \int \phi_j^* e^{-\beta\mathscr{H}} \phi_j \, d\tau = \sum_n e^{-\beta E_n} = \sum_j e^{-\beta E_j} = Q \tag{10–119}$$

which shows that the trace of $\exp(-\beta\mathscr{H})$ is independent of the particular orthonormal set of functions used to compute the matrix elements.

Equation (10–109) gives Q in terms of the quantum mechanical operator $\mathscr{H}$. We now wish to express Eq. (10–101) for $\overline{M}$ in terms of quantum mechanical operators. The ensemble average of M is

$$\overline{M} = \frac{\sum_j M_j e^{-\beta E_j}}{\sum_j e^{-\beta E_j}} \tag{10–120}$$

We first use the fact that M_j is given by

$$M_j = \int \psi_j^* \hat{M} \psi_j \, d\tau \tag{10–121}$$

where $\hat{M}$ denotes the quantum mechanical operator corresponding to M, and ψ_j is the eigenfunction of $\mathscr{H}$ in the jth state. We substitute this into the numerator of Eq. (10–120) and perform a series of elementary manipulations:

$$\begin{aligned} \sum_j M_j e^{-\beta E_j} &= \sum_j e^{-\beta E_j} \int \psi_j^* \hat{M} \psi_j \, d\tau \\ &= \sum_j \int \psi_j^* \hat{M} e^{-\beta E_j} \psi_j \, d\tau = \sum_j \int \psi_j^* \hat{M} e^{-\beta\mathscr{H}} \psi_j \, d\tau \\ &= \sum_j (\hat{M} e^{-\beta\mathscr{H}})_{jj} = \mathrm{Tr}(\hat{M} e^{-\beta\mathscr{H}}) \end{aligned} \tag{10–122}$$

The invariance of the trace with respect to the functions used to calculate the matrix elements says that any convenient orthonormal set of functions could have been used to derive this result.

Using this result for the numerator in Eq. (10–120) gives

$$\overline{M} = \frac{\mathrm{Tr}(\hat{M} e^{-\beta\mathscr{H}})}{\mathrm{Tr}(e^{-\beta\mathscr{H}})} \tag{10–123}$$

The denominator of this expression is a scalar, and hence we can define a new operator $\hat{\rho}$ by

$$\hat{\rho} = \frac{e^{-\beta\mathscr{H}}}{\mathrm{Tr}(e^{-\beta\mathscr{H}})} \tag{10–124}$$

and write for Eq. (10–123)

$$\overline{M} = \mathrm{Tr}(\hat{M}\hat{\rho}) \tag{10–125}$$

The operator $\hat{\rho}$ is the quantum mechanical analog of the equilibrium density of points in the phase space for an canonical ensemble. The matrix corresponding to $\hat{\rho}$ is called the *density matrix*. Equation (10–123) corresponds to the classical expression

$$\overline{M} = \frac{\int \cdots \int dp\, dq M(p, q) e^{-\beta\mathscr{H}(p,q)}}{\int \cdots \int dp\, dq e^{-\beta\mathscr{H}(p,q)}} \tag{10–126}$$

One can prove that the trace operation in Eq. (10–123) goes into the phase space integration in Eq. (10–126) as $h \to 0$.

We have derived the above equations and defined the density matrix ρ only for a canonical ensemble. The density matrix can be defined in a much more general manner, but since we shall need only the results for a canonical ensemble in this book we simply refer to Tolman (see "Additional Reading"), who gives a complete discussion of the density matrix. Tolman shows that ρ can be defined for nonequilibrium systems, giving the quantum mechanical correspondence to the density of phase points in phase space. There is, for example, a quantum mechanical analog of the Liouville equation.

10–7 THE CLASSICAL LIMIT FROM THE QUANTUM MECHANICAL EXPRESSION FOR Q

In this section we shall derive the classical mechanical form of the canonical partition function (an integration over phase space) directly from its quantum mechanical form (a summation over energy states). In other words, we shall show that

$$Q = \sum_j e^{-\beta E_j} \xrightarrow{\hbar \to 0} \frac{1}{N!\,h^{3N}} \int \cdots \int e^{-\beta H}\, d\mathbf{p}_1 \cdots d\mathbf{p}_N\, d\mathbf{q}_1 \cdots d\mathbf{q}_N$$

In Chapter 7 we presented an argument that was meant to make this correspondence plausible, but here we shall present a rigorous treatment due originally to Kirkwood.* In particular, we shall show how to derive an expansion for Q in powers of $\hbar$, with the leading term being the classical limit. Such expansions were derived in Sections 10–1 and 10–3, but there the results were valid only for ideal systems. Here we shall relax this restriction. The quantum mechanical level of this section is somewhat higher than most of the others in this book, and this section can be omitted on first reading, since the material presented here is not necessary in the development of later sections. For those who do choose to read this section, however, Appendix B presents a brief discussion of Fourier transforms, delta functions, and so on.

We start with Eq. (10–116) for a system of N monatomic particles, namely,

$$Q = \sum_j \int \phi_j^* e^{-\beta\mathscr{H}} \phi_j\, d\mathbf{r} \tag{10–127}$$

* J. G. Kirkwood, *Phys. Rev.*, **44** p. 31, 1933; **45**, p. 116, 1934. See also Hill in "Additional Reading."

where we have written $d\mathbf{r} = dx_1\, dy_1 \cdots dz_N$ for $d\tau$. The functions $\phi_j(\mathbf{r}_1, \ldots, \mathbf{r}_N)$ in this case are taken to be the eigenfunctions of the Hamiltonian operator and form a complete orthonormal set. The Hamiltonian operator $\mathscr{H}$ is

$$\mathscr{H} = \mathscr{K} + U = -\frac{\hbar^2}{2m}\sum_{l=1}^{N} \nabla_l^{\,2} + U(\mathbf{r}_1, \ldots, \mathbf{r}_N)$$

We wish to transform Eq. (10–127) into an integration over phase space. To accomplish this, we introduce the eigenfunctions $u(\mathbf{p}_1, \ldots, \mathbf{r}_N)$ of the momentum operator $-i\hbar\nabla$:

$$u(\mathbf{p}_1, \ldots, \mathbf{r}_N) = \exp\left[\frac{i}{\hbar}\sum_{k=1}^{N} \mathbf{p}_k \cdot \mathbf{r}_k\right] \tag{10–128}$$

Note that $-i\hbar\nabla_i u = \mathbf{p}_i u$. We now expand the ϕ_j in Eq. (10–127) in terms of the momentum eigenfunctions

$$\phi_j(\mathbf{r}_1, \ldots, \mathbf{r}_N) = \int \cdots \int A_j(\mathbf{p}_1, \ldots, \mathbf{p}_N)\exp\left[\frac{i}{\hbar}\sum_{k=1}^{N} \mathbf{p}_k \cdot \mathbf{r}_k\right] d\mathbf{p}_1 \cdots d\mathbf{p}_N \tag{10–129}$$

The ϕ_j in Eq. (10–127), being eigenfunctions of $\mathscr{H}$, are either symmetric or antisymmetric in the coordinates of the N particles. Thus they should be expanded in terms of linear combinations of the $u(\mathbf{p}_1, \ldots, \mathbf{r}_N)$ in Eq. (10–128) that themselves are symmetric or antisymmetric in the coordinates of the N particles. We could do this by introducing a permutation operator $\mathscr{P}$, but the resulting equations become fairly involved. For simplicity, we shall ignore this complication and simply use Eq. (10–129). The result of this is that we shall not derive the factor of $N!$ that occurs in the classical partition function. Thus the central result of this section, Eq. (10–133) with Eqs. (10–135), (10–138), through (10–140), is lacking an $N!$ in its denominator. This factor is included, however, in Kirkwood's original paper and in the more pedagogical discussion of Hill.

Equation (10–129) shows $\phi_j(\mathbf{r}_1, \ldots, \mathbf{r}_N)$ as a Fourier transform of A_j, and so we can use the inversion theorem of Fourier transforms to write

$$A_j(\mathbf{p}_1, \ldots, \mathbf{p}_N) = \frac{1}{(2\pi\hbar)^{3N}}\int \cdots \int \phi_j(\mathbf{r}_1, \ldots, \mathbf{r}_N)\exp\left[-\frac{i}{\hbar}\sum_{k=1}^{N} \mathbf{p}_k \cdot \mathbf{r}_k\right] d\mathbf{r}_1 \cdots d\mathbf{r}_N \tag{10–130}$$

We now substitute Eq. (10–129) into Eq. (10–127) to get

$$Q = \sum_j \int \cdots \int \phi_j^*(\mathbf{r}_1, \ldots, \mathbf{r}_N) A_j(\mathbf{p}_1, \ldots, \mathbf{p}_N) e^{-\beta\mathscr{H}} \exp\left[\frac{i}{\hbar}\sum_{k=1}^{N} \mathbf{p}_k \cdot \mathbf{r}_k\right] d\mathbf{p}_1 \cdots d\mathbf{r}_N$$

and then use Eq. (10–130) for A_j (with $\mathbf{r}_1', \ldots, \mathbf{r}_N'$ replacing $\mathbf{r}_1, \ldots, \mathbf{r}_N$ as the variables of integration) to give

$$Q = \frac{1}{h^{3N}}\int \cdots \int \left\{\sum_j \phi_j^*(\mathbf{r}_1, \ldots, \mathbf{r}_N)\phi_j(\mathbf{r}_1', \ldots, \mathbf{r}_N')\right\}\exp\left[-\frac{i}{\hbar}\sum_k \mathbf{p}_k \cdot \mathbf{r}_k'\right]$$
$$\times \exp(-\beta\mathscr{H})\exp\left[\frac{i}{\hbar}\sum_k \mathbf{p}_k \cdot \mathbf{r}_k\right] d\mathbf{p}_1 \cdots d\mathbf{r}_1 \cdots d\mathbf{r}_1' \cdots d\mathbf{r}_N'$$

The summation over j in the braces is $\delta(\mathbf{r}_1 - \mathbf{r}_1', \ldots, \mathbf{r}_N - \mathbf{r}_N')$ (*cf.* Problem B–7), and so the integration over the primed variables simply gives

$$Q = \frac{1}{h^{3N}} \int \cdots \int \exp\left[-\frac{i}{\hbar}\sum_k \mathbf{p}_k \cdot \mathbf{r}_k\right] \exp(-\beta\mathscr{H}) \exp\left[\frac{i}{\hbar}\sum_k \mathbf{p}_k \cdot \mathbf{r}_k\right] d\mathbf{p}_1 \cdots d\mathbf{r}_N \tag{10–131}$$

Note that $\mathscr{H}$ contains the operators ∇_l^2 and so does not commute with

$$\exp\left[\pm\frac{i}{\hbar}\sum_k \mathbf{p}_k \cdot \mathbf{r}_k\right]$$

Thus the order of the terms in the integrand of Eq. (10–131) is important.

Equation (10–131) is in the desired form of an integral over phase space, but the integrand contains the quantum mechanical operator $\exp(-\beta\mathscr{H})$ instead of the classical function $\exp(-\beta H)$. Kirkwood defines a function $w(\mathbf{p}_1, \ldots, \mathbf{r}_N, \beta)$ by the relation

$$\begin{aligned}\exp(-\beta\mathscr{H}) \exp\left[\frac{i}{\hbar}\sum_k \mathbf{p}_k \cdot \mathbf{r}_k\right] &= \exp(-\beta H) \exp\left[\frac{i}{\hbar}\sum_k \mathbf{p}_k \cdot \mathbf{r}_k\right] w(\mathbf{p}_1, \ldots, \mathbf{r}_N, \beta) \\ &= F(\mathbf{p}_1, \ldots, \mathbf{r}_N, \beta)\end{aligned} \tag{10–132}$$

Note that this function $w(\mathbf{p}_1, \ldots, \mathbf{r}_N, \beta)$ has been defined such that Eq. (10–131) becomes

$$Q = \frac{1}{h^{3N}} \int \cdots \int \exp(-\beta H) w(\mathbf{p}_1, \ldots, \mathbf{r}_N, \beta)\, d\mathbf{p}_1 \cdots d\mathbf{r}_N \tag{10–133}$$

Remember that this expression for Q is missing a factor of $N!$ in the denominator since we have chosen to ignore the symmetry properties of the ϕ_j in Eq. (10–127). This equation for Q shows that the quantum corrections to the classical partition function lie in the function $w(\mathbf{p}_1, \ldots, \mathbf{r}_N, \beta)$, and so we must now investigate this function. In particular, we wish to show that $w(\mathbf{p}_1, \ldots, \mathbf{r}_N, \beta) \to 1$ as $\hbar \to 0$.

It is possible to evaluate w by carrying out the operation on the left-hand side of Eq. (10–132) and comparing the result to the right-hand side, but this turns out to be an extremely tedious route. A more convenient way is to differentiate Eq. (10–132) with respect to β to get

$$\begin{aligned}\frac{\partial F}{\partial \beta} &= \frac{\partial}{\partial \beta} \exp(-\beta\mathscr{H}) \exp\left[\frac{i}{\hbar}\sum_k \mathbf{p}_k \cdot \mathbf{r}_k\right] \\ &= \frac{\partial}{\partial \beta}\left(1 - \beta\mathscr{H} + \frac{\beta^2}{2}\mathscr{H}^2 + \cdots\right) \exp\left[\frac{i}{\hbar}\sum_k \mathbf{p}_k \cdot \mathbf{r}_k\right] \\ &= -\mathscr{H}F\end{aligned} \tag{10–134}$$

This differential equation is called a Bloch differential equation and in this case has the boundary condition

$$F(\beta = 0) = \exp\left[\frac{i}{\hbar}\sum_k \mathbf{p}_k \cdot \mathbf{r}_k\right]$$

It is not possible to solve this equation in general (note its similarity with the Schrödinger equation), but it is fairly straightforward to determine the first few coefficients in an expansion of F in powers of $\hbar$. We do this by actually expanding not F but w according to

$$w(\mathbf{p}_1, \ldots, \mathbf{r}_N, \beta) = \sum_{l=0}^{\infty} \hbar^l w_l(\mathbf{p}_1, \ldots, \mathbf{r}_N, \beta) \tag{10–135}$$

This equation defines the functions w_l. We now substitute this expansion into the right-hand side of Eq. (10–132) and that result into Eq. (10–134), and after some amount of work and cancellation of $\exp(-\beta H)$ and

$$\exp\left(\frac{i}{\hbar}\sum_k \mathbf{p}_k \cdot \mathbf{r}_k\right)$$

we get

$$\left\{\frac{\partial w_0}{\partial \beta} + \frac{\hbar\, \partial w_1}{\partial \beta} + O(\hbar^2)\right\} - H\{w_0 + \hbar w_1 + O(\hbar^2)\} = -U\{w_0 + \hbar w_1 + O(\hbar^2)\}$$
$$- K\{w_0 + \hbar w_1 + O(\hbar^2)\} + \frac{i\hbar}{m}\left\{\sum_{i=1}^{N} \mathbf{p}_i \cdot \nabla_i w_0 - \beta w_0 \sum_{i=1}^{N} \mathbf{p}_i \cdot \nabla_i U\right\} + O(\hbar^2) \tag{10–136}$$

where K is the sum of the kinetic energies of all the particles. To arrive at this result, we have used the following relations:

$$\nabla e^{-\beta H} = e^{-\beta K}\, \nabla e^{-\beta U} = -\beta e^{-\beta H}\, \nabla U$$
$$\nabla^2 e^{-\beta H} = -\beta e^{-\beta H}\, \nabla^2 U + \beta^2 e^{-\beta H} (\nabla U)^2$$
$$\nabla_j \exp\left[\frac{i}{\hbar}\sum_k \mathbf{p}_k \cdot \mathbf{r}_k\right] = \frac{i}{\hbar}\mathbf{p}_j \exp\left[\frac{i}{\hbar}\sum_k \mathbf{p}_k \cdot \mathbf{r}_k\right]$$
$$\nabla_j{}^2 \exp\left[\frac{i}{\hbar}\sum_k \mathbf{p}_k \cdot \mathbf{r}_k\right] = -\frac{2m}{\hbar^2} K_j \exp\left[\frac{i}{\hbar}\sum_k \mathbf{p}_k \cdot \mathbf{r}_k\right]$$

where K_j is the kinetic energy of the jth particle.

The coefficient of $\hbar$ to the zero power in Eq. (10–136) gives $\partial w_0/\partial\beta = 0$ or $w_0 =$ constant. The value of the "constant" can be found from the boundary condition

$$F(\beta = 0) = \exp\left[\frac{i}{\hbar}\sum_k \mathbf{p}_k \cdot \mathbf{r}_k\right]$$

From the defining equation for $F(\beta)$, that is, Eq. (10–132), we see that this is equivalent to the condition

$$w(\mathbf{p}_1, \ldots, \mathbf{r}_N, \beta = 0) = 1$$

and so Eq. (10–135) reads

$$1 = w_0(\mathbf{p}_1, \ldots, \mathbf{r}_N, \beta = 0) + \hbar w_1(\mathbf{p}_1, \ldots, \mathbf{r}_N, \beta = 0) + \hbar^2 w^2(\mathbf{p}_1, \ldots, \mathbf{r}_N, \beta = 0) + \cdots$$

Since the w_l are independent of $\hbar$, this implies that

$$\begin{aligned} w_0(\mathbf{p}_1, \ldots, \mathbf{r}_N, \beta = 0) &= 1 \\ w_l(\mathbf{p}_1, \ldots, \mathbf{r}_N, \beta = 0) &= 0 \qquad l \geq 2 \end{aligned} \tag{10–137}$$

But we have seen that w_0 is a constant independent of β, and so we have generally that

$$w_0 = 1 \tag{10–138}$$

and a formal proof that the quantum mechanical partition function goes over into the classical limit according to Eq. (7–9). Note that the classical limit is obtained in two ways. The more obvious one is to let $\hbar \to 0$, but it is also obtained from the limit $\beta = 0$ [*cf.* Eqs. (10–137)].

We can calculate the first quantum correction to $w(\mathbf{p}_1, \ldots, \mathbf{r}_N, \beta)$ by comparing the coefficients of $\hbar$ to the first power in Eq. (10–136), namely,

$$\frac{\partial w_1}{\partial \beta} = -\frac{i\beta}{m} \sum_{j=1}^{N} \mathbf{p}_j \cdot \nabla_j U$$

which upon integration gives

$$w_1 = -\frac{i\beta^2}{2m} \sum_{j=1}^{N} \mathbf{p}_j \cdot \nabla_j U \qquad (10\text{–}139)$$

Since this term is odd in the momenta, its contribution to Q according to Eq. (10–133) will vanish. (See Problem 10–35.) The contribution from w_2 does not vanish, however, and represents the first correction to Q. The evaluation of w_2 follows along the same lines as the evaluation for w_0 and w_1, but is quite a bit more lengthy. The result is

$$w_2 = -\frac{1}{2m}\left\{\frac{\beta^2}{2}\nabla^2 U - \frac{\beta^3}{3}\left[(\nabla U)^2 + \frac{1}{m}(\mathbf{p}\cdot\nabla)^2 U\right] + \frac{\beta^4}{4m}(\mathbf{p}\cdot\nabla U)^2\right\} \qquad (10\text{–}140)$$

where we have used the abbreviated notation

$$\mathbf{p}\cdot\mathbf{a} \equiv \sum_{j=1}^{N} \mathbf{p}_j \cdot \mathbf{a}_j$$

In Chapter 15, Eq. (10–140) is used to calculate the first quantum correction to the thermodynamic properties of imperfect gases.

ADDITIONAL READING

General

HUANG, K. 1963. *Statistical mechanics*. New York: Wiley. Chapters 9, 11, and 12.
ISIHARA, A. 1971. *Statistical physics*. New York: Academic. Chapter 4.
KESTIN, J., and DORFMAN, J. R. 1971. *A course in statistical thermodynamics*. New York: Academic. Chapter 8.
KITTEL, C. 1969. *Thermal physics*. New York: Wiley. Chapters 14, 15, and 17.
KUBO, R. 1965. *Statistical mechanics*. Amsterdam: North-Holland Publishing Co. Chapter 4.
LANDAU, L. D., and LIFSHITZ, E. M. 1958. *Statistical physics*. Oxford: Pergamon. Chapter 5.
MAYER, J. E., and MAYER, M. G. 1940. *Statistical mechanics*. New York: Wiley. Chapter 16.
MÜNSTER, A. 1969. *Statistical thermodynamics*, Vol. I. Berlin: Springer-Verlag. Chapter 2.
SCHRÖDINGER, E. 1952. *Statistical thermodynamics*. Cambridge: Cambridge University Press. Chapter 8.
TER HAAR, D. 1966. *Elements of thermostatics*. London: Oxford University Press. Chapter 4.
TOLMAN, R. C. 1938. *Statistical mechanics*. London: Oxford University Press. Chapter 10.

Fermi-Dirac statistics (electrons in metals)

FOWLER, R. H., and GUGGENHEIM, E. A. 1956. *Statistical thermodynamics*. (Cambridge: Cambridge University Press. Chapter 11.
KITTEL, C. 1967. *Solid state physics*, 3rd ed. New York: Wiley. Chapter 7.
REIF, F. 1965. *Statistical and thermal physics*. New York: McGraw-Hill Book Co. Sections 9–16 and 9–17.

Bose-Einstein statistics

LONDON, F. 1954. *Superfluids* Vol. II. New York: Wiley.

Blackbody radiation

DAVIDSON, N. 1962. *Statistical mechanics.* New York: McGraw-Hill. Chapter 12.
EYRING, H., HENDERSON, D., STOVER, B. J., and EYRING, E. M. 1964. *Statistical mechanics and dynamics.* New York: Wiley. Chapters 5 and 7.
KESTIN, J., and DORFMAN, J. R. 1971. *A course in statistical thermodynamics.* New York: Academic. Chapter 10.
KNUTH, E. 1966. *Statistical thermodynamics.* New York: McGraw-Hill. Chapter 10.
MANDL, F. 1971. *Statistical physics.* New York: Wiley. Chapter 10.
REIF, F. 1965. *Statistical and thermal physics.* New York: McGraw-Hill. Sections 9–13 through 9–15.

Density matrix

EYRING, H., HENDERSON, D., STOVER, B. J., and EYRING, E. M. 1964. *Statistical mechanics and dynamics.* New York: Wiley. Chapter 5.
HILL, T. L. 1956. *Statistical mechanics.* New York: McGraw-Hill. Sections 11 and 12.
ISIHARA, A. 1971. *Statistical physics.* New York: Academic. Chapter 10.
KUBO, R. 1965. *Statistical mechanics.* Amsterdam: North-Holland Publishing Co. Section 2–7 and Example 2–7.
TER HAAR, D. 1966. *Elements of thermostatics.* London: Oxford University Press. Chapter 6.
TOLMAN, R. C. 1938. *Statistical mechanics.* London: Oxford University Press. Chapter 9.

Classical limit of Q

HILL, T. L. 1956. *Statistical mechanics.* New York: McGraw-Hill. Section 16.
MÜNSTER, A. 1969. *Statistical thermodynamics*, Vol. I. Berlin: Springer-Verlag. Section 216.

PROBLEMS

10–1. Referring to the discussion following Eq. (10–5), derive expressions for the thermodynamic functions A, G, μ, and S from E and p as functions of N, V, and T.

10–2. Derive a virial expansion for E, A, and S for a Fermi-Dirac ideal gas.

10–3. Derive a virial expansion for E, A, and S for a Bose-Einstein ideal gas.

10–4. Derive Eqs. (10–11) and (10–12) from Eqs. (10–9) and (10–10).

10–5. Consider the power series

$$z = x + \frac{x^2}{2!} + \frac{x^3}{3!} + \cdots$$

Invert this to find x as a power series in z. Compare your result to the expansion of $\ln(1+z)$, since z above is actually $e^x - 1$.

10–6. Give numerical estimates of the Fermi energy, μ_0, and of $T_F = \mu_0/k$ for (a) electrons in a typical metal such as Ag or Cu, (b) nucleons, for example, neutrons and protons, in a heavy nucleus, (c) He^3 atoms in He^3 gas, in which the volume available to each atom is 50Å^3. Treat the particles as free fermions.

10–7. The density of sodium metal at room temperature is 0.95 g/cm^3. Assuming that there is one conduction electron per sodium atom, calculate the Fermi energy and Fermi temperature of sodium.

10–8. Estimate the pressure of an ideal Fermi-Dirac gas of electrons at 0°K.

10–9. Show that the derivative of $f(\varepsilon)$ in Eq. (10–20) is symmetric about μ and that

$$\int_{-\infty}^{\infty} f'(\varepsilon)\, d\varepsilon = -1$$

where $f'(\varepsilon) = df/d\varepsilon$.

10–10. Prove that the expression $e^x/(1+e^x)^2$ is an even function of x.

10–11. Prove that

$$\int_{-\infty}^{\infty} \frac{x^j e^x\, dx}{(1+e^x)^2} = 0 \qquad j \text{ odd}$$

$$= -2(j!) \sum_{n=1}^{\infty} \frac{(-1)^n}{n^j} \qquad j \text{ even}$$

$$\equiv 2(j!)\eta(j)$$

The summation $\eta(j)$ is closely related to the Riemann zeta function (*cf.* Abromowitz and Stegun) and has the values $\eta(2) = \pi^2/12$, $\eta(4) = 7\pi^4/720, \ldots$.

10–12. Carry Eq. (10–36) one term further and show that

$$\mu = \mu_0\left\{1 - \frac{\pi^2}{12}\left(\frac{kT}{\mu_0}\right)^2 - \frac{\pi^4}{80}\left(\frac{kT}{\mu_0}\right)^4 + \cdots\right\}$$

10–13. Show that for an ideal Fermi-Dirac gas that

$$p = \frac{2}{5}\frac{N\mu_0}{V}\left\{1 + \frac{5\pi^2}{12}\left(\frac{kT}{\mu_0}\right)^2 - \frac{\pi^4}{16}\left(\frac{kT}{\mu_0}\right)^4 + \cdots\right\}$$

10–14. Show that at 3000°K, μ for aluminum differs from μ_0 by less than 0.1 percent ($\mu_0 = 11.7$ eV).

10–15. Show that for an ideal Fermi-Dirac gas, the Helmholtz free energy is

$$A = \tfrac{3}{5}N\mu_0\left\{1 - \frac{5\pi^2}{12}\left(\frac{kT}{\mu_0}\right)^2 + \frac{\pi^4}{48}\left(\frac{kT}{\mu_0}\right)^4 + \cdots\right\}$$

and that the entropy is

$$S = \frac{N\mu_0}{T}\left[\frac{\pi^2}{2}\left(\frac{kT}{\mu_0}\right)^2 - \frac{\pi^2}{20}\left(\frac{kT}{\mu_0}\right)^4 + \cdots\right]$$

10–16. Take Eq. (10–37) one term further and show that

$$E = \tfrac{3}{5}N\mu_0\left\{1 + \frac{5\pi^2}{12}\left(\frac{kT}{\mu_0}\right)^2 - \frac{\pi^4}{16}\left(\frac{kT}{\mu_0}\right)^4 + \cdots\right\}$$

10–17. Show that

$$C_V = \frac{\pi^2}{3}k^2Tf(\mu_0)$$

is the constant volume heat capacity of an ideal Fermi-Dirac gas if $\mu_0 \gg kT$, where $f(\varepsilon)$ is the density of states.

10–18. Consider a system in which the density of states of the electrons $f(\varepsilon)$ is

$$\begin{aligned} f(\varepsilon) &= \text{constant} = D \qquad & \varepsilon > 0 \\ &= 0 & \varepsilon < 0 \end{aligned}$$

Calculate the Fermi energy for this system; determine the condition for the system being highly degenerate; and then show that the heat capacity is proportional to T for the highly degenerate case.

10–19. Derive Eq. (10–71) for the constant volume heat capacity of an ideal Bose-Einstein gas.

10–20. Prove that the heat of transition associated with Bose-Einstein condensation is given by Eq. (10–68).

10–21. Show that $g_n(\lambda)$ defined by Eq. (10–47) obeys the following recursion formula:

$$g_{n-1} = \frac{\partial g_n}{\partial(\ln \lambda)}$$

Also show that for λ close to unity, that

$$g_{5/2}(\lambda) = 2.363(-\ln \lambda)^{3/2} + 1.342 + 2.612 \ln \lambda - 0.730 (\ln \lambda)^2 + \cdots$$

From these two results, show that the discontinuity of $(\partial C_V/\partial T)$ at $T = T_c$ for an ideal Bose-Einstein gas is

$$\left(\frac{\partial C_V}{\partial T}\right)_{T \to T_c + \varepsilon} - \left(\frac{\partial C_V}{\partial T}\right)_{T \to T_c - \varepsilon} = \frac{3.66Nk}{T_c}$$

10–22. Consider an ideal Bose-Einstein gas in which the particles have internal degrees of freedom. Assume for simplicity that only the first excited state need be considered and that this has an energy ε relative to the ground state, taken to be zero. Show that the Bose-Einstein condensation temperature of this system is given by

$$T_c = T_c^0\{1 - 0.255e^{-\varepsilon/kT_c^0} + \cdots\}$$

assuming that $e^{-\varepsilon/kT_c^0} \ll 1$. How are the thermodynamic functions affected by ε.

10–23. Does a two-dimensional Bose-Einstein ideal gas display a condensation as it does in three dimensions?

10–24. Show that in two dimensions the heat capacity C_V of an ideal Fermi-Dirac and Bose-Einstein is the same.

10–25. Derive the Rayleigh-Jeans law

$$\rho(\nu, T)\,d\nu = \frac{8\pi\nu^2}{c^3}\,kT\,d\nu$$

from the Planck radiation law by considering the limit $h\nu \ll kT$. Derive the Wien empirical distribution law

$$\rho(\nu, T)\,d\nu = \frac{8\pi h\nu^3}{c^3}\,e^{-h\nu/kT}\,d\nu$$

by considering the high-frequency limit.

10–26. Show that the Planck blackbody distribution can be written in terms of wavelengths λ rather than frequency:

$$\rho(\lambda, T)\,d\lambda = \frac{8\pi hc}{\lambda^5}\,\frac{d\lambda}{e^{hc/\lambda kT} - 1}$$

where $\rho(\lambda, T)\,d\lambda$ is the amount of energy between wavelength λ and $\lambda + d\lambda$.

10–27. If ω_{max} is the frequency at which $\rho(\omega, T)$ is a maximum, illustrate by maximizing $\ln \rho(\omega, T)$ that ω_{max} is given by

$$\frac{\hbar\omega_{max}}{kT} = 3(1 - e^{-\hbar\omega_{max}/kT})$$

and so

$$\frac{\hbar\omega_{max}}{kT} = 2.82$$

Similarly show that

$$\lambda_{max}T = 0.290 \text{ cm-deg}$$

Calculate the temperature for which λ_{max} is in the red region of the spectrum.

10–28. Derive Eq. (10–93) by evaluating the integral in Eq. (10–98).

10–29. In Problem 7–24, it was shown that the number of molecules striking a surface per unit area per unit time is $\rho\bar{v}/4$. By a similar approach, show that the total energy flux radiated by a blackbody is

$$e(T) = \frac{c}{4}\frac{E}{V} = \sigma T^4$$

where $\sigma = 2\pi^5k^4/15h^3c^3$. This result is known as the Stefan-Boltzmann law, and σ is the Stefan-Boltzmann constant. Verify that σ, a universal constant, equals 5.669×10^{-5} erg/cm^2-deg^4-sec.

10–30. Show that

$$p = \frac{1}{3}\frac{E}{V}$$

for a photon gas and compare this to the analogous result for bosons and fermions with nonzero rest mass.

10–31. It has been stated that in the early stages, a nuclear fission explosion generates a temperature of the order of a million degrees Kelvin over a sphere 10 cm in diameter. Assuming this to be true, estimate the total rate of radiation emitted from the surface of this sphere, the radiation flux a few miles away, and the wavelength corresponding to the maximum in the radiated power spectrum.

10–32. Derive an expression for Planck's blackbody distribution law and Stefan's radiation law for a two-dimensional world.

10–33. Derive all of the principal results for a photon gas by setting $\lambda = 1$ in the Bose-Einstein formulas.

10–34. Prove Eq. (10–112).

10–35. Show that the contribution of the term w_1 [Eq. (10–139)] to the partition function Q vanishes, since it is odd in the momenta.

10–36. Derive Eqs. (10–62).

CHAPTER 11

CRYSTALS

In this chapter we shall discuss the application of statistical thermodynamics to the calculation of the thermodynamic properties of crystals. Unlike dilute gases, the interatomic interactions in a crystal are not negligible, but we shall see that the concept of normal coordinates allows us to treat a crystal as a system of independent "particles." In Section 11–1 we show that all of the thermodynamic properties of a crystal can be expressed in terms of the distribution of its normal vibrational frequencies. This distribution function is difficult to calculate exactly, but in Sections 11–2 and 11–3 we discuss two well-known simple approximations, the first due to Einstein (Section 11–2) and the other due to Debye (Section 11–3). In Section 11–4 we turn to the problem of an exact vibrational analysis of a crystalline solid. We shall determine the exact vibrational spectrum of two types of one-dimensional lattices. Although the results for one-dimensional lattices are not directly applicable to real crystals, the basic ideas and techniques associated with such a calculation serve as an introduction to the field of lattice dynamics, that is, the calculation of the vibrational spectrum of more realistic lattices. The final two sections of the chapter contain a discussion of two important topics in the statistical thermodynamics of crystals. In Section 11–5 we introduce the concept of a phonon, and in Section 11–6 we discuss several of the most important types of defects or imperfections that occur in real crystals.

11–1 THE VIBRATIONAL SPECTRUM OF A MONATOMIC CRYSTAL

In this section we shall derive the partition function for a monatomic crystal. Although we can hardly ignore the interatomic (or intermolecular) interactions in a solid, we shall see that it is, nevertheless, possible to treat a crystalline solid as a system of independent "particles." The crucial point here is the existence of normal coordinates. For many purposes, a crystal may be represented by a system of regularly spaced masses and springs, as illustrated two dimensionally in Fig. 11–1. The springs represent the resultant interatomic force that each atom "sees" about its lattice point. Each

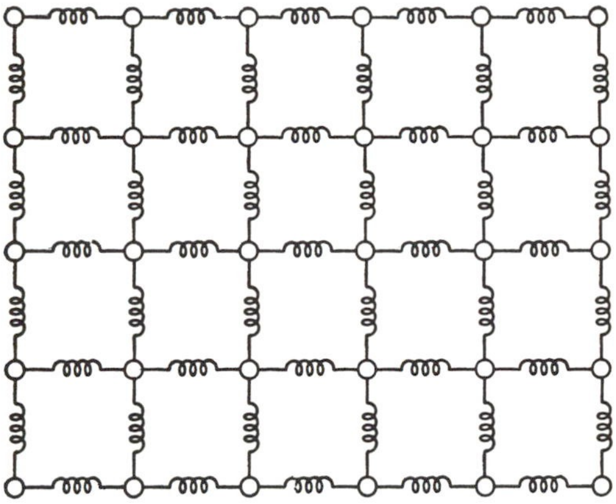

Figure 11–1. **A two-dimensional version of a mass and spring model of a crystalline lattice.**

atom sits in a potential well whose minimum is at a lattice point, and as the atom moves from its equilibrium position, the net force restores it to its equilibrium position. Thus the effect of the intermolecular forces between all the atoms in a crystalline solid may be represented by a system of springs as illustrated in Fig. 11–1. Clearly the force constants that we assign to these springs reflect the interatomic forces involved.

In a typical crystal, the potential well that each atom "sees" is very steep, and so each atom vibrates about its equilibrium lattice point with a small amplitude. This allows us to expand the interatomic potential of the entire crystal in a Taylor series. Consider the one-dimensional example shown in Fig. 11–2. The total potential energy is a function of the displacements of the N atoms from their equilibrium positions, that is, $U = U(\xi_1, \xi_2, \ldots, \xi_N)$. Since the atoms vibrate with small amplitude about their equilibrium positions, we write

$$U(\xi_1, \xi_2, \ldots, \xi_N) = U(0, 0, \ldots, 0) + \sum_{j=1}^{N} \left(\frac{\partial U}{\partial \xi_j}\right)_0 \xi_j + \frac{1}{2} \sum_{i=1}^{N} \sum_{j=1}^{N} \left(\frac{\partial^2 U}{\partial \xi_i \, \partial \xi_j}\right)_0 \xi_i \xi_j + \cdots \tag{11-1}$$

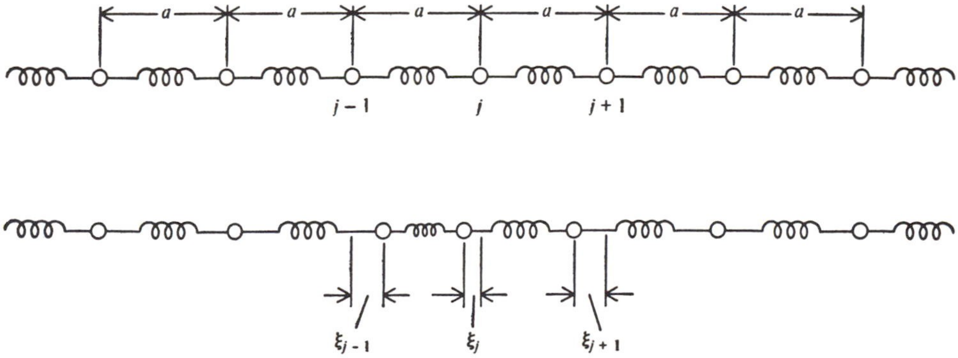

Figure 11–2. **A one-dimensional system of masses and springs. The upper system is in its equilibrium configuration, where all the masses are separated by a distance *a*, the unstrained length of the spring. The lower curve shows an arbitrary configuration which is described by the set of displacements $\{\xi_j\}$ of the atoms from their equilibrium positions.**

where the zero subscript on the derivatives indicated that they are to be evaluated at the point at which all the ξ_j equal zero. Since U is a minimum when the $\xi_j = 0$, the first derivatives in Eq. (11–1) are zero, and so we have

$$U(\xi_1, \xi_2, \ldots, \xi_N) = U(0, 0, \ldots, 0) + \frac{1}{2}\sum_{i,j} \left(\frac{\partial^2 U}{\partial \xi_i \xi_j}\right)_0 \xi_i \xi_j + \cdots$$

$$= U(0, 0, \ldots, 0) + \frac{1}{2}\sum_{i,j} k_{ij}\xi_i \xi_j + \cdots \tag{11–2}$$

where we have introduced the set of force constants k_{ij}. The important result here is that $U(\xi_1, \xi_2, \ldots, \xi_N)$ is a *quadratic* function of the displacements. Note that $U(0, 0, \ldots, 0)$ is a function of the lattice spacing, which in turn is a function of the ratio V/N, or the density. To emphasize this, we shall write $U(0, 0, \ldots, 0)$ as $U(\mathbf{0}; \rho)$. We are taking the zero of energy to be the separated atoms at rest. In addition, the curvature of $U(\xi_1, \xi_2, \ldots, \xi_N)$ at the minimum is a function only of V/N, and so the force constants are functions of V/N only.

Equation (11–2) represents a system of *coupled* harmonic oscillators. We say that they are coupled because of the cross terms in Eq. (11–2). If it were not for these cross terms, Eq. (11–2) would be a sum of independent squared terms, and the Lagrange equations of motion (classically) or the Schrödinger equation (quantum mechanically) would yield N separate or uncoupled harmonic oscillators. This is similar to the situation that occurred when we treated the vibration of a polyatomic molecule in Section 8–1, and, in fact, a crystal containing N atoms can be considered to be just a large polyatomic molecule. We saw in Chapter 8 that the vibrational motion of a polyatomic molecule can be rigorously decomposed into a set of *independent* harmonic oscillators by introducing normal coordinates. In principle, we can apply such a normal coordinate analysis to an entire crystal. If there are N atoms in a monatomic crystal, there are $3N$ degrees of freedom, of which three are associated with the translational motion of the whole crystal, and three more are concerned with the rotation of the crystal. There are then $3N - 6$ vibrational degrees of freedom. But with $N = O(10^{20})$, we can take this number of vibrational degrees to be $3N$, without noticeable error. The result of a normal coordinate analysis would yield $3N - 6 \approx 3N$ vibrational frequencies

$$\nu_j = \frac{1}{2\pi}\left(\frac{k_j}{\mu_j}\right)^{1/2} \qquad j = 1, 2, \ldots, 3N - 6 \approx 3N \tag{11–3}$$

where k_j and μ_j are an effective force constant and an effective reduced mass, respectively. The precise form of k_j and μ_j are not important for our purposes. The important point is that the complicated general vibrational problem can be mathematically reduced to $3N$ (really $3N - 6$) independent harmonic oscillators, each with its own frequency, which is a complicated function of the masses, force constants, and geometry of the lattice. Since the k_j in Eq. (11–3) depends upon the k_{ij} in Eq. (11–2), the k_j and the frequencies ν_j in Eq. (11–3) depend upon V/N rather than V or N separately.

Since the crystal does not translate or rotate,* the complete partition function is given by

$$Q\left(\frac{V}{N}, T\right) = e^{-U(\mathbf{0};\rho)/kT} \prod_{j=1}^{3N-6} q_{\text{vib},j} \tag{11–4}$$

* More precisely, the translational and rotational degrees of freedom of the crystal contribute negligibly to the partition function on a per molecule basis.

where $q_{\text{vib},j}$ is the vibrational partition function associated with the jth vibrational frequency. Note that there is no factor of $N!$ in the denominator of Q. Since each molecule is restricted to the neighborhood of its lattice point, and the lattice points could, in principle, be labeled, the atoms themselves must be considered to be distinguishable. We have already evaluated q_{vib} in our treatment of diatomic gases and found that

$$q_{\text{vib}} = \frac{e^{-h\nu/2kT}}{1 - e^{-h\nu/kT}} \tag{11-5}$$

Therefore the total partition function is given by

$$Q = \prod_{j=1}^{3N} \left(\frac{e^{-h\nu_j/2kT}}{1 - e^{-h\nu_j/kT}} \right) e^{-U(0;\rho)/kT} \tag{11-6}$$

Since there are $3N$ normal frequencies, they are essentially continuously distributed, and we can introduce a function $g(\nu)\,d\nu$, which gives the number of normal frequencies between ν and $\nu + d\nu$. If we introduce this into the logarithm of Eq. (11–6), we have

$$-\ln Q = \frac{U(0;\rho)}{kT} + \int_0^\infty \left[\ln(1 - e^{-h\nu/kT}) + \frac{h\nu}{2kT} \right] g(\nu)\,d\nu \tag{11-7}$$

where

$$\int_0^\infty g(\nu)\,d\nu = 3N \tag{11-8}$$

since there are $3N$ normal frequencies in all. If we can determine the function $g(\nu)$, we can then calculate the thermodynamic properties of the crystal. For example, we have

$$E = U(0;\rho) + \int_0^\infty \left[\frac{h\nu e^{-h\nu/kT}}{(1 - e^{-h\nu/kT})} + \frac{h\nu}{2} \right] g(\nu)\,d\nu \tag{11-9}$$

and

$$C_V = k \int_0^\infty \frac{(h\nu/kT)^2 e^{-h\nu/kT} g(\nu)\,d\nu}{(1 - e^{-h\nu/kT})^2} \tag{11-10}$$

Equations (11–7) through (11–10) are essentially exact. In order to use them, we must know the function $g(\nu)$, and this is, of course, where the difficulty lies. The function $g(\nu)$ is easier to determine than the entire set of individual frequencies, but it is, nevertheless, a very difficult problem. We shall discuss some exact calculations of $g(\nu)$ in Section 11–4, but before that, we shall introduce two useful and well-known approximations to $g(\nu)$. One of these is due to Einstein and says that all the normal frequencies are the same; the other is due to Debye, who treated a crystal as a continuous elastic medium and calculated $g(\nu)$ by studying the elastic waves that can be set up in such a body.

11–2 THE EINSTEIN THEORY OF THE SPECIFIC HEAT OF CRYSTALS

In this section we shall discuss an extremely simple model for the vibrational character of a crystal. It is so simple, in fact, that we should go back in time to the beginning of this century in order to appreciate its great insight and impact. Classical statistical thermodynamics was fairly well developed by the end of the nineteenth century. If it is considered that the N atoms of a crystalline solid behave as harmonic oscillators about their equilibrium positions, classical theory (equipartition) predicts that each atom would contribute R cal/deg-mole for each of its three vibrational degrees of freedom, or that the molar heat capacity at constant volume would be $3Nk = 3R = 6$ cal/deg-mole. This prediction, which is known as the law of Dulong and Petit, is in good agreement with the observed heat capacity of many crystals at high enough temperatures, and often down to room temperature, but the agreement fails completely at low temperatures. For example, the heat capacity C_V of silver is shown in Fig. 11–3. It can be seen that the Dulong and Petit value is approached asymptotically, but that the curve falls rapidly to zero as $T \to 0$. This behavior is observed quite generally, and experimentally it is found that the heat capacity goes to zero as T^3 as $T \to 0$. This is known as the T^3-law and is an experimental observation which any successful theory must reproduce.

At the beginning of this century, deviations from predicted classical behavior were being discovered regularly, and each one was a severe challenge to the physical theories of the time. Einstein was the first to present a theoretical explanation of the low-temperature heat capacity of solids by applying the revolutionary blackbody radiation theory ideas of Planck to the vibrations of atoms in crystals. Einstein assumed that each atom in the crystal vibrates about its equilibrium configuration as a simple harmonic oscillator, so that the entire crystal could be considered to be a set of $3N$ independent harmonic oscillators, each oscillator having the same frequency ν. Physically then, he assumed that each atom of the crystal sees the same environment as any other, and so all N atoms could be treated as independent oscillators in the x-, y-, and z-directions. Classically such an assumption leads to the Dulong and Petit value of $3R$ cal/deg-mole, but Einstein's great contribution (in 1907) was to say that the energy of each of these $3N$ independent oscillators had to be quantized according to the procedure developed by Planck. Thus, with our advantage of using a formalism and notation developed long after the turbulent years of the beginning of the century, we can say that Einstein assumed that the frequency spectrum $g(\nu)$ was a delta function at one frequency

$$g(\nu) = 3N\delta(\nu - \nu_E) \tag{11–11}$$

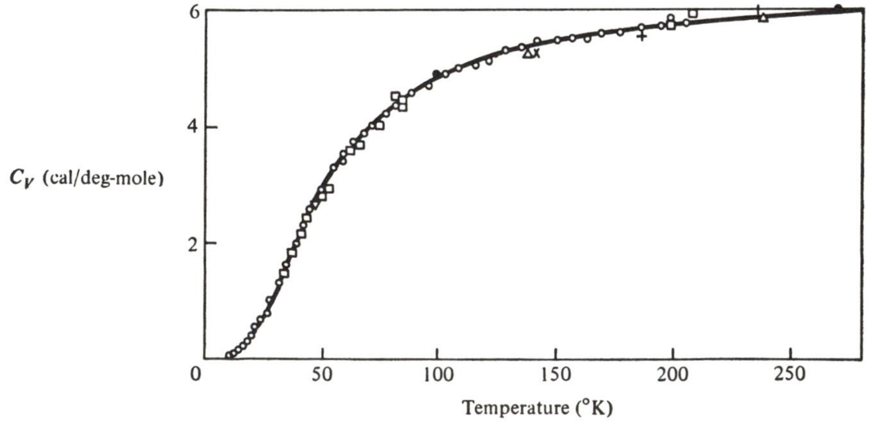

Figure **11–3. The molar heat capacity at constant volume of metallic silver as a function of temperature. (From C. Kittel, *Solid State Physics, 2nd ed.* New York: Wiley, 1956.)**

where the factor $3N$ is included, so that Eq. (11–8) is satisfied, and ν_E is the single frequency assigned to all $3N$ independent oscillators of the crystal. The value of the Einstein frequency ν_E varies from substance to substance and, in some way, reflects the nature of the interatomic interactions for the particular crystal. In the light of the organized development presented in Section 11–1, this may appear to be a terribly gross assumption, but it was a major step forward, and "in a sense the final step,"* in the understanding of the heat capacity of solids.

If we substitute Eq. (11–11) into Eq. (11–10), we find

$$C_V = 3Nk\left(\frac{h\nu_E}{kT}\right)^2 \frac{e^{-h\nu_E/kT}}{(1 - e^{-h\nu_E/kT})^2} \tag{11–12}$$

for the heat capacity. It is customary to define a quantity Θ_E by $h\nu_E/k$, which has units of temperature and is called the Einstein temperature of the crystal. In terms of the Einstein temperature, Eq. (11–12) is

$$C_V = 3Nk\left(\frac{\Theta_E}{T}\right)^2 \frac{e^{-\Theta_E/T}}{(1 - e^{-\Theta_E/T})^2} \tag{11–13}$$

Equation (11–13) contains one adjustable parameter to fit the entire heat capacity curve shown in Fig. 11–3. Figure 11–4 shows a comparison of Eq. (11–13) versus the experimental heat capacity of diamond. This figure is taken from Einstein's original paper and shows the success of such a simple theory. It is easy to show from Eq. (11–13) that C_V approaches the Dulong and Petit value of $3Nk = 3R$ as $T \to \infty$. (See Problem 11–2.)

Although Fig. 11–4 shows that the Einstein model of a crystal is capable of giving an impressive qualitative agreement with experiment, it is not in quantitative agreement. In particular, Eq. (11–13) predicts that the low-temperature heat capacity goes as

$$C_V \to 3Nk\left(\frac{\Theta_E}{T}\right)^2 e^{-\Theta_E/T} \tag{11–14}$$

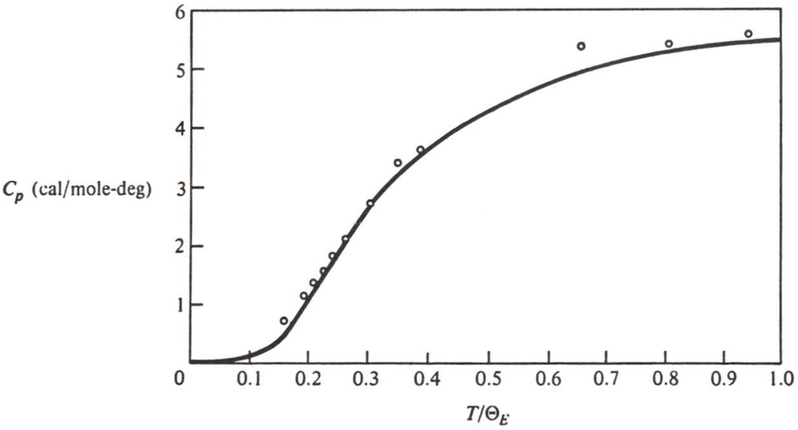

Figure 11–4. **Comparison of the experimental heat capacity of diamond with the prediction based on the Einstein theory with $\Theta_E = 1320°K$.** (From C. Kittel, *Solid State Physics*, 3rd *ed.* New York: Wiley, 1967., after A. Einstein, *Ann. Physik.*, **22**, 180, 1907.)

* From Blackman in "Additional Reading."

instead of T^3 as the experimental data. The low-temperature heat capacity predicted by Eq. (11–14) falls to its zero value more rapidly than T^3-law. We shall see in the next section that the Debye theory, which came a few years after Einstein's, gives a T^3-law as $T \to 0$.

Before going on to the Debye theory, we point out an important feature of Eq. (11–13), which is also exhibited in more rigorous theories. Equation (11–13) predicts that C_V is the same function for all substances if it is plotted versus T/Θ_E. When this happens, we say that C_V is a universal function of T/Θ_E, and that the various crystals obey a *law of corresponding states.* Once the temperature is scaled or "reduced" by a quantity that depends upon the particular substance, the heat capacity versus the reduced temperature will superimpose for all crystals. Although the Einstein model does not quantitatively reproduce experimental data, its prediction of a law of corresponding states is, in fact, correct.

11–3 THE DEBYE THEORY OF THE HEAT CAPACITY OF CRYSTALS

According to the ideas of Planck, the energy of an oscillator is proportional to the frequency, and since it is the lower energies that are populated at low temperatures, we can reason that it is the low-frequency or long-wavelength modes that are most important at low temperatures. The success of the Debye theory is that it treats the long-wavelength frequencies of a crystal in an exact manner, and hence is able to predict the low-temperature heat capacity.

The normal frequencies of a crystal vary from essentially zero to some value of the order of 10^{13} cycles/sec (Hz) or so. Normal frequencies are not due to the vibrations of single atoms, but are a concerted harmonic motion of all the atoms. This concerted motion is called a normal coordinate or a normal mode. Note, for example, that the normal coordinates of CO_2 and H_2O involve the synchronous motion of all the atoms in each molecule. Two extremes of normal modes of a one-dimensional crystal are shown in Fig. 11–5. The upper mode is one in which the atoms vibrate against each other and has a wavelength of $2a$. The lower one is one in which a long row of atoms moves smoothly together to produce a long-wavelength mode of low frequency. It is the long-wavelength modes that Debye was able to treat in a clever manner.

Debye reasoned that those normal modes whose wavelengths are long compared to the atomic spacing do not depend upon the detailed atomic character of the solid and could be calculated by assuming that the crystal is a continuous elastic body. The approximation of the Debye theory is that it treats all the normal frequencies from this point of view.

The distribution of frequencies that can be set up in a solid body is calculated in almost the same way that we calculated the set of standing waves in a blackbody cavity.

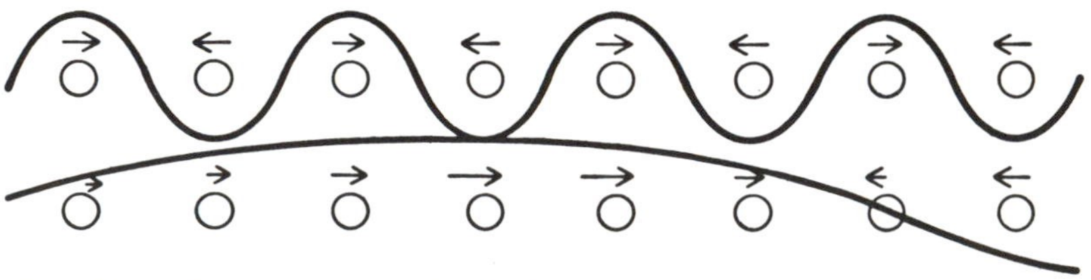

Figure 11–5. **Two types of normal modes in a one-dimensional crystal. The upper one is a high-frequency mode and the bottom is a low-frequency mode.**

The first part of Section 10–5 can be read independently of the rest of the chapter. We state there that the imaginary part of

$$u(\mathbf{r}, t) = Ae^{i(\mathbf{k}\cdot\mathbf{r} - \omega t)} \tag{11–15}$$

represents a wave of amplitude A traveling through a medium in the direction of **k** and with frequency $\omega = 2\pi\nu$. The quantity **k** is called the wave vector, and its magnitude is $2\pi/\lambda$. The velocity of this wave is given by $v = \omega/k = \nu\lambda$. A standing wave can be obtained by superimposing two waves traveling in opposite directions. This gives

$$u = 2Ae^{i\mathbf{k}\cdot\mathbf{r}} \cos \omega t$$

In order that this be a standing wave, we require that the imaginary part of this vanish at the edges of the crystal. If we assume for simplicity that the crystal is a cube of length L, this boundary condition gives that $k_x L = n_x \pi$, $k_y L = n_y \pi$, and $k_z L = n_z \pi$, where k_x, and so on, are the components of **k**, and n_x, n_y, and n_z are positive integers. In vector notation, we have

$$\mathbf{k} = \frac{\pi}{L}\mathbf{n} \tag{11–16}$$

The frequency ν depends upon only the magnitude of **k** (through the relation $\omega = vk$), which is given by

$$k^2 = \left(\frac{\pi}{L}\right)^2 (n_x^{\,2} + n_y^{\,2} + n_z^{\,2}) \tag{11–17}$$

The number of standing waves with wave number between k and $k + dk$ is found by the same method that we used to find the number of translational energy states between ε and $\varepsilon + d\varepsilon$ in Section 1–3. Using Eq. (11–17), the number of standing waves with the wave vector of magnitude less than k is [*cf.* Eq. (10–84)]

$$\Phi(k) = \frac{\pi}{6}\left(\frac{Lk}{\pi}\right)^3 = \frac{L^3k^3}{6\pi^2} = \frac{Vk^3}{6\pi^2}$$

and the number between k and $k + dk$ is [*cf.* Eq. (10–85)]

$$\omega(k)\, dk = \frac{d\Phi}{dk}\, dk = \frac{Vk^2\, dk}{2\pi^2} \tag{11–18}$$

We can convert this into $g(\nu)\, d\nu$ by using the relation $\nu = v/\lambda = vk/2\pi$:

$$g(\nu)\, d\nu = \frac{4\pi V \nu^2}{v^3}\, d\nu \tag{11–19}$$

This is almost the desired result. We must recognize that there are two kinds of waves that can propagate through a continuous medium. These are transverse waves, in which the medium vibrates perpendicular to the direction of propagation (the direction of **k**), and longitudinal waves, in which the medium vibrates in the same direction as the wave is propagated. Since it is possible to draw two independent vectors perpendicular to **k** and only one parallel to **k**, there are two transverse waves and one longitudinal wave. The three of these contribute to $g(\nu)$, and we finally have the complete expression for the Debye approximation to $g(\nu)$, namely,

$$g(\nu)\, d\nu = \left(\frac{2}{v_t^{\,3}} + \frac{1}{v_l^{\,3}}\right) 4\pi V \nu^2\, d\nu \tag{11–20}$$

In this expression, v_t and v_l are the transverse and longitudinal velocities, respectively. It is conventional to introduce a kind of average velocity by means of

$$\frac{3}{v_0^3} \equiv \frac{2}{v_t^3} + \frac{1}{v_l^3} \tag{11–21}$$

so that Eq. (11–20) can be written in the form

$$g(\nu)\,d\nu = \frac{12\pi V}{v_0^3}\,\nu^2\,d\nu \tag{11–22}$$

This expression is exact in the limit of low frequencies or long wavelengths, where the atomic nature of the solid is not important, and the crystal can, in fact, be treated as a continuous elastic body. The Debye theory uses Eq. (11–22) for all the normal frequencies, however. The total number of normal frequencies is $3N$, and so Debye defined a maximum frequency ν_D such that the integral of $g(\nu)\,d\nu$ from 0 to ν_D equals $3N$. Thus

$$\int_0^{\nu_D} g(\nu)\,d\nu = 3N \tag{11–23}$$

which, when Eq. (11–22) is used for $g(\nu)$, gives

$$\nu_D = \left(\frac{3N}{4\pi V}\right)^{1/3} v_0 \tag{11–24}$$

The frequency ν_D is called the Debye frequency. In terms of ν_D, the distribution function $g(\nu)\,d\nu$ is

$$\begin{aligned} g(\nu)\,d\nu &= \frac{9N}{\nu_D^3}\,\nu^2\,d\nu \qquad 0 \le \nu \le \nu_D \\ &= 0 \qquad\qquad\qquad \nu > \nu_D \end{aligned} \tag{11–25}$$

This summarizes the Debye theory of crystals.

We can now substitute Eq. (11–25) for $g(\nu)\,d\nu$ into Eqs. (11–7) through (11–10) to calculate the thermodynamic properties of a crystal according to the Debye theory. The most interesting thermodynamic function is the heat capacity C_V, given by Eq. (11–10) with Eq. (11–25) for $g(\nu)\,d\nu$:

$$C_V = 9Nk\left(\frac{T}{\Theta_D}\right)^3 \int_0^{\Theta_D/T} \frac{x^4 e^x}{(e^x - 1)^2}\,dx \tag{11–26}$$

where we have let $x = h\nu/kT$ and have defined the Debye temperature by

$$\Theta_D = \frac{h\nu_D}{k} \tag{11–27}$$

The integral in Eq. (11–26) cannot be evaluated in terms of simple functions and must be evaluated numerically. Note that the integral is a function of only the upper limit of the integral, that is, a function of Θ_D/T. It is customary to define a function $D(T/\Theta_D)$ by

$$D\left(\frac{T}{\Theta_D}\right) = 3\left(\frac{T}{\Theta_D}\right)^3 \int_0^{\Theta_D/T} \frac{x^4 e^x}{(e^x - 1)^2}\,dx \tag{11–28}$$

so that the heat capacity is

$$C_V = 3NkD\left(\frac{T}{\Theta_D}\right) \tag{11-29}$$

The function $D(T/\Theta_D)$ is called the Debye function. It is a well-tabulated function of T/Θ_D. (See Appendix C.) Figure 11–6 is from Debye's original paper and shows two comparisons of Eq. (11–29) with experimental data. It can be seen that the agreement is very good. The values of Θ_D are those which give the best overall fit to the data. Table 11–1 gives the Debye temperatures for many monatomic solids. Note that most of these values are of the order of a few hundred degrees Kelvin.

Table 11–1. **The Debye temperature of various monatomic solids**

solid	Θ_D(°K)	solid	Θ_D(°K)
Na	150	Fe	420
K	100	Co	385
Cu	315	Ni	375
Ag	215	Al	390
Au	170	Ge	290
Be	1000	Sn	260
Mg	290	Pb	88
Zn	250	Pt	225
Cd	172	C (diam)	1860

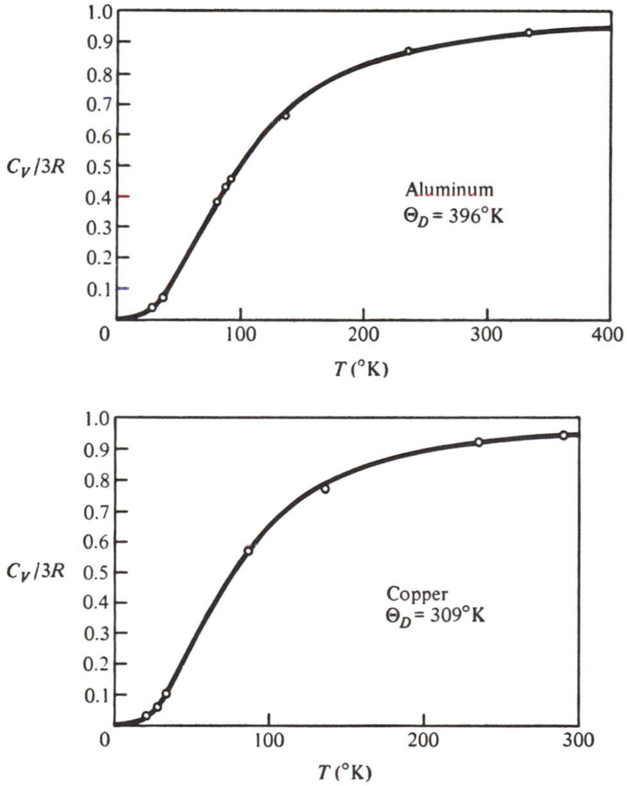

Figure 11–6. **Observed values and curves calculated on the Debye model for the heat capacity of aluminum and copper, taking $\Theta_D = 396$°K and 309°K, respectively. (After P. Debye, *Ann. Physik*, 39,789, 1912. From C. Kittel, *Solid State Physics, 2nd ed.*, New York: Wiley, 1956.)**

Although Eq. (11–29) must be evaluated numerically for arbitrary values of T/Θ_D, it is easy to investigate its high- and low-temperature behavior. At high temperatures, the upper limit in the integral in Eq. (11–28) becomes very small. Hence the range of x is very small and it is legitimate to expand the integrand to get

$$\int_0^{\Theta_D/T} \frac{x^4 e^x}{(e^x-1)^2}\,dx \to \int_0^{\Theta_D/T} \frac{x^4(1+x+\cdots)}{(1+x+\cdots-1)^2}\,dx = \int_0^{\Theta_D/T} x^2\,dx = \frac{1}{3}\left(\frac{\Theta_D}{T}\right)^3$$

Therefore $D(T/\Theta_D) \to 1$, and Eq. (11–29) for C_V becomes

$$C_V \to 3Nk = 3R = 6 \text{ cal/deg-mole} \tag{11–30}$$

which is the classical limiting law of Dulong and Petit.

The low-temperature limit of C_V is more interesting. This can be obtained by letting the upper limit of the integral in $D(T/\Theta_D)$ go to infinity. Then

$$D\left(\frac{T}{\Theta_D}\right) \to 3\left(\frac{T}{\Theta_D}\right)^3 \int_0^{\infty} \frac{x^4 e^x}{(e^x-1)^2}\,dx$$

The integral here is standard and equals $4\pi^4/15$. (See Problem 11–6.) The low-temperature limit of C_V then is

$$C_V \to \frac{12\pi^4}{5} Nk \left(\frac{T}{\Theta_D}\right)^3 \tag{11–31}$$

which is the famous T^3-law. This was the great triumph of the Debye theory. Although it is not obvious from Fig. 11–4, the Einstein heat capacity curve falls much too rapidly as $T \to 0$, and the agreement at low temperatures is very poor. This is more readily seen in Table 11–2, where the Einstein and Debye theories are compared to experimental data for silver. It can be seen that although both theories agree for temperatures greater than approximately 100°K, only the Debye theory is able to be used at lower temperature.

There are several important features of the Debye theory. The one that we have just discussed in some detail is that it predicts a T^3-law for the low-temperature heat capacity. Another is that it predicts a law of corresponding states for the heat capacity. Equation (11–29) clearly shows that if C_V is plotted versus T/Θ_D, all substances will lie on the one curve. Another way of saying this is that C_V is a universal function for all substances, determined by one parameter Θ_D in the form of T/Θ_D. Figure 11–7 shows the heat capacity data for a number of substances plotted on the same graph of C_V versus T/Θ_D. Note that the Debye curve fits all the points over the entire temperature range.

An interesting consequence of Debye's approach is that it is possible to calculate Θ_D in terms of the elastic constants of the solid. We shall not prove it here, but it should be clear that such a thing is possible since the Debye theory is based upon treating a crystal as a continuous elastic body. The elastic constants of a body are quantities such as the compressibility and Young's modulus. Table 11–3 compares the Debye temperatures determined by fitting heat capacity data with those calculated from the elastic properties of the solid. The agreement, although not perfect, is quite good.

Table 11–2. **Heat capacity of silver at different temperatures**

temperatures (°K)	C_V(obs.) (cal/mole-deg)	C_V calculated	
		Einstein	Debye
1.35	0.000254	8.76×10^{-49}	—
2	0.000626	1.39×10^{-32}	—
3	0.00157	6.16×10^{-20}	—
4	0.00303	5.92×10^{-15}	—
5	0.00509	1.62×10^{-11}	—
6	0.00891	3.24×10^{-9}	—
7	0.0151	1.30×10^{-7}	0.0172
8	0.0236	2.00×10^{-6}	0.0257
10	0.0475	1.27×10^{-4}	0.0502
12	0.0830	0.0010	0.0870
14	0.1336	0.0052	0.137
16	0.2020	0.0180	0.207
20	0.3995	0.0945	0.394
28.56	1.027	0.579	1.014
36.16	1.694	1.252	1.69
47.09	2.582	2.272	2.60
55.88	3.186	2.946	3.22
65.19	3.673	3.521	3.73
74.56	4.039	3.976	4.13
83.91	4.326	4.309	4.45
103.14	4.797	4.795	4.86
124.20	5.084	5.124	5.17
144.38	5.373	5.323	5.37
166.78	5.463	5.476	5.51
190.17	5.578	5.581	5.61
205.30	5.605	5.633	5.66

Source: C. Kittel, *Solid State Physics*, 2nd ed. New York: Wiley, 1956.

One cannot expect perfect agreement in Table 11–3 since the Debye theory is, of course, an approximate theory. In fact, as more experimental data became available in the 1920s, the Debye theory was subjected to more and more severe tests, and discrepancies began to appear. For example, one can calculate Θ_D at any particular temperature from Eq. (11–29) if the experimental heat capacity of that temperature is known. If the Debye theory were exact, the value of Θ_D obtained would be independent of the temperature used to calculate it. However, it turns out that the value

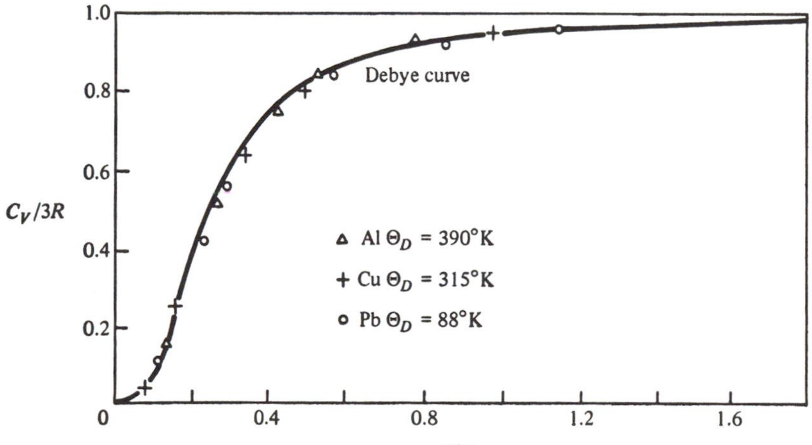

Figure **11–7. An illustration of the law of corresponding states predicted by the Debye theory. (From F. Mandl, *Statistical Physics*. New York: Wiley, 1971).**

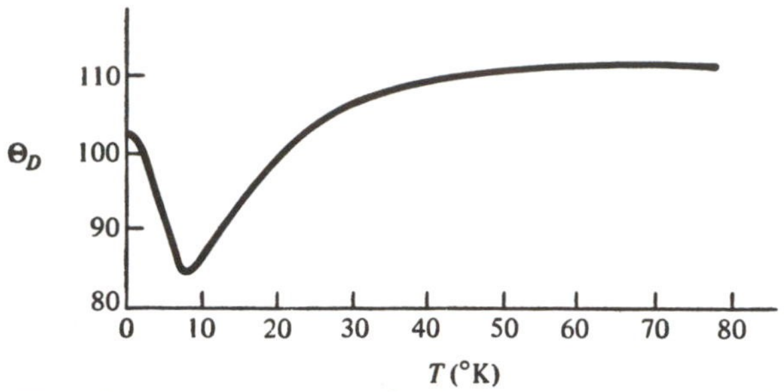

Figure 11–8. **A typical result for the Debye temperature as function of T for a monatomic solid.**

Table 11–3. **Comparison of Debye temperatures determined from elastic constants or heat capacity measurements**

substance	Θ (elastic) °K	Θ (heat capacity) °K
Al	399	396
Cu	329	313
Ag	212	220
Au	166	186
Cd	168	164
Sn	185	165
Pb	72	86
Bi	111	111
Pt	226	220

of Θ_D does depend upon the temperature at which it is evaluated. This can be shown most clearly on a Θ_D versus T plot, where the value of Θ_D calculated at some temperature is plotted against the temperature. Such a Θ_D–T plot is shown in Fig. 11–8. If the Debye theory were exact, Θ_D versus T would be a horizontal straight line, and so a deviation from such a straight line indicates a failing of the Debye theory. In the main, however, these deviations do not exceed 10 percent in most cases, and so we can say that the Debye theory is a successful theory of the thermodynamic properties of crystal lattices. In the next section we shall present the basic ideas of a more rigorous approach to the dynamical properties of lattices, which has been highly developed since the 1930s and constitutes the presently active field of lattice dynamics.

11–4 INTRODUCTION TO LATTICE DYNAMICS

In this section we shall introduce the basic ideas of lattice dynamics by calculating the frequency distribution of two types of one-dimensional lattices: one where all the masses are the same (a one-dimensional model for an elemental crystal such as Zn) and one with two alternating different masses (a one-dimensional model of NaCl, say). These two models have quite different vibrational spectra, both of which are qualitatively observed in real crystals. At the end of the section, we shall simply state the types of results found in two and three dimensions.

The Hamiltonian for the first case is (see Fig. 11–2)

$$H = \sum_{j=1}^{N} \frac{m}{2} \dot{\xi}_j^2 + \sum_{j=2}^{N} \frac{f}{2} (\xi_j - \xi_{j-1})^2 \qquad (11\text{–}32)$$

where we are using f here for the force constants. The equations of motion corresponding to this are (see Problem 11–14)

$$m\ddot{\xi}_j = f(\xi_{j+1} + \xi_{j-1} - 2\xi_j) \tag{11–33}$$

This set of equations represents a set of *coupled* harmonic oscillators. We assume that the time dependence is harmonic and let

$$\xi_j(t) = e^{i\omega t} y_j \tag{11–34}$$

where $\omega = 2\pi\nu$, and y_j is independent of time. Substituting this into Eq. (11–33) gives

$$-m\omega^2 y_j = f(y_{j+1} + y_{j-1} - 2y_j) \tag{11–35}$$

This type of equation is called a difference equation (as opposed to a differential equation); it is a linear difference equation with constant coefficients. Linear difference equations with constant coefficients are solved by $y_j = A^j$ (compare to $y(x) = e^{nx}$ as solutions to linear differential equations with constant coefficients). A little experience with equations of this type would suggest letting A be $e^{i\phi}$. Substituting this into Eq. (11–35) gives

$$\begin{aligned} -m\omega^2 &= f(e^{i\phi} + e^{-i\phi} - 2) \\ &= f(2\cos\phi - 2) \end{aligned}$$

or

$$\omega^2 = \frac{4f}{m}\sin^2\left(\frac{\phi}{2}\right) \tag{11–36}$$

If we note that the maximum value that ω^2 can have is $4f/m$ (since the maximum value that $\sin^2(\phi/2)$ can have is 1), then we can write

$$\omega = \omega_{\text{max}}\left|\sin\left(\frac{\phi}{2}\right)\right| \tag{11–37}$$

The solution to Eq. (11–33) is then

$$\xi_j(t) = e^{i(\omega t + j\phi)}$$

where ϕ is as yet undetermined.

Since this functional form for $\xi_j(t)$ repeats for every $\Delta j = 2\pi/\phi$, there is a wavelength λ equal to $a\,\Delta j = 2\pi a/\phi$, where a is the lattice spacing of the one-dimensional chain. From this, ϕ is given by

$$\phi = \frac{2\pi a}{\lambda} \equiv ka \tag{11–38}$$

where k, which equals $2\pi/\lambda$, is the wave vector of the motion of the chain. We see then that

$$\xi_j(t) = e^{i(jka + \omega t)} \tag{11–39}$$

represents a wave of wavelength $2\pi/k$ and frequency ω traveling along the chain. From the De Broglie relation, $\hbar k$ is the momentum of this wave, or in the language of the next section, $\hbar k$ is the momentum of a phonon associated with this frequency.

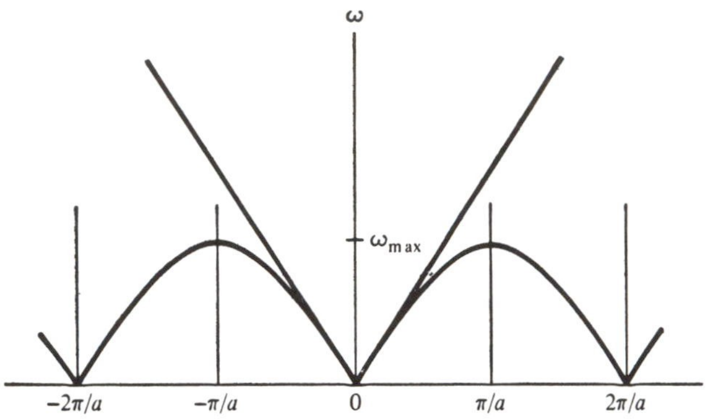

Figure. 11–9 **The dispersion curve for a one-dimensional monatomic lattice. [Eq. (11–40)]. The straight lines correspond to a continuous string.**

The relation between the frequency and the wave vector is called a dispersion curve. The dispersion curve for our simple one-dimensional lattice is [*cf.* Eqs. (11–37) and (11–38)]

$$\omega = \omega_{\max}\left|\sin\left(\frac{ka}{2}\right)\right| \tag{11–40}$$

This is plotted in Fig. 11–9. In the limit of small values of ka (wavelength long compared to the lattice spacing), we have essentially a continuous chain as in the Debye theory, and the dispersion curve is

$$\omega = \omega_{\max}\frac{ka}{2} \tag{11–41}$$

which we write as

$$\frac{\omega}{k} = \lambda\nu = \frac{a\omega_{\max}}{2} = \text{constant velocity} \tag{11–42}$$

Typical lattice spacings are $O(10^{-8}\text{cm})$, and typical maximum frequencies are $O(10^{13}\text{ sec}^{-1})$, giving $O(10^5\text{ cm/sec})$ for the velocity of the wave. This is the order of magnitude for the velocity of sound in solids.

Note that, in general, however, $\lambda\nu$ (or ω/k) is *not* a constant. From Eq. (11–40),

$$\lambda\nu = \frac{\omega}{k} = \frac{\omega_{\max}}{k}\left|\sin\left(\frac{ka}{2}\right)\right| = c(k) \tag{11–43}$$

which shows that the velocity is, in fact, a function of k or λ. Waves actually have two types of velocities associated with them, and the velocity $c(k)$ defined in Eq. (11–43) is called the phase velocity. It is the fact that different wavelengths travel with different velocities, which leads to the dispersion of waves as they pass through a prism, and so $\omega(k)$ is called a dispersion curve. Dispersion curves can be determined experimentally by inelastic neutron scattering measurements.

An important property of Eq. (11–39) for $\xi_j(t)$ is that the substitution $k \to k_n = k + 2\pi n/a$ with $n = \pm 1, \pm 2, \ldots$ leaves $\xi_j(t)$ unchanged. Furthermore, this substitution leaves Eq. (11–40) for ω unchanged. In other words, there is no physical difference between states corresponding to wave vectors k or $k \mp 2\pi n/a$. In order to obtain a unique relationship between the state of vibration of the lattice and the wave vector k, k must be restricted to a range of values $2\pi/a$. Usually one chooses

$$-\frac{\pi}{a} \leq k \leq \frac{\pi}{a}$$

We now show how to express the thermodynamic properties of the lattice in terms of the dispersion curve. Before doing this, however, we must consider the boundary conditions in our lattice. We shall use *periodic boundary conditions*, namely, that $\xi_j(t) = \xi_{j+N}(t)$. One way of thinking about these boundary conditions is to picture the linear chain of N atoms to be bent around into a circle. Clearly, if the chain is long enough, joining its two ends has a negligible effect on the thermodynamic properties of the chain. Applying these boundary conditions to the $\xi_j(t)$ [Eq. (11–39)] gives $\exp(iNka) = 1$, which implies that $k = 2\pi j/Na$, where j is an integer. Now because k is restricted to lie between $\pm\pi/a$, the possible values for j are $\pm 1, \pm 2, \ldots, \pm N/2$. The thermodynamic energy of the crystal (without the zero-point energy) is then

$$\begin{aligned} E &= \sum_j \frac{\hbar\omega_j}{\exp(\beta\hbar\omega_j) - 1} \\ &= \frac{Na}{\pi}\int_0^{\pi/a} \frac{\hbar\omega(k)\,dk}{\exp(\beta\hbar\omega(k)) - 1} \end{aligned} \qquad (11\text{–}44)$$

where the summation over the possible wave vectors defined above has been approximated by an integral and the fact that $\omega(k) = \omega(|k|)$ has been used. We see, then, that if $\omega(k)$, that is, the dispersion curve of the lattice, is known, we can calculate E and other thermodynamic properties of the lattice.

This integral over k can be converted to an integral over ω itself by means of the dispersion curve.

$$\begin{aligned} dk &= \frac{dk}{d\omega}\,d\omega = \frac{d}{d\omega}\left\{\frac{2}{a}\sin^{-1}\left(\frac{\omega}{\omega_{\text{max}}}\right)\right\} d\omega \\ &= \frac{2\,d\omega}{a(\omega_{\text{max}}^2 - \omega^2)^{1/2}} \end{aligned} \qquad (11\text{–}45)$$

Substituting this into Eq. (11–44) gives

$$E = \frac{2N}{\pi}\int_0^{\omega_{\text{max}}} \frac{\hbar\omega\,d\omega}{[\exp(\beta\hbar\omega) - 1][\omega_{\text{max}}^2 - \omega^2]^{1/2}} \qquad (11\text{–}46)$$

If we compare this to Eq. (11–9) (without the zero-point energy $h\nu/2$), we see that

$$g(\nu) = \frac{2N}{\pi}\frac{1}{(\nu_{\text{max}}^2 - \nu^2)^{1/2}} \qquad (11\text{–}47)$$

Problem 11–33 is involved with showing that a one-dimensional Debye approximation agrees with Eq. (11–47) as $\nu \to 0$.

In one dimension, the distribution of frequencies is related to the dispersion curve [*cf.* Eq. (11–47) and Eq. (11–45)] by

$$g(\nu) = \frac{Na}{\pi}\frac{1}{d\nu/dk}$$

Notice that $g(\nu)$ has singularities at the points where $d\nu/dk$ equals zero. For a continuum, $d\nu/dk$ = velocity = constant. In general, however, $d\nu/dk$ is not constant. The derivative dv/dk is called the group velocity of the wave and physically represents the rate-of-energy transmission of the wave. For a continuum, the group velocity and phase velocity [Eq. (11–43)] are both equal to the same constant value.

Now let us consider a one-dimensional lattice with two alternating kinds of atoms with masses m_1 and m_2. This represents a one-dimensional analog of a crystal like sodium chloride. The Hamiltonian for this lattice is

$$H = \sum_{j=1}^{N} \left\{ \frac{m_1}{2} \dot{\xi}_{2j}^2 + \frac{m_2}{2} \dot{\xi}_{2j-1}^2 \right\} + \frac{f}{2} \sum_{j=1}^{N} \{(\xi_{2j} - \xi_{2j-1})^2 + (\xi_{2j+1} - \xi_{2j})^2\}$$

In this case one gets two sets of equations of motion, one for the masses m_1 and one for the masses m_2, which eventually yield (see Problem 11–34)

$$\omega^2 = \omega_0^2 \left\{ 1 \pm \left(1 - \frac{4m_1 m_2 \sin^2 \phi}{(m_1 + m_2)^2} \right)^{1/2} \right\} \tag{11–48}$$

where

$$\omega_0^2 = \frac{f}{\mu} \tag{11–49}$$

In these equations, ϕ = multiple of π/N, and μ is the reduced mass of m_1 and m_2.

According to which sign is chosen, the dispersion curve of Eq. (11–48) yields two branches: a high-frequency branch called the optical branch and a low-frequency branch called the acoustical branch. Figure 11–10(a) shows these two dispersion curves, and Fig. 11–10(b) shows the corresponding frequency distribution $g(\nu)$. In the normal modes belonging to the acoustical branch, neighboring atoms are displaced in the same direction (as in the lower curve in Fig. 11–5) to produce a long wavelength mode. In the optical branch, neighboring atoms are displaced in opposite directions (as in the upper curve in Fig. 11–5).

If the two masses of our lattice are ions of opposite sign (such as NaCl), the vibrational motion in the optical modes produces oscillating dipole moments. An oscillating dipole moment leads to an absorption of infrared radiation. Figure 11–11 shows the infrared spectrum of NaCl. Furthermore, the larger the reduced mass, the lower

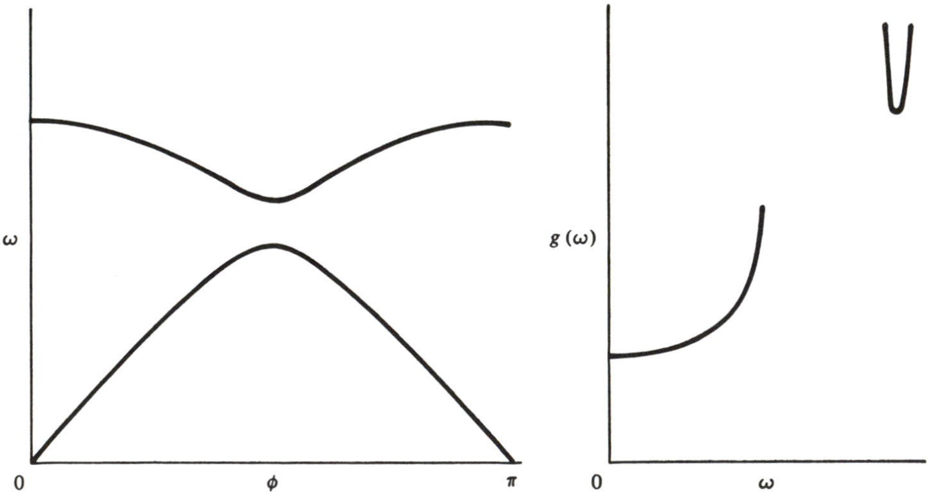

(a) (b)

Figure 11–10. **(a) The angular frequency for a diatomic chain as a function of the phase angle ϕ. (b) Density of normal vibrations of a linear chain with $m_1/m_2 = 3$.**

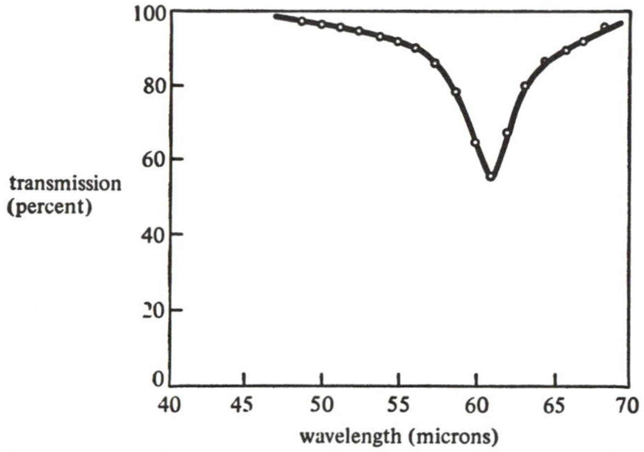

Figure 11–11. **The absorption of infrared radiation through a thin film of sodium chloride.** (From C. Kittel, *Solid State Physics, 3rd ed.*, New York: Wiley, 1967; after R. B. Barnes, *Z. Physik.*, **75**, 723, 1932.)

the frequency of the optical modes, and hence the longer the wavelengths at which the infrared absorption occurs. Figure 11–12 shows the onset of absorption in various alkali halides. Evidently, if we wish to get good transmission in the infrared, we must use a crystal in which the ions are as heavy as possible. Such considerations are important when designing prisms for infrared spectrometers.

The exact lattice dynamic calculations of two- and three-dimensional lattices proceeds in much the same way that we did in the previous section, but the actual calculations are much more difficult. Figure 11–13 shows two experimentally determined vibrational spectra to give an idea of the complexity of the spectra that are obtained for three-dimensional lattices. There have been several exact or near-exact calculations of such spectra, and there is now an extensive literature on lattice dynamics calculations (see "Additional Reading").

The last two sections of this chapter deal with two special topics, namely, the concept of phonons (Section 11–5) and the concept of defects in crystal lattices (Section 11–6).

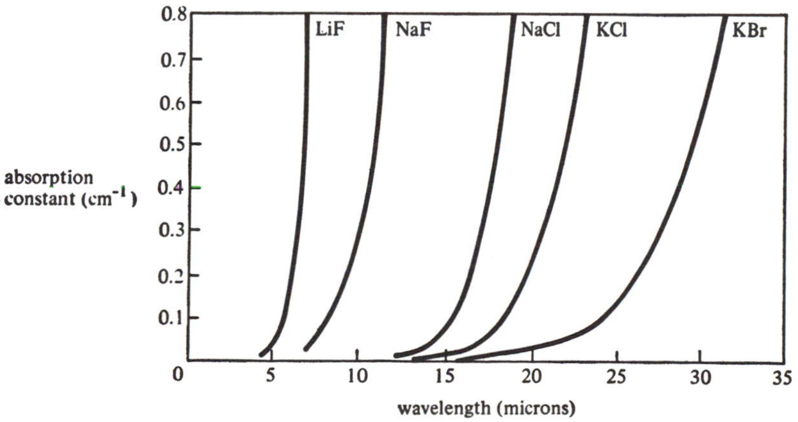

Figure 11–12. **The onset of the absorption of infrared radiation by alkali halide crystals, showing that those alkali halide ion pairs with the heavier reduced masses begin to absorb at longer wavelengths.** (From C. Kittel, *Solid State Physics, 3rd ed.*, New York: Wiley, 1967.)

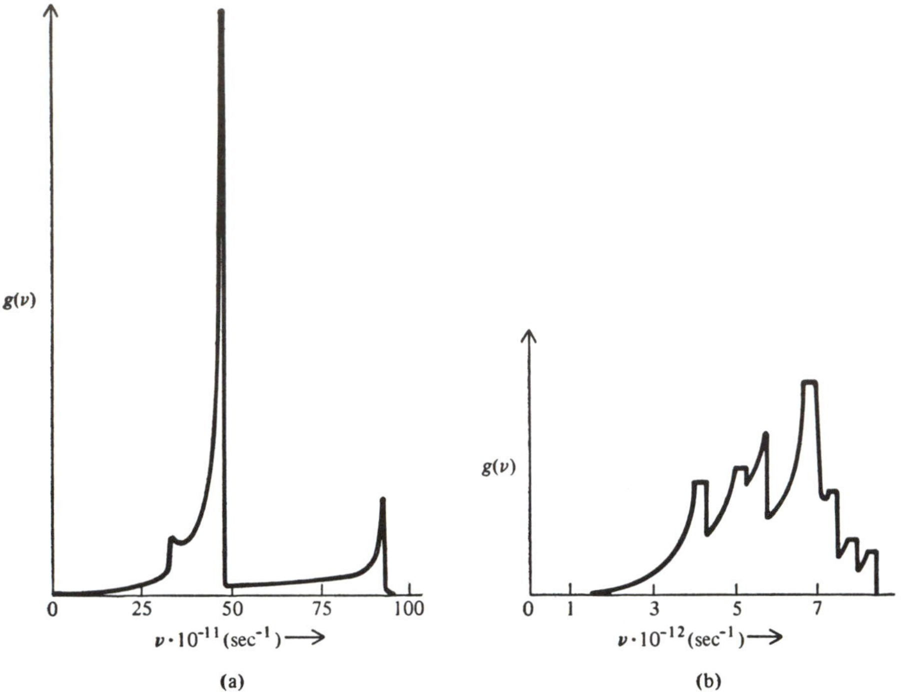

Figure 11–13. **The experimental frequency spectrum of (a) aluminum and (b) iron.** (From M. Blackman in *Encyclopedia of Physics*, vol. VII, pt. 1, ed. by S. Flügge. Berlin: Springer-Verlag, 1955.)

11-5 PHONONS

We have seen in Section 11–1 that if a crystal has N atoms, it has $3N$ normal coordinates, each with its own characteristic frequency $\nu_1, \nu_2, \ldots, \nu_{3N}$. The total energy of the crystal is

$$E(\{n_j\}) = \sum_{j=1}^{3N} h\nu_j(n_j + \tfrac{1}{2}) \tag{11–50}$$

$$= \sum_{j=1}^{3N} h\nu_j n_j + \sum_{j=1}^{3N} \frac{h\nu_j}{2}$$

$$= \sum_{j=1}^{3N} h\nu_j n_j + E_0 \tag{11–51}$$

where we have set E_0 equal to the total zero-point vibrational energy. Since E is a sum of terms, this expression for E can be *interpreted* as the energy of a system of *independent* particles, which occupy the states 1, 2, ..., $3N$ with corresponding energies $h\nu_1, h\nu_2, \ldots, h\nu_{3N}$ with n_1 particles in the first state, n_2 in the second, and so on. Note that the set of numbers $\{n_j\}$, the occupation numbers of the $3N$ states, completely specifies the state of the system. Since it is only the number of "particles" in each state that specifies the system, we can consider these "particles" to be indistinguishable, and furthermore, since there is no restriction on the numbers n_j, that is, $n_j = 0, 1, \ldots, 3N$, these "particles" are bosons.

This is actually a very useful interpretation, since it gives us a system of *non-interacting* bosons, and we can use the Bose-Einstein formulas of Chapter 4 or

Chapter 10. As far as the vibrations of the lattice are concerned then, we shall treat it as if it were an *ideal* Bose-Einstein gas. The "particles" we have invented here are examples of the quasi-particles we mentioned at the beginning of Chapter 4 and are called *phonons*. Phonons are essentially quanta of lattice vibrations, just as photons are quanta of electromagnetic vibrations. Since lattice vibrations are closely related to sound waves passing through the crystals, phonons can be thought of as quanta of sound waves.

We can directly apply the basic Bose-Einstein equations of Chapter 4 or even the formulas developed in Section 10–5 for a photon gas. For instance, the equation for the average occupation of the jth state is [Eq. (4–26)]

$$\bar{n}_j = \frac{\lambda e^{-\beta\varepsilon_j}}{1 - \lambda e^{-\beta\varepsilon_j}} = \frac{1}{\lambda^{-1} e^{\beta\varepsilon_j} - 1} \tag{11–52}$$

As we saw in Chapter 10, these quantum statistical equations are awkward to use because of the presence of λ, but in the case of a phonon gas, we can evaluate λ easily. The reason for this is that the number of phonons is, in fact, not fixed, since clearly it is possible to have a number of sets $\{n_j\}$ in which $E(\{n_j\})$ is fixed but $n = \sum_j n_j$ is different. Thermodynamically, the system is characterized by E and V only. A system of phonons, then, is mathematically identical to a photon gas (photons obey Bose-Einstein statistics and their number is not conserved), and in Chapter 10 we showed that $\mu = 0$ or $\lambda = 1$ for this case. In Section 10–5 we gave a simple thermodynamic proof that $\mu = 0$. In brief, the phonons are likened to a chemical equilibrium of the type $nA \rightleftharpoons mA$, where n and m are integers. Since the number of phonons is not conserved, $n \neq m$. The condition for equilibrium (Chapter 9) is that $(m - n)\mu = 0$, and thus we have $\mu = 0$ and $\lambda = 1$.

With λ set equal to 1, then we have

$$\bar{n}_j = \frac{1}{e^{\beta\varepsilon_j} - 1} \tag{11–53}$$

and

$$\bar{E} = \sum_{j=1}^{3N} \bar{n}_j h\nu_j + E_0 = \sum_{j=1}^{3N} \frac{h\nu_j}{e^{\beta\nu_j} - 1} + E_0 \tag{11–54}$$

If we introduce $g(\nu)$, we can write

$$\bar{E} = E_0 + \int_0^\infty \frac{g(\nu) h\nu \, d\nu}{e^{\beta h\nu} - 1} \tag{11–55}$$

which is the same as Eq. (11–9). Equation (11–54) is the same as Eq. (11–44) without the zero-point energy. The derivative of this gives Eq. (10–10) for the heat capacity. We can thus derive all of the results of Sections 11–2 and 11–3 by treating the lattice vibrations of a crystal as a gas of noninteracting phonons.

Actually, the concept of phonons is much more useful than this. For instance, phonons can be assigned a momentum (just as a photon can), which must be conserved in collisions. The collision of the phonons and the electrons in a metal lead to the electrical resistance of a metal. The inelastic scattering of phonons with photons is known as Brillouin scattering. The vibrational frequency spectrum of a crystal can be determined experimentally by inelastic neutron scattering, which is treated theoretically by phonon–neutron scattering.

11-6 POINT DEFECTS IN SOLIDS

Up to now we have assumed that every atom or ion of a crystal is situated at a lattice site and that every lattice site is occupied by one and only one particle. Such a perfect periodic arrangement is called a perfect crystal. We shall show below, however, that such a perfect arrangement is thermodynamically unattainable. Any deviation from such perfect behavior is called an imperfection or a defect. There are quite a variety of defects that exist in crystals, but in this section we shall study only the most common *point* defects, namely, vacant lattice sites and interstitial atoms (extra atoms not at lattice sites). Although it will turn out that the concentration of such defects will be fairly small, they nevertheless have a profound effect on the properties of crystals. The conductivity of some semiconductors is due entirely to trace amounts of chemical impurities. The color of many crystals is due to imperfections, and the mechanical and elastic properties of crystals depend strongly on the number and type of defects present. The diffusion of atoms through solids is another property that is dependent upon defects.

The simplest type of defect is a lattice vacancy. Such a missing atom or ion is known as a *Schottky defect*. A Schottky defect can be created by transferring an atom or ion from the body of the crystal to its surface. Although it requires energy to do this, there is an increase of entropy. The final equilibrium concentration is found by minimizing the free energy. If we assume that it takes an energy ε_v to bring an atom from an interior lattice site to a surface lattice site and also assume that the concentration of defects is small enough to consider them to be independent, then we can write

$$\begin{aligned} A(n) &= E - TS \\ &= n\varepsilon_v - kT \ln \frac{N!}{n!(N-n)!} \end{aligned} \tag{11-56}$$

The quantity n is the number of vacancies, and the combinatorial factor is just the number of ways of distributing n vacancies over N sites. We now minimize $A(n)$ with respect to n to find the number of vacancies expected. Setting $(\partial A/\partial n)_T$ equal to zero and solving for n give

$$n \simeq N e^{-\varepsilon_v/kT} \tag{11-57}$$

where we have neglected n compared to N. If ε_v is 1 eV and $T = 300°\text{K}$, then $n/N \approx 10^{-17}$. At 1000°K, $n/N \approx 10^{-5}$ (see Problem 11–19). (In ionic crystals such as NaCl, it is usually favorable to form roughly equal numbers of positive and negative defects. This keeps the crystal electrostatically neutral on a local scale.)

The other common type of defect that we shall discuss here is a *Frenkel defect*, in which an atom is displaced from a lattice position to an interstitial position. If we let ε_I be the energy it takes to do this, let N be the number of lattice sites, and let N' be the number of possible interstitial sites, then

$$A(n) = n\varepsilon_I - kT \ln\left\{\frac{N!}{n!(N-n)!} \cdot \frac{N'!}{n!(N'-n)!}\right\} \tag{11-58}$$

In this case the combinatorial is a product of the number of ways of choosing n out of N lattice sites and the number of ways of distributing the n chosen atoms over the N' available interstitial sites. Minimizing $A(n)$ gives (see Problem 11–20)

$$n \approx (NN')^{1/2} e^{-\varepsilon_I/2kT} \tag{11-59}$$

It turns out that the most common type of point defect in alkali halides are Schottky defects, and the most common defect in silver halides are Frenkel defects. This is probably due to the fact that silver ions are smaller than alkali ions and so can fit into interstitial positions more easily. It is possible to determine if one type of defect is predominant by careful measurements of the density. The formation of Schottky defects lowers the density of the crystal since the volume is increased with no change in mass. On the other hand, the formation of Frenkel defects does not change the volume, and so there is no change in density. It is also possible to use ionic conductivity measurements to differentiate between Schottky and Frenkel defects.

Lattice vacancies in controlled concentrations can be produced by the addition of divalent ions. For instance, if a crystal of KCl is grown with controlled amounts of $CaCl_2$, the Ca^{2+} enters the lattice at a normal K^+ site, and the two Cl^- enter at two normal Cl^- sites. The net result of this is the production of vacant positive ion sites. This is shown in Fig. 11–14. The formation of these lattice vacancies can be observed by measuring the density of $CaCl_2$–KCl mixtures. The volume of such a mixture is actually larger than the volume of the separate components.

We said earlier in this section that the rate of diffusion in solids is greatly affected by the presence of defects. As an atom diffuses through a crystal, it must surmount a series of energy barriers presented by its neighbors as it moves from lattice site to lattice site or from interstitial position to interstitial position. Let us consider interstitial diffusion of impurities. If this barrier height is ε, then $\exp(-\varepsilon/kT)$ can be thought of as the fraction of time that the atom will have an energy exceeding ε. If ν is the frequency with which the diffusing atom vibrates around its interstitial position, then the probability per unit time that the atom will be able to pass over the barrier is

$$p \approx \nu e^{-\varepsilon/kT} \tag{11–60}$$

Now consider two parallel planes of impurity atoms in interstitial sites. The planes are separated by the lattice constant a. There will be c impurity atoms on one plane and $c + a(dc/dx)$ on the other. The net number of atoms crossing between these planes per unit time is $pa(dc/dx)$. If n is the concentration of impurity atoms, then $c = an$. The diffusion flux is then

$$j = -pa^2 \frac{\partial n}{\partial x} \tag{11–61}$$

Remember that a flux of a quantity Ψ is the rate of flow of Ψ through a unit area of surface per second. The minus sign occurs because the direction of the diffusion is

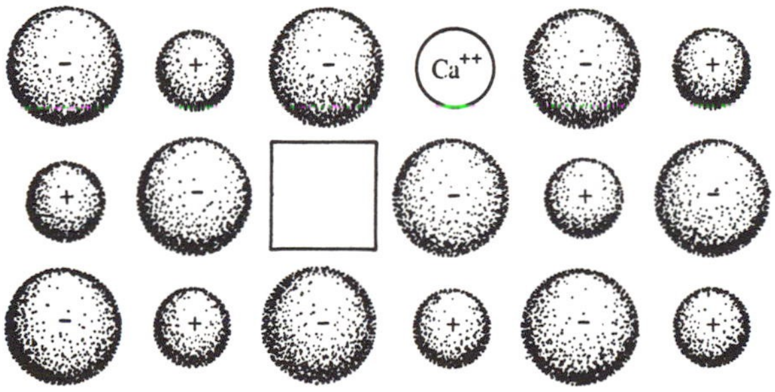

Figure 11–14. **Production of a lattice vacancy by the solution of $CaCl_2$ in KCl: to ensure electrical neutrality a positive ion vacancy is introduced into the lattice with each divalent cation Ca^{2+}. The two Cl^- ions of $CaCl_2$ enter normal negative ion sites. (From C. Kittel, *Solid State Physics*, 3rd ed. New York: Wiley, 1967.)**

opposed to the concentration gradient. We compare this to Fick's law, which simply states that a diffusion flux is proportional to the concentration gradient of the diffusing substance, that is,

$$j = -D\frac{\partial n}{\partial x} \tag{11-62}$$

where D is the diffusion constant. We see then that

$$D = pa^2 \approx \nu a^2 e^{-\varepsilon/kT} \tag{11-63}$$

Figure 11–15 shows the experimentally determined temperature dependence of the diffusion coefficient of carbon in iron. If we let ν be 10^{14} sec^{-1}, a be 3×10^{-8} cm, and ε be 1 eV, then $D \approx 10^{-18}$ cm^2/sec at 300°K and 10^{-6} cm^2/sec at 1000°K. (See Problem 11–23.)

Another property that is much affected by defects is the color of slightly impure (doped) alkali halide crystals. Pure alkali halides are transparent in the visible region, but if a sodium chloride crystal is heated in sodium vapor, it becomes yellow. Similarly, if KCl is heated in potassium vapor, it takes on a magenta color. These same effects can be produced by X-ray, neutron, or electron bombardment or by electrolysis. The color of the doped crystals is due to defects which absorb light and are called *color centers.* A number of different kinds of color centers have been discovered, but here we shall mention only the first discovered and perhaps the simplest, namely, an *F center.* The name comes from the German word for color, *farbe. F* centers have been identified by electron spin resonance to be electrons bound to a negative ion vacancy. This is consistent with the fact that they can be produced by heating the pure crystal in an excess of the metal vapor. When this is done, the metal atoms of the vapor are incorporated into the crystal lattice. The metal atom loses its electron, and the positive ion takes up a lattice site while the electron associates itself with a vacant negative

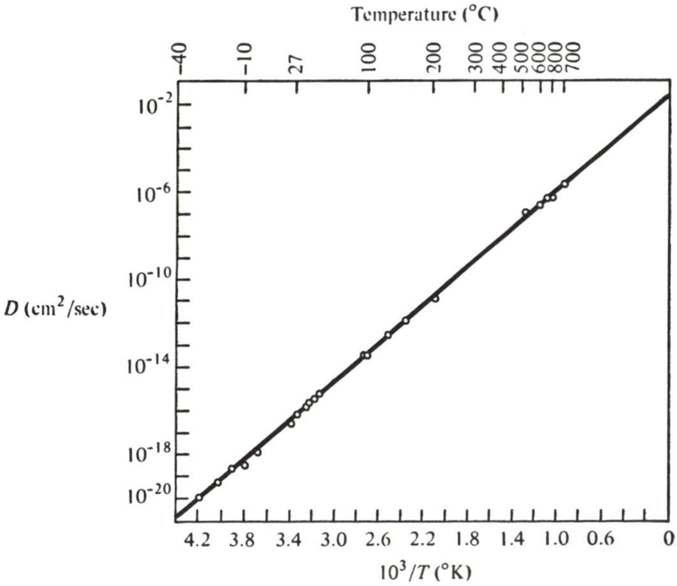

Figure **11–15. Diffusion coefficient of carbon in iron.** (From C. Kittel, *Solid State Physics*, 3rd ed. New York: Wiley, 1967.)

ion lattice site which is lacking an anion. A vacant negative ion lattice site has an effective positive charge, and so the electron behaves somewhat like the electron bound to a nucleus and can absorb light. The model is consistent with a number of experimental facts. For instance, F band absorption is characteristic of the crystal and not of the alkali metal used in the vapor; that is, NaCl will become yellow whether heated in sodium vapor or any other alkali metal vapor. Furthermore, the absorption corresponds quantitatively to the amount of excess metal, and it has been determined that colored alkali halide crystals are less dense than pure alkali halide crystals.

ADDITIONAL READING

General

BLACKMAN, M. 1955. In *Encyclopaedia of physics*, Vol. VII, pt. 1, ed. by S. Flügge. Berlin: Springer-Verlag.

DEKKER, A. J. 1957. *Solid state physics*. Englewood Cliffs, N.J.: Prentice-Hall.

FOWLER, R. H., and GUGGENHEIM, E. A. 1956. *Statistical thermodynamics*. Cambridge: Cambridge University Press. Chapter 4.

KESTIN, J., and DORFMAN, J. R. 1971. *A course in statistical thermodynamics*. New York: Academic. Chapter 9.

KITTEL, C. 1967. *Solid state physics*, 3rd ed. New York: Wiley.

MAYER, J. E., and MAYER, M. G. 1940. *Statistical Mechanics*. New York: Wiley. Chapter 11.

MOORE, W. J. 1967. *Seven solid states*. New York: Benjamin.

WANNIER, G. H. 1966. *Statistical physics*. New York: Wiley. Chapter 13.

Einstein and Debye theories

DAVIDSON, N. 1962. *Statistical mechanics*. New York: McGraw-Hill. Chapter 16.

EYRING, H., HENDERSON, D., STOVER, B. J., and EYRING, E. M. 1964. *Statistical mechanics and dynamics*. New York: Wiley. Chapter 7.

HILL, T. L. 1960. *Statistical thermodynamics*. Reading, Mass.: Addison-Wesley. Chapter 5.

KNUTH, E. 1966. *Statistical thermodynamics*. New York: McGraw-Hill. Chapters 9 and 14.

MANDL, F. 1971. *Statistical physics*. New York: Wiley. Chapter 6.

REIF, F. 1965. *Statistical and thermal physics*. New York: McGraw-Hill. Sections 10–1 and 10–2.

Lattice dynamics

DE LAUNAY, J. 1956. *Solid state physics*, Vol. 2, ed. by F. Seitz and D. Turnbull. New York: Academic.

MARADUDIN, A., *Ann. Rev. Phys. Chem.*, **14**, p. 89, 1963.

MARADUDIN, A. A., MONTROLL, E. W., and WEISS, G. H. 1963. *Theory of lattice dynamics in the harmonic oscillator approximation*. New York: Academic.

MITRA, S. S. 1962. *Solid state physics*, Vol. 13, ed. by F. Seitz and D. Turnbull. New York: Academic.

Defects in crystals

BARR, L. W., and LIDIARD, A. B. 1970. In *Physical chemistry, an advanced treatise*, Vol. X, ed. by H. Eyring, D. Henderson, and W. Jost. New York: Academic.

PROBLEMS

11–1. The difference between the constant pressure and constant volume heat capacities is

$$C_p - C_V = -T\left(\frac{\partial V}{\partial T}\right)_p^2 \left(\frac{\partial p}{\partial V}\right)_T$$

In terms of the volume expansion coefficient $\alpha_V = (1/V)(\partial V/\partial T)_p$ and the isothermal compressibility $\kappa = -(1/V)(\partial V/\partial p)_T$, this difference is

$$C_p - C_V = \alpha_V^2 \frac{TV}{\kappa}$$

which is often rewritten as

$$C_p = C_V(1 + \gamma\alpha_V T)$$

where $\gamma = \alpha_V V/\kappa C_V$ is practically independent of temperature. This is called the Grüneisen constant. Calculate the Grüneisen constant given that $\alpha_V = 6.22 \times 10^{-5}$ deg^{-1} and $\kappa = 12.3 \times 10^{-12}$ cm^2-dyne^{-1} at room temperature.

11–2. Find both the high- and low-temperature limiting forms of the heat capacity according to the Einstein model.

11–3. Derive an expression for the Einstein specific heat for a two-dimensional crystal.

11–4. Determine the various thermodynamic properties of an Einstein crystal.

11–5. Derive Eq. (11–24).

11–6. Prove that

$$\int_0^\infty \frac{x^4 e^x \, dx}{(e^x - 1)^2} = \frac{4\pi^4}{15}$$

Hint: See Problem 1–63.

11–7. Derive an expression for the heat capacity of a Debye crystal as a power series expansion in Θ_D/T.

11–8. Why is the heat capacity of diamond at room temperature far below its Dulong-Petit value?

11–9. Prove that for a monatomic crystal that

$$\int_0^\infty [3Nk - C_V(T)] = E(0)$$

where $E(0)$ is the energy of the crystal at 0°K. Give a graphical interpretation of this equation.

11–10. Starting with a two-dimensional wave equation, derive a Debye type equation for the heat capacity C_V for a two-dimensional crystal, assuming that the transverse and longitudinal velocities are the same.

11–11. Show that in a monatomic crystal, the high-temperature limiting form of the heat capacity depends only on the existence of a cutoff frequency and is given by the law of Dulong and Petit. That is, assume only that the distribution of frequencies

$$g(\nu) = 0 \qquad \text{for } \nu > \nu_{\text{max}}$$

11–12. The potential energy of the atoms of a solid of density N/V in their equilibrium positions is denoted by $U(0; N/V)$. The normal frequencies of vibration of the atoms near their equilibrium positions are functions of the density $\nu_j(N/V)$ where $j = 1, 2, \ldots, 3N - 6$. It is a good approximation that

$$\frac{\partial \log \nu_j}{\partial \log V} = -\gamma \qquad (j = 1, 2, \ldots, 3N - 6)$$

for all frequencies. The constant γ is called the Grüneisen constant. Show that under this approximation, the pressure of the solid is given by

$$p = -\frac{\partial U}{\partial V} + \gamma \frac{E}{V}$$

This equation of state is known as the Mie-Grüneisen equation.

11–13. A modification of the Debye theory was introduced by Born, who proposed a different cutoff for the spectrum of vibrational modes. He proposed that the cutoff be made such that both the longitudinal and transverse modes have a common minimum wavelength. If we denote this common minimum wavelength by λ_m, then $\lambda_m \nu_{\text{long}} = c_{\text{long}}$ and $\lambda_m \nu_{\text{trans}} = c_{\text{trans}}$, Equation (11–23) now becomes

$$4\pi V \left\{ \int_0^{\nu_t} \frac{2}{c_t^3} \nu^2 \, d\nu + \int_0^{\nu_l} \frac{\nu^2 \, d\nu}{c_l^3} \right\} = 3N$$

Show that this leads to the following expression for the specific heat:

$$C_V = R\left[D\left(\frac{\Theta_l}{T}\right) + 2D\left(\frac{\Theta_t}{T}\right)\right]$$

where $D(x)$ is the Debye function

$$D(x) = \frac{3}{x^3}\int_0^x \frac{e^z z^4\, dz}{(e^z - 1)^2}$$

11–14. Derive Eq. (11–33) from (11–32).

11–15. The Debye theory treats the crystal and as a continuum body, and hence the dispersion relation for the phonons is $\omega = ck$, where c is the speed of sound in the body. In a ferromagnetic solid at low temperatures, there exist quantized waves of magnetization called spin waves, and the dispersion relation for these type of waves is $\omega \propto k^2$. Find the low-temperature heat capacity due to spin waves.

11–16. Suppose that wavelike quasi-particles, having the dispersion relation $\omega = Ak^n$, exist in a solid and yield a specific heat when they are excited as thermal motion. Using the relation

$$E = \sum_{\mathbf{k}} \frac{\hbar\omega_{\mathbf{k}}}{\exp(\beta\hbar\omega_{\mathbf{k}}) - 1}$$

$$= \frac{V}{(2\pi)^3}\int \frac{\hbar\omega(\mathbf{k})\, d\mathbf{k}}{\exp(\beta\hbar\omega(\mathbf{k})) - 1} \qquad \text{(three dimensions)}$$

show that the heat capacity is proportional to $T^{3/n}$ at low temperatures. Note that for $n = 1$, that is, $\omega/k = \lambda\nu = A = \text{constant}$, one has the Debye theory.

11–17. Consider a planar square lattice of identical atoms which vibrate perpendicular to the plane of the lattice. If we let u_{lm} be the displacement of the atom in the lth column and mth row, show that the equation of motion is

$$m\left(\frac{d^2 u_{lm}}{dt^2}\right) = f[(u_{l+1,\,m} + u_{l-1,\,m} - 2u_{lm}) + (u_{l,\,m+1} + u_{l,\,m-1} - 2u_{lm})]$$

where m is the mass of an atom, and f is the force constant. By analogy with Eq. (11–39), assume a solution of the form

$$u_{lm} = \exp[i(lk_y a + mk_x a + \omega t)]$$

where a is the nearest-neighbor lattice spacing. Show that the dispersion relation for this system is

$$\omega^2 = \frac{2f}{m}(2 - \cos k_x a - \cos k_y a)$$

For $ka = (k_x^2 + k_y^2)^{1/2}a$ where $a \ll 1$, show that

$$\omega = \left(\frac{fa^2}{m}\right)^{1/2} k$$

11–18. Use the equations $E = h\nu$ and $p = h/\lambda$ to show that the group velocity for a free particle of mass m is p/m.

11–19. Calculate the number of Schottky defects per mole of crystal at 300°K and 1000°K given that it takes 1.0 eV to bring an atom or ion from an interior lattice site to a surface lattice site.

11–20. Show that the number n of Frenkel defects in equilibrium in a crystal having N lattice points and N' possible interstitial positions is given by the equation

$$\varepsilon_1 = k_B T \log\left[\frac{(N-n)(N'-n)}{n^2}\right]$$

which, for $n \ll N, N'$, gives

$$n \simeq (NN')^{1/2} \exp\left(\frac{-\varepsilon_1}{2k_B T}\right)$$

Here ε_1 is the energy necessary to remove an atom from a lattice site to an interstitial position.

11–21. (a) Assuming a simple coulombic interaction between positive and negative ion vacancies, calculate the binding energy between a pair of oppositely charged vacancies in NaCl.

(b) If n_1 represents the number of single positive or negative ion vacancies and n_2 the number of pairs of vacancies, derive an expression for the Helmholtz free energy and show by minimizing this with respect to n_1 and n_2 that

$$\frac{n_2}{n_1} = 6 \exp\left[\frac{(\varepsilon - \frac{1}{2}\phi)}{kT}\right]$$

where ε is the binding energy of a pair of oppositely charged vacancies, and ϕ is the energy required to produce a single positive and negative ion vacancy.

11–22. Treat an F center as a free electron moving in the field of a point charge e in a medium of dielectric constant $\varepsilon = n^2$ where n is the index of refraction. What is the $1s - 2p$ energy difference of F centers in NaCl?

11–23. If a sodium atom next to a vacancy has to move over a potential hill of 0.5 eV, and the atomic vibration frequency is 10^{12} Hz, estimate the diffusion coefficient at room temperature for radioactive sodium in normal sodium. Assume a lattice spacing of 4 Å.

11–24. Consider a closed container containing a small solid and its vapor in equilibrium at temperature T. Assume that the volume of the solid, v, is much less than the volume of the container, V. Let the partition function of the solid be of the form $Q = q_s(T)^{N_s}$, and let there be N_g molecules in the vapor phase. Show that the equilibrium condition is given by

$$N_g = \frac{q_g}{q_s}$$

where q_g is the partition function of a vapor phase molecule. Hint: Minimize the total free energy of the system, that is, $A_g(T, V, N_g) + A_s(T, N_s)$, with respect to N_g keeping $N_g + N_s = N =$ constant.

11–25. Using the result of Problem 11–24, show that the vapor pressure of an Einstein crystal is

$$p = kT\left(\frac{2\pi mkT}{h^2}\right)^{3/2}\left(2 \sinh \frac{\Theta_E}{2T}\right)^3 e^{-\phi/kT}$$

where ϕ is given by $U(0; \rho) = -N_s\phi$.

11–26. The pV term in $G = A + pV$ can be neglected for a condensed phase. Using this fact, the chemical potential of a Debye crystal can be well approximated by $\mu = A/N$ rather than the correct G/N. Show that the vapor pressure of a Debye crystal is given by

$$\ln p = \ln\left[\left(\frac{2\pi mkT}{h^2}\right)^{3/2} kT\right] + \left(\frac{\phi(0)/2 + 9k\Theta_D/8}{kT}\right) - \frac{\pi^4}{5}\left(\frac{T}{\Theta_D}\right)^3$$

Identify the quantity $\phi(0)/2 + 9K\Theta_D/8$.

11–27. Consider a gas in equilibrium with the surface of a solid. Some of the molecules of the gas will be adsorbed onto the surface, and the number adsorbed will be a function of the pressure of the gas. A simple statistical mechanical model for this system is to picture the solid surface to be a two-dimensional lattice of M sites. Each of these sites can be either unoccupied or occupied by at most one of the molecules of the gas. Let the partition function of an unoccupied site be 1 and that of an occupied site be $q(T)$. (We do not need to know $q(T)$ here.) Assuming that molecules adsorbed onto the lattice sites do not interact with each other, the partition function of N molecules adsorbed onto M sites is then

$$Q(N, M, T) = \frac{M!}{N!(M-N)!}[q(T)]^N$$

The binomial coefficient accounts for the number of ways of distributing the N molecules over the M sites. By using the fact the adsorbed molecules are in equilibrium with the gas phase molecules (considered to be an ideal gas), derive an expression for the fractional coverage, $\theta \equiv N/M$, as a function of the pressure of the gas. Such an expression, that is, $\theta(p)$, is called an adsorption isotherm, and this model gives the so-called Langmuir adsorption isotherm.

11–28. The low-temperature constant volume heat capacity of many metals can be written in the form

$$C_V = \gamma T + AT^3$$

where the T^3 term is from the lattice vibrations and the linear term is due to the electrons. In the previous chapter it was shown that C_V for a free electron gas is given by $C_V = \pi^2 NkT/2T_F$, where T_F is the Fermi temperature. Given that the Fermi temperatures of Na and Cu are 3.7×10^4°K and 8.2×10^4°K, respectively, calculate γ and compare it to the experimental values.

11–29. Compare the contributions of the electrons and the lattice vibrations to the heat capacity of sodium at low temperatures.

11–30. Show that the bulk modulus $B = V(\partial P/\partial V)_T$ of an electron gas at 0°K is $B = \frac{5}{3}P = 10E_0/9V$. The valence electron density of potassium is about 1.40×10^{22} cm^{-3}. Calculate B for potassium and compare with the experimental value of 3.66×10^{10} dyne/cm^2 at 4°K.

11–31. Show that the entropy of a Debye crystal at low temperature is given by

$$S = \frac{4\pi^4 Nk}{5}\left(\frac{T}{\Theta_D}\right)^3$$

11–32. The heat capacity of copper at 100°K is 3.85 cal/mole-deg. Using this information, calculate the value of Θ_E and Θ_D for copper. Now calculate the heat capacity at 25°K and compare it to the experimental value of 0.23 cal/mole-deg. Which model gives better results at low temperatures?

11–33. Show that $g(\nu)$ of a one-dimensional Debye crystal agrees with Eq. (11–47) as $\nu \to 0$.

11–34. Derive Eq. (11–48).

CHAPTER 12

TWO SIMPLE THEORIES OF LIQUIDS

In the previous chapter we saw that even though the interatomic or intermolecular interactions cannot be neglected, it is still possible to apply the statistical thermodynamic equations of systems of independent particles to the solid state. In this chapter we shall develop two rather simple theories of liquids that are also based on an independent particle picture. The first of these theories, called the theory of significant structures, specifically recognizes that a liquid is intermediate in structure between a gas and a solid and constructs an approximate partition function for a liquid that is composed of a gaslike part and a solidlike part. This theory is discussed in Section 12–1. Then in Section 12–2 we discuss the Lennard-Jones Devonshire theory of liquids. In this theory, we treat a liquid like a solid in which the molecules no longer vibrate harmonically about their lattice sites, but are allowed to wander throughout a space bounded by their neighboring molecules. This theory starts with a solidlike partition function and modifies it in several ways so that it is representative of a liquid phase.

12–1 THE THEORY OF SIGNIFICANT STRUCTURES

In the 1930s Eyring formulated a hole theory of liquids based on the assumption that a liquid is similar to a solid that has a large number of vacant lattice sites. The basis of his assumption is that the volume of most liquids is greater than that of the solid under similar conditions, and this extra volume exists as holes. The hole theory of liquids has undergone many modifications and extensions since then, and the 1960s saw the latest extension, called the theory of significant structures.

The central idea of the theory of significant structures is that the partition function of a liquid is made up of a gaslike part and a solidlike part. The presence of holes in the liquid is assumed to confer gaslike properties on the molecules next to these holes. If V is the volume of the liquid, and V_s is the volume of the corresponding solid phase, then $V - V_s$ is considered to be the volume available to these gaslike molecules, and

$N(V - V_s)/V$ is said to be the number of such gaslike molecules. The contribution from these gaslike molecules to the partition function of the liquid is then given by an equation like Eq. (5–14), but with V replaced by $V - V_s$ and N replaced by $N(V - V_s)/V$:

$$\left\{\left(\frac{N(V - V_s)}{V}\right)!\right\}^{-1} \left\{\frac{(2\pi mkT)^{3/2}(V - V_s)}{h^3}\right\}^{N(V-V_s)/V} \tag{12–1}$$

The remaining NV_s/V of the molecules of the fluid are assumed to be solidlike, and each one vibrates about some point in space with the same vibrational frequency that it would have in the solid.

We should emphasize here that we are not assuming that some parts of the liquid have a solid structure and some parts have a gaseous structure, but we are assuming that at any instant of time, a particular molecule in the fluid will find itself surrounded by other molecules and thus either will behave like a solidlike molecule or will be next to a hole and thus have a gaslike property. The reader is referred to the many papers by Eyring *et al.*, for a more detailed discussion of the physical basis of these ideas. (See Eyring and Jhon in "Additional Reading.")

The presence of holes in the liquid will not only confer gaslike properties on some of the molecules but will also provide a *positional degeneracy* to a solidlike molecule. This is assumed to be equal to the number of neighboring holes n_h, multiplied by the probability of the molecule and a hole exchanging positions, where this probability is given by a Boltzmann factor $e^{-\varepsilon/kT}$, where ε is the energy required to move into one of these holes. Thus the number of additional sites is $n_h e^{-\varepsilon/kT}$, and so the total number of positions available to a vibrating solidlike molecule is $1 + n_h e^{-\varepsilon/kT}$.

We now assume that the solidlike molecules can be treated by the Einstein approximation of crystals. Equation (11–6) with all the frequencies equal to each other gives that the partition function of an Einstein solid of N_s solidlike atoms ($N_s = NV_s/V$) is

$$Q_s = e^{-U(\mathbf{0};\,\rho)/kT}\left[\frac{e^{-\Theta_E/2T}}{(1 - e^{-\Theta_E/T})}\right]^{3N_s} \tag{12–2}$$

In this equation, $U(\mathbf{0};\,\rho)$ is the total energy of interaction of all the atoms at rest at their lattice sites. If we assume that this total energy is made up of pairwise interaction between all pairs of atoms in the crystal, then $U(\mathbf{0};\,\rho)$ can be written as $N_s\phi(0)/2$, where $\phi(0)$ is the potential energy of an atom situated at a lattice site. The factor of two in the denominator is included to avoid counting all pairwise interactions twice. In terms of $\phi(0)$, Eq. (12–2) can be written

$$Q_s = \left[\frac{e^{-\phi(0)/2kT}e^{-3\Theta_E/2T}}{(1 - e^{-\Theta_E/T})^3}\right]^{N_s} \tag{12–3}$$

Now the numerator here is exactly the exponential of the energy required to take an atom from the crystal and remove it to infinity (remember that $\phi(0)$ is a negative quantity since it is referred to the separated atoms). This energy is just the energy of sublimation E_s, and so we can write Eq. (12–3) as

$$Q_s = \left[\frac{e^{E_s/kT}}{(1 - e^{-\Theta_E/T})^3}\right]^{N_s} \tag{12–4}$$

Thus we have for the solidlike contribution to the partition function of the liquid:

$$\left\{\frac{e^{E_s/kT}}{(1-e^{-\Theta_E/T})^3}(1+n_h e^{-\varepsilon/kT})\right\}^{NV_s/V} \tag{12–5}$$

where we have included the factor $1 + n_h e^{-\varepsilon/kT}$ to take into account the fact that not all of the solidlike atoms are completely surrounded by other atoms; that is, there are likely to be vacancies around any particular atom and hence a positional degeneracy. The total partition function of the liquid is then the product of Eqs. (12–1) and (12–5):

$$Q = \left\{\frac{e^{E_s/kT}}{(1-e^{-\Theta_E/T})^3}(1+n_h e^{-\varepsilon/kT})\right\}^{NV_s/V}\left\{\frac{(2\pi mkT)^{3/2}}{h^3}(V-V_s)\right\}^{N(V-V_s)/V} \times \left\{\left(\frac{N(V-V_s)}{V}\right)!\right\}^{-1} \tag{12–6}$$

This is the partition function of the significant structure theory. When $V = V_s$, the exponent of the second set of brackets and the quantity n_h in the first brackets go to zero, leaving the partition function of a solid. When V is much larger than V_s, the first set of brackets effectively becomes unity, and we are left with the partition function of a gas. All the quantities appearing in Eq. (12–6) are known except for n_h and ε. Eyring and his co-workers do not treat them as completely adjustable parameters, but have tried to relate them to observable quantities. (See Problems 12–3 and 12–4.)

Significant structure theory has been applied to a wide variety of systems such as inert gases, diatomic liquids, organic liquids, molten salts, molten metals, liquid hydrogen, water, solutions, surface tension and transport properties. Table 12–1 shows the type of agreement that is found. The values of E_s, Θ_E, and V_s used were taken from the properties of the solid phase, and the values of n_h and ε were estimated beforehand. Figures 12–1 through 12–3 show some typical results of the theory of significant structures. In particular, Fig. 12–3 shows the results for the vapor pressure of p–H_2, HD, and o–D_2. Remember that liquid hydrogen must be treated as a quantum liquid.

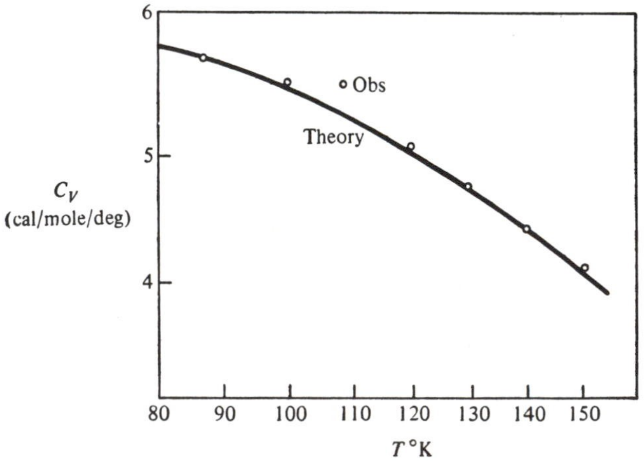

Figure 12–1. **Specific heat of liquid argon. The curve represents theory and the circles represent the experimental data.**

Table 12–1. **Calculated and observed properties of various substances according to the theory of significant structures**

		n–H_2	CH_4	NaCl	KCl	Na	Hg	Ne	Ar	Kr	Xe
T_m(°K)	calc.*	13.94	90.65	1070	1023	371.01	234.29	24.55	83.85	116.0	161.3
	obs.	13.94	90.65	1074	1049	371.01	234.29	24.55	83.85	116.0	161.3
V_m(cm^3-mole^{-1})	calc.	26.093	34.11	38.14	49.06	24.85	14.55	16.43	28.90	33.11	42.30
	obs.	26.108	35.42	37.74	48.80	24.81	14.65	16.15	28.03	34.13	42.68
ΔS_m(cal-mole^{-1}-deg^{-1})	calc.	1.936	2.48	7.75	5.40	1.39	1.99	3.244	3.263	3.456	3.415
	obs.	—	2.48	6.3	5.8	1.68	2.34	3.259	3.35	3.35	3.40
T_b(°K)	calc.	20.70	111.67	1750	1684	1207	648	27.13	87.29	119.28	167.5
	obs.	20.365	111.67	1738	1680	1156.1	630	27.26	87.29	119.93	165.1
V_b(cm^3-mole^{-1})	calc.	28.692	36.35	51.15	71.20	29.17	16.50	17.17	29.33	3331	42.96
	obs.	28.393	37.79	—	—	(30)	15.75	16.80	28.69	—	—
ΔS_b(cal-mole^{-1}-deg^{-1})	calc.	10.564	17.40	21.55	21.63	16.86	20.39	15.68	19.04	19.27	19.43
	obs.	—	17.51	23.5	23. 1	17.84	22.43	15.81	17.85	17.99	18.29
T_c(°K)	calc.	36.2	214.0	3600	3092	4230	1552	46.29	149.7	208.33	287.8
	obs.	33.24	191.04	—	—	—	—	44.47	150.66	210.6	289.8
V_c(cm^3-mole^{-1})	calc.	77.3	103.9	293	431	96.6	63.7	47.11	83.68	88.32	113.52
	obs.	—	99.02	—	—	—	—	41.74	75.26	—	113.8
P_c(atm)	calc.	13.8	63.3	235.5	135.5	1163	610	29.12	52.93	69.68	74.89
	obs.	12.797	45.8	—	—	—	—	26.86	48.00	54.24	58.2

* The melting temperature is used to fix parameters.
Source: H. Eyring, D. Henderson, and T. Ree, *Prog. Int. Res. Therm. Trans. Prop.*, 1962.

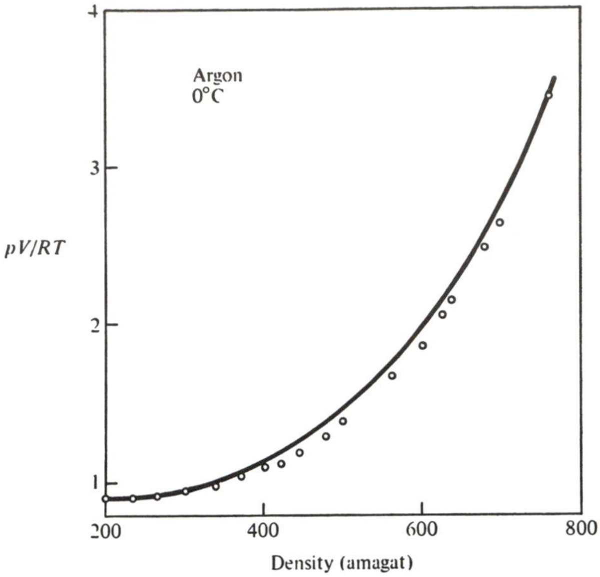

Figure 12–2. **Compressibility factor of argon gas as a function of the density. An amagat is a unit of density used by many high-pressure experimentalists and is the ratio of the volume of the gas at 25°C and 1-atm pressure to the volume of the gas under the actual temperature and pressure.**

A look at the above figures and tables shows that the significant structure theory is capable of producing good agreement with experiment. The theory has been criticized for its artificial assumptions concerning the structure of the liquid state, and some people have referred to it as simply an interpolation scheme since the properties of the limiting gas and solid states are used as input. But regardless, it is probably the most useful simple model theory of liquids today, at least from a practical point of view. Problems 12–8 and 12–9 illustrate some other criticisms of the theory as well.

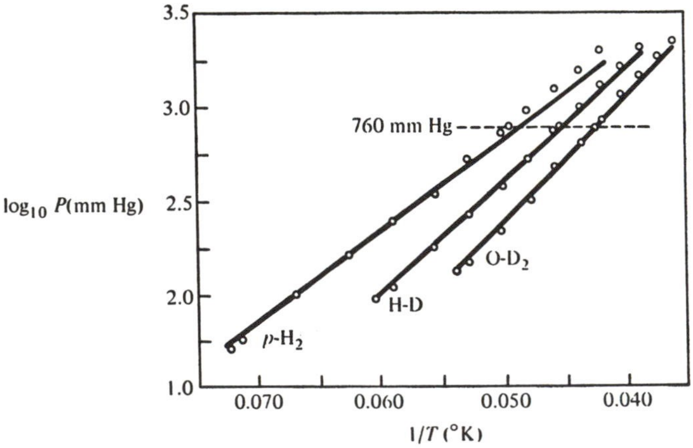

Figure 12–3. **Vapor pressure of liquid p–H_2, H–D, and o–D_2. The curve represents the theoretical values and the circles represent the experimental data.**

12–2 THE LENNARD-JONES DEVONSHIRE THEORY OF LIQUIDS

The theory of significant structures is based upon the assumption that a liquid can be approximated by a solid with many holes in it. This theory belongs to a general class of liquid theories called *hole theories*. In this section we shall discuss one of another general class of liquid theories called *cell theories*. The meaning of this name will be clear as we proceed. The basic cell theory of liquids is due to Lennard-Jones and Devonshire, and it is their theory that we shall treat here.

The physical model behind a cell theory of liquids is the following. Let us assume for the moment that a liquid can be represented by a solid where, instead of just vibrating about its lattice site, each atom is allowed to move within a cage or a cell whose boundaries are defined by the positions of its nearest-neighbor atoms. The atom does not move freely around its cell since it interacts with all of its nearest neighbors. We define an effective or free volume available to the atom by the equation

$$v_f = \int_\Delta e^{-[\phi(\mathbf{r})-\phi(0)]/kT}\, d\mathbf{r} \tag{12–7}$$

where Δ represents the volume of the cell, and $\phi(\mathbf{r})$ is the total interaction of the "central" atom with all of its nearest neighbors when the central atom is at a position $\mathbf{r}$ from its lattice site (*cf.* Fig. 12–5). Note that Eq. (12–7) is an integral of the volume of the cell weighted by the Boltzmann factor $\exp\{-\beta[\phi(\mathbf{r}) - \phi(0)]\}$.

The partition function of any one "central" atom then is given by an expression similar to that for an ideal gas molecule confined to a volume v_f instead of V:

$$q = \left(\frac{2\pi mkT}{h^2}\right)^{3/2} v_f \tag{12–8}$$

Since each central atom is confined to its lattice site (for now), the N atoms of the liquid are distinguishable, and so the partition function of the system of N atoms is given by

$$Q = e^{-N\phi(0)/2kT}\left[\left(\frac{2\pi mkT}{h^2}\right)^{3/2} v_f\right]^N \tag{12–9}$$

The factor of $\exp[-N\phi(0)/2kT]$ has been included to take into account the interaction of all the atoms at their lattice positions relative to a zero of energy where all N atoms are infinitely separated from each other.

Equation (12–9), in fact, is very similar to the partition function of the Einstein model of a crystal,

$$Q = e^{-N\phi(0)/2kT}[q_{\text{vib}}^3]^N \tag{12–10}$$

where q_{vib} is the vibrational partition function of a one-dimensional harmonic oscillator (hence q_{vib}^3 to represent a harmonic oscillator that vibrates independently in each of three directions). The term $(2\pi mkT/h^2)^{3/2}v_f$ arises because the potential $\phi(\mathbf{r})$ is such that the "central" atom can be considered to be an atom moving about the volume of the cell defined by its nearest neighbors instead of simply vibrating about its lattice position. If $\phi(\mathbf{r}) - \phi(0)$, the potential that the atom sees about its lattice position, were a parabola, so that $\phi(\mathbf{r}) - \phi(0) = \frac{1}{2}fr^2$, the motion would be just that of a classical three-dimensional harmonic oscillator. The effective or free volume would be

$$v_f = \int_0^\infty 4\pi r^2 e^{-fr^2/2kT}\, dr = \left(\frac{2\pi kT}{f}\right)^{3/2}$$

and the partition function in Eq. (12–8) would be

$$q = \left(\frac{2\pi mkT}{h^2}\right)^{3/2}\left(\frac{2\pi kT}{f}\right)^{3/2} = \left(\frac{kT}{h\nu}\right)^3 \tag{12–11}$$

where $\nu = (f/m)^{1/2}/2\pi$. Equation (12–11) is the partition function of a three-dimensional classical harmonic oscillator. Thus Eq. (12–9) is simply the classical partition function of a *solid* in which the atoms are not restricted to harmonic vibrations about their lattice points.

We now introduce one more approximation that modifies Eq. (12–9) so that it can be considered to be more applicable to a liquid than a solid. In a solid, each atom is confined to its lattice site and never wanders away from the neighborhood or cell around that position. In a gas, on the other hand, each atom is free to wander over the entire volume of the container. Consequently, a gas has a higher entropy than a solid, and this excess entropy is called the *communal entropy*. We can calculate the magnitude of this communal entropy quite easily.

First consider an ideal gas of N atoms in a volume V. The partition function is given by Eqs. (5–1) and (5–5)

$$Q = \frac{1}{N!}\left(\frac{2\pi mkT}{h^2}\right)^{3N/2} V^N$$

and the entropy is

$$S = Nk + Nk\ln\left[\left(\frac{2\pi mkT}{h^2}\right)^{3/2}\frac{Ve^{3/2}}{N}\right] \tag{12–12}$$

Now consider a system of N atoms, in which the total volume V is divided into N compartments of volume V/N each, and one atom is confined to each compartment. The partition function for this system is

$$Q = \left(\frac{2\pi mkT}{h^2}\right)^{3N/2}\left(\frac{V}{N}\right)^N$$

Note that there is no division by $N!$ here since the compartments, and hence the atoms, can in principle be labeled. The entropy is

$$S = Nk\ln\left[\left(\frac{2\pi mkT}{h^2}\right)^{3/2}\frac{Ve^{3/2}}{N}\right] \tag{12–13}$$

If we compare Eqs. (12–12) and (12–13), we see that the entropy of the gas is Nk larger than that of the solid, and so we say that the communal entropy of a gas is Nk.

A liquid is, in some sense, intermediate between a solid and a gas, and it is not clear how the communal entropy of a liquid varies between its melting point and its boiling point. The Lennard-Jones Devonshire theory, as well as most of its modifications, assumes, however, that a liquid possesses the same amount of communal entropy as a gas. This communal entropy can be introduced into the partition function in Eq. (12–9) by multiplying Q by a factor of e^N. This factor will add an additional contribution of Nk to the entropy (see Problem 12–22). Thus we finally have the basic partition function of the Lennard-Jones Devonshire theory:

$$Q = e^{-N\phi(0)/2kT}\left[\left(\frac{2\pi mkT}{h^2}\right)^{3/2} v_f e\right]^N \tag{12–14}$$

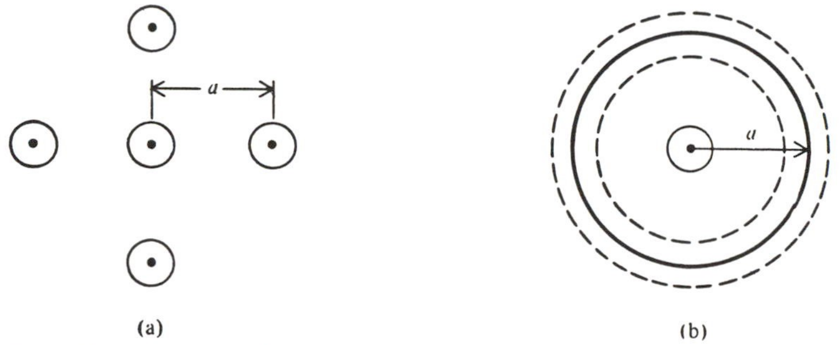

Figure 12–4. **(a) A two dimensional representation of an atom surrounded by its nearest neighbors in the solid, and (b) the cell formed in a liquid by spreading its four nearest neighbors uniformly over a sphere (a circle in two dimensions) of radius *a*.**

where the effective or free volume is given by Eq. (12–7). We emphasize here that $\phi(0)$ is a function of the density, and v_f is a function of the density and the temperature.

In order to apply Eq. (12–14), we must calculate v_f, which in turn requires that we calculate $\phi(\mathbf{r}) - \phi(0)$, the potential that an atom sees as it moves around the cell about its lattice site. We assume that $\phi(\mathbf{r}) - \phi(0)$ can be calculated by treating the C nearest neighbors of the central atom to be distributed uniformly over a sphere of radius a from the lattice site of the central atom, where a is the distance between centers of nearest neighbors in the corresponding solid. A two-dimensional version of this is shown in Fig. 12–4. The distance between nearest-neighbor cells can be related to the number density (N/V) of the crystal by simple geometry. For example, for a face-centered cubic lattice,

$$a^3 = \sqrt{2}\left(\frac{V}{N}\right) = \sqrt{2}\,v \tag{12–15}$$

and the number of nearest neighbors, C, equals 12.

We now refer to Fig. 12–5 to calculate $\phi(\mathbf{r}) - \phi(0)$. In Fig. 12–5 the central atom is at the point P, a distance r from the center of the cell. The area of the ring shown on the surface of the sphere is

$$2\pi a^2 \sin\theta \, d\theta$$

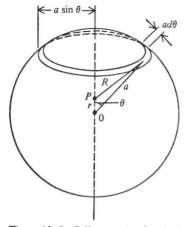

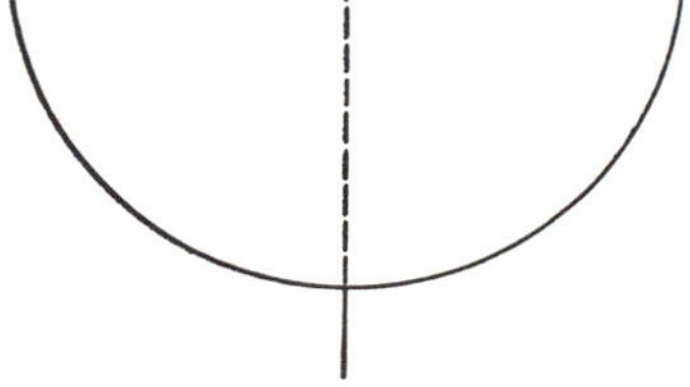

Figure 12–5. **Cell geometry for the Lennard-Jones Devonshire theory.**

The number of nearest neighbors that are uniformly smeared over this area is

$$C \cdot \frac{2\pi a^2 \sin \theta \, d\theta}{4\pi a^2} = \frac{C}{2} \sin \theta \, d\theta \tag{12–16}$$

The potential energy of interaction between the central atom located at P and the nearest neighbors in the ring is the product of Eq. (12–16) and the pairwise potential energy of interaction $u(R)$ between two atoms:

$$u(R) \cdot \frac{C}{2} \sin \theta \, d\theta \tag{12–17}$$

where from the law of cosines

$$R^2 = r^2 + a^2 - 2 \, ar\cos \theta \tag{12–18}$$

The total energy of interaction $\phi(r)$ is obtained by integrating Eq. (12–17) over all values of θ in Fig. 12–5, that is, from 0 to π:

$$\phi(r) = \frac{C}{2} \int_0^{\pi} u(R) \sin \theta \, d\theta \tag{12–19}$$

At this point we must specify $u(R)$. The potential energy of interaction between two atoms is difficult to determine exactly, but it is known to be of the general shape shown in Fig. 12–6. We see that the two atoms repel each other at short distances and attract each other at longer distances.

In Chapter 15 we shall discuss a number of empirical functions that are used to represent the curve shown in Fig. 12–6. The one we shall use here is called the Lennard-Jones potential and is

$$u(R) = \varepsilon \left(\frac{r^*}{R}\right)^{12} - 2\varepsilon \left(\frac{r^*}{R}\right)^{6} \tag{12–20}$$

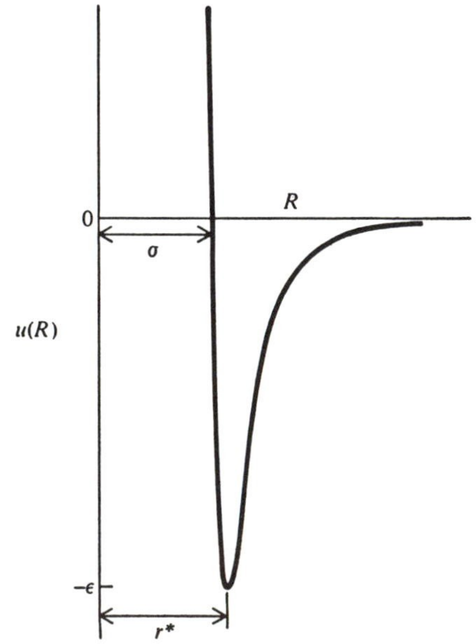

Figure **12–6. The Lennard-Jones intermolecular potential, Eq. (12–20).**

As Fig. 12-6 shows, ε is the depth of the well, and r^* is the separation of which u is a minimum. The parameters r^* and ε depend upon the particular pair of molecules and are tabulated for many molecules. Thus the nature of the liquid to which we apply the Lennard-Jones Devonshire theory is reflected through the choice of r^* and ε. If we substitute Eq. (12–20) into (12–19), the integrations are tedious but straightforward, and the result is that

$$\phi(r) - \phi(0) = C\varepsilon\left[\left(\frac{r^*}{a}\right)^{12} l\left(\frac{r^2}{a^2}\right) - 2\left(\frac{r^*}{a}\right)^{6} m\left(\frac{r^2}{a^2}\right)\right] \tag{12–21}$$

where

$$\phi(0) = C\varepsilon\left[-2\left(\frac{r^*}{a}\right)^{6} + \left(\frac{r^*}{a}\right)^{12}\right] \tag{12–22}$$

$$l(y) = (1 + 12y + 25.2y^2 + 12y^3 + y^4)(1 - y)^{-10} - 1 \tag{12–23}$$

$$m(y) = (1 + y)(1 - y)^{-4} - 1 \tag{12–24}$$

We define a quantity v^* by

$$v^* = \frac{v}{a^3} r^{*3} \tag{12–25}$$

and Eqs. (12–21) and (12–22) become

$$\phi(r) - \phi(0) = C\varepsilon\left[\left(\frac{v^*}{v}\right)^{4} l\left(\frac{r^2}{a^2}\right) - 2\left(\frac{v^*}{v}\right)^{2} m\left(\frac{r^2}{a^2}\right)\right] \tag{12–26}$$

$$\phi(0) = C\varepsilon\left[-2\left(\frac{v^*}{v}\right)^{2} + \left(\frac{v^*}{v}\right)^{4}\right] \tag{12–27}$$

Equation (12–26) is given in Fig. 12–7 for several values of v/v^*. When the density is fairly low ($v/v^* = 3.18$), the potential within the cell is uniform except for a slight dip near the edges of the cell. For a normal crystal density ($v/v^* = 1.21$), the potential is almost parabolic, and the motion would be that of a harmonic oscillator. The case for liquid densities ($v/v^* = 1.82$) is intermediate between the two.

We can now substitute Eq. (12–26) into Eq. (12–7) for v_f. If we then substitute v_f into Eq. (12–14) for Q and differentiate its logarithm with respect to V at constant N and T, we finally get, after lengthy but straightforward algebra (see Problem 12–12),

$$\frac{pv}{kT} = 1 - \frac{2C\varepsilon}{kT}\left[\left(\frac{v^*}{v}\right)^{2} - \left(\frac{v^*}{v}\right)^{4}\right] + \frac{4C\varepsilon}{kT}\left[\left(\frac{v^*}{v}\right)^{4}\frac{g_l}{g} - \left(\frac{v^*}{v}\right)^{2}\frac{g_m}{g}\right] \tag{12–28}$$

where

$$g_l = \int_0^{y_0} \exp\left\{-\frac{C\varepsilon}{Tk}\left[\left(\frac{v^*}{v}\right)^{4} l(y) - 2\left(\frac{v^*}{v}\right)^{2} m(y)\right]\right\} y^{1/2} l(y)\, dy \tag{12–29}$$

$$g_m = \int_0^{y_0} \exp\left\{-\frac{C\varepsilon}{kT}\left[\left(\frac{v^*}{v}\right)^{4} l(y) - 2\left(\frac{v^*}{v}\right)^{2} m(y)\right]\right\} y^{1/2} m(y)\, dy \tag{12–30}$$

and

$$g = \int_0^{y_0} \exp\left\{-\frac{C\varepsilon}{kT}\left[\left(\frac{v^*}{v}\right)^{4} l(y) - 2\left(\frac{v^*}{v}\right)^{2} m(y)\right]\right\} y^{1/2}\, dy \tag{12–31}$$

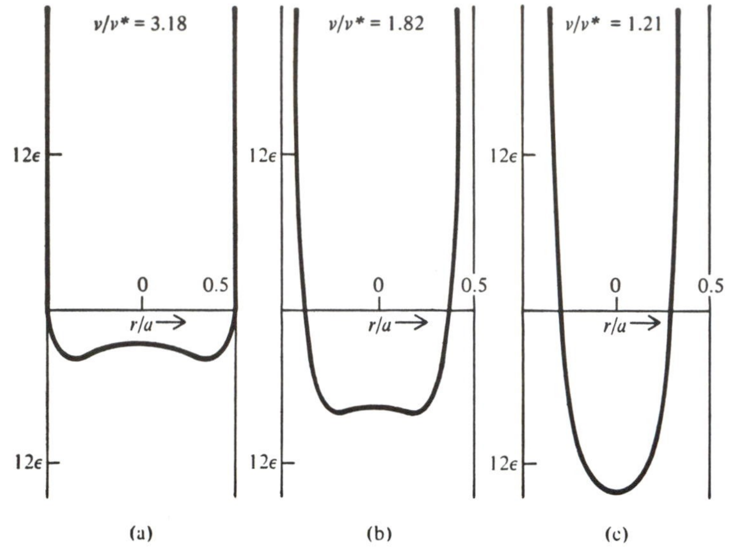

Figure 12–7. **The potential field within a cell for various cell sizes.** (From J. E. Lennard-Jones and A. F. Devonshire, *Proc. Roy. Soc.* (London), **A163**, 53, 1937; taken from J. O. Hirschfelder, C. F. Curtiss and R. B. Bird, *Molecular Theory of Gases and Liquids*, New York: Wiley, 1954.)

In these integrals the upper limit of integration is taken to the edges of the cell. Its precise definition is not important since the integrals are insensitive to the value of y_0 as long as it allows the central atom to wander over the entire cell. The integrals g, g_l, and g_m are complicated and must be evaluated numerically as functions of v/v^* and $T^* = kT/\varepsilon$. There are extensive tabulations of these functions in the literature, and in addition there are extensive tabulations of all the thermodynamic functions calculated from the Lennard-Jones Devonshire theory. (See, e.g., Hirschfelder, Curtiss, and Bird in "Additional Reading.")

Equation (12–28) is the Lennard-Jones Devonshire equation of state of a liquid. The first thing to notice is that it follows a law of corresponding states since pv/kT is a universal function of a reduced volume v/v^* and a reduced temperature $T^* = kT/\varepsilon$. Thus we can write the equation of state in the form

$$\frac{pv}{kT} = 1 + f\left(\frac{kT}{\varepsilon}, \frac{v}{v^*}\right) \tag{12–32}$$

The left-hand side of this equation is dimensionless and so is reduced as it stands, but we can write it in terms of a reduced p, a reduced volume, and a reduced temperature since

$$\frac{pv}{kT} = \frac{\left(\frac{p\sigma^3}{\varepsilon}\right)\left(\frac{v}{\sigma^3}\right)}{kT/\varepsilon} = \frac{\left(\frac{pv^*}{\varepsilon}\right)\left(\frac{v}{v^*}\right)}{kT/\varepsilon} \tag{12–33}$$

We can solve Eq. (12–32) for the reduced pressure pv^*/ε:

$$\frac{pv^*}{\varepsilon} = \left(\frac{kT}{\varepsilon}\right)\left(\frac{v^*}{v}\right)\left[1 + f\left(\frac{kT}{\varepsilon}, \frac{v}{v^*}\right)\right] \tag{12–34}$$

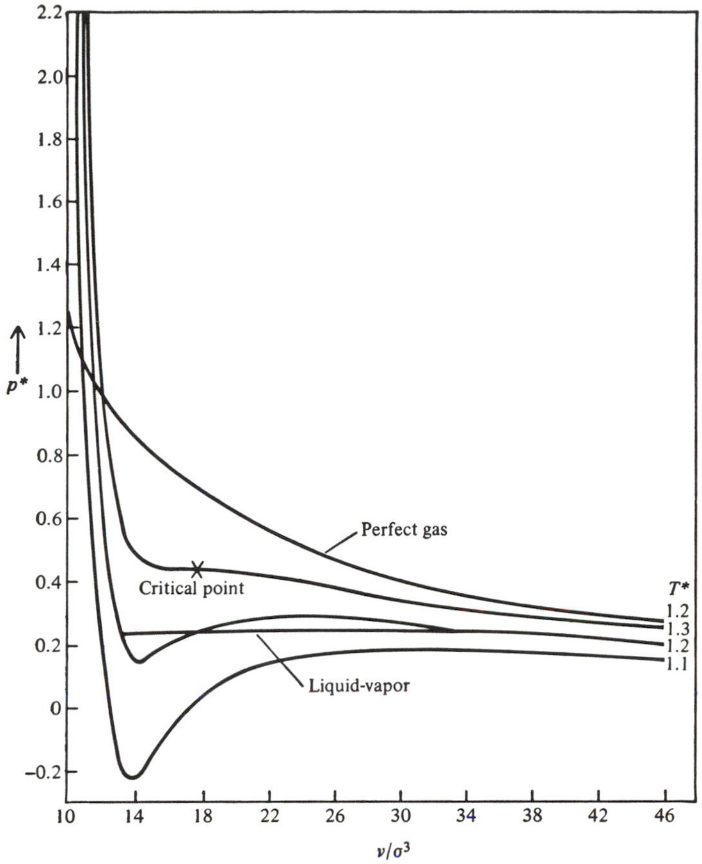

Figure 12–8. **Pressure volume isotherms of the Lennard-Jones Devonshire theory. The pressure is given as $p^* = p\sigma^3/\varepsilon$, and the volume is given as v/σ^3. The temperature $T^* = kT/\varepsilon$.** (From R. H. Wentorf, R. J. Buehler, J. O. Hirschfelder, and C. F. Curtiss, *J. Chem. Phys.*, **18**, 1484, 1950; taken from J. O. Hirschfelder, C. F. Curtiss, and R. B. Bird, *Molecular Theory of Gases and Liquids*. New York: Wiley, 1954.)

In Fig. 12–8, Eq. (12–34) is plotted versus v/v^* for fixed values of $T^* = kT/\varepsilon$. The equation predicts a critical isotherm, and at lower temperatures the p–v curves are S-shaped like the isotherms of the van der Waals equation. These S-shaped isotherms, which are not physically real since they give a region where $(\partial p/\partial v) > 0$, are due to the many approximations introduced. We shall discuss the significance and nature of these van der Waals loops in Chapter 19 (*Statistical Mechanics*).

The critical constants predicted by the Lennard-Jones Devonshire theory can be read off Fig. 12–8. Table 12–2 compares these values with experimental values. This

Table 12–2. **A comparison of the Lennard-Jones Devonshire critical constants with experimental values**

gas	kT_c/ε	v_c/v^*	$p_c v^*/\varepsilon$	$p_c v_c/kT_c$
Ne	1.25	3.33	0.111	0.296
Ar	1.26	3.16	0.116	0.291
Xe	1.31	2.90	0.132	0.293
N_2	1.33	2.96	0.131	0.292
CH_4	1.29	2.96	0.126	0.290
LJD	1.30	1.77	0.434	0.591

shows that the Lennard-Jones Devonshire theory predicts kT_c/ε quite well but the other quantities are in poor agreement with the experimental values. Table 12–3 presents a comparison of the thermodynamic properties of the Lennard-Jones Devonshire theory with experimental values for nitrogen. It is clear that the theory is most satisfactory at low temperatures and high densities. This is particularly so for the entropy. This is to be expected from the nature of the model since, under these conditions, the atoms are more restricted to move in cells about their lattice sites as pictured by the theory.

Table 12–3. **A comparison of the thermodynamic functions of the Lennard-Jones Devonshire theory with experimental data for nitrogen**

T	density*	pv/kT (exp.)	pv/kT (calc.)	E† (exp.)	E† (calc.)	C_V‡ (exp.)	C_V‡ (calc.)	S§ (exp.)	S§ (calc.)
0°C	200	1.04	1.47	−302	−267	0.40	0.31	−1.05	−0.28
	240	1.08	1.62	−360	−316	0.45	0.42	−1.29	−0.66
	280	1.17	1.75	−418	−370	0.49	0.49	−1.54	−1.08
	320	1.26	1.88	−475	−423	0.56	0.55	−1.80	−1.49
	400	1.56	2.14	−588	−531	0.71	0.72	−2.38	−2.34
	480	2.04	2.49	−693	−640	0.91	0.93	−3.05	−3.21
50°C	200	1.12	1.52	−285	−252	0.27	0.29	−1.00	−0.22
	240	1.17	1.72	−340	−295	0.33	0.41	−1.22	−0.59
	280	1.28	1.85	−395	−346	0.38	0.49	−1.46	−1.00
	320	1.39	1.99	−449	−395	0.45	0.55	−1.72	−1.40
	400	1.70	2.29	−554	−495	0.60	0.72	−2.28	−2.22
	480	2.24	2.69	−650	−594	0.78	0.92	−2.91	−3.05
150°C	200	1.18	1.56	−270	−226	0.06	0.24	−0.95	−0.16
	240	1.24	1.77	−312	−254	0.07	0.40	−1.16	−0.48
	280	1.33	1.92	−372	−297	0.10	0.49	−1.39	−0.87
	320	1.48	2.07	−422	−340	0.14	0.55	−1.63	−1.25
	400	1.81	2.39	−514	−424	0.22	0.71	−2.16	−2.03
	480	2.30	2.79	−594	−504	0.33	0.89	−2.74	−2.82

* In units of amagats, a unit of density commonly used in p–V–T work. An amagat is the volume of the gas at 0°C and 1 atm divided by the volume at the temperature and pressure measured.
† In units of cal/mole.
‡ In units of cal/deg-mole.
§ In units of cal/deg-mole.
Source: J. O. Hirschfelder, C. F. Curtiss, and R. B. Bird, *Molecular Theory of Gases and Liquids* (New York: Wiley, 1954).

Many refinements of the Lennard-Jones Devonshire theory have been made. For example, Hirschfelder, Curtiss, and Bird present tables that include the effects of more than a spherical shell of first nearest neighbors. The generalization of allowing 0, 1, 2, ... atoms in a cell has also been worked out in great detail. None of these refinements, however, improve the basic theory presented here a great deal.

A comparison of the agreement found in Tables 12–1 and 12–3 should be made here. It can be seen that the agreement in Table 12–1 is much better than in Table 12–3. It is important to realize, however, that the theory of significant structures and the Lennard-Jones Devonshire theory are quite different in approach. The theory of significant structures contains several parameters which are adjusted or determined from other experimental data, such as the melting point. Once these parameters are determined, a large number of other experimental data can be predicted. The Lennard-Jones Devonshire theory, on the other hand, is formulated in terms of only the two *molecular* parameters that govern the intermolecular potential. These parameters can, in principle, be calculated from quantum mechanics, and so the Lennard-Jones Devonshire theory is in a sense a completely molecular theory. Once the intermolecular interaction between a pair of molecules is specified, all of the thermodynamic properties can be calculated. Thus the Lennard-Jones Devonshire theory can be said to be more fundamental than the theory of significant structures, but the price it pays for this is a poorer overall agreement with experimental data. The Lennard-Jones Devonshire theory is representative of one of the ultimate goals of statistical thermodynamics, namely, the calculation of macroscopic thermodynamic quantities in terms of the properties of the individual molecules of the system and their intermolecular potential. We shall see many more attempts of this in Chapters 15 through 20 (*Statistical Mechanics*), where we discuss in detail systems of molecules in which the intermolecular potential plays a central role.

ADDITIONAL READING

Significant structure theory

EYRING, H., and JHON, M. S. 1969. *Significant liquid structures*. New York: Wiley.

EYRING, H., HENDERSON, D., STOVER, B. J., and EYRING, E. M. 1964. *Statistical mechanics and dynamics*. New York: Wiley. Chapter 12.

JHON, M. S., and EYRING, H., 1971. In *Physical chemistry, an advanced treatise*, Vol. VIIIA, ed. by H. Eyring, D. Henderson, and W. Jost. New York: Academic.

Lennard-Jones Devonshire theory

BARKER, J. A. 1963. *Lattice theories of the liquid state*. New York: Pergamon.

FOWLER, R. H., and GUGGENHEIM, E. A. 1956. *Statistical thermodynamics*. Cambridge: Cambridge University Press. Sections 800 through 812.

HILL, T. L. 1956. *Statistical mechanics*. New York: McGraw-Hill. Chapter 8.

HIRSCHFELDER, J. O., CURTISS, C. F., and BIRD, R. B. 1954. *Molecular theory of gases and liquids*. New York: Wiley. Chapter 4.

LEVELT, J. M. H., and COHEN, E. G. D. 1964. In *Studies in statistical mechanics*, Vol. II, ed. by J. DeBoer and G. E. Uhlenbeck. Amsterdam: North-Holland Publishing Co.

PRIGOGINE, I. 1957. *Molecular theory of solutions*. Amsterdam: North-Holland Publishing Co. Chapters 7, 12, and 13.

PROBLEMS

12–1. It has been argued that n_h in Eq. (12–6) is proportional to the number of vacancies, and so we can write

$$n_h = \frac{n(V - V_s)}{V_s}$$

where n is a proportionality constant. In addition, it can be argued that ε_h should be inversely proportional to the number of vacancies and directly proportional to the sublimation energy of the solid E_s. If we define a proportionality constant a, we can write

$$\varepsilon_h = \frac{aE_sV_s}{V - V_s}$$

Using these expressions and Stirling's approximation for $[N(V - V_s)/V]!$, derive the equation

$$Q = \left\{\frac{e^{E_s/RT}}{(1 - e^{-\Theta_E/T})^3}\left(1 + n\left(\frac{V - V_s}{V_s}\right)\exp\left(-\frac{aE_sV_s}{(V - V_s)RT}\right)\right)\right\}^{NV_s/V}\left\{\frac{(2\pi mkT)^{3/2}}{h^3}\frac{eV}{N}\right\}^{N(V - V_s)/V}$$

This equation is often used as the starting point in calculations. The parameters n and a are discussed in Problem 12–3.

12–2. Using a value of $a = 0.0053$ and $n = 10.7$ in the final equation of Problem 12–1, calculate p and $\Delta S_b = S_{gas} - S_{liquid}$ at the boiling point. The solid-state parameters of argon that are needed can be easily researched; to save time, they are $E_s = 1889$ cal/mole, $V_s = 24.98$ ml, and $\Theta_E = 60.0°K$.

12–3. The parameters n and a introduced in Problem 12–1 can be estimated theoretically. For example, for a simple liquid such as argon near the melting point, the liquid volume is 12 percent larger than the volume of the solid at the melting point. Thus near the melting point, the fraction of nearest-neighbor sites that are unoccupied and therefore available for occupancy is

$$C\left(\frac{V_l - V_s}{V_l}\right) = n\left(\frac{V_l - V_s}{V_s}\right)$$

where C is the number of nearest neighbors in the crystal state, and where V_l is the volume of the liquid near the melting point. This gives that

$$n = \frac{CV_s}{V_l}$$

Using the available data for argon, calculate n and compare it to the value used in Problem 12–2.

12–4. There also exists a method for estimating a theoretically. This is discussed in Eyring and Jhon (pp. 33–35), who argue that a should be given by

$$a = \left(\frac{n-1}{2C}\right)\frac{(V_m - V_s)^2}{V_m V_s}$$

Using this expression, calculate a for argon and compare it to the value used in Problem 12–2.

12–5. Henderson* has attempted to improve the significant structure theory by using the Lennard-Jones Devonshire theory for the solidlike partition function. First show that if this is done, the partition function is

$$Q = \left[\left(\frac{2\pi mkT}{h^2}\right)^{3/2} v_f\right]^{NV_s/V} \exp\left\{-\frac{NV_s\phi(0)}{2kTV}\right\} \times (1 + n_h e^{-\varepsilon_h/kT})^{NV_s/V}$$

where v_f is given by Eq. (12–7). Now show that for the 6–12 potential,

$$\phi(0) = C\varepsilon\left[\frac{1}{\omega^{*4}} - \frac{2}{\omega^{*2}}\right]$$

where $\omega^* = V_s/N\sigma^3$ for a face-centered cubic lattice.

12–6. The significant structure theory requires Θ_E, the Einstein temperature of the solid, a quantity that is not usually available experimentally since the Debye theory is usually used to fit solid-state thermal data. Show that if we require the *average* frequency of the Einstein and Debye distributions to be the same, then we can write

$$\Theta_E \approx \tfrac{3}{4}\Theta_D$$

an approximation often employed by Erying and his co-workers.

12–7. Formulate a two-dimensional version of the significant structure theory of liquids, and show how this reduces to the partition function of a two-dimensional gas.

12–8. Significant structure theory has been applied to diatomic fluids such as liquid chlorine.† The partition function used was

$$Q = \left\{\frac{e^{E_s/RT}}{(1 - e^{-\Theta_E/T})^5}\frac{1}{1 - e^{-h\nu/kT}}(1 + n_h e^{-aE_s/n_h RT})\right\}^{NV_s/V}$$
$$\times \left\{\frac{(2\pi mkT)^{3/2}}{h^3}\frac{8\pi^2 IkT}{2h^2}\frac{1}{1 - e^{-h\nu/kT}}\frac{eV}{N}\right\}^{N(V-V_s)/V}$$

* In *J. Chem. Phys.*, **39**, 1857, 1963.
† See T. R. Thomson, H. Eyring, and T. Ree, *Proc. Natl. Acad. Sci.*, **46**, p. 336, 1960.

Interpret each of the factors in this partition function, particularly the power of 5 in the very first term.

12–9. The theory of significant structures has been applied to simple molten salts such as NaCl by Carlson, Eyring, and Ree.* Given that the vapor phase of salts such as NaCl consists of diatomic covalent NaCl molecules, construct a significant structure partition function and compare your result to that used by Carlson, *et al.* It very likely will differ from theirs, and this has been a common criticism of the significant structure theory, namely, that each class of substances to which it applies requires modification of the basic partition function Eq. (12–6). See also Problem 12–8 for Cl_2.

12–10. Use the tables in Hirschfelder, Curtiss, and Bird to calculate the compressibility factor of N_2 at 60°K and 1 atm according to the Lennard-Jones Devonshire theory. Assume that $\varepsilon/k = 95°K$ and $r^* = 4.15$ Å.

12–11. Show that

$$\frac{1}{2}\int_0^\pi (a^2 + r^2 - 2\,\text{arcos}\,\theta)^{-n/2} \sin\theta\, d\theta = \frac{1}{(2n-4)ar}[(a-r)^{-n+2} - (a+r)^{-n+2}]$$

12–12. Show that the partition function for the Lennard-Jones Devonshire theory is

$$Q = \left[\left(\frac{2\pi mkT}{h^2}\right)^{3/2} e2\pi a^3 g e^{-\phi(0)/2kT}\right]^N$$

where g is given by Eq. (12–31). Derive Eq. (12–28) for the pressure from this.

12–13. Show that if only first nearest neighbors are included in the Lennard-Jones Devonshire theory, the energy of sublimation is

$$E_s = NC\varepsilon\left[1.0\left(\frac{N\sigma^3}{V_s}\right)^2 - 0.5\left(\frac{N\sigma^3}{V_s}\right)^4\right]$$

12–14. Show that E for a monatomic liquid is given by

$$\frac{E}{N} = \tfrac{3}{2}kT - C\varepsilon\left\{1.0\left(\frac{v^*}{v}\right)^2 - 0.5\left(\frac{v^*}{v}\right)^4\right\} + C\varepsilon\left(\frac{v^*}{v}\right)^4 \frac{g_l}{g} - 2C\varepsilon\left(\frac{v^*}{v}\right)^2 \frac{g_m}{g}$$

for the Lennard-Jones Devonshire theory.

12–15. Show that if second and third nearest neighbors are included in the Lennard-Jones Devonshire theory for a face-centered cubic lattice, the functions $l(y)$ and $m(y)$ [Eqs. (12–23) and (12–24)] are replaced by

$$L(y) = l(y) + \frac{1}{128}\,l\left(\frac{1}{2}y\right) + \frac{2}{729}\,l\left(\frac{1}{3}y\right)$$

and

$$M(y) = m(y) + \frac{1}{16}\,m\left(\frac{1}{2}y\right) + \frac{2}{27}\,m\left(\frac{1}{3}y\right)$$

12–16. We can generalize the results of Problem 12–15 to any type of crystal lattice. Let m_n be the number of atoms or molecules in the nth shell, and let a_n be the distance of the nth shell from the central atom. First show that for a face-centered cubic lattice:

shell number n	1	2	3	4	...
distance a_n	a_1	$a_1\sqrt{2}$	$a_1\sqrt{3}$	$a_1\sqrt{4}$	...
number m_n	12	6	24	12	...

* In *Proc. Natl. Acad. Sci.*, **46**, p. 333, 1960.

Now show in general that $l(y)$ and $m(y)$ are replaced by

$$L(y)=\sum_n \left(\frac{m_n}{m_1}\right)\left(\frac{a_1}{a_n}\right)^{12} l\left(\frac{ya_1{}^2}{a_n{}^2}\right)$$

and

$$M(y)=\sum_n \left(\frac{m_n}{m_1}\right)\left(\frac{a_1}{a_n}\right)^{6} m\left(\frac{ya_1{}^2}{a_n{}^2}\right)$$

Show how these reduce to the special case in Problem 12–15.

12–17. For the functions g_l, g_m, and g defined in Eqs. (12–29) to (12–31), show that

$$v^*\left(\frac{\partial g}{\partial v^*}\right)=\frac{4C}{T^*}\left(\frac{g_l}{v^{*4}}-\frac{g_m}{v^{*2}}\right)$$

and

$$T^*\left(\frac{\partial g}{\partial T^*}\right)=\frac{C}{T^*}\left(\frac{g_l}{v^{*4}}-\frac{2g_m}{v^{*2}}\right)$$

12–18. Show that at high densities, the cell potential of the Lennard-Jones Devonshire theory becomes that of a harmonic oscillator, and in fact that the Lennard-Jones Devonshire theory goes over into the classical analog of the Einstein theory of crystals.

12–19. Derive the equation of state for the Lennard-Jones Devonshire theory for a two-dimensional square planar lattice.

12–20. It is easy to derive an equation for the vapor pressure associated with a Lennard-Jones Devonshire liquid. We simply derive an equation for the chemical potential of the liquid and set it equal to the chemical potential of the vapor phase. For the vapor phase, we can use an ideal gas to give Eq. (5–21)

$$\mu_g=-kT\ln\left[\left(\frac{2\pi mkT}{h^2}\right)^{3/2} kT\right]+kT\ln p$$

It is easiest to calculate μ_l by recognizing that A and G differ by pV_l, which can be neglected since typically $pV_l \ll kT$. Thus we use A/N for μ_l to give

$$\mu_l=-kT\ln\left(\frac{2\pi mkT}{h^2}\right)^{3/2}-kT-kT\ln(2\pi a^3 g)+\frac{\phi(0)}{2}$$

Equating μ_l to μ_g, show that

$$\ln p=\ln\left(\frac{kT}{2\pi a^3 g}\right)-\frac{(kT-\phi(0)/2)}{kT}$$

This equation is in fairly reasonable agreement with experiment. (See Fowler and Guggenheim in "Additional Reading," Section 811.)

12–21. The Lennard-Jones Devonshire theory has been applied to fused salts such as NaCl.* Sodium chloride in the crystalline state exists as two interpenetrating simple cubic lattices, one containing only sodium ions and one chloride ions. First convince yourself that the Lennard-Jones Devonshire partition function for this system is

$$Q=\frac{1}{N_+!N_-!}\left[\frac{v_f}{\Lambda_+{}^3}\right]^{N+}\left[\frac{v_f}{\Lambda_-{}^3}\right]^{N-}\exp\left(-\frac{N}{2}\frac{\phi(0;v)}{RT}\right)$$

where $N_\pm$ is the number of $\pm$ ions, $\Lambda_\pm=h/(2\pi m_\pm kT)^{1/2}$, and $N=N_++N_-$. The only complication for fused salts is that since the coulombic potential decreases so slowly with

* See D. A. McQuarrie, *J. Phys. Chem.*, **66**, p. 1508, 1962.

distance, that is, as r^{-1}, all neighbor shells must be considered in calculating v_f. Referring to Fig. 12–5, let $\phi_j(r)$ be the total interaction energy of an ion at P, and all of its C_j jth nearest neighbors in the spherical shell [*cf.* Eq. (12–19)]:

$$\phi_j(r) = \frac{C_j}{2} \int_0^\pi u(R_j) \sin\theta \, d\theta$$

Assume now that $u(R_j)$ has a short-range repulsive part that can be represented for simplicity as a rigid sphere and a long-range coulombic part, so that

$$\begin{aligned} u(R_j) &= \infty \qquad && r > a - \sigma \\ &= \frac{(-1)^j z^2 e^2}{R_j} && r < a - \sigma \end{aligned}$$

where σ is the sum of the radii of the positive and negative ions, and $\pm ze$ is the charge on each. Show that the total interaction energy between an ion at P and all of its shells is

$$\begin{aligned} u(r) &= \frac{e^2}{2} \sum_{j=1}^{\infty} (-1)^j C_j \int_0^\pi \frac{\sin\theta \, d\theta}{(r^2 + a_j^2 - 2ar_j \cos\theta)^{1/2}} \\ &= -\frac{e^2}{2a} \left\{ \frac{6}{\sqrt{1}} - \frac{12}{\sqrt{2}} + \frac{24}{\sqrt{3}} - \cdots \right\} \\ &= -\frac{e^2 \alpha}{2a} \qquad r < a - \sigma \end{aligned}$$

and

$$\phi(r) = \infty \qquad r > a - \sigma$$

Interpret this result from elementary electrostatics. The constant α here is the Madelung constant of the crystal (look this up in almost any physical chemistry text).

Show now that

$$v_f = \frac{4\pi v}{3} \left[1 - \left(\frac{v^*}{v} \right)^{1/3} \right]^3$$

Show that the equation of state is

$$pv = RT \left[1 - \left(\frac{v^*}{v} \right)^{1/3} \right]^{-1} - \frac{e^2 \alpha N^{4/3}}{6 v^{1/3}}$$

and give a physical interpretation of each of the two terms.

Now show that there is a law of corresponding states for fused salts in which the reduced temperature is $\sigma kT/z^2e^2$, and the reduced volume is v/σ^3.* By referring to Fig. 2 in the reference to McQuarrie given in the footnote for Problem 12–21, estimate the critical constants for NaCl.

12–22. Show that a factor of e^N in Q contributes Nk to the entropy.

* Compare H. Reiss, S. W. Mayer, and J. L. Katz, *J. Chem. Phys.*, 35, p. 820, 1961.

CHAPTER 13

IDEAL SYSTEMS IN ELECTRIC AND MAGNETIC FIELDS

In this chapter we shall discuss systems of noninteracting particles in electric and magnetic fields. In Section 13–1 we develop the basic equations for the energy of isolated molecules in an electric field. Then in Section 13–2 we apply the statistical thermodynamic formulas of an ideal gas to a system of noninteracting molecules in an electric field. We derive the equation relating the macroscopic dielectric constant of a dilute gas to the electrical properties of the molecules, namely, the polarizability and the dipole moment. In Section 13–3 we extend this equation to higher densities by applying macroscopic electrostatics to calculate the approximate field in the vicinity of any molecule. This leads to the Clausius-Mossotti equation. Section 13–4 is the magnetic field analog of Section 13–1. In particular, we discuss the origin of magnetic dipoles in atoms and ions and show how they are related to the electronic state of the atom or ion. Lastly, in Section 13–5 we apply the standard statistical thermodynamic equations of ideal systems to systems of noninteracting magnetic dipoles and derive the relation between the macroscopically observable magnetic susceptibility and the electronic state of the atom or ion.

13–1 AN ISOLATED MOLECULE IN AN ELECTRIC FIELD

We shall consider the effect of an electric field on the energy states of an isolated atom or molecule. In the absence of an electric field, the energy eigenvalues are a function of only the volume, and we denote the set of energies by $\{\varepsilon_j^0(V)\}$, where the zero superscript emphasizes that there is no applied field. When we switch on the field, each of the energies is altered, and we denote the new set of energies by $\{\varepsilon_j(V, \mathscr{E}_z^*)\}$, where $\mathscr{E}_z^*$ denotes the magnitude of the electric field, assumed for simplicity to lie in the z-direction. We are using what may appear to be an unnecessarily complicated notation for the electric field, but the reason for this will become clear in the next two sections. This new set of energies can be calculated by quantum mechanical perturbation theory. We shall state and discuss the result of such a calculation later in this section, but the important point is simply that the energies depend upon V and $\mathscr{E}_z^*$.

In Chapter 2 we defined the pressure in the jth quantum state to be given by $p_j = -(\partial E_j/\partial V)_N$. If the particles of the system are independent (which we assume here), the partial pressure due to one particle in its jth quantum state is $-(\partial \varepsilon_j/\partial V)$. The molecular interpretation of the average work done on a system of N independent particles can be written as

$$\delta w = N \sum_j \pi_j \, d\varepsilon_j$$

where π_j is the probability that the particle is in the jth state with energy ε_j. If we write this equation as

$$\delta w = N \sum_j \pi_j \left(\frac{\partial \varepsilon_j}{\partial V}\right) dV = -N \sum_j \pi_j p_j \, dV = -\bar{p} \, dV$$

we see that this definition shows that the work done on a system when we change its volume by dV is $-\bar{p}\,dV$, or $-p\,dV$ when we equate $\bar{p}$ with the thermodynamic pressure.

If we now apply an electric field to a system of independent particles, the energies depend upon both V and $\mathscr{E}_z^*$. Since the system of particles is assumed to be dilute, each one sees the same electric field $\mathscr{E}_z^*$, which can be considered to be applied from outside the system. We define the z-component of the dipole moment of a particle in its jth quantum state to be

$$\mu_{jz} = -\left(\frac{\partial \varepsilon_j}{\partial \mathscr{E}_z^*}\right)_V \tag{13-1}$$

The work that is done on a particle in its jth quantum state when $\mathscr{E}_z^*$ is changed by $d\mathscr{E}_z^*$ is

$$\delta w_j = -\mu_{jz} \, d\mathscr{E}_z^*$$

The average z-component of the dipole moment of a particle is

$$\bar{\mu}_z = \sum_j \pi_j \mu_{jz} \tag{13-2}$$

and the average work that is done on the particle when $\mathscr{E}_z^*$ is changed by $d\mathscr{E}_z^*$ is

$$\delta w = \sum_j \pi_j \left(\frac{\partial \varepsilon_j}{\partial \mathscr{E}_z^*}\right)_V d\mathscr{E}_z^* = -\bar{\mu}_z \, d\mathscr{E}_z^* \tag{13-3}$$

For a system of independent particles, the average z-component of the moment of the entire system $\overline{M}_z$ is N times the average value for each molecule, and so

$$\delta w = -N\bar{\mu}_z \, d\mathscr{E}_z^* = -\overline{M}_z \, d\mathscr{E}_z^* \tag{13-4}$$

represents the electrical work done on the system. We say again that since the system is dilute, each molecule sees only the applied external electric field $\mathscr{E}_z^*$. There are no fields at any molecule due to the dipole moments, say, of the other molecules of the system.

Thus if we know ε_j as a function of $\mathscr{E}_z^*$, we can calculate its average electric moment as a function of the applied field. Since the electric properties of the entire macroscopic system are due to the average electric moments of the individual particles (independent particles), we can calculate the macroscopic electrical properties in terms of molecular properties.

For most molecules, however, it is not necessary to calculate $\varepsilon_j(V, \mathscr{E}_z^*)$ since classical statistics serves as an excellent approximation, and so we need to discuss only the classical energy of a molecule in an electric field. Consider a molecule with dipole moment $\boldsymbol{\mu}$ in an electric field $\mathscr{E}_z^*$ (see Fig. 13–1). We picture the dipole moment as a positive charge and a negative charge of magnitude q separated by a vector $\mathbf{l}$. The field $\mathscr{E}_z^*$ causes the dipole to rotate into a direction parallel with $\mathscr{E}_z^*$. Therefore it requires work to rotate it to an angle θ to $\mathscr{E}_z^*$. The force causing the molecule to rotate is actually a torque (torque is the angular analog of force), and is given by $l/2$ times the force perpendicular to $\mathbf{l}$ at each end of $\mathbf{l}$. The force perpendicular to $\mathbf{l}$ is $q\mathscr{E}_z^* \sin\theta$, and so the torque is $2(l/2)q\mathscr{E}_z^* \sin\theta = \mu\mathscr{E}_z^* \sin\theta$. The energy required to rotate the dipole from 0 to some angle θ is

$$\varepsilon = \int \mu\mathscr{E}_z^* \sin\theta' \, d\theta'$$

The energy of a dipole in an electric field then is

$$\varepsilon = -\mu\mathscr{E}_z^* \cos\theta = -\mu_z\mathscr{E}_z^* \tag{13–5}$$

where μ_z is the z-component of $\boldsymbol{\mu}$. If the field were not necessarily in the z-direction, Eq. (13–5) would be $\varepsilon = -\boldsymbol{\mu}\cdot\boldsymbol{\mathscr{E}}^*$.

Not all molecules possess permanent dipole moments, but all molecules are polarizable, that is, their electronic charge distributions distort in an electric field. This distortion is such that an additional dipole moment is induced by $\mathscr{E}_z^*$, and for ordinary field strengths we can say that the magnitude of the dipole moment induced is proportional to $\mathscr{E}_z^*$. The proportionality constant is called the polarizability of the molecule. If we denote the induced moment by μ_{ind}, we write

$$\mu_{\text{ind}} = \alpha\mathscr{E}_z^* \tag{13–6}$$

where α is the polarizability. Equation (13–6) assumes that the induced dipole lies in the same direction as $\mathscr{E}_z^*$, and so μ_{ind} could be more completely denoted by $\mu_{\text{ind},z}$. The energy required to induce the dipole moment is given by Eq. (13–3)

$$\varepsilon = -\int_0^{\mathscr{E}_z^*} \mu_{\text{ind}} \, d\mathscr{E}_z^* = -\tfrac{1}{2}\alpha\mathscr{E}_z^{*2} \tag{13–7}$$

The total electrical energy of a polarizable dipolar molecule in an external electric field is

$$\varepsilon(\mathscr{E}_z^*) = -\mu_z\mathscr{E}_z^* - \tfrac{1}{2}\alpha\mathscr{E}_z^{*2} \tag{13–8}$$

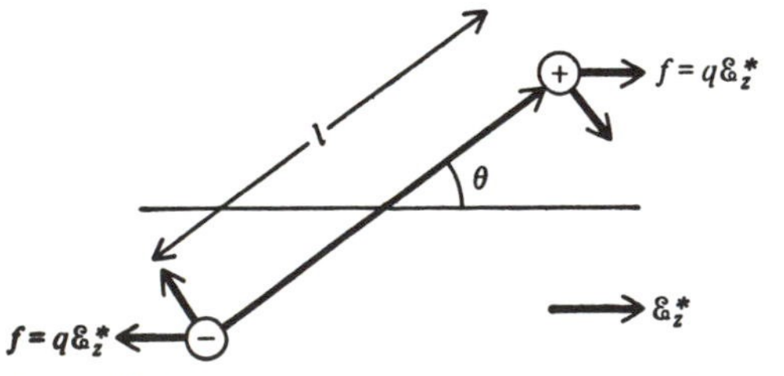

Figure 13–1. **The torque acting upon a dipole moment in an electric field $\mathscr{E}_z^*$. The length of the dipole is l and the charges at each end are $+q$ and $-q$.**

Equations (13–3) and (13–8) allow us to calculate the electrical properties of an ideal gas of such molecules. We shall do that in the next section. Before concluding this section, however, we shall discuss the quantum mechanical calculation of Eq. (13–8). We do this because in Section 13–4 we shall discuss the analogous case of a single atom or molecule in a magnetic field. The magnetic problem must be set up quantum mechanically from the beginning, since the magnetic properties of atoms are related to the stable orbital motion of the electrons within the atom, and these, of course, must be treated by quantum mechanics. In order to compare the electric and magnetic results then, we shall present the quantum mechanical derivation of Eq. (13–8) here.

Consider some molecule with a Hamiltonian operator $\mathscr{H}_0$ and ground-state electronic wave function ψ_0. Since most atoms and molecules are in their ground electronic states at ordinary temperatures, we shall consider only ψ_0. If the molecule is now placed at some angle θ with respect to an electric field $\mathscr{E}_z{}^*$, there is an additional term in the Hamiltonian, given by

$$\mathscr{H}' = -\sum_j q_j z_j \mathscr{E}_z{}^* = -\hat{\mu}_z \mathscr{E}_z{}^* \tag{13–9}$$

where z_j is the z-component of the jth point charge and where the summation is over all the charges in the molecule, that is, over all the electrons and nuclei. The quantity $\hat{\mu}_z$ is the z-component of the quantum mechanical operator corresponding to the dipole moment. The Schrödinger equation is

$$\begin{aligned}(\mathscr{H}_0 + \mathscr{H}')\psi &= \varepsilon\psi \\ (\mathscr{H}_0 - \hat{\mu}_z \mathscr{E}_z{}^*)\psi &= \varepsilon\psi\end{aligned} \tag{13–10}$$

where $\mathscr{H}_0$ is the Hamiltonian of the molecule in the absence of an electric field. This equation cannot be solved exactly, but can be solved by perturbation theory by treating $\mathscr{H}' = -\hat{\mu}_z \mathscr{E}_z{}^*$ as a small perturbation to the original or unperturbed problem. The procedure is to expand ψ and ε as a power series in the perturbation (in this case the field $\mathscr{E}_z{}^*$) and solve the problem term by term. To quadratic terms in $\mathscr{E}_z{}^*$, we get Eq. (13–8) where μ_z is defined by

$$\mu_z = \int \psi_0{}^* \hat{\mu}_z \psi_0 \, d\tau$$

where ψ_0 is the ground-state wave function (assumed known) of the unperturbed problem, and $d\tau$ represents an integration over all the electrons in the molecule. The equation for α is more complicated, but nevertheless it has been used to calculate α in good agreement with experiment. (See Davies in "Additional Reading.")

13–2 AN IDEAL GAS IN AN ELECTRIC FIELD

We now consider an ideal gas of polar molecules in an electric field. Since the gas is dilute enough to consider ideal, the intermolecular interactions can be neglected, and so each molecule of the gas experiences the same electric field $\mathscr{E}_z{}^*$, which in this case is due to external sources such as charges on the plates of a capacitor. The partition function of the ideal gas is

$$Q(N, V, T, \mathscr{E}_z{}^*) = \frac{[q(V, T, \mathscr{E}_z{}^*)]^N}{N!} \tag{13–11}$$

where $q(V, T, \mathscr{E}_z^*)$ is the partition function of a single molecule in the field $\mathscr{E}_z^*$. The form of q depends upon the geometry of the molecule, that is, whether it is linear, a symmetric top, and so on. For simplicity, we shall consider a linear dipolar molecule, but the final result, Eq. (13–16), is more general. If we assume that all the degrees of freedom except vibrational are classical, then the molecular partition function is [*cf.* Eq. (7–15)]

$$q(V, T, \mathscr{E}_z^*) = q_{\text{vib}}(T) \cdot \frac{1}{h^5} \int \cdots \int e^{-H/kT}\, dp_x\, dp_y\, dp_z\, dx\, dy\, dz\, dp_\theta\, dp_\phi\, d\theta\, d\phi \tag{13–12}$$

where

$$H = \frac{1}{2M}(p_x^2 + p_y^2 + p_z^2) + \frac{p_\theta^2}{2I} + \frac{p_\phi^2}{2I\sin^2\theta} - \mu\mathscr{E}_z^* \cos\theta - \tfrac{1}{2}\alpha\mathscr{E}_z^{*2} \tag{13–13}$$

where I is the moment of inertia of the molecule. The integrations over $p_x, p_y, p_z, x, y,$ and z are easy, and we get

$$q(V, T, \mathscr{E}_z^*) = \frac{V}{\Lambda^3} \frac{e^{-\Theta_v/2T}}{(1 - e^{-\Theta_v/T})} \frac{1}{h^2} \int \cdots \int e^{-H'/kT}\, dp_\theta\, dp_\phi\, d\theta\, d\phi \tag{13–14}$$

where $\Lambda = (h^2/2\pi MkT)^{1/2}$ and

$$H' = \frac{p_\theta^2}{2I} + \frac{p_\phi^2}{2I\sin^2\theta} - \mu\mathscr{E}_z^* \cos\theta - \tfrac{1}{2}\alpha\mathscr{E}_z^{*2} \tag{13–15}$$

The integrals over $dp_\theta\, dp_\phi$ are also easy, and Eq. (13–14) becomes

$$q(V, T, \mathscr{E}_z^*) = \frac{V}{\Lambda^3}\left(\frac{8\pi^2 IkT}{h^2}\right)\left(\frac{e^{-\Theta_v/2T}}{1 - e^{-\Theta_v/T}}\right) \times \frac{1}{4\pi}\int_0^{2\pi}\int_0^{\pi} e^{\mu\mathscr{E}_z^* \cos\theta/kT} e^{\alpha\mathscr{E}_z^{*2}/2kT} \sin\theta\, d\theta\, d\phi$$

$$= q(V, T, 0) \times \frac{1}{2}\int_0^{\pi} e^{\mu\mathscr{E}_z^* \cos\theta/kT} e^{\alpha\mathscr{E}_z^{*2}/2kT} \sin\theta\, d\theta \tag{13–16}$$

Notice that the partition function factors into the partition function in the absence of an electric field and a function of $\mathscr{E}_z^*$, $f(\mathscr{E}_z^*)$,

$$f(\mathscr{E}_z^*) = \frac{1}{2}\int_0^{\pi} e^{\mu\mathscr{E}_z^* \cos\theta/kT} e^{\alpha\mathscr{E}_z^{*2}/2kT} \sin\theta\, d\theta \tag{13–17}$$

Of course, $f(\mathscr{E}_z^*)$ becomes unity when $\mathscr{E}_z^* = 0$. Equation (13–17) can be easily integrated by letting $x = \cos\theta$:

$$f(\mathscr{E}_z^*) = e^{\alpha\mathscr{E}_z^{*2}/2kT}\left\{\frac{\sinh\left(\dfrac{\mu\mathscr{E}_z^*}{kT}\right)}{\left(\dfrac{\mu\mathscr{E}_z^*}{kT}\right)}\right\} \tag{13–18}$$

We have derived Eq. (13–16) here for a diatomic gas only, but it is more general. For a diatomic molecule, the total partition function is

$$q(V, T, \mathscr{E}_z^*) = \frac{V}{\Lambda^3}\left(\frac{8\pi^2 IkT}{h^2}\right)\frac{e^{\alpha\mathscr{E}_z^{*2}/2kT} e^{-\Theta_v/2T}}{(1 - e^{-\Theta_v/T})} \frac{\sinh\left(\dfrac{\mu\mathscr{E}_z^*}{kT}\right)}{\left(\dfrac{\mu\mathscr{E}_z^*}{kT}\right)} \tag{13–19}$$

We showed in Eq. (13–4) that the electrical work done on a system of independent particles is $-\overline{M}_z \, d\mathscr{E}_z{}^*$, where $\overline{M}_z = N\bar{\mu}_z$ is the total dipole moment of the macroscopic system. We can then write the first law of thermodynamics as

$$dE = T\,dS - \bar{p}\,dV - \overline{M}_z\,d\mathscr{E}_z{}^* \tag{13–20}$$

By subtracting $d(TS)$ from both sides of this equation, we get

$$dA = -S\,dT - \bar{p}\,dV - \overline{M}_z\,d\mathscr{E}_z{}^* \tag{13–21}$$

from which we write

$$p = -\left(\frac{\partial A}{\partial V}\right)_{N,\mathscr{E}_z{}^*,T} = kT\left(\frac{\partial \ln Q}{\partial V}\right)_{N,\mathscr{E}_z{}^*,T} \tag{13–22}$$

and

$$\overline{M}_z = -\left(\frac{\partial A}{\partial \mathscr{E}_z{}^*}\right)_{N,V,T} = kT\left(\frac{\partial \ln Q}{\partial \mathscr{E}_z{}^*}\right)_{N,V,T} \tag{13–23}$$

From Eqs. (13–11), (13–19), and (13–22) we see that $pV = NkT$, that is, the equation of state is unchanged by the introduction of an external electric field. The average moment of the system is

$$\overline{M}_z = N\bar{\mu}_z = N(\alpha\mathscr{E}_z{}^* + \mu\mathscr{L}(y)) \tag{13–24}$$

where $y = \mu\mathscr{E}_z{}^*/kT$ and $\mathscr{L}(y)$ is the Langevin function

$$\mathscr{L}(y) = \coth y - \frac{1}{y} \tag{13–25}$$

The Langevin function is shown in Fig. 13–2. It can be seen that $\mathscr{L}(y)$ approaches unity asymptotically. Physically, $\mu\mathscr{L}(\mu\mathscr{E}_z{}^*/kT)$ is the average z-component of $\boldsymbol{\mu}$ since

$$\bar{\mu}_z = \mu\,\overline{\cos\theta} = \mu\,\frac{\int_0^\pi e^{\mu\mathscr{E}_z{}^*\cos\theta/kT}\cos\theta\sin\theta\,d\theta}{\int_0^\pi e^{\mu\mathscr{E}_z{}^*\cos\theta/kT}\sin\theta\,d\theta}$$

$$= \mu\mathscr{L}\left(\frac{\mu\mathscr{E}_z{}^*}{kT}\right) \tag{13–26}$$

Thus the Langevin function describes the tendency of the dipolar molecules to line up in the external electric field. This tendency is opposed by the thermal agitation of the molecules, but as the field increases, the molecules line up almost completely and $\overline{\cos\theta} \to 1$. In the absence of an external field, the dipoles have no preferred orientation and hence $\bar{\mu}_z \to 0$ as $\mathscr{E}_z{}^* \to 0$.

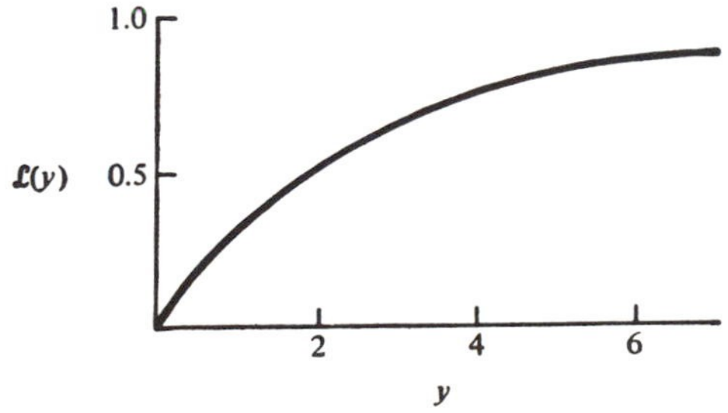

Figure **13–2. The Langevin function $\mathscr{L}(y) = \coth y - 1/y$ versus y.**

In practice, the argument of the Langevin function $\mu \mathscr{E}_z^*/kT$ is much less than unity. For example, the permanent dipole moment of a molecule is of the order of 10^{-18} esu-cm, that is, 1 debye, and a fairly strong electric field is $0(10^5)$ volts/cm (which in cgs units is $0(10^5/300)$ statvolts/cm). Thus at 300°K, the ratio

$$\frac{\mu \mathscr{E}_z^*}{kT} = \frac{10^{-18} \times (10^5/300)}{1.38 \times 10^{-16} \times 300} = 8.0 \times 10^{-3}$$

Electromagnetic units can become extremely confusing, but we shall use only cgs here, and a short table of cgs electric and magnetic units is given in Table 13–1.

Table 13–1. **A brief tabulation of the cgs electromagnetic units used in the chapter**

property	symbol	units	comments
electric field	$\mathscr{E}$	statvolts/cm	1 statvolt = 300 practical volts 10^4 statvolts/cm is a large electric field
charge	q	esu	charge on the electron = 4.803×10^{-10} esu
electric dipole moment	μ	debye	1 debye = 10^{-18} esu-cm
dielectric constant	ε	unitless	$\varepsilon \geq 1$
potential	ϕ	statvolts	
magnetic field	B	gauss	10^5 gauss is a strong magnetic field
magnetic dipole moment	μ	ergs/gauss	given in terms of the Bohr magneton
Bohr magneton	β	ergs/gauss	$\beta = 9.273 \times 10^{-21}$ erg/gauss

Since $\mu \mathscr{E}_z^*/kT$ is much smaller than unity, we may use the first term of the expansion of $\mathscr{L}(y)$ for small y (see Problem 13–5)

$$\mathscr{L}(y) = \frac{y}{3} - \frac{y^3}{45} + \cdots \tag{13–27}$$

The total dipole moment in the macroscopic system is

$$\overline{M}_z = N\left(\alpha + \frac{\mu^2}{3kT}\right)\mathscr{E}_z^* \tag{13–28}$$

This is an experimentally measurable quantity, since it can be directly related to the dielectric constant ε of the system.

In order to develop the connection between $\overline{M}_z$ and ε, consider a parallel plate capacitor with the region between the plates evacuated. The plates are large enough so that any end effects can be neglected. Let the charge density on one plate be $+\sigma$ esu/cm^2 and the charge density on the other plate be $-\sigma$ esu/cm^2. The electric field $\mathscr{E}_0$ due to the surface charges is perpendicular to the plates and is equal to $4\pi\sigma$ (see Problem 13–2). The capacitance is defined as the ratio of the charge on one of the plates divided by the potential difference between them. The potential difference is $\mathscr{E}_0$ times the distance between the plates d, and so

$$C_0 = \frac{\text{charge}}{\text{voltage}} = \frac{\sigma A}{4\pi\sigma d} = \frac{A}{4\pi d} \tag{13–29}$$

where A is the area of one of the plates.

Now fill the region between the plates with some dielectric (nonconducting) material, such as a gas or a liquid. The electric field acting between the plates causes a separation of positive and negative charges within the dielectric material. This *polarization* is

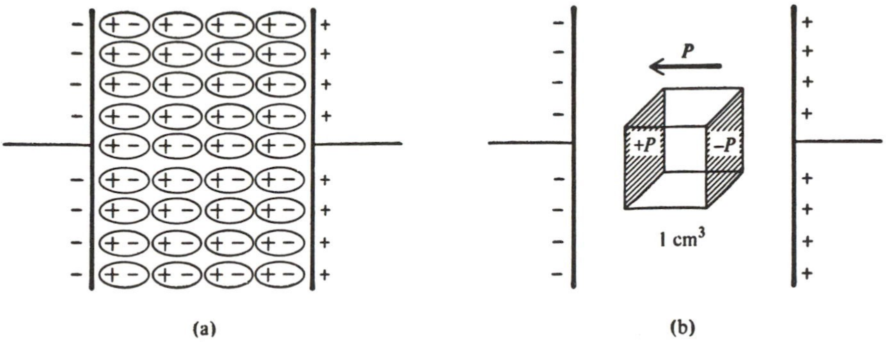

Figure 13–3. **(a) The polarization charges induced in a dielectric material and (b) the definition of the polarization vector P.**

depicted in Fig. 13–3. This polarization can be due to induced dipoles or the orientation of permanent dipoles or both. In either case, we get the effect shown in Fig. 13–3(a).

If the surface charge on the plates is held fixed and a dielectric substance is introduced between the plates of the capacitor, the capacitance will increase by a factor ε, called the dielectric constant. According to the definition of capacitance given in Eq. (13–29), the voltage, and hence the electric field between the plates, is reduced by a factor ε, so that $\mathscr{E} = \mathscr{E}_0/\varepsilon$. The field is reduced because the material between the plates is polarized by the charge on the plates, and its net dipole moment acts against the field $\mathscr{E}_0$. In Fig. 13–3(b) is shown a unit cube of dielectric material between the plates of a capacitor. We define a vector **P**, called the polarization vector, as the dipole moment per unit volume. It is equivalent to a charge $+P$ on one face of the unit dielectric cube and $-P$ on the opposite face. These charges partially cancel the charges on the plates so that the average macroscopic electric field within the dielectric is

$$\mathscr{E}_z = 4\pi(\sigma - P_z) \tag{13–30}$$

We define a new quantity $D_z = 4\pi\sigma$ perpendicular to the plates and pointing toward the negative plate. Thus we write

$$D_z = \mathscr{E}_z + 4\pi P_z \tag{13–31}$$

Furthermore, since D_z is what we previously called $\mathscr{E}_0$, we can write

$$D_z = \varepsilon\mathscr{E}_z \tag{13–32}$$

From Eqs. (13–31) and (13–32) we have the relation

$$P_z = \frac{\overline{M}_z}{V} = \frac{(\varepsilon - 1)}{4\pi}\mathscr{E}_z \tag{13–33}$$

where remember that $\mathscr{E}_z$ is the average macroscopic field within the dielectric material. Equation (13–28) tells us that

$$\overline{M}_z = N\left(\alpha + \frac{\mu^2}{3kT}\right)\mathscr{E}_z^* \tag{13–34}$$

where $\mathscr{E}_z^*$ is the field within the vicinity of each molecule. The immediate problem is to relate $\mathscr{E}_z^*$, the *local field,* to $\mathscr{E}_z$, the average macroscopic field. In the case of an ideal gas, the local field acting upon any one molecule is the field produced by the charges fixed on the plates of the capacitor. This field is $4\pi\sigma$, or in the notation introduced after Eq. (13–30), we write that $\mathscr{E}_z^* = D_z = \varepsilon\mathscr{E}_z$. If we substitute this into Eq. (13–34) and that into Eq. (13–33), we have finally that

$$\frac{\varepsilon - 1}{\varepsilon} = 4\pi\rho\left(\alpha + \frac{\mu^2}{3kT}\right) \tag{13–35}$$

which gives the desired relation between the dielectric constant ε and the electrical properties of the individual molecules, namely, α and μ.

Equation (13–35) predicts that for molecules with a permanent dipole moment, a plot of $(\varepsilon - 1)/\varepsilon$ versus $1/T$ yields a straight line whose slope is a measure of the square of the dipole moment of the molecule. This is a standard physical chemical method for determining dipole moments, and some typical plots of Eq. (13–35) are given in Fig. 13–4.

It is important to get an appreciation of the types of numbers involved in Eq. (13–35). The polarizability of an atom or a molecule is approximately the volume of the atom or molecule, so that α is typically of the order of 10^{-23} or 10^{-24} cm^3. The term $\mu^2/3kT$ is of the order of $10^{-36}/(3 \times 1.38 \times 10^{-16} \times 300) = 0(10^{-23})$, since dipole moments are $0(10^{-18}$ esu-cm), or 1 debye. Thus both terms in parentheses in Eq. (13–35) are of the order of the volume of an atom or molecule. If this is multiplied by the number density ρ, we see that the right-hand side of Eq. (13–35) is roughly the volume of the liquid or solid divided by the volume of the gas, which typically is a number of the order of 10^{-4} or so. We see then that the dielectric constant of 1 mole of dilute gas is a number like $1.000n$, where n is some number. Because of this, Eq. (13–35) is essentially

$$\varepsilon - 1 = 4\pi\rho\left(\alpha + \frac{\mu^2}{3kT}\right) \tag{13–36}$$

This equation is often used in place of Eq. (13–35). Table 13–2 gives the molar dielectric constant at room temperature, the polarizability, and the permanent dipole moment of various atoms and molecules.

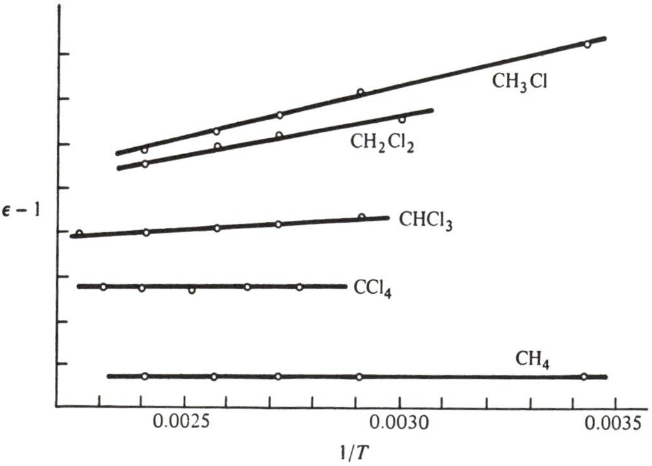

Figure 13–4. **The dielectric constant of the chlorinated methanes versus the reciprocal of the temperature.** (From P. Debye, *Polar Molecules*, New York: Dover.)

Table 13–2. **The contribution of the polarizability term and the dipole moment term to the dielectric constant of several gases at 298°K and 1 atm***

gas	$\alpha \times 10^{25}$(cm^3)	μ(debyes)	$4\pi\rho\alpha$ ($\times 10^5$)	$4\pi\rho\mu^2/3kT$ ($\times 10^5$)	$[\varepsilon - 1] \times 10^5$(calc.)
Ne	3.93	0.00	12.2	0.0	12.2
Ar	16.23	0.00	50.2	0.0	50.2
Kr	24.6	0.00	76.1	0.0	76.1
N_2	17.6	0.00	54.5	0.0	54.5
CO_2	26.5	0.00	82.0	0.0	82.0
CH_4	26.0	0.00	80.5	0.0	80.5
CCl_4	105.0	0.00	325.0	0.0	325.0
CO	19.5	0.12	60.4	3.6	64.0
N_2O	30.0	0.166	92.8	6.9	99.7
HCN	25.9	2.98	80.2	2227.0	2307.0
NH_3	22.6	1.47	69.9	542.0	612.0
$CHCl_3$	82.3	1.01	255.0	256.0	511.0
CH_2Cl_2	64.8	1.58	201.0	626.0	827.0
CH_3Cl	45.6	1.87	141.0	877.0	1018.0

* The experimental values are almost identical with the calculated values since α and μ are determined from experimental dielectric constant data; compare Eq. (13–36).

One can use the partition function in Eq. (13–19) to calculate all the thermodynamic functions of an ideal gas in an external electric field. For example, the thermodynamic energy is

$$\frac{E}{NkT} = \frac{E(0)}{NkT} - \frac{1}{kT}\left(\frac{\alpha}{2} + \frac{\mu^2}{3kT}\right)\mathscr{E}_z^{*2}$$

where $E(0)$ is the energy of the ideal gas with no electric field. Typically, the additional term due to $\mathscr{E}_z^*$ is $0(10^{-6})$ compared to 0(1) for the leading term. It is easy to show that the above equation is simply $E(0)/NkT$ plus $\langle\varepsilon(\mathscr{E}_z^*)\rangle kT$ (see Problem 13–11). Other thermodynamic functions are left to the problems.

13–3 THE CLAUSIUS-MOSSOTTI EQUATION

Equation (13–35) is valid only for dilute gases. Not only did we assume that the total dipole moment of the system was $N\bar{\mu}$, but we also assumed that the local electric field acting upon any single molecule was due only to the charges fixed on the plates of a capacitor, so that $\mathscr{E}_z^* = \varepsilon\mathscr{E}_z$. A rigorous extension of Eq. (13–35) to more concentrated systems requires that we consider the interaction of any molecule with its neighbors, so that the local field includes a contribution from its neighboring molecules as well as from the externally fixed charges. On strictly molecular grounds, such an extension is fairly complicated, but there exists a well-known and useful semimolecular theory due to Lorentz *et al.*, which we present in this section. This theory introduces the important concept of the local field, which is a central concept of many statistical mechanical theories of dielectrics.

Equation (13–28) shows that the average z-component of a dipole moment due to an electric field $\mathscr{E}_z^*$ in the z-direction is

$$\bar{\mu}_z = \left(\alpha + \frac{\mu^2}{3kT}\right)\mathscr{E}_z^* \tag{13–37}$$

It is important to understand here that $\mathscr{E}_z^*$ is the electric field acting upon the molecule. In the previous section, we argued that for a dilute gas, $\mathscr{E}_z^*$ was equal to the field due to the charges fixed on the plates of the capacitor, and so we were able to equate

$\mathscr{E}_z^*$ with $\varepsilon\mathscr{E}_z$, where $\mathscr{E}_z$ is the average macroscopic field in the dielectric material. For a more dense system, however, the local field $\mathscr{E}_z^*$ is not just due to the charges on the plates of the capacitor, but also has a contribution from the other molecules in the system. The essence of the problem we are setting up then is the following. The general macroscopic equations of classical electrostatics give that the polarization is

$$P_z = \frac{(\varepsilon - 1)}{4\pi}\mathscr{E}_z \tag{13–38}$$

and Eq. (13–37) says that

$$P_z = \frac{N\bar{\mu}_z}{V} = \rho\left(\alpha + \frac{\mu^2}{3kT}\right)\mathscr{E}_z^* \tag{13–39}$$

The problem is to relate the local field $\mathscr{E}_z^*$ to the macroscopic field so that we eliminate it between Eqs. (13–38) and (13–39) and obtain an expression for ε in terms of α and μ. In the previous section, we set $\mathscr{E}_z^* = \varepsilon\mathscr{E}_z$ to derive Eq. (13–35), but this is not valid for a dense system. The determination of the relation between $\mathscr{E}_z^*$ and $\mathscr{E}_z$ is called the local field problem and is a central problem of dielectric theory.

In order to calculate the local field acting on a molecule in a dense system, consider a parallel plate capacitor filled with the dielectric material. Let us now construct a sphere of radius a around this central atom, where the radius of this sphere is large compared to molecular dimensions, but small compared to the separation between the plates. We shall treat the molecules within the sphere molecularly, but we shall treat those outside of the sphere as a continuous dielectric medium. In a sense, then, we are formulating a semimolecular theory, where the nearby molecules are considered as molecules, and the rest of the dielectric is treated as a continuum. The surface charges on the plates of the capacitor induce polarization charges on all the surfaces of the dielectric, and this is shown in Fig. 13–5. The magnitude of the polarization charges on the surfaces of the dielectric is given by the component of the polarization vector **P** normal to the surface. Thus the surface charge on the spherical surface is **P** $\cos\theta$. We point out that the charges on the left-hand planar surface of the dielectric and the right half of the spherical surface are both negative, since the polarization vector **P** is entering the dielectric at these surfaces.

The local field acting upon a molecule at the center of the sphere can be written as

$$\mathscr{E}_z^* = E_0 + E_1 + E_2 + E_3 \tag{13–40}$$

The first contribution is the field due to the real fixed charges on the plates of the capacitor. This field is $4\pi\sigma$. The term E_1 is the field due to the polarization charges on the surface of the dielectric next to the plates. If we denote the polarization charge density there by σ', this field is $-4\pi\sigma'$. The first two contributions to $\mathscr{E}_z^*$ give

$$E_0 + E_1 = 4\pi(\sigma - \sigma') = \mathscr{E}_z \tag{13–41}$$

where we have recognized that $4\pi\sigma - 4\pi\sigma' = D_z - 4\pi P_z = \mathscr{E}_z$.

The term E_2 in Eq. (13–40) is the field at the center of the sphere due to the polarization charges on the spherical surface. The charge density there is equal to the component of P normal to the spherical surface, or $P\cos\theta$, where the angle θ is shown in Fig. 13–5. The area of the spherical ring shown in the figure is $2\pi a^2 \sin\theta\, d\theta$, and so

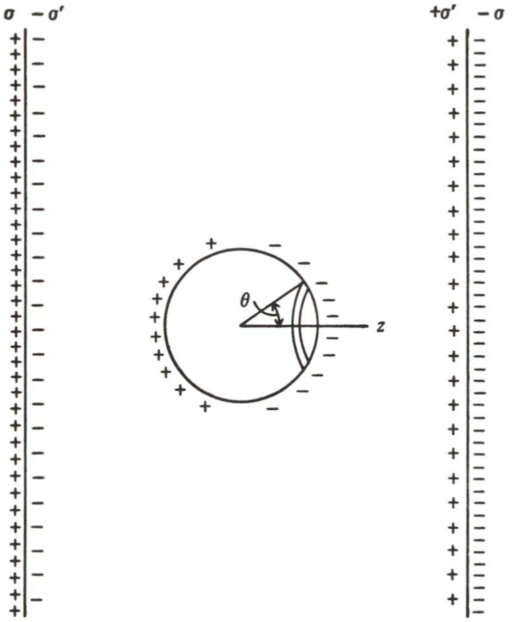

Figure 13–5. **The sphere hollowed out from the dielectric material between the plates of a parallel plate capacitor. The plates have a fixed surface charge density σ which induces charges on the surfaces of the dielectric.**

the charge in that area is $dq = (P \cos \theta)(2\pi a^2 \sin \theta \, d\theta)$. By symmetry, there is a resultant field only in the z-direction, and this is given by

$$dE_2 = \frac{dq}{a^2} \cos \theta = 2\pi P \cos^2 \theta \sin \theta \, d\theta$$

which gives for E_2

$$E_2 = 2\pi P \int_0^\pi \cos^2 \theta \sin \theta \, d\theta = \frac{4\pi P}{3} \tag{13–42}$$

Up to this point we have made no approximations. If the sphere is large compared to molecular dimensions, it is possible to treat the fluid beyond the central sphere as a continuous dielectric. It is in the calculation of E_3, the field due to the molecules within the sphere, that we shall have to make an approximation. We shall, in fact, make a very severe approximation, namely, that $E_3 = 0$. The reasoning behind setting $E_3 = 0$ is that Lorentz showed that $E_3 = 0$ for simple cubic, body-centered cubic, and face-centered cubic crystals. It is also equal to zero for an ideal gas, where there is no interaction between the molecules. Although it is clearly incorrect to set $E_3 = 0$ for a dense gas or liquid, it is hoped that it is not a bad approximation, and so we set

$$E_3 = 0 \tag{13–43}$$

The total local field in Eq. (13–40), then, is

$$\mathscr{E}_z^* = \mathscr{E}_z + \frac{4\pi}{3} P_z \tag{13–44}$$

This expression for the local field is called the Lorentz-Lorenz field. According to the discussion following Eqs. (13–38) and (13–39), our aim is to write the local field

$\mathscr{E}_z^*$ in terms of the macroscopic field $\mathscr{E}_z$, and this can now be done by eliminating P_z between Eqs. (13–38) and (13–44):

$$\mathscr{E}_z^* = \frac{(\varepsilon + 2)}{3}\mathscr{E}_z \tag{13–45}$$

If we substitute Eq. (13–38) into the left-hand side of Eq. (13–39) and Eq. (13–45) into the right-hand side, we get the Clausius-Mossotti equation:

$$\frac{\varepsilon - 1}{\varepsilon + 2} = \frac{4\pi\rho}{3}\left(\alpha + \frac{\mu^2}{3kT}\right) \tag{13–46}$$

which is applicable to much higher densities than Eq. (13–35). As the density becomes that of gas, $\varepsilon \to 1$, and Eq. (13–35) results. Also, as $\varepsilon \to 1$, the local field $\mathscr{E}_z^*$ approaches the macroscopic field $\mathscr{E}_z$ [*cf.* Eq. (13–45)].

The Clausius-Mossotti equation is not applicable to polar liquids since there is strong interaction between neighboring dipoles, and Eq. (13–43) is not valid. For non-polar substances, on the other hand, it serves as a useful equation up to very high densities. Table 13–3 gives the Clausius-Mossotti function over a wide range of densities for argon and carbon dioxide.

Table 13–3. **The variation of the Clausius-Mossotti function, $(\varepsilon - 1)/(\varepsilon + 2)\rho$, with density for argon and carbon dioxide at 100°C***

	density (mole/liter)	$(\varepsilon - 1)/\rho$	$(\varepsilon - 1)/\varepsilon\rho$	$(\varepsilon - 1)/(\varepsilon + 2)\rho$
argon	0.01	12.414	12.414	4.138
	0.10	12.420	12.405	4.138
	0.50	12.443	12.366	4.139
	1.00	12.472	12.319	4.140
	2.00	12.531	12.225	4.142
	3.00	12.591	12.132	4.145
	4.00	12.651	12.041	4.147
carbon dioxide	0.01	22.036	22.037	7.346
	0.10	22.065	22.016	7.350
	0.50	22.179	21.936	7.366
	1.00	22.324	21.836	7.386
	2.00	22.617	21.638	7.427
	3.00	22.916	21.442	7.468

* For comparison, the functions $(\varepsilon - 1)/\rho$ (see Eq. (13–36)) and $(\varepsilon - 1)/\varepsilon\rho$ (see Eq. (13–35)) are given also.

13–4 AN ISOLATED ATOM IN A MAGNETIC FIELD

This section will be similar to Section 13–1 in which we discussed an isolated molecule in an electric field. Here we shall discuss the analogous magnetic case. We point out that although atoms do not possess permanent electric dipole moments, we shall see that atoms and ions can possess magnetic dipole moments. For simplicity then, we shall consider only atoms and ions in magnetic fields. First let us consider some general results.

Denote the set of energies of the atom (or ion) in the absence of a magnetic field by $\{\varepsilon_j^0(V)\}$. When we switch on the magnetic field, each of the energies is altered, and we have a new set $\{\varepsilon_j(V, B_z^*)\}$, where B_z^* denotes the magnetic field (taken in the z-direction). Again we use an asterisk to denote the local field acting on the atom. In analogy with Eq. (13–1), we define the z-component of the magnetic moment in the jth state by

$$\mu_{jz} = -\left(\frac{\partial \varepsilon_j}{\partial B_z^*}\right)_V \tag{13–47}$$

We are using the same symbol for the electric moment and the magnetic moment, but this should cause no confusion since we shall never have both in the same equation. The work that is done on an atom in the jth quantum state when B_z^* is changed by dB_z^* is

$$\delta w_j = -\mu_{jz}\, dB_z^* \tag{13–48}$$

The average z-component of a magnetic moment of a particle is

$$\bar{\mu}_z = \sum_j \pi_j \mu_{jz} \tag{13–49}$$

where π_j is the probability that the particle is in the jth quantum state.

The average work that is done on a particle when B_z^* is changed by dB_z^* is

$$\delta w = \sum_j \pi_j \left(\frac{\partial \varepsilon_j}{\partial B_z^*}\right) dB_z^* = -\bar{\mu}_z\, dB_z^* \tag{13–50}$$

For a system of independent dipoles, the average z-component of the magnetic moment of the entire system $\mathscr{M}_z$ is N times the average value for each atom, and so

$$\delta w = -N\bar{\mu}_z\, dB_z^* = -\mathscr{M}_z\, dB_z^* \tag{13–51}$$

represents the work done of the system. We must now explore the origin of the magnetic moments of atoms and ions.

It is shown in elementary electricity and magnetism that a magnetic dipole is associated with the motion of an electric charge around a closed loop. This is the basis for using a solenoid to generate a magnetic field. The magnitude of the magnetic dipole is iA/c, where i is the electric current around the loop; A is its area; and c is the velocity of light.

The orbital motions of the electrons around the nucleus of an atom or ion act as tiny electric currents, and hence produce a magnetic dipole moment. Consider the motion of a charged particle in a circular orbit. Let the radius of the orbit be r, the velocity of the charge v, and its charge be q. The electric current generated by this moving charge is the charge per second moving past a given point and is given by $qv/2\pi r$. Thus the magnetic dipole μ due to this motion is

$$\mu = \frac{iA}{c} = \frac{qv\pi r^2}{c2\pi r} = \frac{qvr}{2c} \tag{13–52}$$

If the orbit is not circular, the generalization of Eq. (13–52) is

$$\boldsymbol{\mu} = \frac{q}{2c}(\mathbf{r} \times \mathbf{v}) \tag{13–53}$$

which also serves to show that the magnitude of $\boldsymbol{\mu}$ is perpendicular to the plane of orbital motion if the motion takes place in a single plane. Equation (13–53) can also be written in terms of the angular momentum $\mathbf{l}$ of the charge, since $\mathbf{l} = m\mathbf{r} \times \mathbf{v}$:

$$\boldsymbol{\mu} = \frac{q}{2mc}\mathbf{l} \tag{13–54}$$

For an electron, $q = -|e|$; and the magnetic moment is in the opposite direction to the angular momentum. There is also a magnetic dipole associated with a spin of an electron since a spinning electron is like an electric current, but we shall consider only the magnetic dipole associated with the orbital motion for now.

The expression for the energy of interaction of a magnetic dipole with an external magnetic field B_z^* is given by an equation similar to Eq. (13–9), namely,

$$-\mu_z B_z^* = \frac{|e|}{2c}(x\dot{y} - y\dot{x})B_z^* \tag{13–55}$$

where we have used Eq. (13–53) for μ_z. Notice that we have *not* denoted this by $\mathscr{H}'$, the symbol for the perturbation to the Hamiltonian operator. Remember that the classical Hamiltonian is defined so that it is a function of the momenta and coordinates. The term in Eq. (13–55) is given in terms of the velocities and coordinates. This term properly is an addition to the Lagrangian of the system, since it is the Lagrangian that is the classical mechanical function of the velocities and coordinates. Thus we write

$$L(\dot{x}, \dot{y}, \dot{z}, x, y, z) = \frac{m}{2}(\dot{x}^2 + \dot{y}^2 + \dot{z}^2) - U(x, y, z) - \frac{|e|B_z^*}{2c}(x\dot{y} - y\dot{x}) \tag{13–56}$$

where $U(x, y, z)$ is the potential energy of the electron in the field of the nucleus. The momentum is defined by

$$p_x = \frac{\partial L}{\partial \dot{x}} = m\dot{x} + \frac{|e|yB_z^*}{2c} \tag{13–57}$$

and so on, and the Hamiltonian is (see Problem 13–24)

$$H = \frac{1}{2m}(p_x^2 + p_y^2 + p_z^2) + U(x, y, z) + \frac{|e|B_z^*}{2mc}(xp_y - yp_y) + \frac{e^2(x^2 + y^2)B_z^{*2}}{8mc^2} \tag{13–58}$$

The perturbation term in the Hamiltonian of an atom in an external magnetic field, then, is

$$\mathscr{H}' = \frac{|e|B_z^*}{2mc}(xp_y - yp_x) + \frac{e^2(x^2 + y^2)B_z^{*2}}{8mc^2} \tag{13–59}$$

To quadratic terms in B_z^*, the energy is

$$\varepsilon(B_z^*) = \varepsilon_0 - \mu_z B_z^* + \frac{\alpha}{2}B_z^{*2} \tag{13–60}$$

where

$$\mu_z = -\int \psi_0^* \frac{|e|}{2mc}(xp_y - yp_x)\psi_0 \, d\tau \tag{13–61}$$

and α consists of a first-order perturbation theory term from the quadratic term in Eq. (13–59) and a second-order perturbation theory term from the first term. Unlike the electric field case, the systems of most interest here are those that possess a permanent magnetic moment given by Eq. (13–61). Such substances are called paramagnetic, and consequently we shall consider only paramagnetic substances in this and the next section.

Equation (13–59) is the perturbation Hamiltonian for a one-electron atom or ion. The generalization to many electrons simply involves a summation over all the electrons, but in order to keep the equations simple, we shall discuss only the hydrogen atom (without spin) and then state the results for other atoms or ions. The basic ideas are the same for any other system as for hydrogen.

The operator $xp_y - yp_x$ appearing in Eq. (13–61) is the operator corresponding to the z-component of the angular momentum of an isolated hydrogen atom. If we denote this by $\hat{l}_z$, then

$$\mu_z = \frac{-|e|}{2mc} \int \psi_0{}^* \hat{l}_z \psi_0 \, d\tau \tag{13–62}$$

But remember that for the hydrogen atom, the square of the orbital angular momentum is quantized by $l(l+1)\hbar^2 (l = 0, 1, 2, \ldots)$ and that its z-component is $m_l \hbar$, where $-l \le m_l \le l$. This allows us to express the magnetic moment of the hydrogen atom in terms of the two quantum numbers l and m_l:

$$\mu^2 = \frac{e^2\hbar^2}{4m^2c^2} l(l+1) = \beta_0{}^2 l(l+1) \tag{13–63}$$

$$\mu_z = -\frac{|e|\hbar}{2mc} m_l = -\beta_0 m_l \qquad -l \le m_l + l \tag{13–64}$$

where β_0 is defined as $|e|\hbar/2mc$.

Equations (13–63) and (13–64) do not take into account the spin of the electron. Furthermore, they are restricted to only one orbital electron. For a many-electron atom, the important angular momentum operator is the total angular momentum operator $\hat{J}$, given the sum of the operators corresponding to the total orbital angular momentum $\hat{L}$ and the total spin angular momentum $\hat{S}$. It is shown in Section 5–4 that, strictly speaking, it is the total angular momentum J and not the orbital or spin orbital momentum separately that is conserved. The mean-square angular momentum is $J(J+1)\hbar^2$, and its z-component is $M_J\hbar$, where $-J \le M_J \le J$. Since the total angular momentum includes spin, J can be integral or half-integral, depending upon whether the atom has an even number or an odd number of electrons, respectively. The generalization of Eqs. (13–63) and (13–64) is

$$\mu^2 = g^2 \beta_0{}^2 J(J+1) \tag{13–65}$$

$$\mu_z = -g\beta_0 M_J \qquad -J \le M_J \le J \tag{13–66}$$

The factor g is a number called the Lande g factor. It is difficult to determine g exactly, but within the approximation of Russell-Saunders coupling, it can be shown that

$$g = 1 + \frac{J(J+1) + S(S+1) - L(L+1)}{2J(J+1)} \tag{13–67}$$

where L and S are the total orbital and total spin quantum numbers. The Lande g factor can be calculated from the atomic term symbol, since the term symbol is simply a notation for J, L, and S. The Lande g factor can be measured experimentally by magnetic resonance methods, and it is found that Eq. (13–67) is often an accurate approximation. Table 13–4 gives the term symbol and the g factor for a number of ions.

Table 13–4. **A comparison of $(3kT\chi/N)^{1/2}$ and $[g\beta_0 J(J+1)]^{1/2}$ according to Eq. (13–80)**

ion	no. of 4f electrons	normal state	S	L	J	g	calc. (Hund)	calc. (Van Vleck)	$(3kT\chi/N)^{1/2}$
La^{3+}	0	$^1S_0$	0	0	0	—	0	0.00	0
Ce^{3+}	1	$^2F_{5/2}$	$\frac{1}{2}$	3	$\frac{5}{2}$	$\frac{6}{7}$	2.54	2.56	2.4
Pr^{3+}	2	$^3H_4$	1	5	4	$\frac{4}{5}$	3.58	3.62	3.5
Nd^{3+}	3	$^4I_{9/2}$	$\frac{3}{2}$	6	$\frac{9}{2}$	$\frac{8}{11}$	3.62	3.68	3.5
Pm^{3+}	4	$^5I_4$	2	6	4	$\frac{3}{5}$	2.68	2.83	—
Sm^{3+}	5	$^6H_{5/2}$	$\frac{5}{2}$	5	$\frac{5}{2}$	$\frac{2}{7}$	0.84	1.55, 1.65	1.5
Eu^{3+}	6	$^7F_0$	3	3	0	—	0	3.40, 3.51	3.6
Gd^{3+}	7	$^8S_{7/2}$	$\frac{7}{2}$	0	$\frac{7}{2}$	2	7.94	7.94	8.0
Tb^{3+}	8	$^7F_6$	3	3	6	$\frac{3}{2}$	9.72	9.7	9.5
Ds^{3+}	9	$^6H_{15/2}$	$\frac{5}{2}$	5	$\frac{15}{2}$	$\frac{4}{3}$	10.65	10.6	10.7
Ho^{3+}	10	$^5I_8$	2	6	8	$\frac{5}{4}$	10.61	10.6	10.3
Er^{3+}	11	$^4I_{15/2}$	$\frac{3}{2}$	6	$\frac{15}{2}$	$\frac{6}{5}$	9.58	9.6	9.5
Tu^{3+}	12	$^3H_6$	1	5	6	$\frac{7}{6}$	7.56	7.6	7.3
Yb^{3+}	13	$^2F_{7/2}$	$\frac{1}{2}$	3	$\frac{7}{2}$	$\frac{8}{7}$	4.54	4.5	4.5
Lu^{3+}	14	$^1S_0$	0	0	0	—	0	0.00	0

Source: J. Van Vleck, *The Theory of Electric and Magnetic Susceptibilities* (Oxford: Oxford University Press, 1932).

The quantity β_0 that we defined above is the natural unit of the electronic magnetic moment and is called the Bohr magneton. In cgs units

$$\beta_0 = \frac{|e|\hbar}{2mc} = 9.274 \times 10^{-21} \text{ erg/gauss} \tag{13–68}$$

There is also a magnetic moment associated with the spin of the nucleus, which is given by equations similar to Eq. (13–65) and Eq. (13–66), but where J is replaced by the spin quantum number I of the nucleus, where M_J is replaced by M_I, and most importantly, where β_0 is replaced by the nuclear magneton. The nuclear magneton is defined by an equation like Eq. (13–68), but with the mass of the nucleus replacing the electron mass. Since nuclei are of the order of 10^3 times more massive than electrons, the splitting of the nuclear energy states is much smaller than the splitting for electronic states, and so we shall ignore this splitting.

Equation (13–65) shows that μ^2 will be nonzero for atoms or ions with nonzero total electronic angular momentum. This includes all atoms and ions possessing an odd number of electrons and all atoms or ions with a partly filled inner shell, such as the transition metal ions or rare earth ions.

13–5 SYSTEMS OF NONINTERACTING PARAMAGNETIC ATOMS OR IONS

From Chapter 5, the partition function of an atom contains a translational part, an electronic part, and a nuclear part. We adopted the convention in Chapter 5 of setting $q_{nucl} = 1$, and since the nuclear states are split very little by the magnetic field, we can still use this convention and write the partition function of an atom (or an ion) in a magnetic field as

$$q(V, T, B_z^*) = \left(\frac{2\pi mkT}{h^2}\right)^{3/2} V q_{elec}(T, B_z^*) \tag{13–69}$$

We saw in Chapter 5 that the excited electronic states of most atoms have energies that are large compared to kT, and so in the absence of a magnetic field q_{elect} is simply the degeneracy of the ground electronic state, which is $2J + 1$, where J is the total electronic (orbital and spin) angular momentum of the atom. In the presence of a magnetic field, this $(2J + 1)$-fold degeneracy is split according to Eq. (13–66) into $2J + 1$ states with energies

$$\varepsilon_{M_J} = g\beta_0 B_z^* M_J \qquad -J \le M_J \le J \tag{13–70}$$

where β_0 is the Bohr magneton, and g is the Lande g factor.

In the presence of the magnetic field then, the electronic partition function is

$$q_{\text{elec}}(T, B_z^*) = \sum_{M_J=-J}^{+J} e^{g\beta_0 B_z^* M_J/kT} \tag{13–71}$$

This can be summed exactly. First let $\exp(g\beta_0 B_z^*/kT) = x$. Equation (13–71) becomes

$$\begin{aligned} q_{\text{elec}}(T, B_z^*) &= \sum_{M_J=-J}^{+J} x^{M_J} \\ &= x^{-J} \sum_{M_J=-J}^{+J} x^{(M_J+J)} = x^{-J} \sum_{m=0}^{2J} x^m \end{aligned} \tag{13–72}$$

This summation is a truncated geometric series

$$\sum_{n=0}^{N} z^n = \frac{1 - z^{N+1}}{1 - z} \tag{13–73}$$

which can be proved by straightforward division of the right-hand side. Note that if $N \to \infty$, the right-hand side becomes $1/(1 - z)$ if $z < 1$. Applying this to Eq. (13–72), we get (see Problem 13–23)

$$q_{\text{elec}}(T, B_z^*) = \frac{\sinh[(2J + 1)y/2]}{\sinh(y/2)} \tag{13–74}$$

where $y = g\beta_0 B_z^*/kT$. When $B_z^* \to 0$, $y \to 0$, and $q_{\text{elec}} \to 2J + 1$.

The magnetic analog of Eq. (13–21) is

$$dA = -S\,dT - p\,dV - \mathcal{M}_z\,dB_z^* \tag{13–75}$$

where $\mathcal{M}_z$ is the average z-component of the total magnetic moment of the system. It is easy to show that the ideal gas equation of state is unaltered by the external magnetic field. Since the magnetic dipoles are assumed to be noninteracting,

$$\begin{aligned} \mathcal{M}_z = N\bar{\mu}_z &= kT\left(\frac{\partial \ln Q}{\partial B_z}\right)_{N,V,T} \\ &= NkT\left(\frac{\partial \ln q_{\text{elec}}}{\partial B_z}\right)_{V,T} \end{aligned} \tag{13–76}$$

where we have dropped the asterisk of B_z since we assume that each dipole sees the same magnetic field and since the dipoles of the system do not interact with each other. By substituting Eq. (13–74) into Eq. (13–76) we get

$$\frac{\bar{\mu}_z}{g\beta_0} = \frac{\bar{\mu}}{g\beta_0} = \frac{(2J + 1)}{2} \coth \frac{(2J + 1)y}{2} - \frac{1}{2} \coth \frac{y}{2} \tag{13–77}$$

The right-hand side is called the *Brillouin function* and is the magnetic analog of the Langevin function. The parameter y is approximately $10^{-4}B_z/T$. In cgs units, magnetic fields are expressed in gauss, and a large, but conventional magnetic field strength, is 10^4 gauss. (See Table 13–1.) For all but the very lowest temperatures then, y is much smaller than unity, and we can expand the Brillouin function in powers of y. Equation (13–77) becomes

$$\bar{\mu} = \frac{g^2\beta_0{}^2J(J+1)}{3kT}B_z$$

$$= \frac{\mu^2 B_z}{3kT} \tag{13–78}$$

Equation (13–78) is the magnetic analog of Eq. (13–28) (where $\alpha = 0$). The experimentally measurable property associated with $\bar{\mu}$ is the magnetic susceptibility, defined for dilute systems by

$$\chi = \frac{\mathcal{M}_z}{B_z} \tag{13–79}$$

The magnetic susceptibility can be measured from the force on the sample in a nonuniform magnetic field. From Eqs. (13–78) and (13–79), the magnetic susceptibility is

$$\chi = \frac{N\mu^2}{3kT} = \frac{Ng^2\beta_0{}^2J(J+1)}{3kT} \tag{13–80}$$

Table 13–4 compares Eq. (13–80) to experimental data taken on the ions of the rare earth metals. Because of the incomplete $4f$ shell, these ions are paramagnetic. Table 13–4 gives the electronic term symbol of the ion, the values of S, L, J, and g, the value of $[g\beta_0 J(J+1)]^{1/2}$, and the experimental value of $(3kT\chi/N)^{1/2}$, which is equal to $[g\beta_0 J(J+1)]^{1/2}$ according to Eq. (13–80). The agreement between Eq. (13–80) is very good except for Sm^{3+} and Eu^{3+}. It has been shown by VanVleck* that good agreement can be obtained by including higher electronic states. Equation (13–80) is derived under the assumption that only the ground electronic state contributes to $q_{elec}(T, B_z)$. Such good agreement is not found in the case of the transition metal ions, however.

Equation (13–80) predicts that the paramagnetic contribution to χ varies as $1/T$. This was discovered experimentally by P. Curie and is known as Curie's law. An experimental verification of Curie's law is shown in Fig. 13–6.

These data are taken on a powdered crystal, which is a random arrangement like an ideal gas. Many of the data that are available for paramagnetic systems, however, are taken on crystals containing paramagnetic ions. The condition of independence of the individual magnetic dipoles requires that the ions be far enough apart that their energy of interaction is small compared to kT. Magnetic dipoles are of the order of a few multiples of a Bohr magneton, or a few multiples of 10^{-20} erg/gauss. The energy of interaction between two dipoles separated by a distance a is of the order of μ^2/a^3. Thus we have the condition that $Ta^3 \gg 1$, where T is in degrees Kelvin and a is in angstroms. This condition is readily satisfied even at low temperature by hydrated crystals of salts of paramagnetic ions, such as the transition metal ions or the rare

* See J. H. Van Vleck, *The Theory of Electric and Magnetic Susceptibilities* (Oxford: Oxford University Press, 1932).

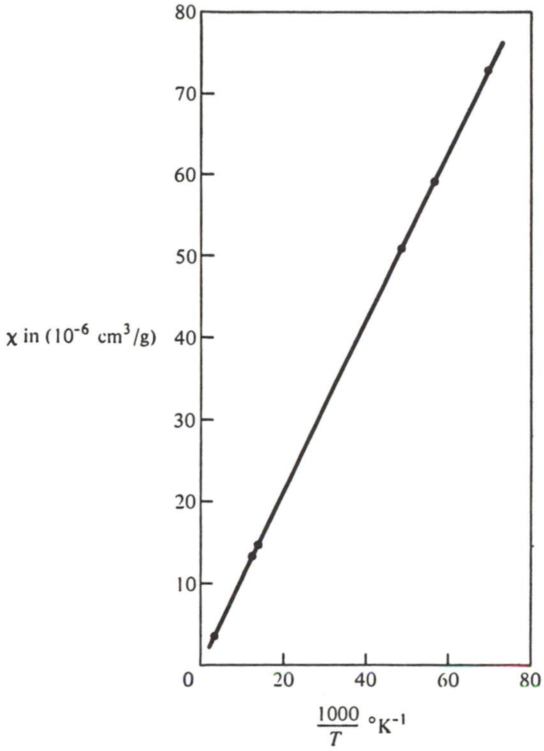

Figure 13–6. **Plot of susceptibility per gram versus reciprocal temperature for powdered $CuSO_4 \cdot K_2SO_4 \cdot 6H_2O$, showing the Curie law temperature dependence.** (From C. Kittel, *Solid State Physics*, 3rd ed. New York: Wiley, 1967.)

earth ions. Particularly useful systems are the class of compounds called alums, such as potassium chromium alum $KCr(SO_4)_2 \cdot 12H_2O$, and iron ammonium alum $FeNH_4(SO_4)_2 \cdot 12H_2O$. Since there are many data on dilute magnetic crystals, we wish to derive here the partition function for a lattice of noninteracting magnetic dipoles.

Consider a lattice of N equivalent noninteracting magnetic dipoles. Since the dipoles are restricted to lattice sites, they can be treated as distinguishable, and so the partition function of the lattice system is

$$Q(N, T, B_z) = [q_{\text{int}}(T)q_{\text{elec}}(T, B_z)]^N \tag{13–81}$$

where $q_{\text{int}}(T)$ is the partition function of the atom or ion exclusive of its electronic part. In Eq. (13–69), which is applicable to an atom or ion in an ideal gas or an ideal solution, $q_{\text{int}}(T)$ is a translational partition function. In this case, however, q_{int} is a partition function of a particle restricted to a lattice site and is more apt to be described by an Einstein model or a cell model, for example. The exact form of $q_{\text{int}}(T)$ will not be important here. The electronic partition function is still given by Eq. (13–74), and Eq. (13–77) is still valid. Thus Eq. (13–77) applies to a system of indistinguishable or distinguishable atoms or ions. Figure 13–7 compares Eq. (13–77) with some experimental data on two types of alum and another highly hydrated crystal. The agreement is very good over the complete range of $\bar{\mu}$ from 0 to its highest possible value.

In Chapter 18 (*Statistical Mechanics*) we shall study lattice systems in which the

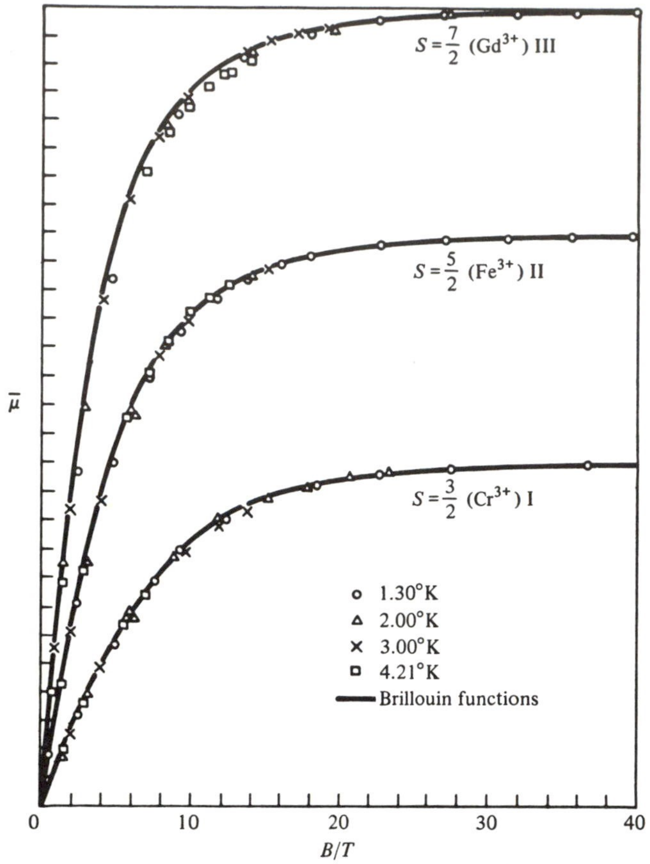

Figure 13–7. **Plot of magnetic moment versus B/T for spherical samples of (I) potassium chromium alum, (II) ferric ammonium alum, and (III) gadolinium sulfate octahydrate. Over 99.5 percent magnetic saturation is achieved at 1.3°K and about 50,000 Gauss.** (From C. Kittel, *Solid State Physics*, 3rd ed. New York: Wiley,1967; after J. C. Jupse, *Physica*, 9, 633, 1942.)

interactions are not negligible. An example of such a system is a ferromagnet. In preparation for this, it is useful to derive Eq. (13–81) in a manner that is more readily generalized to the case of interacting particles. For simplicity we shall consider a system of noninteracting magnetic dipoles which can point in one of only two directions, say in the direction of the external magnetic field B_z or against the field. This corresponds to an atom with only spin angular momentum, so that $L = 0$ and $J = S$ in Eq. (13–67), and $g = 2.00$. Consequently we take the energy to be

$$\varepsilon = \beta_0 m B_z \qquad m = \pm 1 \tag{13–82}$$

Let N_+ be the number of dipoles with $m = +1$ (aligned against the field) and N_- be the number with $m = -1$ (i.e., aligned with the field). For any pair of values of N_+ and N_-, the total magnetic energy of the system is

$$E(N_+, N_-) = \beta_0(N_+ - N_-)B_z \tag{13–83}$$

Since $N_+ + N_- = N$, only one of them is independent. Take N_+ to be the independent one and write

$$E(N_+) = \beta_0(2N_+ - N)B_z \tag{13–84}$$

There are $N!/N_+!(N - N_+)!$ possible states with the energy $E = \beta_0(2N_+ - N)B_z$, and so

$$Q(N, T, B_z) = q^N_{\text{int}} \sum_{N_+=0}^{N} \frac{N!}{N_+!(N - N_+)!} e^{-\beta_0(2N_+ - N)B_z/kT} \tag{13–85}$$

where $q_{\text{int}}(T)$ is the partition function of the ion in the absence of B_z. This summation can be easily evaluated by recognizing it as essentially a binomial expansion, and so

$$Q(N, T, B_z) = [q_{\text{int}}(T)(e^{\beta_0 B_z/kT} + e^{-\beta_0 B_z/kT})]^N \tag{13–86}$$

in agreement with Eq. (13–81) for a two-state system. It is Eq. (13–85) that we shall generalize in Chapter 18 (*Statistical Mechanics*).

ADDITIONAL READING

General

BROWN, Jr., W. F. 1967. In *Physical chemistry, an advanced treatise*, Vol. II, ed. by H. Erying, D. Henderson, and W. Jost. New York: Academic.

———. 1956. In *Encyclopedia of physics*, Vol. XVII, ed. by S. Flügge. Berlin: Springer-Verlag.

DAVIDSON, N. 1962. *Statistical mechanics*. New York: McGraw-Hill. Chapters 17 through 19.

EYRING, H., HENDERSON, D., STOVER, B. J., and EYRING, E. M. 1964. *Statistical mechanics and dynamics*. New York: Wiley. Chapter 8.

FOWLER, R. H., and GUGGENHEIM, 1956. *Statistical thermodynamics*, Cambridge: Cambridge University Press, Chapter 14.

MAYER, J. E., and MAYER, M. G., 1940. *Statistical mechanics*, New York: Wiley. Chapter 15.

Electric and magnetic properties of molecules

BUCKINGHAM, A. D. 1970. In *Physical chemistry, an advanced treatise*, Vol. IV, ed. by H. Eyring, D. Henderson, and W. Jost. New York: Academic.

DAVIES, D. W. 1967. *The theory of the electric and magnetic properties of molecules*. New York: Wiley.

VANVLECK, J. H. 1932. *The theory of electric and magnetic susceptibilities*. London: Oxford University Press.

Dielectric properties

DEBYE, P. 1929. *Polar molecules*. New York: Dover.

FRÖHLICH, H. 1958. *Theory of dielectrics*, 2nd ed. London: Oxford University Press.

HILL, T. L. 1960. *Statistical thermodynamics*. Reading, Mass.: Addison-Wesley. Chapter 12.

MOORE, W. J. 1972. *Physical chemistry*, 4th ed. Englewood Cliffs, N.J.: Prentice-Hall. Sections 15–15 to 15–19.

RUSHBROOKE, G. S. 1949. *Statistical mechanics*. London: Oxford University Press. Chapter 10.

Magnetic properties

KESTIN, J., and DORFMAN, J. R. 1971. *A course in statistical thermodynamics*. New York: Academic. Chapter 11.

KITTEL, C. 1967. *Solid state physics*, 3rd ed. New York: Wiley. Chapter 14.

WANNIER, G. H. 1966. *Statistical physics*. New York: Wiley. Chapter 15.

PROBLEMS

13–1. Prove that the potential energy of a dipole moment $\boldsymbol{\mu}$ in an electric field $\mathscr{E}_0$ is $-\boldsymbol{\mu} \cdot \mathscr{E}_0$ and that the torque acting on $\boldsymbol{\mu}$ is $\boldsymbol{\mu} \times \mathscr{E}_0$.

13–2. Using either Coulomb's law or Gauss' law, prove that the electric field $\mathscr{E}_0$ between two parallel plates with surface charge density $+\sigma$ on one plate and $-\sigma$ on the other is equal to $4\pi\sigma$. This is most easily shown by considering one plate at a time since the $4\pi\sigma$ arises from a contribution of $2\pi\sigma$ from each plate.

13–3. The dielectric constant of a gas at 300°K is $\varepsilon - 1 = 5.26 \times 10^{-3}$ and at 400°K is $\varepsilon - 1 = 3.56 \times 10^{-3}$. Calculate the polarizability and the dipole moment of this substance.

13–4. Show that

$$\bar{\mu} = \mu_0 \mathscr{L}\left(\frac{\mu_0 \mathscr{E}^*}{kT}\right)$$
$$= \frac{\mu_0{}^2 \mathscr{E}^*}{3kT} + \cdots$$

13–5. Show that the Langevin function has the following expansion for small x:

$$\mathscr{L}(x) \to \frac{x}{3} - \frac{x^3}{45} + \cdots$$

13–6. Calculate $\mu_0 \mathscr{E}^*/kT$ for $\mu_0 = 1$ debye, $\mathscr{E}^* = 1000$ volts/cm and $T = 300°K$. Is this small enough to linearize the Langevin function?

13–7. Prove that the work required to create an induced dipole moment is $-\alpha\mathscr{E}^{*2}/2$.

13–8. Plot $\bar{\mu}/\mu_{max} = \bar{\mu}/\mu_0 = \mathscr{L}(\beta\mu_0\mathscr{E}^*)$ versus $\beta\mu_0\mathscr{E}^*$. Calculate $\bar{\mu}/\mu_{max}$ for $\mu_0 = 5$ debyes and an external field of 1000 volts/cm at $T = 300°K$.

13–9. The dipole moment of $CHCl_3$ as measured in the gas phase is 1.01 debyes, and the polarizability is 82.3×10^{-25} cm^3-mole^{-1}. The density of chloroform at 300°K is 1.48 g/cm^3. Calculate the dielectric constant of the gas at 300°K and 1-atm pressure and of the liquid under the same conditions.

13–10. Show that

$$\frac{S}{Nk} = \frac{S(0)}{Nk} - \frac{1}{6}\left(\frac{\mu_0 \mathscr{E}^*}{kT}\right)^2 + \cdots$$

and

$$\mu = \mu(0) - \left(\alpha + \frac{\mu_0{}^2}{3kT}\right)\frac{\mathscr{E}^{*2}}{2kT}$$

where $S(0)$ and $\mu(0)$ are the values of S and μ in the absence of an external field. Show that for ordinary fields, the magnitude of these field-dependent terms is quite small.

13–11. Derive the equation

$$\frac{E}{NkT} = \frac{E(0)}{NkT} - \left(\frac{\alpha}{2} + \frac{\mu^2}{3kT}\right)\frac{\mathscr{E}^{*2}}{kT}$$

by calculating $N\langle\varepsilon(\mathscr{E}_z^*)\rangle/kT$ and adding it to $E(0)/NkT$.

13–12. Hollow out an infinitely-long right cylindrical cavity parallel to the plates of a parallel plate capacitor and show that the Claussius-Mossotti equation of a nonpolar fluid for this system is

$$\frac{\varepsilon - 1}{\varepsilon + 1} = 2\pi\, \rho\alpha$$

This equation has been used in the study of two-dimensional dielectrics.

13–13. When an external electric field D is applied to a real fluid at constant volume, the pressure and density charge. Calculate $(\partial p/\partial D)_{T,\mu,V}$ and $(\partial p/\partial D)_{T,N,V}$ for an ideal gas with permanent dipole moment μ_0 and polarizability α. Note that the two quantities to be calculated are the pressure change for an open and closed system, respectively. This effect is called electrostriction. Some thermodynamic equations are

$$dA = -S\,dT - p\,dV - M\,dD + \mu\,dN$$
$$d(pV) = S\,dT + p\,dV + M\,dD + N\,d\mu$$

Your answer should be expressed first in terms of μ_0, α, ρ, and D and then in terms of ε and D only.

13–14. An alternative formulation of the Claussius-Mossotti equation says that for a nonpolarizable molecule, that is, $\alpha = 0$,

$$\varepsilon - 1 = \frac{4\pi\rho}{kT}\langle\mu_z{}^2\rangle$$

The average here is in the absence of any external field, that is, $D = 0$. Show how this can be used to derive the C–M equation.

13–15. Solve Laplace's equation for a conducting sphere in a uniform external electric field $\mathscr{E}_0$ and show that the polarizability of a conducting sphere of radius a is a^3. This is a physical basis for relating the polarizability of an atom or molecule to its size. (Remember that the electric field inside a conductor is zero.)

13–16. Calculate the strength of an electric field required to achieve 1 percent of the saturation value of the orientational polarization of a dipolar gas at room temperature, assuming that the molecules have a dipole moment of 1 debye.

13–17. Consider a system of N electron spins in an external field B_0. If $B_0 = 10^4$ gauss and $T = 300°K$, calculate the excess number of spins aligned with the field. Repeat this calculation at 1°K.

13–18. Calculate the magnetic susceptibility per mole of gadolinium sulfate $Gd_2(SO_4)_3 \cdot 8H_2O$ at 4°K (liquid helium temperature), given that the density of the crystal is 3.00 g/cm^3 at this temperature.

13–19. A K-39 nucleus has a spin of $\frac{3}{2}$ and a nuclear g factor of 0.2606. What are the allowed orientations of this nucleus in a magnetic field? Calculate the energy levels in a field of 10^3 gauss.

13–20. A single electron in a magnetic field has a potential energy $\pm\beta_0 B_z$. Show that at 0°K the total magnetic moment of a free electron gas is given by

$$M = -\beta_0(N_+ - N_-)$$
$$= \beta_0 \frac{4\pi V}{3h^3}[\{2m(\mu_F + \beta_0 B_z)\}^{3/2} - \{2m(\mu_F - \beta_0 B_z)\}^{3/2}]$$

where μ_F is the Fermi energy.

Show that if $\mu_F \gg \beta_0 B_z$, we have

$$M = \frac{3}{2}\frac{\beta_0{}^2 N}{\mu_F} B_z + \cdots$$

where Eq. (10–22) has been used for N. From this result we see that the susceptibility is

$$\chi = \frac{3}{2}\frac{N\beta_0{}^2}{\mu_F}$$

Hint: Use the fact that at 0°K the kinetic energy of an electron with $\pm$ spin ranges from 0 to $\mu_F \mp \beta_0 B_z$.

13–21. The Bohr–van Leeuwen theorem states that the magnetic susceptibility of a system obeying classical mechanics and classical statistics is always equal to zero. Prove this by showing that the classical Hamiltonian

$$H = \sum_{j=1}^{N} \frac{1}{2m_j}\left\{\mathbf{p}_j + \frac{e_j}{c}A(\mathbf{r}_j)\right\}^2 + U(\mathbf{r}_1, \ldots, \mathbf{r}_N)$$

(in terms of the vector potential **A**) when substituted into the classical partition function gives a partition function that is independent of **A** and hence magnetic field B_z.

13–22. Under what conditions does the Brillouin function, Eq. (13–77), go over into the Langevin function, Eq. (13–25)? Interpret this limit physically.

13–23. Derive Eq. (13–74) from (13–72).

13–24. Derive the Hamiltonian given in Eq. (13–58).

POLYMERS

In this chapter we shall discuss some applications of statistical thermodynamics to polymer molecules. Polymers are high-molecular-weight molecules made up of the repeated addition of some fundamental unit called a monomer. The simplest such molecule is polyethylene:

$$\diagdown CH_2 \diagup^{CH_2} \diagdown CH_2 \diagup^{CH_2} \diagdown CH_2 \diagup^{CH_2} \diagdown CH_2 \diagup^{CH_2} \diagdown$$

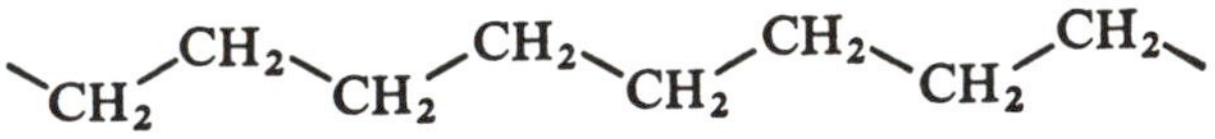

In a typical system of a synthetic polymer, such as polyethylene, there are molecules of all lengths or molecular weights, but in this elementary introduction to polymer physics, we shall ignore that complication and assume that all the molecules of the system have the same molecular weight.

In Sections 14–1 and 14–2 we shall introduce a model of a polymer molecule which is mathematically simple and yet displays many of the observed physical characteristics of polymers. In Section 14–3 we shall discuss the role of light scattering to deduce the average size and shape of macromolecules. Section 14–4 will involve a treatment of one of the most striking properties of many polymers, namely, that of their high elasticity. Lastly, Section 14–5 is a more rigorous and somewhat more mathematically advanced treatment of the physical model presented in Section 14–1.

14–1 THE END-TO-END DISTRIBUTION OF POLYMER CHAINS

Ordinarily the chemical formula of a polymer molecule is written across a page as though it were a rigid linear molecule. For many polymer molecules, however, this is a gross misrepresentation of the actual configuration. Consider a molecule of polyethylene. There is free or almost free rotation about each carbon–carbon bond so that the molecule is able to take on a great many configurations. A two-dimensional version of some typical configuration of such a molecule is shown in Fig. 14–1(a). Note that the molecule is not extended as a rigid rod, but is in the configuration of a random coil.

We shall now introduce a mathematical model of a polymer chain, which at first

(a) (b)

Figure 14–1. **(a) A two-dimensional version of an instantaneous configuration of a polymer molecule. There is a fixed bond angle, but there is rotation about the bonds. (b) A two-dimensional version of a random flight.**

may appear to be too simple, but which leads to a great understanding of the general properties of polymer molecules. We shall later modify the model to account for some of its obvious inadequacies.

We imagine a configuration of a polymer to be generated by starting at an origin and drawing a vector of length l in an arbitrary random direction. Then we do the same thing starting from the end of the first vector, then the same from the end of the second vector, and so on. Such a process is called a random flight, and a two-dimensional version is shown in Fig. 14–1(b). Certainly if we consider each random flight vector to be a carbon–carbon bond in polyethylene, the model is an oversimplification of the real molecule. For example, the direction of successive vectors must conform to the tetrahedral valence angles of the carbon atom. Furthermore, the random flight model does not account for the fact that the atoms have a finite size and exclude other atoms in the chain from occupying the same position. Nevertheless, we shall see that the random flight model of a polymer molecule captures much of the physics. In a sense it is the ideal gas analog of polymer molecules. Since the direction of each bond is assumed to be completely arbitrary and random, this model is often called the freely jointed chain model.

In order to discuss the mathematics of a random flight, it is convenient to treat first a simple one-dimensional version of this process called a random walk. Consider a process (a walk) in which a particle (or walker) starts at the origin and moves to the left or to the right at successive intervals of time with equal probability $p = \frac{1}{2}$. After N such steps, the particle could be found at the points $-N, -N+1, \ldots, -1, 0, 1, 2, \ldots, N$. We wish to calculate $W(m, N)$, the probability that the particle is at the position m after N steps.

In order to reach m after N steps, the particle must have made $(N+m)/2$ steps in the positive direction and the others, that is, $(N-m)/2$, in the negative direction. It can be seen by simple numerical examples that if N is odd, m must be odd, and that m must be even if N is even. The number of different ways that $(N+m)/2$ out of N steps can be taken to the right and that $(N-m)/2$ out of N steps can be taken to the left is

$$\text{number of ways of reaching } m \text{ in } N \text{ steps} = \frac{N!}{\left(\frac{N+m}{2}\right)!\left(\frac{N-m}{2}\right)!}$$

Each of the N steps has probability $p = \frac{1}{2}$ associated with it, and so

$$W(m, N) = \frac{N!}{\left(\frac{N+m}{2}\right)!\left(\frac{N-m}{2}\right)!}\left(\frac{1}{2}\right)^N \qquad (14\text{–}1)$$

Equation (14–1) is exact but difficult to use in practice. It can be seen immediately, however, that the average value of m is zero since $W(m, N)$ is symmetric in $\pm m$. This result, of course, is also expected on physical grounds, since the probabilities of moving in either direction are the same.

On the average, then, the particle is found at the origin and is found near the origin most of the time. Thus we can limit Eq. (14–1) to small values of m/N. Furthermore, we shall be interested in only large values of N (polymers), and so we may apply Stirling's approximation to the factorials in Eq. (14–1). Taking the logarithm of Eq. (14–1), then we get

$$\begin{aligned}
\ln W &\approx N \ln N - N - \left(\frac{N+m}{2}\right)\ln\left(\frac{N+m}{2}\right) + \left(\frac{N+m}{2}\right) \\
&\qquad - \left(\frac{N-m}{2}\right)\ln\left(\frac{N-m}{2}\right) + \left(\frac{N-m}{2}\right) - N\ln 2 \\
&\approx N\ln N - \left(\frac{N+m}{2}\right)\ln\left(\frac{N+m}{2}\right) - \left(\frac{N-m}{2}\right)\ln\left(\frac{N-m}{2}\right) - N\ln 2 \\
&\approx N\ln N - \frac{N}{2}\left(1+\frac{m}{N}\right)\ln\left[\frac{N}{2}\left(1+\frac{m}{N}\right)\right] - \frac{N}{2}\left(1-\frac{m}{N}\right)\ln\left[\frac{N}{2}\left(1-\frac{m}{N}\right)\right] - N\ln 2 \\
&\approx -\frac{N}{2}\ln\left(1+\frac{m}{N}\right) - \frac{m}{2}\ln\left(1+\frac{m}{N}\right) - \frac{N}{2}\ln\left(1-\frac{m}{N}\right) + \frac{m}{2}\ln\left(1-\frac{m}{N}\right)
\end{aligned}$$

We now use the fact that $\ln(1 \pm x) = \pm x - x^2/2 + \cdots$ to get

$$\ln W(m, N) \approx -\frac{m^2}{2N}$$

or

$$W(m, N) \approx e^{-m^2/2N} \tag{14–2}$$

Before discussing the significance of this result, introduce the variable $\xi = mh$, where h is the length of each step. The quantity ξ, then, is the distance of the particle from the origin. Equation (14–2) becomes

$$W(\xi) \approx e^{-\xi^2/2Nh^2}$$

Apart from a normalization factor, $W(\xi)$ is the probability that the system be found a distance ξ from the origin after N displacements of length h. Since N is large, we can treat ξ as a continuous variable and normalize $W(\xi)$ to get

$$W(\xi) = \frac{1}{(2\pi N h^2)^{1/2}} e^{-\xi^2/2Nh^2} \tag{14–3}$$

Equation (14–3) has been derived on the basis that each of the N steps is of length h. If all the steps are not of the same length, the result is

$$W(\xi) = \frac{1}{(2\pi N \sigma^2)^{1/2}} e^{-\xi^2/2N\sigma^2} \tag{14–4}$$

where $\sigma^2 = \overline{h^2}$.

It is now a simple matter to derive a similar result for random flights. Any component, say the z-component, of a random flight will obey an equation like Eq. (14–4). In this case $\xi = z$ and $h = l \cos \theta$, where θ is the angle between the random flight vector and the z-axis. The angle θ may take on all values with equal probabilities and hence σ^2 in Eq. (14–4) is given by $\overline{l^2 \cos^2 \theta} = l^2/3$. Thus the z-component of a random flight is given by

$$W(z)\,dz = \left(\frac{3}{2\pi N l^2}\right)^{1/2} e^{-3z^2/2Nl^2}\,dz \tag{14–5}$$

But since there is no preferred direction in the random-flight process we are considering here, both the x- and y-components will be given by an equation identical to Eq. (14–5).

Since the x-, y-, and z-directions are independent, the probability of an overall distance of r between the beginning and the end of a random flight is given by

$$W(x, y, z)\,dx\,dy\,dz = W(x)W(y)W(z)\,dx\,dy\,dz = \left(\frac{3}{2\pi N l^2}\right)^{3/2} e^{-3r^2/2Nl^2}\,dx\,dy\,dz$$

where $r^2 = x^2 + y^2 + z^2$. We can convert this to spherical coordinates to get

$$W(r)4\pi r^2\,dr = 4\pi\left(\frac{3}{2\pi N l^2}\right)^{3/2} e^{-3r^2/2Nl^2} r^2\,dr \tag{14–6}$$

This is the equation for the end-to-end distribution in the model of a freely jointed polymer chain. The average value of r is given by $(8Nl^2/3\pi)^{1/2}$ (see Problem 14–3). The mean-square end-to-end distance, however, is a more fundamental property of polymers, and this is given by

$$\overline{r^2} = \int_0^\infty W(r)4\pi r^4\,dr = Nl^2 \tag{14–7}$$

The root-mean-square length then is $(\overline{r^2})^{1/2} = N^{1/2}l$. Since the fully extended length of a polymer (called the contour length) is Nl, this result indicates that a polymer chain will be tightly coiled rather than extended. Since N is proportional to the molecular weight, Eq. (14–7) says that the root-mean-square end-to-end distance is proportional to $M^{1/2}$. We shall see that this is closely confirmed experimentally.

The form of Eq. (14–6) is valid even if the valence angle requirements of the chain are considered. The only modification is that the quantity l^2 must be redefined. To consider more realistic models of polymer chains, we can write Eq. (14–6) in terms of $\overline{r^2}$

$$W(r)4\pi r^2\,dr = 4\pi\left(\frac{3}{2\pi \overline{r^2}}\right)^{3/2} e^{-3r^2/2\overline{r^2}} r^2\,dr \tag{14–8}$$

and concentrate on the determination of $\overline{r^2}$ itself.

14–2 THE ROOT-MEAN-SQUARE END-TO-END DISTANCE AND THE RADIUS OF GYRATION OF POLYMER CHAINS

In this section we shall focus on a calculation of $\overline{r^2}$ and the radius of gyration for chains in which valence angles are taken into account. First, however, we shall calculate $\overline{r^2}$ directly for a freely jointed chain. Let $\mathbf{l}_j$ be the vector of length l_j that describes the length and orientation of the jth bond. The vector from the beginning of the chain to the end is

$$\mathbf{r} = \sum_{j=1}^{N} \mathbf{l}_j \tag{14–9}$$

and its mean square is

$$\overline{r^2} = \overline{\left(\sum_{i=1}^{N} \mathbf{l}_i\right) \cdot \left(\sum_{j=1}^{N} \mathbf{l}_j\right)} = \sum_{i=1}^{N} \sum_{j=1}^{N} \overline{\mathbf{l}_i \cdot \mathbf{l}_j} \tag{14–10}$$

Let θ be the angle between the positive directions of any two successive bonds. Then $\mathbf{l}_j \cdot \mathbf{l}_{j+1} = l_j l_{j+1} \cos\theta$. Since θ may take on all values with equal probabilities in a freely jointed chain, $\overline{\cos\theta}$ and hence $\overline{\mathbf{l}_j \cdot \mathbf{l}_{j+1}} = 0$. Similarly, all other terms in Eq. (14–10) with $i \neq j$ vanish. The only terms that survive are those in which $i = j$. There are N such terms, and so

$$\overline{r^2} = \sum_{j=1}^{N} l_j^{\,2} = N l_{\text{av}}^{\;2} \tag{14–11}$$

where $l_{\text{av}}^{\;2}$ is the average square of a bond length in the chain. If the bonds are all of the same length, this reduces to Eq. (14–7).

Polymer chains such as polyethylene and its derivatives consist of chains of carbon atoms with identical bond lengths $l = 1.54$ Å and a fixed angle $\theta = 180° - 109°28' = 70°32'$, $109°28'$ being the tetrahedral angle included between successive bonds. We shall now calculate $\overline{r^2}$ for a chain with a fixed angle θ between successive bonds but with free rotation about the bond. Figure 14–2 shows the angles involved. Equation

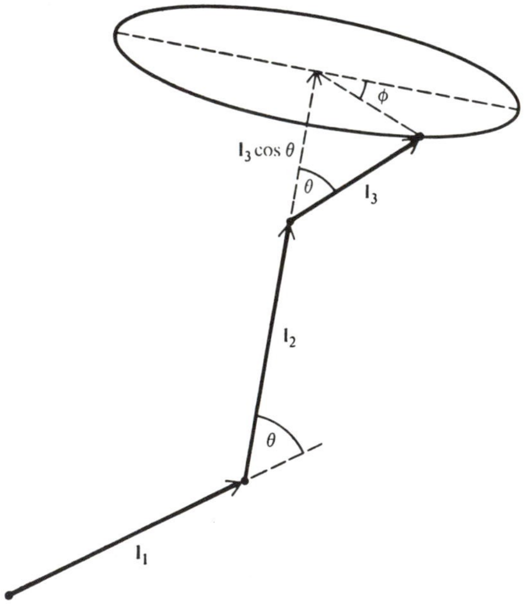

Figure 14–2. **Three successive bonds of a polyethylene chain. The first two are in the plane of the figure. The terminus of l_3 may lie anywhere on the circle due to the assumed free rotation about the carbon-carbon bond.** (From C. Tanford, *Physical Chemistry of Macromolecules.* New York: Wiley, 1961.)

(14–10) is still valid, but most of the terms no longer vanish. Let us write the double summation in Eq. (14–10) in terms of $\mathbf{l}_i \cdot \mathbf{l}_{i+n}$, that is, in terms of bond vectors separated by n intermediate bonds. This is (see Problem 14–5)

$$\overline{r^2} = \sum_{j=1}^{N} \overline{l_j^2} + 2\sum_{j=1}^{N-1} \overline{\mathbf{l}_j \cdot \mathbf{l}_{j+1}} + 2\sum_{j=1}^{N-2} \overline{\mathbf{l}_j \cdot \mathbf{l}_{j+2}} + \cdots + 2\sum_{j=1}^{2} \overline{\mathbf{l}_j \cdot \mathbf{l}_{j+N-2}} + \overline{2\mathbf{l}_1 \cdot \mathbf{l}_N}$$

where the bar indicates an average over the rotation (assumed free) about the carbon–carbon bonds. The terms under each summation are all equal, and so we have

$$\overline{r^2} = Nl^2 + 2(N-1)\overline{\mathbf{l}_1 \cdot \mathbf{l}_2} + 2(N-2)\overline{\mathbf{l}_1 \cdot \mathbf{l}_3} + \cdots + 2(N-(N-2))\overline{\mathbf{l}_1 \cdot \mathbf{l}_{N-1}} + 2\overline{\mathbf{l}_1 \cdot \mathbf{l}_N} \qquad (14\text{–}12)$$

To evaluate this, we need to evaluate terms such as $\overline{\mathbf{l}_1 \cdot \mathbf{l}_2}$, $\overline{\mathbf{l}_1 \cdot \mathbf{l}_3}$, and so on. The term $\mathbf{l}_1 \cdot \mathbf{l}_2 = l^2 \cos\theta$ since θ is the angle between $\mathbf{l}_1$ and $\mathbf{l}_2$. In order to evaluate $\mathbf{l}_1 \cdot \mathbf{l}_3$, we refer to Fig. 14–2. As $\mathbf{l}_3$ is rotated around the angle ϕ, its components perpendicular to $\mathbf{l}_2$ will vanish. Its component parallel to $\mathbf{l}_2$ is $l\cos\theta$, so that effectively the vector can be replaced by a vector parallel to $\mathbf{l}_2$ and of magnitude $l\cos\theta$. Mathematically, we write this as $l\cos\theta(\mathbf{l}_2/l) = \mathbf{l}_2\cos\theta$. But $\mathbf{l}_1 \cdot \mathbf{l}_2 = l^2\cos\theta$, and so $l_1 \cdot l_3 = l^2\cos^2\theta$. In a similar way, we can see that $\overline{\mathbf{l}_1 \cdot \mathbf{l}_{1+n}} = l^2\cos^n\theta$. Thus Eq. (14–12) becomes

$$\overline{r^2} = l^2[N + 2(N-1)\cos\theta + 2(N-2)\cos^2\theta + 2(N-3)\cos^3\theta + \cdots + 2\cos^{N-1}\theta] \qquad (14\text{–}13)$$

For a chain such as polyethylene, $\theta = 70°32'$ and $\cos\theta = \frac{1}{3}$. Thus the terms in Eq. (14–13) decrease very rapidly. Under these conditions then, we may ignore the small integers compared to N and approximate Eq. (14–13) by an infinite series (see Problem 14–6)

$$\overline{r^2} = l^2N(1 + 2\cos\theta + 2\cos^2\theta + 2\cos^3\theta + \cdots) \qquad (14\text{–}14)$$

$$= Nl^2\frac{1+\cos\theta}{1-\cos\theta} \qquad (14\text{–}15)$$

where we have recognized the sum in Eq. (14–14) as a geometric series. The terms in which the "small" integers cannot be ignored with respect to N contain the $\cos\theta$ to such a high power that they do not contribute to Eq. (14–14).

For polyethylene, $\cos\theta = \frac{1}{3}$ and $\overline{r^2} = 2Nl^2$. In this case then, the effect of fixing the bond angle θ, instead of allowing it to take on any value with equal probability, is to introduce a factor of 2 into $\overline{r^2}$. Equation (14–14) shows that, in general, the effect is to introduce a numerical factor into $\overline{r^2}$ without altering the general form $\overline{r^2} = \beta^2N$.

The rotation about carbon–carbon bonds is, of course, not free as we have assumed. This topic has been discussed in Section 8–4, and we have seen that there is a potential energy associated with such internal rotation. If this is taken into account, then Eq. (14–15) is modified to

$$\overline{r^2} = Nl^2\left(\frac{1+\cos\theta}{1-\cos\theta}\right)\left(\frac{1+\overline{\cos\phi}}{1-\overline{\cos\phi}}\right) \qquad (14\text{–}16)$$

The cos ϕ here appears as an average, since this angle is not fixed as θ is. In fact,

$$\overline{\cos \phi} = \frac{\int_0^{2\pi} \cos \phi e^{-u(\phi)/kT}\, d\phi}{\int_0^{2\pi} e^{-u(\phi)/kT}\, d\phi}$$

where $u(\phi)$ is the potential energy of internal rotation. Note that $\overline{r^2}$ is still of the form $\overline{r^2} = \beta^2 N$. Regardless of the complicated restrictions that may be introduced by the valence angles of a chain, $\overline{r^2}$ will have the form $\beta^2 N$, where β reflects the restriction on the angles. Thus we write

$$\overline{r^2} = \beta^2 N \tag{14–17}$$

where β is a function of the bonds lengths and bond angles within the chain and N is the number of units in the chain. Equation (14–17) implies that flexible chains restricted by rotation may be treated as freely jointed chains with a bond length β instead of l. The effect of restricted rotation is to increase the effective bond length. We shall see that β can be determined experimentally and that the ratio β/l can be used to determine the degree of restricted rotation or stiffness of the polymer chain. For a carbon–carbon chain with free rotation about the carbon–carbon bond $\beta/l = 2^{1/2}$, and as more rotational restrictions are imposed on the chain, this ratio increases [*cf.* Eq. (14–16)].

It should be noted that Eq. (14–17) is not necessarily restricted to the extreme ends of the chain. Since the chain is freely jointed or can be considered to be freely jointed with an effective bond length β, Eq. (14–17) can be applied also to the mean-square distance between any two segments in the chain, in which case N becomes the number of links between the two segments. For example, if the two segments are the ith and the jth, then N is replaced by $|j - i|$.

The result that $\overline{r^2}$ is proportional to N is a fundamental result. This can be tested experimentally by measuring the dipole moment of a polymer whose monomeric units can be considered to have dipole moments within the chain. An example is the polymer polyethylene glycol, a polymer with the repeating unit

$$-CH_2-CH_2-O-$$

The dipole moment of the entire polymer chain is given by an equation similar to Eq. (14–9):

$$\boldsymbol{\mu} = \sum_{j=1}^{N} \boldsymbol{\mu}_j$$

But it is μ^2 that is determined in dipole moment measurements and so by the same argument as leading to Eq. (14–17), we have

$$\overline{\mu^2} = N{\mu_1}^2$$

where μ_1 is the effective dipole moment of the repeating unit. In samples of polyethylene glycol of various molecular weights, the square of the dipole moment was found to vary linearly with the molecular weight, that is, with N. This gives experimental support to the validity of a model as abstract as a flight or a freely jointed chain model.

There is another measure of the extent of a polymer chain that is perhaps more important than $\overline{r^2}$ since, as we shall see in Section 14–3, it can be measured directly without the assumption of a model for the polymer chain. This quantity is the radius of gyration. The radius of gyration of an arbitrary assembly of points (see Fig. 14–3) is given by

$$R_G^2 = \frac{\sum_{j=1}^{N} m_j d_j^2}{\sum_{j=1}^{N} m_j} \tag{14–18}$$

where m_j is the mass of the jth point, and d_j is the distance of the jth point to the center of mass of the assembly. For simplicity, we shall assume that all the masses are the same, so that

$$R_G^2 = \frac{1}{N} \sum_{j=1}^{N} d_j^2 \tag{14–19}$$

The center of mass of the system is found from

$$\sum_{j=1}^{N} m_j \mathbf{d}_j = 0 \tag{14–20}$$

Figure 14–3 gives these $\mathbf{d}_j$ in terms of a vector $\boldsymbol{\alpha}$, which locates the center of mass from some arbitrary origin, and the vectors $\mathbf{r}_j$, which locate the jth particle with respect to the same arbitrary origin. Clearly, the location of this arbitrary origin can have no effect on the final results and must eventually drop out of the problem.

In terms of the vectors $\mathbf{r}_j$, the $\mathbf{d}_j$ are given by

$$\boldsymbol{\alpha} + \mathbf{d}_j = \mathbf{r}_j \qquad j = 1, 2, \ldots, N \tag{14–21}$$

Thus we have from Eq. (14–20) (with equal masses)

$$\boldsymbol{\alpha} = \frac{1}{N} \sum_{j=1}^{N} \mathbf{r}_j \tag{14–22}$$

From Eq. (14–19), the square of the radius of gyration is

$$\begin{aligned} R_G^2 &= \frac{1}{N} \sum_{j=1}^{N} \mathbf{d}_j \cdot \mathbf{d}_j = \frac{1}{N} \sum_{j=1}^{N} (\mathbf{r}_j - \boldsymbol{\alpha}) \cdot (\mathbf{r}_j - \boldsymbol{\alpha}) \\ &= \frac{1}{N} \left\{ \sum_{j=1}^{N} r_j^2 - 2\boldsymbol{\alpha} \cdot \sum_{j=1}^{N} \mathbf{r}_j + N\alpha^2 \right\} \end{aligned} \tag{14–23}$$

If we substitute Eq. (14–22) for $\boldsymbol{\alpha}$ into Eq. (14–23), we get

$$R_G^2 = \frac{1}{N} \sum_{j=1}^{N} r_j^2 - \frac{1}{N^2} \sum_{i=1}^{N} \sum_{j=1}^{N} \mathbf{r}_i \cdot \mathbf{r}_j \tag{14–24}$$

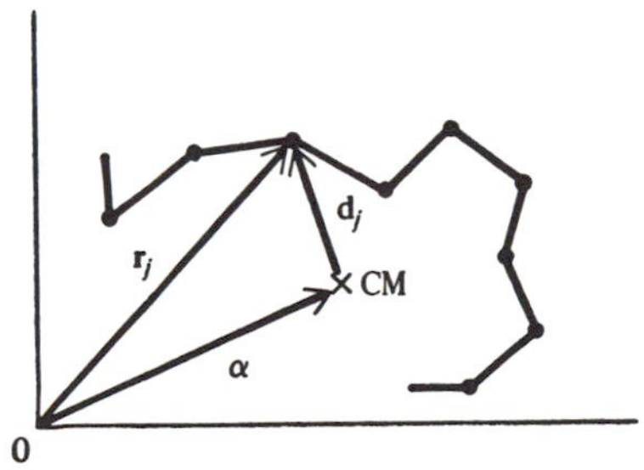

Figure 14–3. **An assembly of N points showing the vector $\boldsymbol{\alpha}$ that locates center of mass of the system with respect to some arbitrary origin 0, the vector $\mathbf{d}_j$, locating the jth point from the center of mass, and the vector $\mathbf{r}_j$, locating the jth point from some arbitrary origin.**

To evaluate the double summation, we apply the law of cosines to a triangle formed by the points O, i, and j:

$$\begin{aligned} r_{ij}^2 &= r_i^2 + r_j^2 - 2r_i r_j \cos \theta_{ij} \\ &= r_i^2 + r_j^2 - 2\mathbf{r}_i \cdot \mathbf{r}_j \end{aligned}$$

We use this equation to eliminate $\mathbf{r}_i \cdot \mathbf{r}_j$ in Eq. (14–24) to get

$$\begin{aligned} R_G^2 &= \frac{1}{N} \sum_{j=1}^{N} r_j^2 - \frac{1}{2N^2} \sum_{i=1}^{N} \sum_{j=1}^{N} (r_i^2 + r_j^2 - r_{ij}^2) \\ &= \frac{1}{2N^2} \sum_{i=1}^{N} \sum_{j=1}^{N} r_{ij}^2 \end{aligned} \tag{14–25}$$

Note that only the distance between all pairs of particles appears in Eq. (14–25), that is, it does not depend upon the location of the arbitrary origin in Fig. 14–3.

Equation (14–25) is the square of the radius of gyration for a fixed configuration. If we average over all the configurations of the flexible chain, we have

$$R_G^2 = \frac{1}{2N^2} \sum_{i=1}^{N} \sum_{j=1}^{N} \overline{r_{ij}^2} \tag{14–26}$$

The average of r_{ij}^2 is the average square distance between the ith and jth elements of a flexible chain or the average square end-to-end distance of a chain with $|j - i|$ elements. From Eq. (14–17), this is

$$\overline{r_{ij}^2} = \beta^2 |j - i| \tag{14–27}$$

Thus Eq. (14–26) becomes

$$R_G^2 = \frac{\beta^2}{2N^2} \sum_{i=1}^{N} \sum_{j=1}^{N} |j - i| \tag{14–28}$$

To evaluate this double summation, first consider the sum over j. In order to handle the absolute-value sign in this summation, we must split it into two parts: one in which $j > i$ and one in which $j < i$. This allows us to write

$$\sum_{j=1}^{N} |j - i| = \sum_{j=1}^{i} (i - j) + \sum_{j=i+1}^{N} (j - i) \tag{14–29}$$

This incidentally is the standard way that one manipulates summations or integrals involving absolute values. The summations in Eq. (14–29) now are easy. Using the fact that the sum of the first n integers is $n(n + 1)/2$, Eq. (14–29) becomes

$$\begin{aligned} \sum_{j=1}^{N} |j - i| &= i^2 - \tfrac{1}{2}i(i + 1) + \tfrac{1}{2}[N(N + 1) - i(i + 1)] - i(N - i) \\ &= i^2 - i(N + 1) + \tfrac{1}{2}N(N + 1) \end{aligned}$$

We now perform the summation over i in Eq. (14–28). To do this, we need the fact that the sum of the squares of the first n integers is $n(n + 1)(2n + 1)/6$. Thus we get

$$\begin{aligned} R_G^2 &= \frac{\beta^2}{2N^2} \left\{ \frac{N(N + 1)(2N + 1)}{6} - \frac{N(N + 1)^2}{2} + \frac{N^2(N + 1)}{2} \right\} \\ &\approx \frac{\beta^2 N}{6} = \frac{\overline{r^2}}{6} \end{aligned} \tag{14–30}$$

In Eq. (14–30) we have ignored terms of $0(N)$ in the braces. Equation (14–30) shows that the radius of gyration of a flexible chain is directly related to the mean-square end-to-end distance. We shall see in the next section that the radius of gyration can be determined directly from light-scattering measurements.

In Section 14–3 we shall need expressions for the radius of gyration of assemblies of points other than just a flexible chain. We shall now calculate the radius of gyration of a uniform sphere and a uniform right cylinder. By a uniform sphere we mean a sphere in which the density is constant throughout. The center of mass of a sphere is obviously its center, and so from the continuum analog of Eq. (14–18), we have

$$
\begin{aligned}
R_G{}^2 &= \frac{\int_{\text{sphere}} r^2\, dm}{\int_{\text{sphere}} dm} \\
&= \frac{1}{M}\int_0^R \int_0^\pi \int_0^{2\pi} r^2 \rho r^2 \sin\theta \, d\phi \, d\theta \, dr \\
&= \frac{4\pi\rho R^5}{5M} = \frac{3}{5} R^2
\end{aligned}
\tag{14–31}
$$

The radius of gyration of a uniform cylinder of radius R and length L is (see Problem 14–12)

$$
R_G{}^2 = \frac{R^2}{2} + \frac{L^2}{12} \approx \frac{L^2}{12} \tag{14–32}
$$

where the last term assumes that the length of the cylinder is much larger than the radius.

We see that the values of $R_G{}^2$ differ greatly for a given mass of polymer molecule in the shape of a flexible chain [Eq. (14–30)], uniform sphere [Eq. (14–31)] and a uniform right cylinder [Eq. (14–32)] (see Problem 14–13). By comparing such calculated $R_G{}^2$ values to the experimental value obtained directly from light scattering, it is possible to determine the configuration of a polymer molecule or to at least eliminate certain possibilities. We shall now discuss the application of light scattering to polymer molecules.

14–3 LIGHT SCATTERING

The measurement of the scattering of electromagnetic radiation by systems of particles is one of the most powerful probes into the configurations of the particles. For example, we shall see in Chapter 16 (*Statistical Mechanics*) that the scattering of X-rays by liquids can be used to determine the local structure in a liquid. In this section we shall discuss the scattering of radiation by polymer molecules.

Consider a polymer molecule to be made up of N single scatterers connected by bonds or links that are freely jointed. An oscillating electric field that is incident to a single scattering element induces an oscillating electric dipole in that element. The oscillating dipole moment emits electromagnetic radiation in the form of a spherical wave, that is, in all directions, and this radiation is said to be scattered spherically. Each of the N scattering elements of the freely jointed chain scatters the incident radiation in all directions. The total intensity of scattered radiation at the position of some observer is obtained by considering the radiation scattered from each of the elements in the following way.

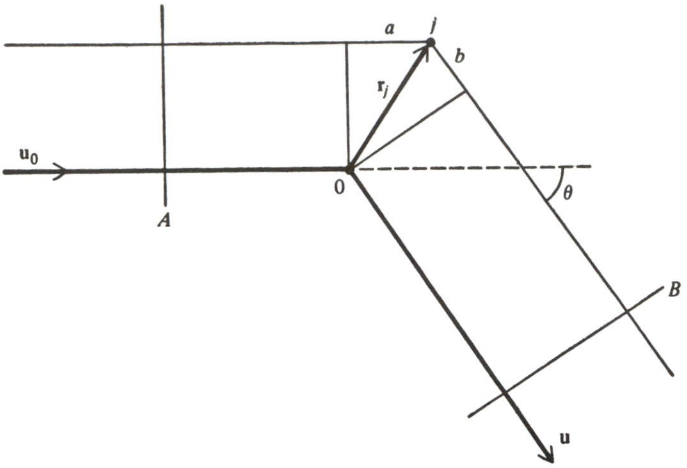

Figure 14–4. **The geometry used to calculate the intensity of radiation scattered through an angle θ. The plane A represents the incident radiation, and B represents the plane of an observer.**

Figure 14–4 shows the jth scattering element referred to some arbitrary origin O. Just as in the previous section, the location of this origin can have no effect on the final results. The incident oscillating electric field is propagated in the direction given by the unit vector $\mathbf{u}_0$ and has the same phase in the plane A. The plane B represents the position of an observer of the scattered radiation, which is propagated in the direction of the unit vector $\mathbf{u}$. The radiation arriving at B will, in general, no longer be in phase because of the extra distance $a + b$ traveled by the upper beam in the figure. Hence the intensity of the radiation arriving at B will be less than the incident intensity.

We defined a quantity $P(\theta)$ by

$$P(\theta) = \frac{\text{scattered intensity from a collection of scatters}}{\text{initial intensity}}$$

Note that as $\theta \to 0$ in Fig. 14–4, the distance $a + b \to 0$ and hence the intensity of scattered radiation equals the intensity of incident radiation. Thus

$$P(\theta) = \frac{\text{intensity of radiation scattered through an angle } \theta}{\text{intensity of radiation scattered through an angle } \theta \to 0}$$

$$= \frac{I(\theta)}{I(0)} \tag{14–33}$$

where $I(\theta)$ is the intensity of radiation scattered through θ. The quantity $P(\theta)$, which can be determined experimentally, yields much information concerning the configuration of the collection of scatterers.

From Fig. 14–4, we see that

$$a = r_j \cos(\mathbf{r}_j, \mathbf{u}_0) \tag{14–34}$$

where $(\mathbf{r}_j, \mathbf{u}_0)$ represents the angle between $\mathbf{r}_j$ and $\mathbf{u}_0$. But $r_j \cos(\mathbf{r}_j, \mathbf{u}_0) = \mathbf{r}_j \cdot \mathbf{u}_0$ since $\mathbf{u}_0$ is a unit vector, and so

$$a = \mathbf{r}_j \cdot \mathbf{u}_0 \tag{14–35}$$

Similarly, $b = r_j \cos(\pi - (\mathbf{r}_j, \mathbf{u}))$, and this can be written as

$$\begin{aligned} b &= -r_j \cos(\mathbf{r}_j, \mathbf{u}) \\ &= -\mathbf{r}_j \cdot \mathbf{u} \end{aligned} \tag{14–36}$$

since $\mathbf{u}$ also is a unit vector. The extra distance traveled by the radiation scattered by the jth element is

$$a + b = \mathbf{r}_j \cdot (\mathbf{u}_0 - \mathbf{u}) \tag{14–37}$$

Although $\mathbf{u}_0$ and $\mathbf{u}$ are unit vectors, their difference is not. The square of $\mathbf{u}_0 - \mathbf{u}$ is $u_0{}^2 + u^2 - 2\mathbf{u}_0 \cdot \mathbf{u} = 2 - 2\cos\theta$, and hence the length of $\mathbf{u}_0 - \mathbf{u} = [2(1 - \cos\theta)]^{1/2} = 2 \sin \theta/2$. Let $\mathbf{n}$ be a unit vector in the direction of $\mathbf{u}_0 - \mathbf{u}$. The vector $\mathbf{u}_0 - \mathbf{u}$ then equals $2\mathbf{n} \sin \theta/2$ and Eq. (14–37) is

$$a + b = 2\mathbf{r}_j \cdot \mathbf{n} \sin\frac{\theta}{2} \tag{14–38}$$

The phase difference ϕ_j due to this extra difference is $(a + b)/\lambda$, where λ is the wavelength of the radiation, and so

$$\phi_j = \mathbf{r}_j \cdot \mathbf{n}\left(\frac{2}{\lambda} \sin\frac{\theta}{2}\right) \tag{14–39}$$

The electric field at B due to the scattering by the element at j is

$$\mathscr{E}_j = A \cos[2\pi(\nu t - \phi_j)] \tag{14–40}$$

where ν is the frequency of the radiation, and A can be considered to be simply a constant. The electric field from the collection of N scatterers is

$$\mathscr{E}_s = \sum_{j=1}^{N} A \cos[2\pi(\nu t - \phi_j)] \tag{14–41}$$

The intensity is the energy which falls on 1 cm^2 of area per second and can be obtained by averaging $\mathscr{E}_s{}^2$ over one period:

$$I(\theta) = \frac{\int_0^{1/\nu} dt \mathscr{E}_s{}^2}{\int_0^{1/\nu} dt} = \nu A^2 \int_0^{1/\nu} dt \left(\sum_{j=1}^{N} \cos\alpha_j\right)^2 \tag{14–42}$$

where $\alpha_j = 2\pi(\nu t - \phi_j)$. To evaluate the integral in Eq. (14-42), first write

$$\begin{aligned} \left(\sum_{j=1}^{N} \cos\alpha_j\right)^2 &= \sum_{i=1}^{N} \sum_{j=1}^{N} \cos\alpha_i \cos\alpha_j \\ &= \sum_{i=1}^{N} \sum_{j=1}^{N} \{\tfrac{1}{2}\cos(\alpha_i - \alpha_j) + \tfrac{1}{2}\cos(\alpha_i + \alpha_j)\} \end{aligned}$$

Then

$$I(\theta) = \nu A^2 \int_0^{1/\nu} dt \sum_{i=1}^{N} \sum_{j=1}^{N} \{\tfrac{1}{2}\cos[2\pi(\phi_i - \phi_j)] + \tfrac{1}{2}\cos[2\pi(2\nu t - (\phi_i + \phi_j))]\}$$

The second integral here vanishes since we are integrating over one period, and we are left with

$$I(\theta) = \frac{A^2}{2} \sum_{i=1}^{N} \sum_{j=1}^{N} \cos 2\pi(\phi_i - \phi_j) \tag{14–43}$$

As $\theta \to 0$, each $\phi_i \to 0$, and so $I(\theta) \to A^2N^2/2$. The ratio of $I(\theta)$ to $I(0)$ then is

$$P(\theta) = \frac{1}{N^2} \sum_{i=1}^{N} \sum_{j=1}^{N} \cos[2\pi(\phi_i - \phi_j)]$$
$$= \frac{1}{N^2} \sum_{i=1}^{N} \sum_{j=1}^{N} \cos[s\mathbf{n} \cdot (\mathbf{r}_i - \mathbf{r}_j)] \tag{14–44}$$

where

$$s = \frac{4\pi}{\lambda} \sin \frac{\theta}{2} \tag{14–45}$$

Again note that only the difference in the positions of particles appears in Eq. (14–44), and that the arbitrarily placed origin in Fig. 14–4 drops out of the equations.

Equation (14–44) is for a collection of scatterers in a fixed orientation with respect to the incident radiation. We must now average Eq. (14–44) over all orientations. This can be done by averaging over all orientations of $\mathbf{n}$. To do this, take a spherical coordinate system with the z-axis to be along $\mathbf{r}_i - \mathbf{r}_j$ with the angle between $\mathbf{n}$ and this z-axis denoted by α and the other angle in the spherical coordinate system denoted by β. Then $d\mathbf{n} = \sin\alpha \, d\alpha \, d\beta$ (Problem 14–34) and

$$\overline{\cos[s\mathbf{n} \cdot \mathbf{r}_{ij}]} = \frac{1}{4\pi} \int_0^{2\pi} \int_0^{\pi} \cos(sr_{ij} \cos\alpha) \sin\alpha \, d\alpha \, d\beta$$
$$= \frac{\sin sr_{ij}}{sr_{ij}} \tag{14–46}$$

The scattering function $P(\theta)$ is given by

$$P(\theta) = \frac{1}{N^2} \sum_{i=1}^{N} \sum_{j=1}^{N} \frac{\sin sr_{ij}}{sr_{ij}} \tag{14–47}$$

An important property of $P(\theta)$ is that for small θ, it becomes independent of any assumption regarding the configuration of the collection of scatterers. Under this limiting condition, it becomes a direct measure of the radius of gyration. No other physical measurement provides a measure of the dimensions of a macromolecule without any assumption regarding its general form.

The relation between $P(\theta)$ and the radius of gyration is obtained by expanding each term in Eq. (14–47). Since $\sin x = x - x^3/3! + x^5/5! \ldots$, we have

$$P(\theta) = \frac{1}{N^2} \sum_{i=1}^{N} \sum_{j=1}^{N} \left(1 - \frac{s^2 r_{ij}{}^2}{3!} + \frac{s^4 r_{ij}{}^4}{5!} \cdots\right)$$
$$= 1 - \left(\frac{s^2}{3!\,N^2}\right) \sum_{i=1}^{N} \sum_{j=1}^{N} r_{ij}{}^2 + \cdots \tag{14–48}$$

According to Eq. (14–25), which is quite general, the double summation here is just $2N^2R_G{}^2$. Equation (14–25) is applicable to a rigid configuration of point masses. For a flexible chain, the quantity $r_{ij}{}^2$ in the expansion of $P(\theta)$ becomes $\overline{r_{ij}{}^2}$, which Eq. (14–26) shows to be also $2N^2R_G{}^2$. Thus Eq. (14–48) is

$$P(\theta) = 1 - \frac{s^2 R_G{}^2}{3} + \cdots$$
$$= 1 - \frac{16\pi^2 R_G{}^2}{3\lambda^2} \sin^2 \frac{\theta}{2} + \cdots \tag{14–49}$$

A plot of $P(\theta)$ versus $\sin^2 \theta/2$ gives $R_G{}^2$ without any assumptions regarding the configuration of the system of particles. It is customary to plot the reciprocal of $P(\theta)$ versus $\sin^2 \theta/2$. Since Eq. (14–49) is an expansion for small values of θ, we can write $1/(1-x)$ as $1+x$ to get

$$\frac{1}{P(\theta)} = 1 + \frac{16\pi^2 R_G{}^2}{3\lambda^2} \sin^2 \frac{\theta}{2} + \cdots \tag{14–50}$$

A plot of $P^{-1}(\theta)$ versus $\sin^2 \theta/2$ is indeed a straight line for small values of $\sin^2 \theta/2$, from which one immediately obtains $R_G{}^2$.

The determination of the radius of gyration is a very powerful probe into the average configuration of a polymer molecule. Suppose we have a polymer of molecular weight = 500,000, a density of 1.33 g/cm^3, a monomeric weight of 150, and a value of β estimated to be 5 Å. Using Eqs. (14–30) through (14–32), we can calculate what the radius of gyration would be, assuming the molecule to be a flexible chain, a solid sphere, and a solid cylinder of various lengths. Table 14–1 shows the results of such a

Table 14–1. **The radius of gyration as a function of the assumed configuration of a polymer of molecular weight 500,000, density 1.33 g/cm^3, monomeric weight 150, and a value of β estimated to be 5 Å**

solid sphere	41 Å
flexible chain	118 Å
solid cylinder	
(diameter 20 Å)	570 Å
(diameter 15 Å)	1000 Å
(diameter 10 Å)	2300 Å

calculation (see Problem 14–13). It can be seen that the radius of gyration is a strong function of the assumed configuration of the molecule. The assumptions used to calculate these numbers are indeed crude, but the general conclusions are not altered. Although there are many other configurations that would be included in Table 14–1, by comparing these numbers to the experimentally observed radius of gyration, one can at least eliminate certain possibilities. Table 14–2 shows some experimentally observed radii of gyration compared to the radii of gyration, assuming that the molecules are solid spheres. Regardless of the crudeness of the assumption that the molecule is a solid sphere, it can be seen that only three of the substances listed are at all close to being solid spheres. The other molecules must all have a much greater extension in space. For example, it is known from electron microscopy that the tobacco mosaic virus is a long rod about 3000 Å long. If it is assumed that the experimental value listed in Table 14–2 is that of a rod, then one can calculate the length

Table 14–2. **Radii of gyration from light scattering or X-ray scattering**

		calculated result unsolvated sphere		experimental result
	mol. wt.	density (g/cc)	R_G (Å)	R_G (Å)
serum albumin	66,000	1.33	21	29.8
catalase	225,000	1.37	31	39.8
myosin	493,000	1.37	41	468
DNA	4×10^6	1.8	74	1170
bushy stunt virus	10.6×10^6	1.35	113	120
tobacco mosaic virus	39×10^6	1.33	175	924

Source: C. Tanford, *Physical Chemistry of Macromolecules* (New York: Wiley, 1961).

from Eq. (14–32) (neglecting the small term $R^2/2$). The result is $L = 3200$ Å, in excellent agreement with the electron microscopic value.

We have shown that the function $P(\theta)$ at small angles gives the radius of gyration directly. The complete expression for $P(\theta)$ is a function of the assumed configuration of the system of scatterers, and the determination of $P(\theta)$ as a function of θ can be used to distinguish between geometric forms. We shall calculate $P(\theta)$ for two specific models and compare the results in Fig. 14–5. The two models will be a long thin rod and a flexible chain.

Consider $N + 1$ scattering elements in a straight line. If the distance between adjacent elements is l, the total length of the rod is Nl. We wish to evaluate Eq. (14–47) for this configuration. For mathematical convenience, we shall consider the limit in which N is large and l is small such that $L = Nl$ is fixed. By the same argument that we used in going from Eq. (14–10) to (14–12), we have

$$P(\theta) = \frac{1}{(N+1)^2} \sum_{k=0}^{N} 2(N-k)\frac{\sin(slk)}{slk} - \frac{N}{(N+1)^2} \tag{14–51}$$

The second term here, $-N(N+1)^2$, arises because there are only N terms for which $k = 0$, but we have included $2N$ of them in the summation. There is no $\sin(slk)/slk$ factor in this term since $\sin(slk/slk) = 1$ when $k = 0$. We can convert the summation in Eq. (14–51) to an integral by taking N to be large and l to be small such that $L = Nl$ is fixed. Physically this corresponds to taking the rod to be a linear array of many closely separated scatterers. In this limit then, Eq. (14–51) becomes

$$P(\theta) = \frac{2}{N}\int_0^N \frac{\sin(slk)}{slk}\,dk - \frac{2}{slN^2}\int_0^N \sin(slk)\,dk - \frac{1}{N} \tag{14–52}$$

where we have neglected 1 with respect to N in the denominators. The second integral in this equation is easy, but the first cannot be expressed in terms of elementary functions (see Problem 14–14) and must be evaluated numerically. By letting $x = slk$, the two integrals become functions of only sL, which is finite, but the last term is still $1/N$ and so can be ignored. Thus Eq. (14–52) becomes

$$P(\theta) = \frac{2}{sL}\int_0^{sL} \frac{\sin x}{x}\,dx - \left[\frac{\sin(sL/2)}{(sL/2)}\right]^2 \tag{14–53}$$

This function is shown in Fig. 14–5.

Debye was the first to calculate $P(\theta)$ for a flexible chain. Equation (14–47) must be averaged over all the allowed configurations of a polymer chain, that is, we must average each of the terms $\sin(sr_{ij})/sr_{ij}$. The probability that the ith and jth elements of a flexible chain are separated by a distance r is given by Eq. (14–6) with N set equal to $|i-j|$. The quantity $P(\theta)$ becomes

$$\begin{aligned} P(\theta) &= \frac{1}{N^2}\sum_{i=1}^{N}\sum_{j=1}^{N}\int d\mathbf{r}W_{ij}(r)\frac{\sin sr}{sr} \\ &= \frac{1}{N^2}\sum_{i=1}^{N}\sum_{j=1}^{N}\left(\frac{3}{2\pi|i-j|l^2}\right)^{3/2}\int_0^\infty 4\pi r^2\,dre^{-3r^2/2|i-j|l^2}\frac{\sin sr}{sr} \\ &= \frac{1}{N^2}\sum_{i=1}^{N}\sum_{j=1}^{N}\exp\left(\frac{-s^2l^2|i-j|}{6}\right) \end{aligned} \tag{14–54}$$

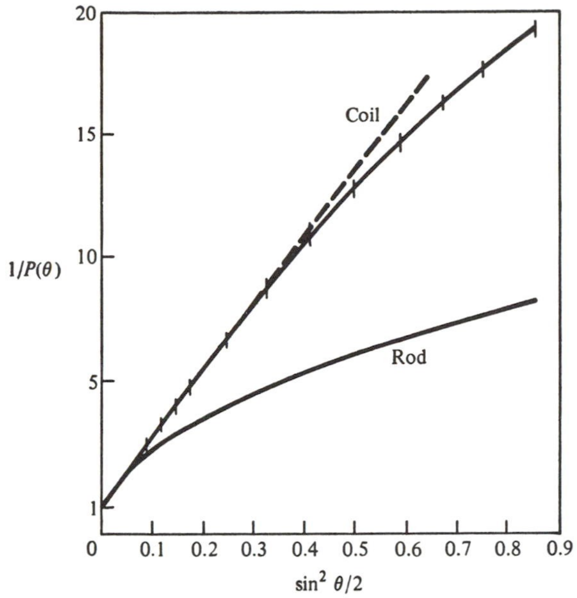

Figure 14–5. **The angular dependence of light scattering from a DNA solution. Experimental values of $1/P(\theta)$ are compared with theoretical curves for a flexible coil and for a rigid rod.** (From C. Tanford, *Physical Chemistry of Macromolecules.* New York: Wiley, 1961; after P. Doty and B. H. Bunce, *J. Am. Chem, Soc.*, **74**, 5029, 1952.)

where we have used the integral

$$\int_0^\infty x \sin axe^{-bx^2}\,dx = \frac{a\pi^{1/2}}{4b^{3/2}}\exp\left(-\frac{a^2}{4b}\right)$$

The summation in Eq. (14–54) may be replaced by integrals:

$$P(\theta) = \frac{2}{N^2}\int_0^N dx \int_0^x dy \exp\left(\frac{-s^2l^2|x-y|}{6}\right)$$

Let $x - y = z$ and $x = x$ to get

$$P(\theta) = \frac{2}{N^2}\int_0^N dx \int_0^x dze^{-s^2l^2z/6}$$

$$= \frac{2}{\omega^2}(e^{-\omega} + \omega - 1) \qquad (14\text{–}55)$$

where $\omega = Ns^2R_G{}^2$. This is Debye's equation for the scattering due to a flexible chain. The reciprocal of Eq. (14–55) is shown in Fig. 14–5, along with experimental data from a DNA solution. The figure indicates that the nucleic acid resembles a flexible coil more closely than a rod. Since DNA is known to have a double-stranded helical structure, the data imply that occasional points of flexibility or kinks must interrupt this structure.

Light scattering can be used to study a number of other static properties of macromolecules such as molecular weights, molecular-weight distributions, and the interaction of macromolecules in solution. In all these cases the quantity that is measured is the total frequency integrated absorption as a function of scattering angle. These are known as static measurements. Before the development of lasers, it was not possible to measure the frequency dependence since the width of the scattered spectral distribution is usually very small. Now, however, it is possible to measure the frequency dependence of the scattering. We shall see in Chapter 29 (*Statistical Mechanics*) that this technique, light-scattering spectroscopy, can be used to measure dynamical properties such as translational and rotational diffusion coefficients and chemical reaction rate constants. The additional reading by Pecora is a review of light-scattering spectroscopy as a tool for studying macromolecular dynamics.

In this section and the previous two sections, we have studied the configuration of single polymer molecules. In the next section we shall study a network of polymer molecules connected to each other as in a rubber. We shall see that the simple random flight model of a polymer chain can be used to describe the general elastic properties of rubber.

14-4 RUBBER ELASTICITY

Rubbers consist of polymer chains connected to each other at various points to form a network. In order to calculate the elastic properties of such a network, we must calculate the free energy of the network as a function of its distortion. If L represents the length of the network (say a rubber band for concreteness), then the Helmholtz free energy is given by

$$A(T, V, L) = E(T, V, L) - TS(T, V, L) \tag{14-56}$$

The reversible work associated with an infinitesimal deformation of a rubber is

$$dA = dE - T\,dS - S\,dT \tag{14-57}$$

where

$$dE = T\,dS - p\,dV + f\,dL \tag{14-58}$$

and where f is the force exerted.

The volume change associated with an isothermal deformation of rubber is negligible, and so we may ignore the $p\,dV$ term in Eq. (14-58). Substitution of Eq. (14-58) into Eq. (14-57) gives

$$dA = -S\,dT + f\,dL \tag{14-59}$$

from which we write

$$f = \left(\frac{\partial A}{\partial L}\right)_T = \left(\frac{\partial E}{\partial L}\right)_T - T\left(\frac{\partial S}{\partial L}\right)_T \tag{14-60}$$

Experimentally, the internal energy is almost independent of L at constant T, so that the term $(\partial E/\partial L)_T \approx 0$. Note that this condition is analogous to the fact that $(\partial E/\partial V)_T = 0$ for an ideal gas. Furthermore, it is observed experimentally that the force is proportional to the absolute temperature. This implies that the entropy of the network does not depend upon the temperature, or in other words is a function only of L. This also is analogous to an ideal gas, in which case the entropy consists of two parts, one associated with the heat capacity of the gas but making no contribution to $(\partial S/\partial V)_T$, and the other associated with the number of configurations available to the system, independent of its composition, and making a contribution to $(\partial S/\partial V)_T$, which is independent of T (see Problem 14-18).

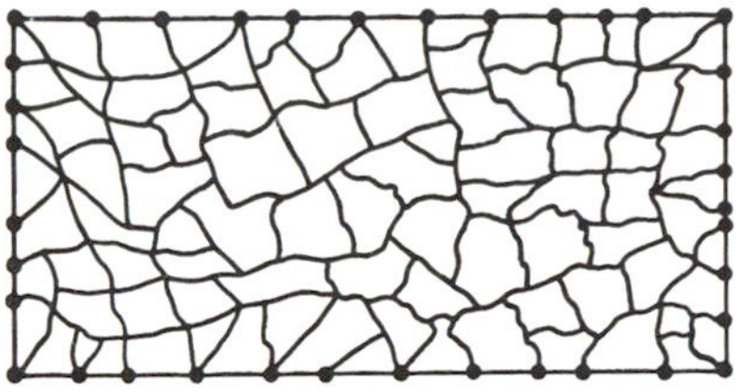

Figure 14–6. **A two-dimensional version of a polymer network, illustrating fixed points (on the edges) and junction points (in the interior).**

The force f can be determined as a function of L from the equation

$$f = -T\left(\frac{\partial S}{\partial L}\right)_T \tag{14–61}$$

if we can determine the entropy of the network as a function of a deformation.

Figure 14–6 shows a two-dimensional version of a polymer network. The chains of the network are connected throughout the network by points. We shall distinguish between fixed points and junction points. Fixed points are points at which the network is localized at the surface of the rubber. Specification of the fixed points corresponds to the specification of the surface of the network. The junctions are points at which chains are connected in the network. These are not necessarily fixed in space, but exhibit a Brownian motion compatible with the constraints of the network. We shall assume that the configurations of the chains between any two points (fixed or junction) are governed by the results of Section 14–1, in particular the result that says the probability that two points of a freely jointed chain will be separated by a vector with components x, y, and z is

$$W(x, y, z) = \left(\frac{3}{2\pi N l^2}\right)^{3/2} \exp\left(-\frac{3(x^2 + y^2 + z^2)}{2Nl^2}\right) \tag{14–62}$$

where N is the number of links between the two points. This quantity may also be reinterpreted to be proportional to the number of configurations available to a chain of N links, whose end points are separated by a vector $\mathbf{r}$ with components x, y, and z. The quantity $W(\mathbf{r})$ then is analogous to a microcanonical partition function (the fixed energy here is zero), and so the entropy is given by $S = k \ln(cW)$, where c is a constant of proportionality which is independent of the configuration of the molecule.

Since the statistics of a freely jointed chain depend only upon the number of links between the two points of interest, the probability of the configuration shown in Fig. 14–7 is (see Problem 14–27)

$$\begin{aligned} W(\mathbf{r}_i, \mathbf{r}_j, \mathbf{r}_k) &= W(\mathbf{r}_i, \mathbf{r}_j)W(\mathbf{r}_j, \mathbf{r}_k) \\ &= W(|\mathbf{r}_i - \mathbf{r}_j|)W(|\mathbf{r}_j - \mathbf{r}_k|) \end{aligned} \tag{14–63}$$

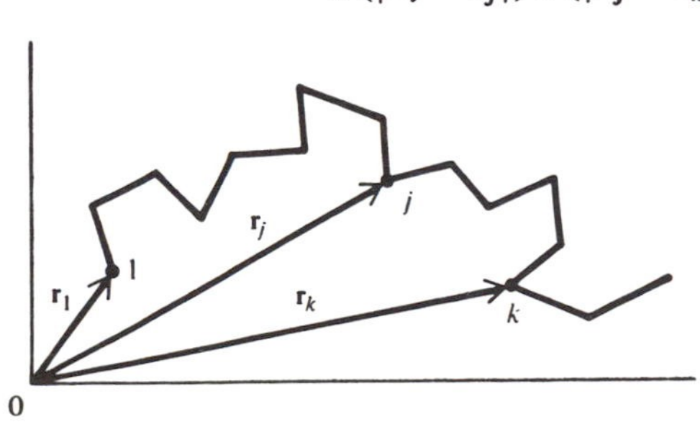

Figure 14–7. **An illustration of the quantities appearing in Eq. (14–63).**

A similar equation applies to the entire network. Let us denote the fixed points of the network by subscripts α, β, ..., the junctions by numerical subscripts, and the points of either types by subscripts τ, ν, ... The positions of the various points are indicated by vectors $\mathbf{r}_\alpha, \mathbf{r}_\beta, \mathbf{r}_1, \mathbf{r}_2, \ldots, \mathbf{r}_\tau, \mathbf{r}_\nu, \ldots$. By the generalization of Eq. (14–63), the number of configurations for the network is equal to the product of the number of configurations possible for all the individual chains

$$W(\mathbf{r}_\alpha, \mathbf{r}_\beta, \ldots, \mathbf{r}_1, \mathbf{r}_2, \ldots) = \prod_{\tau,\nu} W(|\mathbf{r}_\tau - \mathbf{r}_\nu|) \tag{14–64}$$

The quantity $W(|\mathbf{r}_\tau - \mathbf{r}_\nu|)$ is proportional to the number of configurations for a chain of $N_{\tau\nu}$ links of equal length l.

The total number of configurations of the network, consistent with given positions of the fixed points only is obtained by integrating $W(\mathbf{r}_\alpha, \mathbf{r}_\beta, \ldots, \mathbf{r}_1, \mathbf{r}_2, \ldots)$ over all possible sets of positions of the junctions:

$$W(\mathbf{r}_\alpha, \mathbf{r}_\beta, \ldots) = \int d\mathbf{r}_1 \int d\mathbf{r}_2 \cdots W(\mathbf{r}_\alpha, \mathbf{r}_\beta, \ldots, \mathbf{r}_1, \mathbf{r}_2, \ldots) \tag{14–65}$$

Since the $W(|\mathbf{r}_\tau - \mathbf{r}_\nu|)$ in Eq. (14–64) and hence in Eq. (14–65) are given by

$$W(|\mathbf{r}_\tau - \mathbf{r}_\nu|) = C_{\tau\nu} \exp\left(-\frac{3}{2N_{\tau\nu} l^2} |\mathbf{r}_\tau - \mathbf{r}_\nu|^2\right)$$

the integral in Eq. (14–65) is of the form

$$I(x_1, x_2, \ldots, x_n) = \int_{-\infty}^{\infty} \cdots \int \exp\left\{-\sum_{i=1}^{N} \sum_{j=1}^{N} a_{ij} x_i x_j\right\} dx_{n+1}\, dx_{n+2} \cdots dx_N$$

Such integrals can be handled quite generally, but the manipulations are lengthy.* All we need to know, however, is that $I(x_1, x_2, \ldots, x_n)$ is of the form

$$I(x_1, x_2, \ldots, x_n) = c \exp\left\{-\sum_{i=1}^{n} \sum_{j=1}^{n} b_{ij} x_i x_j\right\}$$

where c and the b_{ij} can be expressed in terms of the a_{ij}. In other words, if one integrates over a subset of the variables of a multidimensional Gaussian function, the result is a Gaussian function of the variables that were not integrated.

Applying this result to the integral of Eq. (14–65), we get

$$W(\mathbf{r}_\alpha, \mathbf{r}_\beta, \ldots) = c \exp\left\{-\frac{3}{2l^2} \sum\sum \frac{1}{N_{\alpha\beta}'} |\mathbf{r}_\alpha - \mathbf{r}_\beta|^2\right\} \tag{14–66}$$

The form of this result is independent of the "topology" of the network, which affects only the constants C and $N_{\alpha\beta}'$.

Consider now the special case of a unit cube of a polymer, subjected to uniform stretching parallel to the edges until these have lengths L_x, L_y, and L_z. If all the external forces are applied normal to the surfaces, the fixed points will lie on these surfaces in positions which change in the same proportion as L_x, L_y, and L_z. Taking three of the faces of the cube as coordinate planes, we have

$$x_\alpha = x_\alpha^{(0)} L_x, \quad y_\alpha = y_\alpha^{(0)} L_y \quad z_\alpha = z_\alpha^{(0)} L_z \tag{14–67}$$

* H. M. James, *J. Chem. Phys.*, **15**, p. 651, 1947.

where $x_\alpha^{(0)}$, $y_\alpha^{(0)}$, $z_\alpha^{(0)}$ are the coordinates of the αth fixed point in the undeformed state. Using Eqs. (14–67), $W(\mathbf{r}_\alpha, \mathbf{r}_\beta, \ldots)$ can be written as a function of L_x, L_y, L_z and the initial undeformed state of the network.

By substituting Eqs. (14–67) into Eq. (14–66), we get for the number of configurations available to the network:

$$W(L_x, L_y, L_z) = C \exp-\{(K_x L_x^2 + K_y L_y^2 + K_z L_z^2)\} \tag{14–68}$$

where

$$K_x = \frac{3}{2l^2} \sum_{\alpha>\beta}\sum \frac{1}{N_{\alpha\beta}'}(x_\alpha^{(0)} - x_\beta^{(0)})^2 \tag{14–69}$$

with similar forms for K_y and K_z. If the polymer network in the undeformed state is isotropic on the average, then $K_x = K_y = K_z = K$, and Eq. (14–68) becomes

$$W(L_x, L_y, L_z) = C \exp\{-K(L_x^2 + L_y^2 + L_z^2)\} \tag{14–70}$$

The entropy is given by

$$S(L_x, L_y, L_z) = k \ln W(L_x, L_y, L_z) + \text{constant} \tag{14–71}$$

For a uniform stretch in the z-direction of an incompressible polymer, L_x and L_y must decrease since the volume $1 = L_x L_y L_z$ must be preserved. If we let $L = L_z$, then $L_x^2 = L_y^2 = 1/L$. The entropy is a function of only L and is

$$S(L) = \text{constant} - kK\left(L^2 + \frac{2}{L}\right)$$

Finally, the force required to stretch the network to a length L is given by Eq. (14–61) and is

$$f = 2KkT\left(L - \frac{1}{L^2}\right) \tag{14–72}$$

Equation (14–72) is the fundamental equation of rubber elasticity and is analogous to the equation $pV = NkT$ of ideal gases. It is, in a sense, the ideal equation of state of rubber elasticity. Figure 14–8 shows the force-extension curve predicted by Eq. (14–72) and some typical experimental data. It can be seen that the agreement is

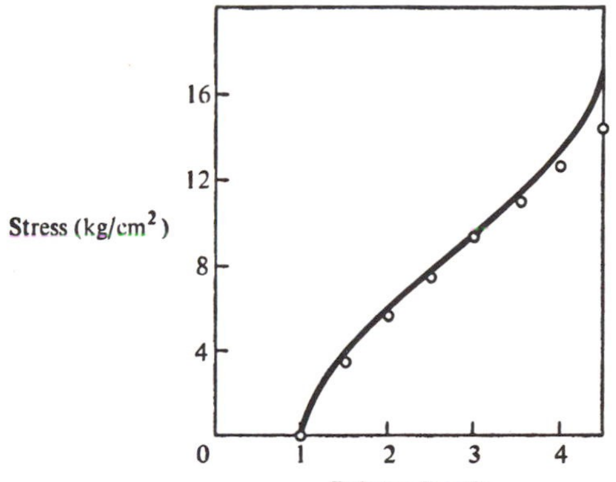

Figure 14–8. **The force length curve of rubber elasticity. The solid curve represents the experimental data and the circles are calculated values. The agreement becomes progressively worse for greater values of the relative extension.**

quite good up to extensions of the order of 300 percent. Furthermore, it predicts the experimental observation that the force is proportional to the absolute temperature at fixed extension, or that a rubber band held under constant tension contracts when heated. Although Eq. (14–72) is based on the simple model of a network of freely jointed chains, it does reflect the basic features of rubber elasticity.

14-5 THE SOLUTION TO THE GENERAL PROBLEM OF RANDOM FLIGHTS

In the first section of the chapter, we considered a random walk in one dimension and then generalized the results to the problem of a random flight. In this section we shall treat the random-flight problem in a more elegant manner and derive Eq. (14–6) by a powerful mathematical technique which has a quite general applicability. Although this section is somewhat of a mathematical disgression, the technique introduced here can be very useful in many problems. We shall solve the general problem of random flights by a method due to Markov.

In the general problem of a random flight, the position $\mathbf{R}$ of a particle after N displacements is given by (see Chandrasekhar in "Additional Reading")

$$\mathbf{R} = \sum_{j=1}^{N} \mathbf{r}_j \tag{14–73}$$

where the $\mathbf{r}_j$ are the N displacements. Let the probability that the jth displacement lie between $\mathbf{r}_j$ and $\mathbf{r}_j + d\mathbf{r}_j$ be given by

$$\tau_j(x_j, y_j, z_j)\, dx_j\, dy_j\, dz_j = \tau_j(\mathbf{r}_j)\, d\mathbf{r}_j \tag{14–74}$$

We wish to find the probability $W_N(\mathbf{R}_0)\, d\mathbf{R}_0$ that the resultant vector in Eq. (14–73) lie in the interval $\mathbf{R}_0$ and $\mathbf{R}_0 + d\mathbf{R}_0$. Since each of the N displacements is assumed to be independent of each other, we can write down a *formal* expression for $W_N(\mathbf{R}_0)$:

$$W_N(\mathbf{R}_0)\, d\mathbf{R}_0 = \int \underset{*}{\cdots} \int \prod_{j=1}^{N} \tau_j(x_j, y_j, z_j)\, dx_j\, dy_j\, dz_j \tag{14–75}$$

where the asterisk denotes the condition that the limits of the integral are such that the variables $x_1, y_1, z_1, \ldots, x_N, y_N, z_N$ are confined to a region such that

$$\mathbf{R}_0 - \tfrac{1}{2}\, d\mathbf{R}_0 \le \mathbf{R} \le \mathbf{R}_0 + \tfrac{1}{2}\, d\mathbf{R}_0 \tag{14–76}$$

We can now *formally* eliminate this awkward restriction on the limits of the integral in Eq. (14–75) by introducing a factor $\Delta(x_1, y_1, \ldots, z_N)$ having the property that

$$\begin{aligned} \Delta(x_1, y_1, \ldots, z_N) &= 1 \qquad \text{whenever } \mathbf{R}_0 - \tfrac{1}{2}\, d\mathbf{R}_0 \le \mathbf{R} \le \mathbf{R}_0 + \tfrac{1}{2}\mathbf{R}_0 \\ &= 0 \qquad \text{otherwise} \end{aligned} \tag{14–77}$$

This allows us to write

$$W_N(\mathbf{R}_0)\, d\mathbf{R}_0 = \int \cdots \int \Delta(\mathbf{r}_1, \mathbf{r}_2, \ldots, \mathbf{r}_N) \prod_{j=1}^{N} \tau_j(\mathbf{r}_j)\, d\mathbf{r}_j \tag{14–78}$$

where, in contrast to Eq. (14–75), the integration is carried out over all values of $\mathbf{r}_1$ through $\mathbf{r}_N$. Clearly, the introduction of the function $\Delta(\mathbf{r}_1, \ldots, \mathbf{r}_N)$ is at this point only a formal device to extend the limits of integration of Eq. (14–75) over all space, but the essence of the Markov method is that an explicit expression for this factor can be given.

To see this, consider the integrals

$$\delta_k = \frac{1}{\pi}\int_{-\infty}^{\infty} \frac{\sin \alpha_k \rho_k}{\rho_k} \exp(i\rho_k \gamma_k)\, d\rho_k \qquad k = x, y, z \tag{14–79}$$

This integral is called Dirichlet's discontinuous integral and has the important property Problem 14–35 that

$$\begin{aligned} \delta_k &= 1 \qquad \text{whenever } -\alpha_k < \gamma_k < \alpha_k \\ &= 0 \qquad \text{otherwise} \quad (k = x, y, z) \end{aligned} \tag{14–80}$$

We can utilize this result to construct Δ by letting

$$\begin{aligned} \alpha_x &= \tfrac{1}{2}\, dX_0; \qquad \gamma_x = \left(\sum_{j=1}^{N} x_j\right) - X_0 \\ \alpha_y &= \tfrac{1}{2}\, dY_0; \qquad \gamma_y = \left(\sum_{j=1}^{N} y_j\right) - Y_0 \end{aligned} \tag{14–81}$$

and

$$\alpha_z = \tfrac{1}{2}\, dZ_0; \qquad \gamma_z = \left(\sum_{j=1}^{N} z_j\right) - Z_0$$

where X_0, Y_0, Z_0 are the components of $\mathbf{R}_0$. By substituting these definitions into Eq. (14–80), we see that

$$\Delta(x_1, y_1, \ldots, z_n) = \delta_x\, \delta_y\, \delta_z \tag{14–82}$$

We now substitute Eqs. (14–82), (14–81), and (14–79) into Eq. (14–78) to obtain

$$W_N(\mathbf{R}_0)\, d\mathbf{R}_0 = \frac{1}{\pi^3}\int \cdots \int \left\{\prod_{j=1}^{N} \tau_j(\mathbf{r}_j)\, d\mathbf{r}_j\right\} \left\{\frac{\sin\left(\dfrac{\rho_x\, dX_0}{2}\right)\sin\left(\dfrac{\rho_y\, dY_0}{2}\right)\sin\left(\dfrac{\rho_z\, dZ_0}{2}\right)}{\rho_x \rho_y \rho_z}\right\}$$
$$\times \exp\left\{-i\left(\boldsymbol{\rho}\cdot\mathbf{R}_0 - \sum_{j=1}^{N} \boldsymbol{\rho}\cdot\mathbf{r}_j\right)\right\} d\rho_x\, d\rho_y\, d\rho_z$$

Since the arguments of the sines are infinitesimal quantities, we can replace $\sin(\rho_x\, dX_0/2)$ by $\rho_x\, dX_0/2$, and so on, and get

$$W_N(\mathbf{R}_0)\, d\mathbf{R}_0 = \frac{d\mathbf{R}_0}{(2\pi)^3}\int \cdots \int \exp(-i\boldsymbol{\rho}\cdot\mathbf{R}_0) A_N(\boldsymbol{\rho})\, d\boldsymbol{\rho} \tag{14–83}$$

where $d\mathbf{R}_0 = dX_0\, dY_0\, dZ_0$, $d\rho = d\rho_x\, d\rho_y\, d\rho_z$, and

$$A_N(\boldsymbol{\rho}) = \prod_{j=1}^{N} \int \cdots \int \exp(i\boldsymbol{\rho}\cdot\mathbf{r}_j)\tau_j(\mathbf{r}_j)\, d\mathbf{r}_j \tag{14–84}$$

Most often all the functions τ_j are identical, and Eq. (14–84) becomes

$$A_N(\boldsymbol{\rho}) = \left[\int \cdots \int \exp(i\boldsymbol{\rho}\cdot\mathbf{r})\tau(\mathbf{r})\, d\mathbf{r}\right]^N \tag{14–85}$$

Thus we see that if the probability function of each individual displacement is given, it is possible to calculate $A_N(\boldsymbol{\rho})$ and hence the desired function $W_N(\mathbf{R})\, d\mathbf{R}$, the probability that the resultant vector lies in the interval $\mathbf{R}$ and $\mathbf{R} + d\mathbf{R}$.

There are many choices for τ that have physical interest, but we shall consider only two here, namely, each displacement of constant length but random direction (the model of the first section) and a general spherically symmetric displacement.

If the displacements are of constant length l and random directions, the probability distribution $\tau_j(\mathbf{r}_j)$ is given by

$$\tau_j = \frac{1}{4\pi}\,\delta(|\xi_j|^2 - 1) \tag{14–86}$$

where the variable $\xi_j = r_j/l$ has been introduced so that the argument of the Dirac delta function is unitless. The factor $1/4\pi$ is a normalization factor such that

$$\int d\mathbf{r}_j\,\tau(\mathbf{r}_j) = \int d\boldsymbol{\xi}_j\,\tau_j(\boldsymbol{\xi}_j) = \int_0^\infty \int_0^\pi \int_0^{2\pi} \tau_j(\xi_j)\xi^2 \sin\theta\, d\xi\, d\theta\, d\phi = 1$$

Substitution of Eq. (14–86) into the integrand of Eq. (14–85) yields

$$\frac{1}{4\pi}\int\cdots\int \exp(il\boldsymbol{\rho}\cdot\boldsymbol{\xi})\tau(\boldsymbol{\xi})\,d\boldsymbol{\xi} \tag{14–87}$$

This may be evaluated by introducing a polar coordinate system with the z-axis in the direction of $\boldsymbol{\rho}$. This gives the integral

$$\frac{1}{4\pi}\int_0^\infty \int_0^\pi \int_0^{2\pi} \exp(il\rho\xi\cos\theta)\,\delta(\xi^2 - 1)\xi^2\,d\xi\,\sin\theta d\theta d\varphi$$

$$= \frac{1}{\rho l}\int_0^\infty \sin(l\rho\xi)\xi\,\delta(\xi^2 - 1)\,d\xi$$

$$= \frac{\sin(\rho l)}{\rho l}$$

If this is used in Eq. (14–85) for $A_N(\boldsymbol{\rho})$,

$$W_N(\mathbf{R}) = \frac{1}{(2\pi)^3}\int\cdots\int \exp(-i\boldsymbol{\rho}\cdot\mathbf{R})\left(\frac{\sin(\rho l)}{\rho l}\right)^N d\boldsymbol{\rho} \tag{14–88}$$

Again, choosing a spherical coordinate system but with the z-axis in the direction of $\mathbf{R}$ this time, we have

$$W_N(\mathbf{R}) = \frac{1}{8\pi^3}\int_0^\infty \int_0^\pi \int_0^{2\pi} \exp(-i\rho R\cos\theta)\left(\frac{\sin\rho l}{\rho l}\right)^N \rho^2 \sin\theta\, d\theta\, d\rho\, d\phi$$

$$= \frac{1}{2\pi^2 R}\int_0^\infty \sin(\rho R)\left(\frac{\sin\rho l}{\rho l}\right)^N \rho\, d\rho \tag{14–89}$$

For arbitrary values of N, this integral is difficult to evaluate in general, but for large N, which is most often the case of interest, we can write

$$\lim_{N\to\infty}\left(\frac{\sin\rho l}{\rho l}\right)^N \approx \exp\left(\frac{-Nl^2\rho^2}{6}\right) \tag{14–90}$$

[This result is most easily proved by expanding $(\sin \rho l/\rho l)^N$ in powers of ρl and retaining only the highest powers of N in each term (see Problem 14–24).] When Eq. (14–90) is used in Eq. (14–89), we have

$$W_N(\mathbf{R}) = \frac{1}{2\pi^2 R}\int_0^\infty \exp\left(\frac{-Nl^2\rho^2}{6}\right)\rho \sin(\rho R)\, d\rho$$
$$= \left(\frac{3}{2\pi N l^2}\right)^{3/2} \exp\left(\frac{-3R^2}{2Nl^2}\right) \tag{14–91}$$

which is exactly Eq. (14–6). It should be noted, however, that for arbitrary values of N Eq. (14–89) must be used.

The other case we wish to discuss is that in which each displacement is governed by a distribution τ_j that is spherically symmetric and identical for all displacements, that is,

$$\tau_j(\mathbf{r}_j) = \tau(|\mathbf{r}_j|) \qquad j = 1, 2, \ldots, N \tag{14–92}$$

In this case

$$A_N(\boldsymbol{\rho}) = \left[\int \cdots \int \exp(i\boldsymbol{\rho}\cdot\mathbf{r})\tau(|\mathbf{r}|)\, d\mathbf{r}\right]^N$$

By introducing spherical coordinates with the z-axis in the direction of $\boldsymbol{\rho}$, $A_N(\boldsymbol{\rho})$ becomes

$$A_N(\boldsymbol{\rho}) = \left[4\pi \int_0^\infty \frac{\sin(\rho r)}{\rho r} r^2 \tau(r)\, dr\right]^N \tag{14–93}$$

If we expand $\sin(\rho r)$ in terms of ρr:

$$A_N(\boldsymbol{\rho}) = \left[4\pi \int_0^\infty (1 - \tfrac{1}{6}\rho^2 r^2 + \cdots) r^2 \tau(r)\, dr\right]^N$$
$$= [1 - \tfrac{1}{6}\rho^2 \langle r^2\rangle_{\text{av}} + \cdots]^N$$

For large N, we can write this as [*cf.* Eq. (14–90)]

$$\approx \exp\left(\frac{-N\rho^2\langle r^2\rangle_{\text{av}}}{6}\right) \qquad \text{(large } N) \tag{14–94}$$

where $\langle r^2\rangle_{\text{av}}$ is the mean-square displacement to be expected on any occasion. The proof of this equation is exactly the same as the proof of Eq. (14–90). Using Eq. (14–94) for $A_N(\boldsymbol{\rho})$ gives (see Problem 14–22)

$$W_N(\mathbf{R}) = \frac{1}{8\pi^3}\int \cdots \int \exp\left(-i\boldsymbol{\rho}\cdot\mathbf{R} - \frac{N\langle r^2\rangle_{\text{av}}\rho^2}{6}\right) d\boldsymbol{\rho}$$
$$= \left(\frac{3}{2\pi N\langle r^2\rangle_{\text{av}}}\right)^{3/2} \exp\left(\frac{-3R^2}{2N\langle r^2\rangle_{\text{av}}}\right) \tag{14–95}$$

This result was used in Section 14–1, but not proved until now.

Markov's method is actually more general than we have indicated in this section, and the reader is referred to Chandrasekhar under "Additional Reading" for a thorough discussion of not only Markov's method but random-flight problems in general.

ADDITIONAL READING

General

FREED, K. F., *Adv. chem. phys.*, **22**, p. 1, 1972.
FLORY, P. J. 1953. *Principles of polymer chemistry*. Ithaca, N.Y.: Cornell University Press.
MOORE, W. J. 1972. *Physical chemistry*, 4th ed. Englewood Cliffs, N.J.: Prentice-Hall. Chapter 20.
TANFORD, C. 1961. *Physical chemistry of macromolecules*. New York: Wiley.

Chain statistics

BIRSHTEIN, T. M., and PTITSYN, O. B. 1966. *Conformations of macromolecules*. New York: Interscience.
VOLKENSTEIN, M. V. 1963. *Configurational statistics of polymeric chains*. New York: Interscience.

Light scattering

Light scattering from dilute polymer solutions, ed. by D. McIntyre and F. Gornick. 1964. New York: Gordon and Breach.
PECORA, R. 1970. In *Photochemistry of macromolecules*, ed. by R. F. Reinisch. New York: Plenum.
ROWELL, R. L., and STEIN, R. S. 1967. *Electromagnetic scattering*. New York: Gordon and Breach.

Rubber elasticity

CIFERRI, A. *J. polym. sci.*, **54**, p. 149, 1961.
JAMES, H. M., and GUTH, E. *J. chem. phys.*, **11**, p. 455, 1943; *J. polym. sci.*, **4**, p. 153, 1949.
KRIGBAUM, W. R., and ROE, R-J. *Rubber chemistry and technology*, **38**, p. 1039, 1965.
TER HAAR, D. 1954. *Elements of statistical mechanics*. New York: Rinehart. Chapter 15.
TRELOAR, L. R. G. 1958. *The physics of rubber elasticity*. London: Oxford University Press.

Markov's method

CHANDRASEKHAR, S. *Rev. mod. phys.*, **15**, p. 1, 1943.

PROBLEMS

14–1. Compare Eq. (14–1) to Eq. (14–3) for the cases $N = 5$, 10, and 20.

14–2. Generalize Eq. (14–3) to the case in which the probability of taking a step to the right is p and to the left is $1 - p$.

14–3. Show that Eq. (14–6) is normalized, that $\bar{r} = (8Nl^2/3\pi)^{1/2}$, and that $\overline{r^2} = Nl^2$.

14–4. Calculate the average x-, y-, or z-component or the end-to-end separation of a freely jointed chain. Calculate the average of $(x^2 + y^2 + z^2)^{1/2}$.

14–5. Derive Eq. (14–12).

14–6. Evaluate Eq. (14–13) exactly for the case in which $\cos\theta = \frac{1}{3}$ and compare this result to Eq. (14–15). How do these two equations compare for $N = 50$?

14–7. Show that the mean-square distance of the ith segment from the center of mass of the polymer chain is

$$\langle s_i^2 \rangle = \frac{Nl^2}{3}[u^3 + (1 - u)^3]$$

where $u = i/N$.

14–8. The end-to-end distribution function of a real polymer chain is not Gaussian because of the so-called excluded volume effect. One of a number of proposed modifications has been presented by Mazur,* who by means of computer generation of real polymer chains suggested that

$$w(r)\,dr = \left[\Gamma\left(\frac{3}{t}\right)\right]^{-1} t\alpha^{3/t} r^2 \exp(-\alpha r^t)\,dr$$

with $t = 3.2$ and α being a parameter. First show that this distribution is normalized, and then show that $\overline{r^2} = \Gamma(5/t)/\alpha^{2/t}\Gamma(3/t)$.

* In *J. Res. Natl. Bur. Standards*, **69A**, p. 355, 1965.

14–9. Consider a freely jointed chain subjected to a stretching force F applied to the ends of the chain. A segment in a freely jointed chain has no preferred direction when there is no force, but when there is an applied force, each segment will tend to align with the force. Assume now that the potential energy of a segment is $-Fl\cos\theta$, where θ is the angle between F and $\mathbf{l}$. Show that component of $\mathbf{l}$ in the direction F, say the x-direction, is

$$\langle l_x\rangle = \frac{\int_0^\pi (l\cos\theta)(2\pi l^2 \sin\theta\, d\theta)\exp(Fl\cos\theta/kT)}{\int_0^\pi 2\pi l^2 \sin\theta\, d\theta \exp(Fl\cos\theta/kT)}$$

$$= l\mathscr{L}\left(\frac{lF}{kT}\right)$$

where $\mathscr{L}$ is the Langevin function (*cf.* Eq. (13–25)), and so

$$F \approx \left(\frac{3kT}{Nl^2}\right)r$$

for small separations. What is this force law called? Discuss the quantity $(3kT/Nl^2)$ physically, particularly with respect to its temperature dependence. What force is necessary to stretch the chain completely?

14–10. The result presented in Problem 14–9 can be used to derive an end-to-end distribution that is valid for all extensions. From Problem 14–9, the force required to achieve an end-to-end separation r is

$$F = \frac{kT}{l}\mathscr{L}^{-1}\left(\frac{r}{Nl}\right)$$

The energy stored in such a chain is the work done in stretching it:

$$V = \int_0^r F\,dr = \frac{kT}{l}\int_0^r \mathscr{L}^{-1}\left(\frac{r}{Nl}\right) dr \equiv kT\phi(r)$$

The probability that a chain has an end-to-end distance r then is

$$P(r) \propto \exp\left(\frac{-V}{kT}\right) = \exp(-\phi(r))$$

Show that this reduces to Eq. (14–6) for small displacements. What is the probability that $r > Nl$? Is this result obtained with Eq. (14–6)?

14–11. The end-to-end distance distribution function of a freely jointed chain can be derived from a partial differential equation. Let $P(\mathbf{r}, n)$ be the end-to-end distribution function. If we add one more segment, we have $P(\mathbf{r}, n+1)$, but this is equivalent to $P(\mathbf{r}+l, n)$. We now expand $P(\mathbf{r}+\mathbf{l}, n)$ about $\mathbf{r}$ to get

$$P(\mathbf{r}+\mathbf{l}, n) = P(\mathbf{r}, n) + \left(\frac{\partial P}{\partial x}l_x + \frac{\partial P}{\partial y}l_y + \frac{\partial P}{\partial z}l_z\right) + \frac{1}{2}\left[\frac{\partial^2 P}{\partial x^2}l_x^2 + \frac{\partial^2 P}{\partial y^2}l_y^2 + \frac{\partial^2 P}{\partial z^2}l_z^2 + \frac{\partial^2 P}{\partial x\partial y}l_x l_y + \cdots\right]$$

In considering an ensemble average of freely jointed chains, terms linear in l_x, l_y, or l_z and cross products will vanish. Using this, show that

$$\langle P(\mathbf{r}+\mathbf{l}, n)\rangle = \langle P(\mathbf{r}, n)\rangle + \frac{l^2}{6}\left[\frac{\partial^2 P}{\partial x^2} + \frac{\partial^2 P}{\partial y^2} + \frac{\partial^2 P}{\partial z^2}\right]$$

This in turn is equal to $\langle P(\mathbf{r}, n+1)\rangle$. Convince yourself that this implies

$$\frac{\partial P}{\partial n} = \frac{l^2}{6}\nabla^2 P$$

This is the diffusion equation. Show that the solution to this equation under the boundary conditions appropriate to this problem is

$$P(\mathbf{r}, N) = \left(\frac{3}{2\pi l^2 N}\right)^{3/2} \exp\left(\frac{-3r^2}{2Nl^2}\right)$$

This is an important approach to the statistics of polymer chains, since it is possible to extend it to include excluded volume effects.

14–12. Show that the radius of gyration of a uniform right cylinder of radius R and length L is

$$R_G{}^2 = \frac{R^2}{2} + \frac{L^2}{12}$$

and show that this is numerically approximately $L^2/12$ for cylindrical molecules.

14–13. Suppose we have a polymer of molecular weight 500,000, a density 1.33 g/cm³, a monomeric weight of 150, and a value of β estimated to be 5 Å. Verify the values of the radius of gyration given in Table 14–1 for various assumed geometries.

14–14. Derive Eq. (14–53) from Eq. (14–52). By referring to any set of mathematical tables (such as Abramowitz and Stegun), see that

$$\int_0^x \frac{\sin t}{t}\, dt$$

is a well-tabulated function called the sine integral. Plot $1/P(\theta)$ and verify the curves shown in Fig. 14–5.

14–15. Derive the Debye scattering function $P(\theta)$ given by Eq. (14–55).

14–16. Doty, Bradbury, and Holtzer* have determined the radius of gyration of various molecular weight samples of poly-γ-benzyl-L-glutamate in chloroform-formaldehyde at 25°C. Their data are

Mol. wt. × 10^{-5}	R_G(Å)
2.62	528
2.08	408
1.30	263

On the basis of these data, would you conclude that poly-γ-benzyl-L-glutamate is a freely jointed chain in the above solvent? Would you say that it was fully extended? Similarly, Outer, Carr, and Zimm† give the following results for polystyrene in butanone at 22°C:

Mol. wt. × 10^{-6}	R_G(Å)
17.70	437
16.30	414
13.20	367
9.40	306
5.24	222
2.30	163

Is polystyrene a freely jointed chain under the above experimental conditions?

14–17. Consider a solution of polystyrene in cyclohexane at 33°C. Given that the polystyrene fraction has a molecular weight of 3.2×10^6 and a specific gravity of 1.0, calculate the radius of gyration, assuming a spherical shape, and compare this to the experimental value of 494 Å.‡ What does this tell you about the shape of polystyrene in cyclohexane?

* In *J. Am. Chem. Soc.*, **78**, 947, 1956.
† In *J. Chem. Phys.*, **18**, 830, 1950.
‡ See W. R. Krigbaum and K. D. Carpenter, *J. Chem. Phys.*, **59**, 1166, 1955.

14–18. Show that the entropy of an ideal gas consists of two parts, one associated with the heat capacity of the gas, but making no contribution to $(\partial S/\partial V)_T$, and the other associated with the number of configurations available to the system, independent of its composition, and making a contribution to $(\partial S/\partial V)_T$, which is independent of T.

14–19. Derive the thermodynamic equation

$$\left(\frac{\partial E}{\partial L}\right)_T = f - T\left(\frac{\partial f}{\partial T}\right)_L$$

This equation provides a method for experimentally determining the change in energy with elongation.

14–20. Show that the temperature change upon stretching is given by

$$\Delta T = -\frac{T}{C_L}\int_{l_0}^{l}\left(\frac{\partial S}{\partial L}\right)_T dL$$

where C_L, the heat capacity at constant L, is assumed to be independent of L. What does this equation tell you about the sign of the change in temperature observed when rubber is stretched.

14–21. Consider a random flight for which the individual displacement probability distribution, τ_j of Eq. (14–74), is Gaussian:

$$\tau_j = \left(\frac{3}{2\pi l_j^2}\right)^{3/2} \exp\left(\frac{-3|\mathbf{r}_j|^2}{2l_j^2}\right)$$

Show that

$$W_N(\mathbf{R}) = \frac{1}{(2\pi N\langle l^2\rangle/3)^{3/2}} \exp\left(\frac{-3|\mathbf{R}|^2}{2N\langle l^2\rangle}\right)$$

where

$$\langle l^2\rangle = \frac{1}{N}\sum_{j=1}^{N} l_j^2$$

14–22. Derive Eq. (14–95).

14–23. Rederive the results of Section 14–5 for a random walk, that is, a random fligh· constrained to lie in a plane.

14–24. Prove Eq. (14–90); that is, prove

$$\left(\frac{\sin x}{x}\right)^N \approx \exp\left(\frac{-Nx^2}{6}\right)$$

for large N.

14–25. Using Markov's method (Section 14–5), show that the distribution of the distance of the ith segment from the center of mass of the chain is given by

$$P(s_i) = \left(\frac{\phi}{\pi}\right)^{3/2} \exp(-\phi s_i^2)$$

where

$$\phi = \frac{9N^2}{\{2l^2[i^3 + (N-i)^3]\}}$$

Show that $\langle s_i^2\rangle = Nl^2[u^3 + (1-u)^3]/3$, where $u = i/N$. (See Problem 14–7.)

14–26. Use Markov's method to solve the problem of a random flight with a bias in one direction, say the z-direction. In particular, let τ_j be given by

$$\tau_j = \frac{1}{(2\pi l_x^2/3)(2\pi l_z^2/3)^{1/2}} \exp\left[-\frac{3}{2}\left(\frac{x_j^2}{l_x^2} + \frac{y_i^2}{l_x^2} + \frac{z_j^2}{l_z^2}\right)\right]$$

This corresponds to a random flight with a drift in the z-direction.

14–27. Show that

$$W(\mathbf{r}_1, \mathbf{r}_j, \mathbf{r}_k) = W(\mathbf{r}_1, \mathbf{r}_j)W(\mathbf{r}_j, \mathbf{r}_k)$$

is not normalized.

14–28. Show that the probability of a given position of the ith segment of a chain having ends fixed at $x_1 = y_1 = z_1 = 0$ and $x_N = y_N = 0$, $z_N = z$ is

$$W(\mathbf{r}_k; z) = \left(\frac{3N}{2\pi k(N-k)l^2}\right)^{3/2} \times \exp\left\{-\frac{3N}{2l^2(N-k)}\left[x_k^2 + y_k^2 + \left(z_k - \frac{kz}{N}\right)^2\right]\right\}$$

and show that

$$\bar{x}_k = \bar{y}_k = 0 \quad \text{and} \quad \bar{z}_k = \frac{kz}{N}$$

Also show that the mean-square distance of the kth atom from the z-axis, that is, the line connecting the two end points, is

$$\overline{r_k^2} = \overline{x_k^2} + \overline{y_k^2} = \frac{2}{3} l^2 \left(\frac{N-k}{N}\right) k$$

14–29. Show that the distribution function for r_k defined in Problem 14–28 is

$$P(r_k) = \frac{3N}{k(N-k)l^2} \exp\left[-\frac{3Nr_k^2}{2k(N-k)l^2}\right]$$

From this show that the maximum value of $\overline{r_k^2} = Nl^2/6$. This is another measure of the size or extent of a freely jointed chain.

14–30. Equation (14–88) for $W_N(\mathbf{R})$ is also the distribution function for the jth segment from the ith segment, call it $\rho_j^i(r)$. Thus we can write

$$\rho_j^i(r) = (2\pi)^{-3} \int \left(\frac{\sin \rho l}{\rho l}\right)^{|i-j|} \exp(-i\boldsymbol{\rho} \cdot \mathbf{r})\, d\boldsymbol{\rho}$$

Substitute this into the expression for $P(\theta)$ given in Problem 14–31 and derive

$$P(\theta) = \frac{1}{N^2} \sum_{i,j} \left[\frac{\sin(sl)}{sl}\right]^{|i-j|}$$

Show that

$$\sum_{i,j}^{N} x^{|i-j|} = N + \frac{2x[(N-1)(1-x) - x(1-x^{N-1})]}{(1-x)^2}$$

Using this result, show that*

$$P(\theta) = N^{-1} + 2\Phi^{-2}\left[1 - \frac{\Phi}{N}\right]\left\{\Phi - 1 + \left[1 - \frac{\Phi}{N}\right]^N\right\}$$

where

$$\Phi = (1 - S(s))N$$

* See A. Isihara, *J. Chem. Phys.*, **40**, p. 1137, 1964.

and

$$S(s) = \frac{\sin (sl)}{sl}$$

Lastly show that if Φ remains finite as $N \to \infty$, then $P(\theta)$ becomes

$$P(\theta) = \left(\frac{2}{\Phi^2}\right)(\Phi - 1 + e^{-\Phi})$$

Compare this to the Debye formula.

14–31. The scattering function $P(\theta)$ is often expressed in the form

$$P(\theta) = \int \rho(r) \exp(i\mathbf{s} \cdot \mathbf{r})\, d\mathbf{r}$$

where $\mathbf{s} = s(\mathbf{u} - \mathbf{u}_0)$ (*cf.* Section 14–3), $\mathbf{r} = \mathbf{r}_i - \mathbf{r}_j$, and $\rho(r)$ is the pair segment distribution function. Show that this is equivalent to Eq. (14–47) and evaluate $P(\theta)$ for a freely jointed chain.

14–32. Debye has shown* that

$$P^{1/2}(\theta) = N^{-1} \int w(r) \exp(i\mathbf{s} \cdot \mathbf{r})\, d\mathbf{r}$$

where $w(r)$ is the distribution function of segments about the center of mass. Using the result of Problem 14–25, show that

$$P^{1/2}(\theta) = \left(\frac{\pi}{x}\right)^{1/2} e^{-x/12} \operatorname{erf}\left(\frac{x^{1/2}}{2}\right)$$

where

$$x = \frac{8\pi^2}{3} \frac{Nl^2}{\lambda^2} \sin^2 \frac{\theta}{2}$$

and

$$\operatorname{erf}(z) = 2\pi^{-1/2} \int_0^z e^{-t^2}\, dt$$

And show that

$$P(\theta) \to 1 - \frac{x}{3} + 0(x^2)$$

for small x.

14–33. Consider a freely jointed chain with one end fixed at the origin O and the other confined to some small volume $\Delta\tau$ located a distance r from O. The number of configurations of the chain that are consistent with the fixed ends is proportional to $\exp(-3r^2/2Nl^2)\Delta\tau$. The energy of each of these configurations is zero, and so an ensemble of such chains can be treated as a microcanonical ensemble. Show that the entropy is given by

$$S = \text{const} - \frac{3k(x^2 + y^2 + z^2)}{2Nl^2}$$

Now consider a network of chains in the form of a rectangular parallelepiped on lengths a, b, and c. Deform this parallelepiped such that $a \to \lambda_1 a$, $b \to \lambda_2 b$, and $c \to \lambda_3 c$. Making the severe assumption that the components of the end-to-end distance of each and every chain transforms in the same manner as the bulk, show that the entropy change of any chain is

$$\Delta S = -3k[(\lambda_1^2 - 1)x^2 + (\lambda_2^2 - 1)y^2 + (\lambda_3^2 - 1)z^2]/2Nl^2$$

* Compare A. Isihara, *J. Chem. Phys.*, **40**, p. 1137, 1964.

Average this over all configurations and all chains in the network to derive

$$\Delta S = -\frac{Nk}{2}(\lambda_1^2 + \lambda_2^2 + \lambda_3^2 - 3)$$

where N is the number of segments between junctions. Using the fact that the volume of an elastomer does not change appreciably upon stretching, that is, $\lambda_1\lambda_2\lambda_3 = 1$, and the symmetry argument that $\lambda_2 = \lambda_3$ for elongation in the x-direction show that

$$f = NkT\left(\lambda - \frac{1}{\lambda^2}\right)$$

where $\lambda = \lambda_1$.

14–34. Verify Eq. (14–46).

14–35. Prove Eq. (14–80).

CHAPTER 15

IMPERFECT GASES

In the limit of low densities, all gases approach perfect-gas behavior, or in other words, they obey the well-known equation of state

$$p = \rho kT \tag{15–1}$$

This equation was derived in Chapter 5 for a monatomic gas in which the intermolecular potential could be ignored. Physically, this means that the particles spend most of their time far away from each other and so do not "feel" any intermolecular potential. To see again how Eq. (15–1) arises from statistical thermodynamics, consider the classical canonical partition function of N monatomic particles contained in a volume V at temperature T:

$$Q = \frac{1}{N!\,h^{3N}} \int \cdots \int e^{-\beta H}\, d\mathbf{p}_1 \cdots d\mathbf{p}_N\, d\mathbf{r}_1 \cdots d\mathbf{r}_N \tag{15–2}$$

Since H is of the form

$$H = \frac{1}{2m} \sum_{n=1}^{N} (p_{xn}^{\ 2} + p_{yn}^{\ 2} + p_{zn}^{\ 2}) + U(x_1, y_1, \ldots, z_N)$$

we can immediately integrate over the momenta to get

$$Q = \frac{1}{N!} \left(\frac{2\pi mkT}{h^2} \right)^{3N/2} Z_N \tag{15–3}$$

where Z_N is the configuration integral

$$Z_N = \int \cdots \int e^{-U_N/kT}\, d\mathbf{r}_1\, d\mathbf{r}_2 \cdots d\mathbf{r}_N \tag{15–4}$$

If we can neglect U_N in the configuration integral, then $Z_N = V^N$ and $Q = q^N/N!$ where $q(V, T) = (2\pi mkT/h^2)^{3/2} V$. The key point is that it is q being of the form $f(T)V$ that leads directly to the ideal gas equation of state. Although we have discussed only the case of monatomic gases here, the same result holds true for polyatomic gases. This is easily proved at the expense of introducing a number of angular integrations (*cf.* Chapters 6 and 8).

As the density of a gas is increased, the particles are closer on the average, and the intermolecular potential becomes nonnegligible. Thus the configuration integral is no longer simply V^N, and the ideal gas equation of state is not obtained as the equation of state of the gas. Of course, it is well known experimentally that real gases exhibit deviations from ideal gas behavior as the density is increased. A large number of empirical and semiempirical equations of state have been constructed to describe the deviations from the simple ideal gas law. The most fundamental of these, in the sense that it has the most sound theoretical foundation, is the so-called virial equation of state, originally proposed by Thiesen and developed by Kamerlingh-Onnes. The virial equation of state expresses the deviations from ideal behavior as an infinite power series in ρ:

$$\frac{p}{kT} = \rho + B_2(T)\rho^2 + B_3(T)\rho^3 + \cdots \tag{15-5}$$

The quantities $B_2(T)$, $B_3(T)$, ... are called the second, third, ... virial coefficients, respectively, and depend only on the temperature and on the particular gas under consideration, but are independent of density or pressure. The primary goal of this chapter is to derive expressions for B_2, B_3, and so on, in terms of intermolecular potentials.

We shall show in Section 15–1 that the jth virial coefficient can be calculated in terms of the interactions of j molecules in a volume V. This is proved most readily by means of the grand canonical partition function. Thus we shall show that the N-body problem of an imperfect gas can be reduced to a series of one-body, two-body, three-body problems, and so on. The initial deviations from ideality (up to 10 atm, say) rest in $B_2(T)$, which we shall see is easy to calculate since it involves only two-body interactions. The derivation presented in Section 15–1 is valid for any one-component gas, including polyatomic quantum-mechanical gases whose intermolecular forces are not pair-wise additive. Then in Section 15–2 we shall specialize these results to a classical monatomic gas whose intermolecular potential is pairwise additive. The next two sections are somewhat detailed discussions of the experimental and theoretical second and third virial coefficients. Section 15–5 is devoted to the calculation of virial coefficients higher than the third; Section 15–6 discusses quantum corrections to the second virial coefficient; Section 15–7 discusses the law of corresponding states; and lastly, Section 15–8 is a general discussion.

Before going on, however, it is helpful to consider Table 15–1, which gives the contribution of the first few terms of the virial expansion to $p/\rho kT$. The data are for argon at 25°C. The contributions of all the remaining terms are shown in the parentheses. It can be seen that the second and third virial coefficients alone give most of $p/\rho kT$ up to pressures approaching 100 atm.

Table 15–1. **The contribution of the first few terms in the virial expansion of $p/\rho kT$ for argon at 25°C**

p(atm)	$p/\rho kT$
	$1 + B_2\rho \quad + B_3\rho^2 \quad + \quad$ remainder
1	$1 - 0.00064 + 0.00000 + \cdots (+0.00000)$
10	$1 - 0.00648 + 0.00020 + \cdots (-0.00007)$
100	$1 - 0.06754 + 0.02127 + \cdots (-0.00036)$
1000	$1 - 0.38404 + 0.68788 + \cdots (+0.37232)$

Source: E. A. Mason and T. H. Spurling, *The Virial Equation of State* (New York: Pergamon, 1969).

15–1 THE VIRIAL EQUATION OF STATE FROM THE GRAND PARTITION FUNCTION

The grand partition function is [*cf.* Eq. (3–15)]

$$\Xi(V, T, \mu) = \sum_{N=0}^{\infty} Q(N, V, T)\lambda^N \tag{15–6}$$

where $\lambda = \exp(\beta\mu)$. When $N = 0$, the system has only one state with $E = 0$, and so $Q(N = 0, V, T) = 1$. This allows us to write Eq. (15–6) as

$$\Xi(V, T, \mu) = 1 + \sum_{N=1}^{\infty} Q_N(V, T)\lambda^N \tag{15–7}$$

where we have written $Q_N(V, T)$ for $Q(N, V, T)$. The characteristic thermodynamic function associated with Ξ is pV according to the relation

$$pV = kT \ln \Xi \tag{15–8}$$

The average number of molecules in the system is given by [*cf.* Eq. (3–33)]

$$N = kT\left(\frac{\partial \ln \Xi}{\partial \mu}\right)_{V,T} = \lambda\left(\frac{\partial \ln \Xi}{\partial \lambda}\right)_{V,T} \tag{15–9}$$

Thus we have the pressure and essentially the density in terms of Ξ. The standard procedure to eliminate Ξ between these two quantities is to obtain a power series for $\ln \Xi$ in some convenient parameter and then to eliminate this parameter between Eqs. (15–8) and (15–9). The most obvious choice for this expansion parameter is λ, since we already have Ξ as a power series in λ in Eq. (15–7). It is more convenient, however, though not at all necessary, to define a new activity z, proportional to λ, such that $z \to \rho$ as $\rho \to 0$. By taking the limit $\lambda \to 0$ in Eq. (15–9), we find that

$$N = \lambda\left(\frac{\partial \ln \Xi}{\partial \lambda}\right)_{V,T} = \lambda Q_1 \qquad (\lambda \to 0)$$

Thus as $\lambda \to 0$, the density $\rho \to \lambda Q_1/V$, and so we set $z = \lambda Q_1/V$. In terms of this new activity, then

$$\Xi(V, T, \mu) = 1 + \sum_{N=1}^{\infty} \left(\frac{Q_N V^N}{Q_1{}^N}\right) z^N \tag{15–10}$$

It is convenient to define a quantity Z_N by

$$Z_N = N!\left(\frac{V}{Q_1}\right)^N Q_N \tag{15–11}$$

It will turn out that the classical limit of the Z_N defined here is just the configuration integral given in Eq. (15–4). With these definitions, Eq. (15–7) becomes

$$\Xi = 1 + \sum_{N=1}^{\infty} \frac{Z_N(V, T)}{N!} z^N \tag{15–12}$$

This gives us Ξ as a power series in z.

We now *assume* that the pressure can be expanded in powers of z according to

$$p = kT \sum_{j=1}^{\infty} b_j z^j \tag{15–13}$$

where we wish now to determine the unknown coefficients b_j in terms of the Z_N of Eq. (15–11). This can be done directly by substituting Eq. (15–13) into $\Xi = \exp(pV/kT)$, expanding the exponential, collecting like powers of z, equating the coefficients to those of Eq. (15–12), and finally solving for the b_j in terms of the Z_N. The result of this straightforward algebra is (see Problem 15–5)

$$\begin{aligned}
b_1 &= (1!\,V)^{-1}Z_1 = 1 \\
b_2 &= (2!\,V)^{-1}(Z_2 - Z_1{}^2) \\
b_3 &= (3!\,V)^{-1}(Z_3 - 3Z_2Z_1 + 2Z_1{}^3) \\
b_4 &= (4!\,V)^{-1}(Z_4 - 4Z_3Z_1 - 3Z_2{}^2 + 12Z_2Z_1{}^2 - 6Z_1{}^4) \\
&\cdots
\end{aligned} \tag{15–14}$$

It is possible to write down a general formula for b_j in terms of Z's, but we shall not need it. We now know the b_j in Eq. (15–13) in terms of the Z_N. Notice that the calculation of b_2, for example, involves the calculation of only Z_2 and Z_1, that is, essentially the partition functions of two and one particle, respectively. Similarly, b_3 involves the determination of a partition function for three particles at the most. Thus we see that we have reduced the original N-body problem to a series of few-body problems. This was accomplished by using the grand partition function.

We are not finished yet, however, since we really want an expansion of the pressure in terms of the density ρ and not some activity z. But we not only have the pressure in terms of z, we also have the density since the pressure and density are connected by $\ln \Xi$ through Eqs. (15–8) and (15–9). Thus we can write

$$\begin{aligned}
\rho = \frac{N}{V} &= \frac{\lambda}{V}\left(\frac{\partial \ln \Xi}{\partial \lambda}\right)_{V,T} = \frac{z}{V}\left(\frac{\partial \ln \Xi}{\partial z}\right)_{V,T} \\
&= \frac{z}{kT}\left(\frac{\partial p}{\partial z}\right)_{V,T}
\end{aligned}$$

from which we have

$$\rho = \sum_{j=1}^{\infty} j b_j z^j \tag{15–15}$$

Now we have both p and ρ as power series in the activity z. The problem is to eliminate z between the two equations. There is a general mathematical technique to accomplish this (by means of complex variable theory), but we can determine the first few terms in a direct algebraic way (*cf.* Section 10–1). We simply write

$$z = a_1\rho + a_2\rho^2 + a_3\rho^3 + \cdots$$

Substitute this into Eq. (15–15) and equate like powers of ρ on the two sides of the equation to get

$$\begin{aligned}
a_1 &= 1 \\
a_2 &= -2b_2 \\
a_3 &= -3b_3 + 8b_2{}^2 \\
&\cdots
\end{aligned}$$

Now that we have z as a power series in ρ. We substitute this into the equation for p as a power series in z [Eq. (15–13)] to get the desired final result, namely, p as a power series in the density:

$$\frac{p}{kT} = \rho + B_2(T)\rho^2 + B_3(T)\rho^3 + \cdots$$

where

$$B_2(T) = -b_2 = -(2!\,V)^{-1}(Z_2 - Z_1^{\,2}) \tag{15–16}$$

$$B_3(T) = 4b_2^{\,2} - 2b_3$$

$$= -\frac{1}{3V^2}\,[V(Z_3 - 3Z_2Z_1 + 2Z_1^{\,3}) - 3(Z_2 - Z_1^{\,2})^2] \tag{15–17}$$

$$\cdots$$

The equations become increasingly complicated as one goes to higher virial coefficients, but Table 15–1 shows that the first few virial coefficients suffice to account for p–V–T data up to hundreds of atmospheres.

In the next section we shall apply these general equations to the important case of a classical monatomic gas. Before doing this, however, several comments are in order. We have expressed the virial coefficients in terms of the b_j's which, in turn, are given in terms of the Z_N. The b_j's then appear to be simply intermediate quantities with no physical significance. They do have an interesting physical significance, however, that is lost in the grand canonical ensemble derivation presented here. The first statistical mechanical development of the virial expansion is due to Mayer, who used the canonical ensemble as his starting point. In this approach, the b_j's turn out to be related in a way to clusters of j molecules and play a more central role in the development of the final equations. (See Mayer and Mayer in "Additional Reading.")

Another point involves the expansion of p or $\ln \Xi$ in powers of z. We have tacitly assumed that such an expansion exists in the first place. This assumption is correct for all the cases we shall discuss in this book, but it is not obvious that it is generally valid. It is possible that special cases of highly degenerate or strongly interacting systems do not allow such a power series expansion. For example, we shall see that the intermolecular potential between a pair of particles must go to zero more rapidly than r^{-3} in order that the second virial coefficient be finite. (See Problem 15–6.) Thus we see that a virial expansion for a fully ionized gas, that is, a plasma, is invalid.

In summary then, if the virial expansion exists, we have a method for calculating the virial coefficients from partition functions. Furthermore, this series can be shown to converge in some nonzero region for a large class of intermolecular potentials. At one time there was much effort to relate this radius of convergence to the point at which condensation sets in, but it is now felt unlikely that there is any such relation.

15–2 VIRIAL COEFFICIENTS IN THE CLASSICAL LIMIT

For simplicity we shall consider only monatomic gases in this section. The extension to include molecules with internal degrees of freedom is straightforward, but the equations are more complicated. (See Problems 15–29 through 15–32.) In the classical limit, $Q(N, V, T)$ is given by Eq. (15–2). In particular, for $N = 1$, in the absence of external forces we have $U = 0$, and so we find that

$$Q_1(V, T) = \left(\frac{2\pi mkT}{h^2}\right)^{3/2} V = \frac{V}{\Lambda^3} \tag{15–18}$$

For $N > 1$ we can still integrate over the momenta as we did to get Eq. (15–3), giving

$$Q_N = \frac{Z_N}{N!\Lambda^{3N}}$$

where Z_N is the configuration integral given by Eq. (15–4). But according to Eq. (15–18), we can write this as

$$Q_N = \frac{1}{N!}\left(\frac{Q_1}{V}\right)^N Z_N \tag{15–19}$$

Comparing this to Eq. (15–11), we see that the quantity Z_N defined there is just the configuration integral. This, of course, is why we used the symbol Z_N for both quantities.

In order to calculate the second and third virial coefficients then, we need

$$Z_1 = \int d\mathbf{r}_1 = V \tag{15–20}$$

$$Z_2 = \iint e^{-U_2/kT}\, d\mathbf{r}_1\, d\mathbf{r}_2 \tag{15–21}$$

and

$$Z_3 = \iiint e^{-U_3/kT}\, d\mathbf{r}_1\, d\mathbf{r}_2\, d\mathbf{r}_3 \tag{15–22}$$

To calculate the second virial coefficient, we need U_2. For monatomic particles it is reasonable to assume that $U_2(\mathbf{r}_1, \mathbf{r}_2)$ depends only upon the separation of the two particles, so that $U_2 = u(r_{12})$, where $r_{12} = |\mathbf{r}_2 - \mathbf{r}_1|$. The intermolecular potential between two particles can be calculated, at least in principle, from quantum mechanics. We shall discuss this in Section 15–3. Figure 15–1(a) shows a typical intermolecular potential between two spherically symmetric molecules. We can derive an equation for $B_2(T)$ in terms of $u(r_{12})$ by substituting Z_2 and Z_1 into Eq. (15–16) to get

$$B_2(T) = -\frac{1}{2V}(Z_2 - Z_1^{\,2})$$

$$= -\frac{1}{2V}\iint [e^{-\beta u(r_{12})} - 1]\, d\mathbf{r}_1\, d\mathbf{r}_2 \tag{15–23}$$

For neutral molecules in their ground electronic states, $u(r_{12})$ goes to zero fairly rapidly, say in a few molecular diameters, and so the integrand in B_2 is zero unless the volume elements $d\mathbf{r}_1$ and $d\mathbf{r}_2$ are near each other. Thus it is possible to change the variables of integration to $\mathbf{r}_1$ and the relative coordinates $\mathbf{r}_{12} = \mathbf{r}_2 - \mathbf{r}_1$ and write

$$B_2(T) = -\frac{1}{2V}\int d\mathbf{r}_1 \int [e^{-\beta u(r_{12})} - 1]\, d\mathbf{r}_{12} \tag{15–24}$$

Now the integration over the relative separation of the two particles is independent of where the pair is in the volume V, except for when the pair of particles is near the

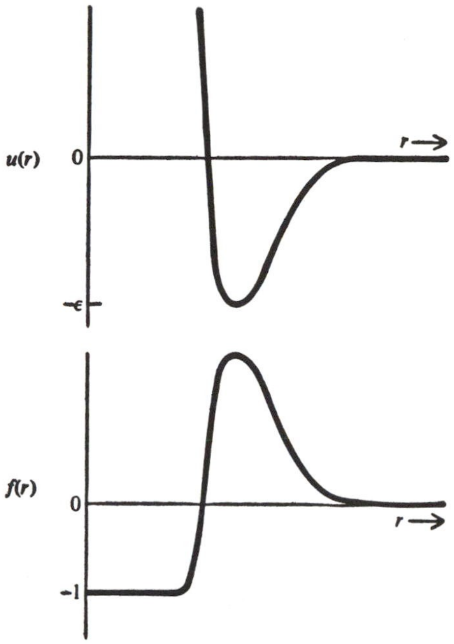

Figure 15–1. **(a) A typical intermolecular potential $u(r)$ and (b) a Mayer f-function $e^{-\beta u(r)} - 1$ plotted versus the intermolecular separation r.**

walls of the container. But thermodynamically we are interested only in the case that $V \to \infty$, and so this "surface effect" becomes entirely negligible. Thus we can carry out the integration over $d\mathbf{r}_1$ separately to give a factor of V and also write $4\pi r^2\,dr$ for $d\mathbf{r}_{12}$ to obtain the final result

$$B_2(T) = -2\pi \int_0^\infty [e^{-\beta u(r)} - 1]r^2\,dr \tag{15–25}$$

This gives the second virial coefficient as a simple quadrature of $u(r)$. We shall give a number of applications of this formula in Section 15–3. The limit of integration has been formally extended to infinity rather than just the walls of the container, since the integrand falls essentially to zero for larger than a few molecular diameters. Note that B_2 is independent of V.

The term in brackets in the integrand of Eq. (15–25) appears throughout the equations of imperfect gas theory and is commonly denoted by $f(r)$,

$$f_{ij} = f(r_{ij}) = e^{-u(r_{ij})/kT} - 1 \tag{15–26}$$

Since $u(r) \to 0$ as r increases, we see that $f(r) \to 0$ as r increases. The importance of the function $f(r)$ was exploited by Mayer and is now called a Mayer f-function. Figure 15–1(b) shows a typical $f(r)$ plotted versus r.

To obtain the third virial coefficient, we need the potential $U_3(\mathbf{r}_1, \mathbf{r}_2, \mathbf{r}_3)$. Here is where the question of the pairwise additivity of intermolecular forces first arises. For many years it was common to assume that the intermolecular potential of a group of three molecules was the sum of the potentials of the three pairs taken one at a time, that is, to assume that

$$U_3(\mathbf{r}_1, \mathbf{r}_2, \mathbf{r}_3) \approx u(r_{12}) + u(r_{13}) + u(r_{23}) \tag{15–27}$$

This apparently can be used as a good approximation, but eventually suffers from close scrutiny. In general, we should write

$$U_3 = u(r_{12}) + u(r_{13}) + u(r_{23}) + \Delta(r_{12}, r_{13}, r_{23})$$

or in an obvious notation which we shall use often

$$U_3 = u_{12} + u_{13} + u_{23} + \Delta_3 \tag{15–28}$$

where Δ represents the deviation from pairwise additivity. We shall neglect Δ in what we do below, but Problem 15–26 involves the generalization to include a non-additive contribution to U_3. We shall now derive an expression for $B_3(T)$ under the assumption (apparently mild) of pairwise additivity. This derivation involves manipulations that may at first be unfamiliar, but are actually quite simple and, furthermore, are common in the statistical mechanical theories of gases and liquids.

In order to calculate B_3, we must calculate b_3, which is given by $6Vb_3 = Z_3 - 3Z_2Z_1 + 2Z_1^3$ [*cf.* Eq. (15–14)]. For Z_3 we have

$$Z_3 = \iiint (1 + f_{12})(1 + f_{13})(1 + f_{23})\, d\mathbf{r}_1\, d\mathbf{r}_2\, d\mathbf{r}_3$$

$$= \iiint [f_{12}f_{13}f_{23} + f_{12}f_{13} + f_{12}f_{23} + f_{13}f_{23} + f_{12} + f_{13} + f_{23} + 1]\, d\mathbf{r}_1\, d\mathbf{r}_2\, d\mathbf{r}_3 \tag{15–29}$$

The next step is to subtract $3Z_2Z_1$ from this. Since $Z_1 = V$, we write

$$Z_1Z_2 = V \iint (f_{12} + 1)\, d\mathbf{r}_1\, d\mathbf{r}_2 = \iiint (f_{12} + 1)\, d\mathbf{r}_1\, d\mathbf{r}_2\, d\mathbf{r}_3 \tag{15–30}$$

where we have taken the volume in under the integral sign in order to obtain $d\mathbf{r}_1\, d\mathbf{r}_2\, d\mathbf{r}_3$ as in Z_3 above. Rather than to just subtract three times Eq. (15–30) from Z_3, however, we recognize that Z_1Z_2 can also be written as

$$Z_1Z_2 = V \iint (f_{13} + 1)\, d\mathbf{r}_1\, d\mathbf{r}_3 = \iiint (f_{13} + 1)\, d\mathbf{r}_1\, d\mathbf{r}_2\, d\mathbf{r}_3$$

or equivalently

$$Z_1Z_2 = V \iint (f_{23} + 1)\, d\mathbf{r}_2\, d\mathbf{r}_3 = \iiint (f_{23} + 1)\, d\mathbf{r}_1\, d\mathbf{r}_2\, d\mathbf{r}_3$$

So instead of subtracting three times any one of these expressions for Z_1Z_2 from Z_3, we subtract each once and get

$$Z_3 - 3Z_1Z_2 = \iiint [f_{12}f_{23}f_{13} + f_{12}f_{13} + f_{12}f_{23} + f_{13}f_{23} - 2]\, d\mathbf{r}_1\, d\mathbf{r}_2\, d\mathbf{r}_3$$

For the $2Z_1^3$ that must be added to this to get $6Vb_3$, we add

$$2 \iiint d\mathbf{r}_1\, d\mathbf{r}_2\, d\mathbf{r}_3$$

to get

$$6Vb_3 = \iiint [f_{12}f_{13}f_{23} + f_{12}f_{13} + f_{12}f_{23} + f_{13}f_{23}]\, d\mathbf{r}_1\, d\mathbf{r}_2\, d\mathbf{r}_3 \tag{15–31}$$

Recall now that $B_3(T) = 4b_2^2 - 2b_3$, or more conveniently (in order to keep the factor of $6V$ with b_3)

$$B_3(T) = -\frac{1}{3V}(6Vb_3 - 12Vb_2^2)$$

The term $6Vb_3$ is given by Eq. (15–31), and so we need only write $12Vb_2^2$ in some convenient form and subtract it from $6Vb_3$. Since

$$b_2 = \frac{1}{2}\int f_{12}\, d\mathbf{r}_{12}$$

we can write

$$4b_2^2 = \left[\int f_{12}\, d\mathbf{r}_{12}\right]^2 = \left[\int f_{12}\, d\mathbf{r}_{12}\right]\left[\int f_{13}\, d\mathbf{r}_{13}\right]$$

or

$$4Vb_2^2 = \int d\mathbf{r}_1 \int f_{12}\, d\mathbf{r}_{12} \int f_{13}\, d\mathbf{r}_{13}$$

$$= \iiint f_{12} f_{13}\, d\mathbf{r}_1\, d\mathbf{r}_2\, d\mathbf{r}_3$$

where the last line is just the inverse of the procedure of converting to relative coordinates. Clearly the subscripts on the f-functions are arbitrary, and we can readily derive two alternative expressions for $4Vb_2^2$, one with $f_{13}f_{23}$ and one with $f_{12}f_{23}$ in the integrands. The three of these make $12Vb_2^2$, the quantity to be subtracted from $6Vb_3$ to give B_3. Thus we have finally

$$B_3(T) = -\frac{1}{3V}\iiint f_{12} f_{13} f_{23}\, d\mathbf{r}_1\, d\mathbf{r}_2\, d\mathbf{r}_3 \tag{15–32}$$

Let us examine the integrand of B_3 more closely. It involves three particles, and since $f_{ij} \to 0$ as particles i and j are separated, the product $f_{12}f_{13}f_{23}$ will vanish unless all the three particles are simultaneously close to one another. We can represent the integrand of B_3 pictorially in the following way. We draw a numbered circle for each different subscript appearing in the product and a line between each pair of particles connected by an f-function. For example, the integrands in the second and third virial coefficients can be represented schematically as shown in Fig. 15–2(a). Thus we see that all three particles in the integrand of B_3 are connected by f-functions. Since this represents a cluster of particles, we call diagrams as in Fig. 15–2 cluster diagrams. The only other cluster diagrams for three particles are shown in Fig. 15–2(b).

There is a general result of imperfect gas theory whose proof is too complicated to present here, but can be readily appreciated in terms of cluster diagrams. This result states that the virial coefficients are given by

$$B_{j+1} = \frac{-j}{j+1}\beta_j \tag{15–33}$$

where

$$\beta_j = \frac{1}{j!\,V}\int \cdots \int S'_{1,2,\ldots,j+1}\, d\mathbf{r}_1\, d\mathbf{r}_2 \cdots d\mathbf{r}_{j+1} \tag{15–34}$$

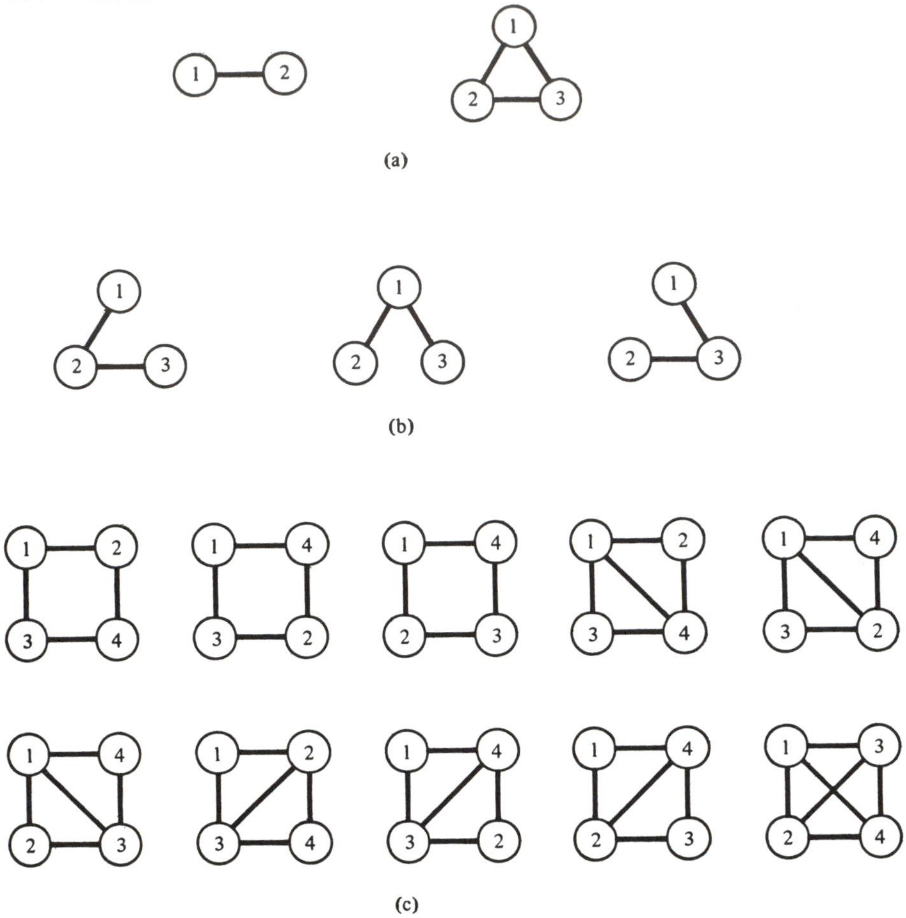

Figure 15–2. **Some examples of cluster diagrams. (a) The integrands of the second and third virial coefficients. (b) The three topologically equivalent diagrams of three particles that are singly connected. (c) The three topologically different stars or doubly connected diagrams for four particles. Compare the degeneracies of these with Eq. (15–35) and Table 15–2.**

where $S'_{1,2,\ldots,j+1}$ is the sum of all products of f-functions that connect molecules $1, 2, \ldots, j+1$ such that the clusters are connected in such a way that the removal of *any* point, together with all of the lines associated with that point, still results in a connected graph, that is, all the particles connected to one another. Notice that all the graphs appearing in Fig. 15–2(a) are connected in this way, while those in Fig. 15–2(b) are not. Such diagrams are also called *doubly connected* since each particle is connected to any other by two independent paths. Doubly connected diagrams are also called *stars*.

All the stars of up to seven particles are listed in Appendix 1 of an article by Hoover and DeRocco.* Table 15–2 lists the number of topologically distinct connected and

Table 15–2. **The number of topologically different connected graphs $C(n)$ and star graphs $S(n)$ for $n \leq 7$**

n:	2	3	4	5	6	7
$C(n)$:	1	2	6	21	112	853
$S(n)$:	1	1	3	10	56	468

Source: W. G. Hoover and A. G. DeRocco, *J. Chem. Phys.*, **36**, p. 3141, 1962.

* W. G. Hoover and A. G. DeRocco, *J. Chem. Phys.*, **36**, p. 3141, 1962.

(p, k)	graph	n
2,1		1
3,2		3
3,3		1
4,3	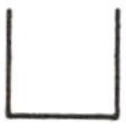	12
		4
4,4		12
		3
4,5		6
4,6		1
5,4		5
		60
		60
5,5		60
		60

(p, k)	graph	n
		30
		60
		12
5,6		60
		60
		15
		60
		10
5,7		20
		10
		60
		30
5,8		30
		15
5,9		10
5,10		1

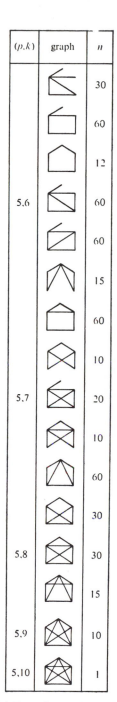

Figure **15–3. The graphs of p points and k lines of up to five particles. n is the number of topologically equivalent graphs of that type.** (From G. E. Uhlenbeck and G. W. Ford, in *Studies in Statistical Mechanics* ed. by J. DeBoer and G. E. Uhlenbeck. New York: North-Holland Publishing Co., 1962.)

star graphs for $n \leq 7$. For example, for $n = 3$ there are two topologically different connected graphs, namely, ∠ and △, and only one star graph, namely, △. Figure 15–2(c) shows the three different types of star diagrams that occur for $n = 4$. The three connected diagrams that are not stars for $n = 4$ are of the form ⊔, ⊼, and ⊠. Note that clusters like ⋁ and ⋉ are really equivalent (topologically) to ⊔. Figure 15–3 gives all the distinct connected graphs and star graphs of up to five particles. Note how Fig. 15–3 is consistent with Table 15–2.

In summary, then, the virial coefficients are integrals over sums of stars, and for $n = 2$, 3, and 4 these sums are given by

$$S'_{1,2} = —$$
$$S'_{1,2,3} = \triangle$$
$$S'_{1,2,3,4} = 3\square + 6\boxslash + \boxtimes \qquad (15\text{–}35)$$

The coefficients in $S'_{1,2,3,4}$ represent the number of equivalent stars of that form [see Fig. 15–2(c)]. We shall now discuss the calculation of $B_2(T)$ and $B_3(T)$ for a number of intermolecular potential functions.

15–3 SECOND VIRIAL COEFFICIENT

Equation (15–25) shows that once the intermolecular potential $u(r)$ is known, the second virial coefficient can be calculated as a function of temperature. Second virial coefficients can be measured experimentally over a large temperature range to within a few percent. Figure 15–4 shows some experimental second virial coefficients versus

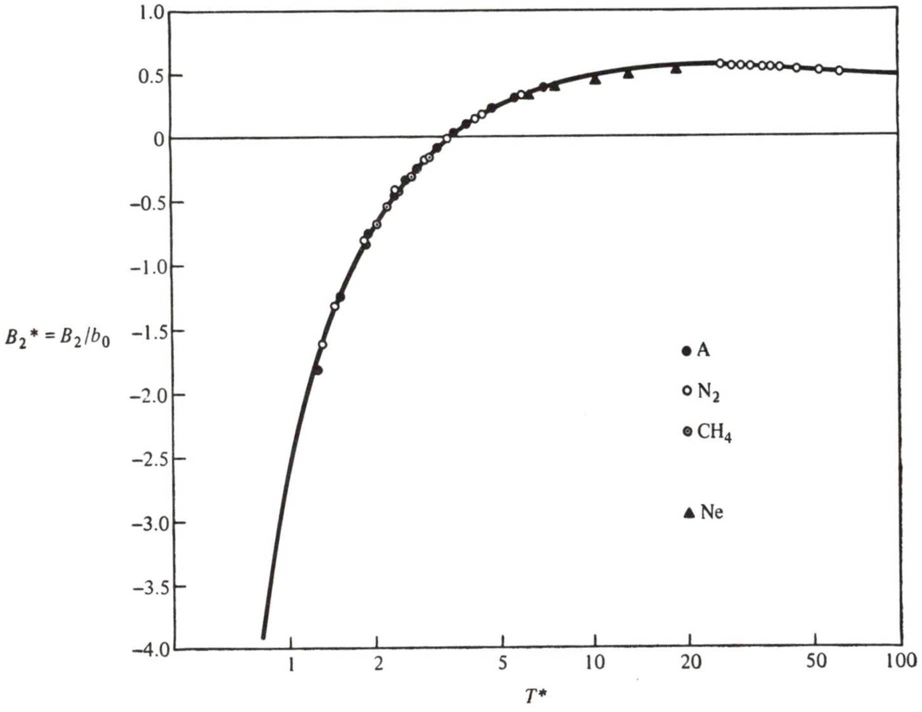

Figure 15–4. **The solid line is the reduced virial coefficient for the Lennard-Jones 6–12 potential as a function of the reduced temperature T^*. Experimental data of a number of substances are also given. (From J. O. Hirschfelder, C. F. Curtiss, and R. B. Bird, *Molecular Theory of Gases and Liquids*. New York: Wiley, 1954.)**

temperature. It is the goal of this section to understand this curve from a theoretical point of view.

In principle, $u(r)$ can be obtained from quantum mechanics, but this is a very difficult numerical problem and has not been done for anything more complicated than H_2.* One can show from perturbation theory, however, that asymptotically

$$u(r) \rightarrow -C_6 r^{-6} \tag{15-36}$$

Usually one uses simple analytical expressions with adjustable parameters for $u(r)$ that go asymptotically as r^{-6}. These adjustable parameters can then be varied to fit experimental data. Perhaps the most well-used form is

$$u(r) = \frac{n\varepsilon}{n-6}\left(\frac{n}{6}\right)^{6/(n-6)}\left\{\left(\frac{\sigma}{r}\right)^{n} - \left(\frac{\sigma}{r}\right)^{6}\right\} \tag{15-37}$$

where σ is the distance at which $u(r) = 0$, and ε is the depth of the well. The exponent n is usually taken to be an integer between 9 and 15, but for historical reasons, 12 is still the most popular value. The r^{-6} is included in Eq. (15–37) so that $u(r)$ has a correct asymptotic form. For $n = 12$, $u(r)$ is called the Lennard-Jones 6-12 potential:

$$u(r) = 4\varepsilon\left\{\left(\frac{\sigma}{r}\right)^{12} - \left(\frac{\sigma}{r}\right)^{6}\right\} \tag{15-38}$$

When plotted versus r, this potential is similar to that in Fig. 15–1(a). There are many other potentials in use nowadays, but they are all similar to this. We shall discuss some of these shortly.

It is not possible to integrate $B_2(T)$ analytically if Eq. (15–38) or any other realistic potential is used, and so before discussing this, however, let us consider some simpler but less realistic forms that have the advantage of allowing $B_2(T)$ to be integrated analytically.

A. HARD-SPHERE POTENTIAL

The hard-sphere potential has the form

$$u(r) = \begin{cases} \infty & r < \sigma \\ 0 & r > \sigma \end{cases} \tag{15-39}$$

This potential has no attractive part, but does simulate the steep repulsive part of realistic potentials. This is the simplest potential used and is the only potential for which the first seven virial coefficients have been calculated. (See, however, Problem 15–46.) It is the potential often used by theorists to try to understand things in a general way (we shall come back to this later). A system of particles with this potential is called a hard-sphere gas or fluid. With the hard-sphere potential, $B_2(T)$ is

$$B_2(T) = -\frac{1}{2}\int_0^{\sigma} (-)4\pi r^2\, dr = \frac{2\pi\sigma^3}{3} \tag{15-40}$$

Note that this is four times the volume of a sphere and also is independent of temperature.

* W. Kolos and L. Wolniewicz, *J. Chem. Phys.*, **41**, p. 3663, 1964.

B. SQUARE-WELL POTENTIAL

An extension of the hard-sphere potential that includes an attractive term and yet is simple enough to handle analytically is the square-well potential:

$$u(r) = \begin{cases} \infty & r < \sigma \\ -\varepsilon & \sigma < r < \lambda\sigma \\ 0 & r > \lambda\sigma \end{cases} \tag{15–41}$$

λ, the range of the attractive well, is usually taken to be between 1.5 and 2.0. If this potential is substituted into Eq. (15–25), one gets

$$B_2(T) = b_0\{1 - (\lambda^3 - 1)(e^{\beta\varepsilon} - 1)\} \tag{15–42}$$

The quantity b_0 is the hard-sphere second virial coefficient $2\pi\sigma^3/3$. Note that as $\lambda \to 1$ or $\varepsilon \to 0$, Eq. (15–42) reduces to the hard-sphere result. Equation (15–42) can be used to fit experimental data very well, at least at ordinary temperatures. It does not pass through a maximum, however. (Why not?) Figure 15–5 shows the calculated and experimental second virial coefficient for argon, and Table 15–3 gives the square-well parameters for a number of molecules.

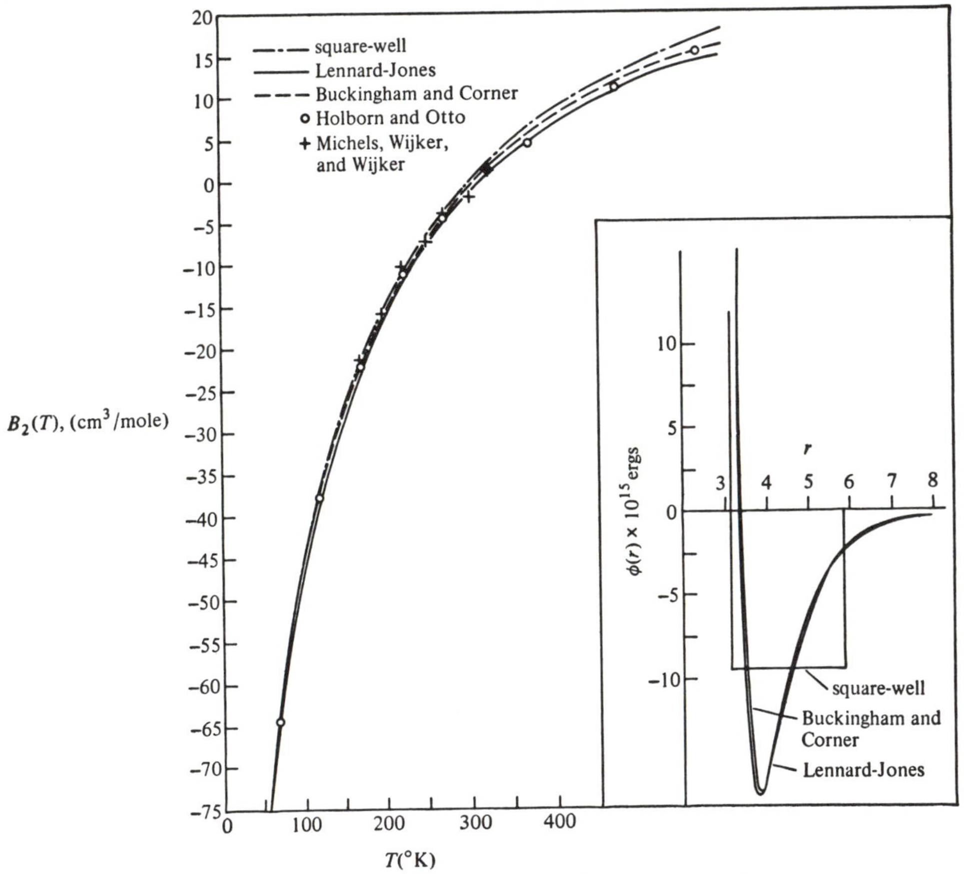

Figure 15–5. **Second virial coefficients for argon calculated for several molecular models. The potential functions obtained from the experimental $B_2(T)$ data are also shown.** (The experimental data are those of L. Holborn and J. Otto, *Z Physik*, 33, 1, 1925, and A. Michels, Hub. Wijker, and Hk. Wijker, *Physica*, **15** 627, 1949, from J. O. Hirschfelder, C. F. Curtiss, and R. B. Bird, *Molecular Theory of Gases and Liquids*, New York: Wiley 1954.)

Table 15–3. **Potential parameters determined from second virial coefficient data**

substance	potential	λ	σ(Å)	ε/k(°K)
argon	sw	1.70	3.067	93.3
	LJ		3.504	117.7
krypton	sw	1.68	3.278	136.5
	LJ		3.827	164.0
methane	sw	1.60	3.355	142.5
	LJ		3.783	148.9
xenon	sw	1.64	3.593	198.5
	LJ		4.099	222.3
tetrafluoromethane	sw	1.48	4.103	191.1
	LJ		4.744	151.5
neopentane	sw	1.45	5.422	382.6
	LJ		7.445	232.5
nitrogen	sw	1.58	3.277	95.2
	LJ		3.745	95.2
carbon dioxide	sw	1.44	3.571	283.6
	LJ		4.328	198.2
n-pentane	sw	1.36	4.668	612.3
	LJ		8.497	219.5
benzene	sw	1.38	4.830	620.4
	LJ		8.569	242.7

Source: A. E. Sherwood and J. M. Prausnitz, *J. Chem. Phys.*, **41**, p. 429, 1964.

C. LENNARD-JONES POTENTIAL*

The hard-sphere and square-well potentials are simple enough to yield analytic expressions for virial coefficients, but presumably do not represent an actual intermolecular potential function. As we mentioned above, the Lennard-Jones potential [Eq. (15–38)] is the most commonly used form that does qualitatively represent the behavior in Fig. 15–1(a). The second virial coefficient for this potential is

$$B_2(T) = -\frac{1}{2}\int_0^\infty \left[\exp\left\{-\frac{4\varepsilon}{kT}\left[\left(\frac{\sigma}{r}\right)^{12} - \left(\frac{\sigma}{r}\right)^6\right]\right\} - 1\right] 4\pi r^2\, dr \tag{15–43}$$

This must be evaluated numerically. Before doing this, however, it is convenient to write Eq. (15–43) in a reduced form by defining a reduced distance, $x = r/\sigma$, and a reduced temperature $T^* = kT/\varepsilon$. If we define a reduced second virial coefficient $B_2^* = B_2/b_0$, then

$$B_2^*(T^*) = -3\int_0^\infty \left[\exp\left\{-\frac{4}{T^*}(x^{-12} - x^{-6})\right\} - 1\right] x^2\, dx \tag{15–44}$$

The reduced second virial coefficient B_2^* is a well-tabulated function. (See Hirschfelder, Curtiss, and Bird in "Additional Reading.")

There are a number of procedures for determining the "best" values of σ and ε from experimental second virial coefficient data. One way is to choose any two temperatures, say T_1 and T_2, and calculate the ratio

$$k_B = \left.\frac{B_2(T_2)}{B_2(T_1)}\right|_{\text{exptl}}$$

and set this equal to

$$k_B = \frac{B_2^*(kT_2/\varepsilon)}{B_2^*(kT_1/\varepsilon)}$$

* Section 3–6 of Hirschfelder, Curtiss, and Bird (in "Additional Reading") has an excellent discussion of virial coefficients and the Lennard-Jones potential.

Then hunt for an ε by trial and error in the table of $B_2{}^*(T^*)$ until the calculated ratio agrees with the experimental ratio. This gives ε. Then calculate σ from

$$\frac{B_2(T_1)}{B_2{}^*(T_1{}^*)} = b_0 = \frac{2\pi\sigma^3}{3}$$

Hopefully, σ and ε will be independent of the choice of T_1 and T_2 and would be if the Lennard-Jones potential were really the "exact" potential. Table 15–3 gives the Lennard-Jones parameters for a number of molecules, and Fig. 15–4 shows the reduced second virial coefficient as a function of T^* for several gases. Figure 15–5 shows B_2 for the Lennard-Jones potential and the square-well potential. Notice that the two curves for B_2 are almost indistinguishable, although the insert shows that the two potentials themselves are quite different. It has often been said that the second virial coefficient is extremely insensitive to the potential used, but there is some indication that, if taken over a sufficiently large temperature range, it can be used to select potential functions.*

D. OTHER POTENTIALS

There are a number of other realistic potentials besides the Lennard-Jones, and B_2 has been tabulated and used for all of them. Fitts' review article† discusses several of these and their comparison with experimental data. One of the most successful of these simply includes an r^{-8} term in the potential given in Eq. (15–37) to give

$$u(r) = \frac{A}{r^n} - \frac{C_6}{r^6} - \frac{C_8}{r^8} \tag{15–45}$$

In terms of the usual parameters of ε, the depth of the potential well, and of σ, the distance at which the potential equals zero, $u(r)$ becomes

$$\frac{u(r)}{\varepsilon} = \frac{(6+2\gamma)}{n-6}\left(\frac{d\sigma}{r}\right)^n - \frac{[n-\gamma(n-8)]}{n-6}\left(\frac{d\sigma}{r}\right)^6 - \gamma\left(\frac{d\sigma}{r}\right)^8 \tag{15–46}$$

where $d = r_m/\sigma$ and $\gamma = C_8/(\varepsilon r_m)^8$ and r_m is the distance at which the potential is a minimum. Since this potential has two more parameters than the Lennard-Jones 6-12 potential, it is able to give much better agreement with experimental data. Nevertheless, it has been subjected to a fairly critical test and has performed quite well.‡ Mason and Spurling discuss a large number of other intermolecular potentials. (See Mason and Spurling under "Additional Reading.")

15–4 THIRD VIRIAL COEFFICIENT

The third virial coefficient can also be measured experimentally, although not as accurately as the second virial coefficient. Equation (15–32) for $B_3(T)$ is

$$\begin{aligned} B_3(T) &= -\frac{1}{3V}\iiint\limits_V f_{12} f_{13} f_{23}\, d\mathbf{r}_1\, d\mathbf{r}_2\, d\mathbf{r}_3 \\ &= -\frac{1}{3}\iint f_{12} f_{13} f_{23}\, d\mathbf{r}_{12}\, d\mathbf{r}_{13} \end{aligned} \tag{15–47}$$

We shall show how to calculate B_3 analytically for the hard-sphere and square-well potentials in the next section. For the Lennard-Jones potential, we can measure all

* A. E. Kingston, *J. Chem. Phys.*, **42**, p. 719, 1965.
† D. D. Fitts, *Ann. Rev. Phys. Chem.*, **17**, p. 59, 1966.
‡ H. J. M. Hanley and M. Klein, *J. Phys. Chem.*, **76**, p. 1743, 1972.

distances in Eq. (15–47) in terms of σ and define a reduced temperature T^*, and so on, to give

$$B_3{}^*(T^*) = \frac{B_3(T)}{b_0{}^2} = \frac{-3}{4\pi^2} \iint f_{12} f_{13} f_{23} \, d\mathbf{r}_{12}{}^* \, d\mathbf{r}_{13}{}^* \tag{15–48}$$

which, of course, must be evaluated numerically. This gives $B_3/b_0{}^2$ as a universal function of kT/ε, and this is shown as a function of T^* in Fig. 15–6. Although the figure caption suggests reasons for the discrepancy between the experimental points and the theoretical curve, the principal reason is probably due to the nonadditivity of the three-body potential.

Third virial coefficients, including a nonadditive contribution, have been calculated for several other potentials by Sherwood and Prausnitz.* They use the square-well, Kihara, and exp-6 potentials and make a very thorough comparison to experimental data. They find the inclusion of the nonadditive part of $U_3(\mathbf{r}_1, \mathbf{r}_2, \mathbf{r}_3)$ gives much better agreement with experimental data. This is shown in Fig. 15–7. Their article provides a readable and extensive study of virial coefficients and intermolecular potentials.

It becomes increasingly difficult to either determine higher virial coefficients experimentally or to calculate them theoretically for any realistic potential. A great deal of work has been done, however, in calculating higher virial coefficients for simple potentials such as the hard-sphere potential, not to compare such results to experiment, but to investigate the structure of the statistical mechanical equations themselves. For this reason we shall discuss some of the hard-sphere calculations of B_3 through B_7. Most of the results in this area are due to Ree and Hoover.†

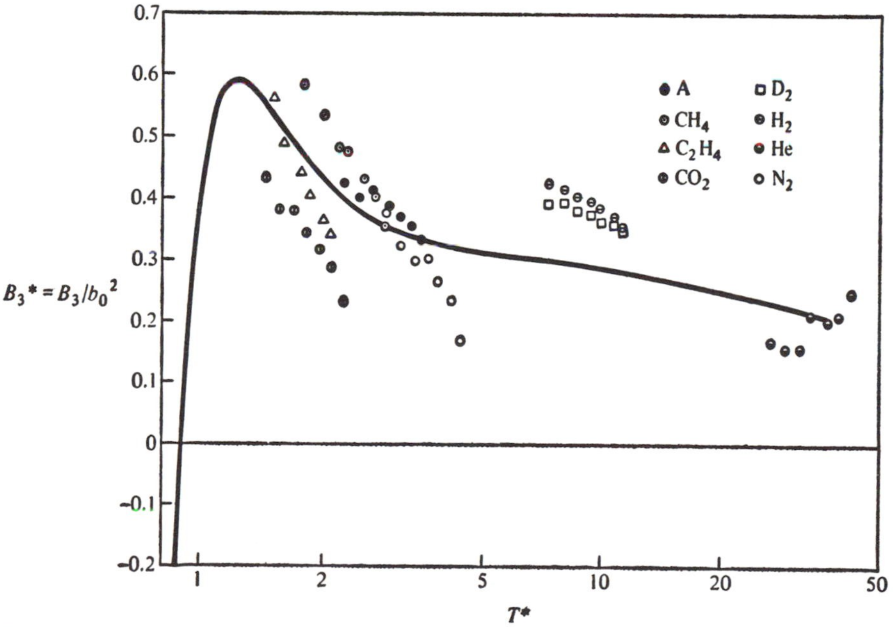

Figure 15–6. **The reduced third virial coefficient for the Lennard-Jones potential. The experimental values have been reduced using values of ε and σ determined by fitting the second virial coefficient to experimental data. The nonspherical molecules (carbon dioxide and ethylene) deviate markedly from the calculated curve. Also the light gases (hydrogen, deuterium and helium) exhibit different behavior because of quantum effects. (From R. B. Bird, E. L. Spotz, and J. O. Hirschfelder, *J. Chem. Phys.* 18, 1395, 1950.)**

* A. E. Sherwood and J. M. Prausnitz, *J. Chem. Phys.*, 41, p. 429, 1964.
† F. H. Ree and W. G. Hoover, *J. Chem. Phys.*, 40, p. 939, 1964.

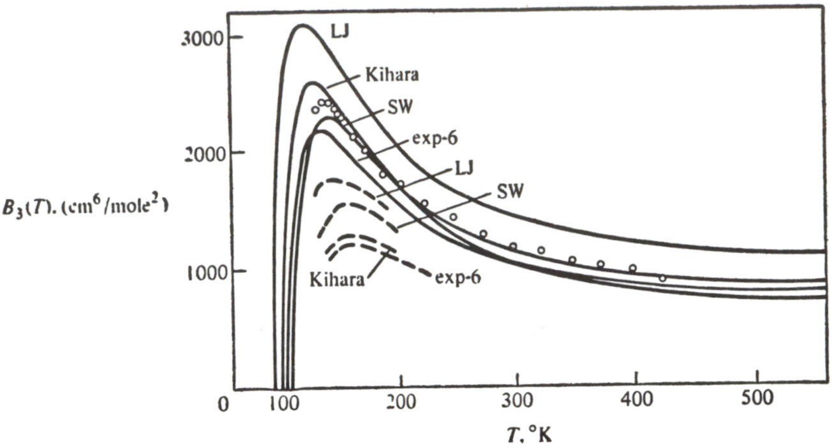

Figure 15–7. **Comparison of observed and calculated values of the third virial coefficient of argon. Solid lines include a nonadditivity correction; dashed lines show a portion of the additive third virial coefficient curve.** (From Sherwood and Prausnitz, *J. Chem. Phys.* **41**, 429, 1964.)

15–5 HIGHER VIRIAL COEFFICIENTS FOR THE HARD-SPHERE POTENTIAL

We have already seen that $B_2 = 2\pi\sigma^3/3$ for the hard-sphere potential. The third virial coefficient can be calculated by a geometrical consideration of the overlapping of three spheres, but this is a laborious and demanding route. We shall calculate B_3 here by a method due to Katsura* using Fourier transforms. Let $\boldsymbol{\rho}_j = \mathbf{r}_j - \mathbf{r}_1$ in Eq. (15–32). Then

$$B_3 = -\frac{1}{3}\iint f(|\boldsymbol{\rho}_2|)f(|\boldsymbol{\rho}_3|)f(|\boldsymbol{\rho}_3 - \boldsymbol{\rho}_2|)\, d\boldsymbol{\rho}_2\, d\boldsymbol{\rho}_3 \tag{15–49}$$

Now let the Fourier transform of $f(|\boldsymbol{\rho}|)$ be (*cf.* Appendix B)

$$\gamma(t) = \gamma(|\mathbf{t}|) = (2\pi)^{-3/2}\int f(|\boldsymbol{\rho}|)e^{-i\mathbf{t}\cdot\boldsymbol{\rho}}\, d\boldsymbol{\rho} \tag{15–50}$$

$$= \left(\frac{2}{\pi}\right)^{1/2}\int_0^\infty \rho f(\rho)\frac{\sin t\rho}{t}\, d\rho \tag{15–51}$$

In Eq. (15–51), ρ is the scalar magnitude of $\boldsymbol{\rho}$. The second expression for $\gamma(t)$ is obtained by calling $\mathbf{t}$ the z-axis of a spherical coordinate system and measuring $\boldsymbol{\rho}$ relative to $\mathbf{t}$. Then we can set $\mathbf{t}\cdot\boldsymbol{\rho} = t\rho\cos\theta$, $d\boldsymbol{\rho} = \rho^2\, d\rho \sin\theta\, d\theta\, d\phi$, and integrate over θ and ϕ (see Problem 15–47). Equation (15–50) can be inverted to give

$$f(\rho) = (2\pi)^{-3/2}\int \gamma(|\mathbf{t}|)e^{i\mathbf{t}\cdot\boldsymbol{\rho}}\, d\mathbf{t}$$

$$= \left(\frac{2}{\pi}\right)^{1/2}\int_0^\infty t\gamma(t)\frac{\sin\rho t}{\rho}\, dt \tag{15–52}$$

For hard spheres,

$$\begin{aligned} f(\rho) &= -1 \qquad 0 < \rho < 1 \\ &= 0 \qquad \rho < 1 \end{aligned} \tag{15–53}$$

* S. Katsura, *Phys. Rev.*, **115**, p. 1417, 1959.

and so according to Eq. (15–51)

$$\gamma(t) = \sigma^3\left(\frac{2}{\pi}\right)^{1/2}\left\{\frac{\cos(\sigma t)}{(\sigma t)^2} - \frac{\sin(\sigma t)}{(\sigma t)^3}\right\}$$
$$= -\sigma^3\frac{J_{3/2}(\sigma t)}{(\sigma t)^{3/2}} \qquad (15\text{–}54)$$

where $J_{3/2}(t)$ is a Bessel function.

The trick now is to realize that Eq. (15–52) shows that

$$f(|\mathbf{\rho}_3 - \mathbf{\rho}_2|) = (2\pi)^{-3/2}\int \gamma(t)e^{i\mathbf{t}\cdot\mathbf{\rho}_3 - i\mathbf{t}\cdot\mathbf{\rho}_2}\,d\mathbf{t} \qquad (15\text{–}55)$$

and substitute this into Eq. (15–49) for B_3 to get

$$B_3 = -\frac{(2\pi)^{-3/2}}{3}\iiint \gamma(t)f(\rho_2)f(\rho_3)e^{i\mathbf{t}\cdot\mathbf{\rho}_3 - i\mathbf{t}\cdot\mathbf{\rho}_2}\,d\mathbf{\rho}_2\,d\mathbf{\rho}_3\,d\mathbf{t}$$
$$= -\frac{(2\pi)^{3/2}}{3}\int d\mathbf{t}\gamma^3(t)$$
$$= \frac{(2\pi)^{3/2}}{3}\sigma^9\int_0^\infty 4\pi t^2\frac{[J_{3/2}(\sigma t)]^3}{(\sigma t)^{9/2}}\,dt$$
$$= \frac{4\pi(2\pi)^{3/2}\sigma^6}{3}\int_0^\infty dx[J_{3/2}(x)]^3x^{-5/2} \qquad (15\text{–}56)$$

where we have converted the variable of integration from t to $x = \sigma t$.

The integral in Eq. (15–56) is not standard, but Katsura* has evaluated many integrals of the form

$$\int_0^\infty x^\lambda J_\alpha(x)J_\beta(x)J_\gamma(x)\,dx$$

of which Eq. (15–56) is a special case. Using Katsura's paper, Eq. (15–56) becomes (see Problem 15–48)

$$B_3 = \frac{4\pi(2\pi)^{3/2}\sigma^6}{3}\cdot\frac{5}{48(2\pi)^{1/2}} = \frac{5\pi^2\sigma^6}{18} = \frac{5}{8}b_0{}^2 \qquad (15\text{–}57)$$

where b_0 is the second hard-sphere virial coefficient.

This is a technique which can be used in many applications involving r_{ij}-type integrands. The integration of the most highly connected stars such as ⊠ cannot be done in this way and are usually done numerically by Monte Carlo methods.†

The fourth virial coefficient B_4 involves the sum of three types of cluster integrals and is given by

$$B_4 = -\frac{1}{8V}\iiiint\{3\square + 6\boxslash + \boxtimes\}\,d\mathbf{r}_1\,d\mathbf{r}_2\,d\mathbf{r}_3\,d\mathbf{r}_4$$
$$= (-0.9714 + 1.4167 - 0.1584)b_0{}^3$$
$$= 0.2869b_0{}^3 \qquad (15\text{–}58)$$

Ree and Hoover‡ have evaluated B_5 through B_7 by a variety of numerical methods.

* S. Katsura, *ibid.*

† G. W. Bern in *Modern Mathematics for the Engineer*, ed. by E. F. Bechenbach (New York: McGraw-Hill, 1956). B. J. Alder and W. G. Hoover, in *Physics of simple liquids*, ed. by H. N. V. Temperly, J. S. Rowlinson, and G. S. Rushbrooke. Amsterdam: North-Holland Publishing Co., 1968.

‡ F. H. Ree and W. G. Hoover, *J. Chem. Phys.*, **46**, p. 4181, 1967.

The number of integrals becomes quite large, there being 468 in the calculation of B_7. They find that

$$\frac{B_5}{b_0^4} = 0.1097 \pm 0.0003$$

$$\frac{B_6}{b_0^5} = 0.0386 \pm 0.0004$$

$$\frac{B_7}{b_0^6} = 0.0138 \pm 0.0004 \tag{15–59}$$

Not even the diagrams have ever been generated for B_8. These calculations were made to study virial coefficient equations themselves and also to determine how much one can learn about the liquid state from the first few virial coefficients. They found, for instance, that using just the truncated seventh degree polynomial in the density, they obtained 10 percent agreement with certain "computer experiments" on hard-sphere liquids. (We shall discuss these "computer experiments" later.) A virial expansion cannot really be expected to describe a liquid though, since the series itself diverges at certain densities and temperatures. Much work has been done examining this series and its convergence. This involves quite a sophisticated mathematical discussion of the β_j for large j, particularly their volume dependence. One of the questions that at one time was investigated is whether the phenomenon of condensation occurs at the density and temperature at which the virial expansion diverges. We shall see in Chapter 19 (*Statistical Mechanics*) that these two are not related to each other.

Up to now we have concentrated on a one-component, classical, monatomic gas. It is straightforward to extend these results to two components (see Problems 15–17 and 15–18) and to polyatomic molecules (see Problems 15–29 through 15–32.) The main difference in the polyatomic-molecule case is that the intermolecular potential now depends upon the relative orientation of the molecules. This is discussed in Sections 3–4 and 3–8 in Hirschfelder, Curtiss, and Bird (see "Additional Reading") We can also extend these classical mechanical formulas to include quantum effects. To calculate the second virial coefficient in a complete quantum-mechanical scheme requires the detailed scattering state wave functions and phase shifts for $u(r)$, and this is very difficult. Chapter 6 of Hirschfelder, Curtiss, and Bird discusses such calculations for several potentials, such as the square well and Lennard-Jones. It is possible, however, to make an expansion in powers of h, and the first few terms not only give a measure of the degree of importance of quantum effects, but are also quite tractable. In the next section we calculate the first quantum correction to our classical $B_2(T)$.

15–6 QUANTUM CORRECTIONS TO $B_2(T)$

Recall Eq. (10–116) for Q_N, namely,

$$Q_N = \sum_m \int \psi_m^* e^{-\beta \mathscr{H}} \psi_m \, d\mathbf{r}_1 \cdots d\mathbf{r}_N$$

In Section 10–7 we showed that Q_N could be expanded in a power series in h, with the leading term being the classical partition function (without the $N!$ since we did not consider the symmetry requirements on the wave functions). We found that

$$Q_N = \frac{1}{h^{3N}} \int \cdots \int \exp(-\beta H) w(\mathbf{p}_1, \ldots, \mathbf{r}_N, \beta) \, d\mathbf{p}_1 \cdots d\mathbf{r}_N$$

where

$$w(\mathbf{p}_1, \ldots, \mathbf{r}_N, \beta) = \sum_{l=0}^{\infty} \hbar^l w_l(\mathbf{p}_1, \ldots, \mathbf{r}_N, \beta)$$

with

$$w_0 = 1,$$

$$w_1 = -\frac{i\beta^2}{2m} \sum_{j=1}^{N} \mathbf{p}_j \cdot \nabla_j U$$

and

$$w_2 = -\frac{1}{2m}\left\{\frac{\beta^2}{2}\sum_{k=1}^{N}\nabla_k^2 U - \frac{\beta^3}{3}\left[\sum_{k=1}^{N}(\nabla_k U)^2 + \frac{1}{m}\left(\sum_{k=1}^{N}\mathbf{p}_k\cdot\nabla_k\right)^2 U\right] + \frac{\beta^4}{4m}\left(\sum_{k=1}^{N}\mathbf{p}_k\cdot\nabla_k U\right)^2\right\}\cdots$$

We can substitute this series into Q_N and integrate over the momenta to get

$$Q_N = \frac{(2\pi mkT)^{3N/2}}{h^{3N}} \iint e^{-\beta U}\left\{1 - \frac{\hbar^2\beta^2}{12m}\sum_{k=1}^{N}\left(\nabla_k^2 U - \frac{\beta}{2}(\nabla_k U)^2\right) + \cdots\right\} d\mathbf{r}_1 \cdots d\mathbf{r}_N \tag{15-60}$$

The contribution from w_1 vanishes upon integration, since it is an odd function of the momenta. We now divide this by $N!$ and substitute it into Eq. (15–11) and then that into Eq. (15–16) to get

$$B_2 = -b_2 = \frac{-1}{2V}(Z_2 - Z_1^2)$$

$$= -2\pi\int_0^\infty [e^{-\beta u(r)} - 1]r^2\,dr + \frac{h^2}{24\pi m(kT)^3}\int_0^\infty e^{-\beta u(r)}\left(\frac{du}{dr}\right)^2 r^2\,dr + O(h^3) \tag{15-61}$$

Table 15–4 shows the magnitude of the various quantum contributions to $B_2(T)$ for several gases at various temperatures. It lists not only the first correction which we have derived above, but also the second, which goes as h^4, and an h^3-term that arises

***Table* 15–4. Contribution of various quantum terms to the second virial coefficient**

gas	T(°K)	B_{class}	h^2-term	h^4-term	ideal quantum gas
He^4	27.3	−4.87	9.16	−4.05	0.50
	83.5	8.87	1.82	−0.19	0.093
	256.0	11.13	0.48	−0.01	0.017
H_2	49.2	−47.1	20.68	−8.63	0.57
	182.8	7.55	2.26	−0.19	0.080
	592.0	15.7	0.49	−0.01	0.014
D_2	37.0	78.94	19.60	−10.21	0.31
	182.8	7.55	1.13	−0.05	0.029
	592.0	15.7	0.25	0	0.004
Ne	35.6	−66.2	3.80	−0.47	0.03
	95.0	−6.23	0.55	−0.01	0.007
	392.0	12.1	0.07	0	0.0008

Source: J. O. Hirschfelder, C. F. Curtiss, and R. B. Bird, *Molecular Theory of Gases and Liquids* (New York: Wiley, 1954).

when the symmetry requirement of the wave functions is taken into account. This h^3-term is just the quantum-mechanical ideal gas second virial coefficient, Eqs. (10–15) and (10–48). You can see from Table 15–4 that this term is not numerically important. Clearly the quantum corrections become less important as the temperature increases. Table 15–4 indicates that the expansion may not be used below about 40°K for helium and 75°K for hydrogen. Of course, it is in the case of light molecules like He and H_2 that quantum effects are most important.

15–7 THE LAW OF CORRESPONDING STATES

In Sections 15–3 and 15–4 we discussed the second and third virial coefficients for several intermolecular potentials. For the Lennard-Jones potential, for example, we found that the virial coefficients could be written in a reduced form such that $B_2(T)/b_0$ or $B_3(T)/b_0^2$ is a function of only the reduced temperature $T^* = kT/\varepsilon$. Furthermore, assuming that all molecules interact through a Lennard-Jones potential (but with different parameters σ and ε), this function of T^* would be the same for all systems. This is illustrated in Fig. 15–4 and is an example of the law of corresponding states. We have seen examples of this law previously in Chapters 11 and 12, where we discussed the Debye theory of crystals and the Lennard-Jones Devonshire theory of liquids. In this section we shall discuss a more general version of the law of corresponding states for classical monatomic systems.

We shall assume that the total intermolecular potential can be written in the form

$$U = \sum_{i,j} u(r_{ij}) = \sum_{i,j} \varepsilon\phi\left(\frac{r_{ij}}{\sigma}\right) \tag{15–62}$$

In particular, we are assuming pairwise additivity and that the pair potential can be written as an energy parameter ε times a function of only the reduced distance r/σ. This pair potential is quite general and need not be a Lennard-Jones potential. However, we do assume that the pair potential ϕ is the same function for all substances. With these assumptions, then, the configuration integral is

$$\begin{aligned} Z_N &= \int \cdots \int e^{-U/kT} d\mathbf{r}_1 \cdots d\mathbf{r}_N \\ &= \sigma^{3N} \int \cdots \int \exp\left\{-\frac{\varepsilon}{kT}\sum_{i,j}\phi\left(\frac{r_{ij}}{\sigma}\right)\right\} d\left(\frac{\mathbf{r}_1}{\sigma^3}\right) \cdots d\left(\frac{\mathbf{r}_N}{\sigma^3}\right) \\ &= \sigma^{3N} f\left(T^*, \frac{V}{\sigma^3}, N\right) \end{aligned}$$

where the function f is the same for all molecules. We have written V/σ^3 as one of the variables of f since all distances have been reduced by σ.

The N dependence of f is restricted by the fact that the Helmholtz free energy A is an extensive thermodynamic quantity. To see this, we use the fact that since A is extensive, A/N is intensive and, consequently, must be a function of only $v = V/N$ and T. In terms of Z_N, A/N is

$$\begin{aligned} \frac{A}{NkT} &= -\frac{1}{N}\ln Q = -\frac{1}{N}\ln\frac{Z_N}{N!\Lambda^{3N}} \\ &= -\frac{1}{N}\ln\frac{Z_N}{N!} + 3\ln\Lambda \end{aligned}$$

Since A/NkT is a function of v and T only, $N^{-1} \ln (Z_N/N!)$ is a function of v and T only, which says that

$$\frac{Z_N}{N!} = \frac{\sigma^{3N}}{N!} f\left(T^*, \frac{V}{\sigma^3}, N\right)$$

$$= \sigma^{3N}\left[g\left(T^*, \frac{v}{\sigma^3}\right)\right]^N \tag{15-63}$$

where g is the same function for all molecules (since f is).

Thus the partition function is of the form

$$Q(N, V, T) = \left[\frac{\sigma^3 g(kT/\varepsilon, v/\sigma^3)}{\Lambda^3}\right]^N \tag{15-64}$$

The pressure is

$$p = kT\left(\frac{\partial \ln Q}{\partial V}\right)_{N,T}$$

$$= \frac{kT}{N\sigma^3}\left(\frac{\partial \ln Q}{\partial(v/\sigma^3)}\right)_{N,T}$$

$$= \frac{kT}{\sigma^3}\left(\frac{\partial \ln g}{\partial(v/\sigma^3)}\right)_{T^*}$$

which shows that

$$\frac{pv}{kT} = \left(\frac{v}{\sigma^3}\right)\left(\frac{\partial \ln g}{\partial(v/\sigma^3)}\right)_{T^*} \tag{15-65}$$

must be the same function of $T^* = kT/\varepsilon$ and v/σ^3 for all substances. Table 15–5 shows the reduced critical parameters for a number of substances.

A look at Table 15–5 shows that the reduced critical constants of Ne, Ar, Kr, Xe, N_2, and CH_4 are quite similar, and hence these substances obey the law of corresponding states as we have formulated it. The other substances do not conform to it for various obvious reasons. For example, we assumed that classical statistics was applicable, and this is not valid for He at 5°K. We also assumed that the intermolecular interactions were spherically symmetric or at least effectively spherically symmetric.

Table 15–5. **The critical constants and reduced critical constants, reduced by means of the Lennard-Jones parameters**

gas	T_c(°K)	v_c(cm^3/mole)	p_c(atm)	T_c^*	$V_c/N\sigma^3$	$p_c v_c/kT_c$
He	5.3	57.8	2.26	0.52	5.75	0.300
Ne	44.5	41.7	25.9	1.25	3.33	0.296
Ar	151	75.3	48.0	1.28	2.90	0.292
Kr	209	91.3	54.3	1.27	2.71	0.289
Xe	290	118.7	58.0	1.30	2.86	0.289
N_2	126	90.0	33.5	1.32	2.84	0.292
CO_2	304	94.0	72.8	1.53	1.93	0.274
CH_4	191	100.0	45.8	1.28	3.07	0.292
n-pentane	470	310.3	33.3	2.14	0.84	0.268
neopentane	434	302.5	31.6	1.87	1.21	0.268
benzene	563	260.0	48.6	2.32	0.68	0.274

Source: J. O. Hirschfelder, C. F. Curtiss, and R. B. Bird, *Molecular Theory of Gases and Liquids* (New York: Wiley, 1954).

Table 15–5 shows that this assumption is not good for molecules like pentane and benzene. Similar disagreement would have been observed if we had included polar molecules. The law of corresponding states can be extended to include polar substances and quantum effects, but we shall not do so here.*

Note that the law of corresponding states provides a useful method for estimating Lennard-Jones parameters. For example, for those substances that do obey a law of corresponding states, T_c^* is about 1.3, and v_c/σ^3 is 2.7. Thus we have

$$\frac{\varepsilon}{k} \approx \tfrac{3}{4} T_c$$

and

$$Nb_0 = \frac{2\pi N\sigma^3}{3} \approx \tfrac{3}{4} v_c \tag{15–66}$$

15–8 CONCLUSION

This almost concludes our discussion of gases. We have been able to derive many both useful and exact results (something which is quite rare) by appealing to the grand canonical ensemble. The area of imperfect gas theory is quite well understood.

Before leaving this, however, let us go back to B_5 for hard spheres and look at its calculation a little more closely. B_5 is a sum of ten types of cluster integrals, whose values are†

$B_5 = -\dfrac{1}{30V}\displaystyle\int \cdots \int \{12$  $+ 60$  $+ 10$  $+ 60$  $+ 30$  $+ 10$ 

$+ 15$  $+ 30$  $+ 10$  $+$  $\}\, d\mathbf{r}_1 \cdots d\mathbf{r}_5$ (15–67)

$$= -\frac{1}{30}(-45.70 + 152.72 + 23.43 - 114.28 - 47.55 - 20.55$$

$$+ 17.15 + 39.93 - 9.17 + 0.73)b_0^4$$

$$= -\frac{1}{30}(-237.25 + 233.96)b_0^4 = \frac{3.29}{30} b_0^4 = 0.11 b_0^4 \tag{15–68}$$

It would appear that if one must add ten difficult-to-calculate terms, each of the order of ten or a hundred, to get 0.11, something is wrong somewhere. Is it possible to calculate 0.11 directly? Maybe our whole approach to gas theory is the wrong way to go about things, in spite of its rigor and physical appeal.

Along this same line, the determination of virial coefficients from experimental p–V–T data is not at all trivial. The usual method is to curve-fit the data to a polynomial, but it is well known that the coefficients depend upon the degree of the polynomial used. One usually uses a greater and greater degree polynomial and waits for the lower coefficients to settle down.‡ A much more satisfactory method is to expand p/kT not in a power series, but in terms of orthogonal polynomials in the density, since then the coefficients are independent of the number of terms used. To

* R. W. Hakala, *J. Chem. Phys.*, **71**, p. 1880, 1967. K. S. Pitzer, *J. Am. Chem. Soc.*, **77**, p. 3427, 1955.
† J. S. Rowlinson, *Proc. Roy. Soc.*, **A279**, p. 147, 1964.
‡ See, for example, K. R. Hall and F. B. Canfield, *Physica*, **33**, p. 481, 1967.

see this, consider the set of orthonormal polynomials $\{\phi_j(\rho)\}$ with weighting function $w(\rho)$.

Then

$$\frac{p}{kT} = \sum_{j=1} C_j(T)\phi_j(\rho)$$

and so

$$C_j(T) = \int \left(\frac{p}{kT}\right)\phi_j(\rho)w(\rho)\, d\rho$$

which is independent of the other $C_j(T)$ and any truncation of the series for the pressure. Is it possible to find a $\{\phi_j(\rho)\}$ and $w(\rho)$ such that the $C_j(T)$ can be written as integrals over physically appealing molecular aggregates and not involve the delicate cancellation that appears in the Mayer theory?

ADDITIONAL READING

General

FOWLER, R. H., and GUGGENHEIM, E. A. 1956. *Statistical thermodynamics.* Cambridge: Cambridge University Press. Chapter 7.

HILL, T. L. 1956. *Statistical mechanics.* New York: McGraw-Hill. Chapter 5.

HIRSCHFELDER, J. O., CURTISS, C. F., and BIRD, R. B. 1954. *Molecular theory of gases and liquids.* New York: Wiley. Chapters 3 and 6.

KAHN, B. 1965. In *Studies in statistical mechanics*, Vol. III, ed. by J. DeBoer and G. E. Uhlenbeck. Amsterdam: North-Holland Publishing Co.

KESTIN, J., and DORFMAN, J. R. 1971. *A course in statistical thermodynamics.* New York: Academic. Chapter 7.

KIHARA, T. *Rev. Mod. Phys.*, **25**, p. 831, 1953.

KILPATRICK, J. E. *Adv. Chem. Phys.*, **20**, p. 39, 1971.

MASON, E. A., and SPURLING, T. H. 1969. *The Virial equation of state.* New York: Pergamon.

MAYER, J. E. 1958. In *Encyclopedia of physics*, Vol. XII, ed. by S. Flügge. Berlin: Springer-Verlag.

MAYER, J. E., and MAYER, M. G. 1940. *Statistical mechanics.* New York: Wiley. Chapter 13.

MÜNSTER, A. 1969. *Statistical thermodynamics*, Vol. I. Berlin: Springer-Verlag. Chapters 8 and 9.

ROWLINSON, J. 1958. In *Encyclopedia of physics*, Vol. XII, ed. by S. Flügge. Berlin: Springer-Verlag.

TER HAAR, D. 1954. *Elements of statistical mechanics.* New York: Rinehard. Chapter 8.

UHLENBECK, G. E., and FORD, G. W. 1962. In *Studies in statistical mechanics*, Vol. I, ed. by J. DeBoer and G. E. Uhlenbeck. Amsterdam: North-Holland Publishing Co.

PROBLEMS

15–1. The usual form of a virial expansion is

$$\frac{pV}{RT} = 1 + \frac{B(T)}{V} + \frac{C(T)}{V^2} + \cdots$$

Some workers, however, prefer to express their data by expanding the compressibility factor in a power series in the pressure

$$\frac{pV}{RT} = 1 + B'(T)p + C'(T)p^2 + \cdots$$

Find the relations between the two sets of virial coefficients.

15–2. Find the second virial coefficient from the Dieterici equation:

$$p(v - b) = kTe^{-a/kTv}$$

15–3. Sketch the cluster diagrams corresponding to the following products of f-functions:

(a) $f_{12}f_{23}f_{34}f_{45}f_{51}f_{14}f_{25}$

(b) $f_{12}f_{23}f_{13}f_{34}f_{45}f_{46}f_{56}$

15–4. Find the second virial coefficient and the Boyle temperature for the Berthelot equation of state

$$\left(P+\frac{N^2A}{V^2T}\right)(V-NB)=NkT$$

where A and B are constants.

15–5. Derive Eqs. (15–14).

15–6. Show that the intermolecular potential must vanish more rapidly than r^{-3} in order for $B_2(T)$ to exist. Do this by breaking up the integral in B_2 into two regions, say 0 to L and L to ∞. Choose L large enough so that the exponential in B_2 can be expanded, and investigate this convergence.

15–7. Show that the second virial coefficient for a Lennard-Jones potential can be written in the form

$$\frac{B_2(T)}{b_0}=\sum_{j=0}^{\infty} b^{(j)}T^{*-(2j+1)/4}$$

with

$$b^{(j)}=-\frac{2^{j+1/2}}{4j!}\Gamma\left(\frac{2j-1}{4}\right)$$

The quantity $\Gamma(x)$ is the gamma function.

15–8. Show that the second virial coefficient for the Sutherland potential, defined by

$$\begin{aligned} u(r)&=\infty & r&<\sigma \\ &=-cr^{-\gamma} & r&>\sigma \end{aligned}$$

is

$$B_2(T)=-\frac{2\pi\sigma^3}{3}\sum_{j=0}^{\infty}\frac{1}{j!}\left(\frac{3}{j\gamma-3}\right)\left(\frac{c}{\sigma^\gamma kT}\right)^j$$

The parameter γ is usually taken to be 6.

15–9. Calculate the second virial coefficient for the triangle potential*

$$\begin{aligned} u(r)&=\infty & r&<\sigma \\ &=\frac{\varepsilon}{\sigma(\lambda-1)}\{r-\lambda\sigma\} & \sigma&<r<\lambda\sigma \\ &=0 & r&>\lambda\sigma \end{aligned}$$

15–10. Show that

$$B_2=-\frac{1}{6kT}\int_0^\infty r\frac{du(r)}{dr}e^{-u(r)/kT}4\pi r^2\,dr$$

is equivalent to

$$B_2=-\frac{1}{2}\int_0^\infty (e^{-\beta u(r)}-1)4\pi r^2\,dr$$

State the condition on $u(r)$ that is necessary.

15–11. Find an exact B_2 for the potential $u(r)=\alpha/r^n$ for all r with $n>3$, using the results of Problem 15–10. Carry the integration to the point of having a well-known function in your result for B_2.

* See M. J. Feinberg and A. G. DeRocco, *J. Chem. Phys.*, **41**, p. 3439, 1964.

15–12. Using the tables in Hirschfelder, Curtiss, and Bird (see "Additional Reading"), plot the second virial coefficient of argon versus temperature from its boiling point, 84°K, to 500°K, assuming a Lennard-Jones potential with parameters given in Table 15–3. Compare your result to the experimental values.

15–13. Using the data in F. Whalley and W. G. Schneider, *J. Chem. Phys.*, **23**, p. 1644, 1955; A. E. Kingston, *J. Chem. Phys.*, **42**, p. 719, 1965; B. E. F. Fender and G. D. Halsey, *J. Chem. Phys.*, **36**, p. 1881, 1962; and R. D. Weir, I. Wynn Jones, J. S. Rowlinson, and G. Saville, *Trans. Far. Soc.*, **63**, p. 1320, 1967, show that the second virial coefficient satisfies the law of corresponding states for argon, krypton, and xenon.

15–14. Using the experimental data in any of the references in Problem 15–13, determine ε and σ for the Lennard-Jones potential.

15–15. Calculate the compressibility factor for argon at 10 atm and 0°C.

15–16. Show that the virial expansion for the thermodynamic energy is

$$\frac{E}{NkT} = \frac{3}{2} - T\sum_{j=1}^{\infty} \frac{1}{j} \frac{dB_{j+1}}{dT} \rho^j$$

and that for entropy is

$$\frac{S}{Nk} = \frac{S_{\text{ideal}}}{Nk} - \sum_{j=1}^{\infty} \frac{1}{j} \frac{\partial}{\partial T}(TB_{j+1})\rho^j$$

15–17. Show that for a binary mixture,

$$\frac{p}{kT} = \rho_1 + \rho_2 + B_{20}(T)\rho_1^2 + B_{11}(T)\rho_1\rho_2 + B_{02}(T)\rho_2^2 + \cdots$$

and derive expressions for the second virial coefficients. Show that if this virial expansion is written in the form

$$\frac{p}{kT} = \rho + B_2(T)\rho^2 + \cdots$$

with $\rho = \rho_1 + \rho_2$, then

$$B_2(x_1, T) = x_1^2 B_{20}(T) + x_1(1 - x_1)B_{11}(T) + (1 - x_1)^2 B_{02}(T)$$

where x_1 is the mole fraction of component 1.

15–18. Show that for a multicomponent mixture,

$$\frac{p}{kT} = \rho + B_2\rho^2 + B_3\rho^3 + \cdots$$

where $\rho = \rho_1 + \rho_2 + \cdots + \rho_n$ and

$$B_2(T) = \sum_{i=1}^{n} \sum_{j=1}^{n} B_{ij}(T) x_i x_j$$

$$B_3(T) = \sum_{i=1}^{n} \sum_{j=1}^{n} \sum_{k=1}^{n} B_{ijk} x_i x_j x_k$$

where the x's are mole fractions.

15–19. Using tables for the Lennard-Jones potential, calculate the volume change when 5 liters of N_2 and 2 liters of Ar are mixed at 10 atm pressure and 300°K.

15–20. Derive the first few terms of a virial expansion for the fugacity f.

15–21. For an ideal gas,

$$C_p - C_V = R$$

Derive a virial expansion for this difference for a real gas.

15–22. The Joule-Thomson coefficient μ is defined by

$$\mu = \left(\frac{\partial T}{\partial p}\right)_H = C_p^{-1}\left[T\left(\frac{\partial V}{\partial T}\right)_p - V\right]$$

Derive a density expansion for μ. At high temperatures μ is negative, and for sufficiently low temperatures, it is positive. The temperature at which μ is zero is called the inversion temperature. Show that for not too dense gases, the inversion temperature is given by $d(B_2/T)/dT = 0$.

15–23. Show that

$$\mu_0 C_p{}^0 = -2\pi N\beta \int_0^\infty e^{-\beta u}\left(u - \frac{r}{3}\frac{du}{dr}\right) r^2\, dr$$

where μ_0 and $C_p{}^0$ are the zero pressure limits of the Joule-Thomson coefficient and C_p.

15–24. Using the result of Problem 1–35, derive an expression for the first nonideal correction to the speed of sound in a gas.

15–25. Calculate the third virial coefficient for the potential*

$$u(r) = \begin{cases} \infty & r < \sigma \\ -\varepsilon & \sigma < r < 2\sigma \\ 0 & r > 2\sigma \end{cases}$$

15–26. Derive an expression for the third virial coefficient without assuming that the intermolecular potential is pairwise additive. Let $U_3(\mathbf{r}_1, \mathbf{r}_2, \mathbf{r}_3)$ be

$$U_3(\mathbf{r}_1, \mathbf{r}_2, \mathbf{r}_3) = u(r_{12}) + u(r_{13}) + u(r_{23}) + \Delta(r_{12}, r_{13}, r_{23})$$

15–27. Derive Eq. (15–51) from Eq. (15–50).

15–28. Evaluate the ring contribution to B_4 by the method of Fourier transformations in Section 15–5.

15–29. The Hamiltonian for a gas composed of N rigid rotors is

$$H = \sum_j \frac{1}{2m} p_j{}^2 + \sum_j \frac{1}{2I}\left(p_{\theta_j}{}^2 + \frac{p_{\phi_j}{}^2}{\sin^2\theta_j}\right) + U(\mathbf{q}_1, \ldots, \mathbf{q}_N)$$

where $\mathbf{q}_j$ is the set of five coordinates x_j, y_j, z_j, θ_j, and ϕ_j needed to specify the location and orientation of a molecule. Show that the second virial coefficient for such a system is given by

$$B_2(T) = -\frac{1}{32\pi^2 V}\iint f_{12}\, d\mathbf{q}_1\, d\mathbf{q}_2$$

where

$$d\mathbf{q}_j = dx_j\, dy_j\, dz_j \sin\theta_j\, d\theta_j\, d\phi_j$$

15–30. The potential between two dipolar molecules (see Fig. 15–8) can be approximated by

$$\begin{aligned} u(r, \theta_1, \theta_2, \phi_1 - \phi_2) &= \infty & r < \sigma \\ &= -\frac{\mu^2}{r^3} g(\theta_1, \theta_2, \phi_2 - \phi_1) & r > \sigma \end{aligned}$$

where

$$g(\theta_1, \theta_2, \phi_1 - \phi_2) = 2\cos\theta_1\cos\theta_2 - \sin\theta_1\sin\theta_2\cos(\phi_2 - \phi_1)$$

Show that the second virial coefficient for this potential can be expressed as

$$B_2(T) = b_0\left[1 - \frac{1}{3}\left(\frac{\mu^2}{\sigma^3 kT}\right)^2 - \frac{1}{75}\left(\frac{\mu^2}{\sigma^3 kT}\right)^4 + \cdots\right]$$

* See S. Katsura, *Phys. Rev.*, **115**, p. 1417, 1959.

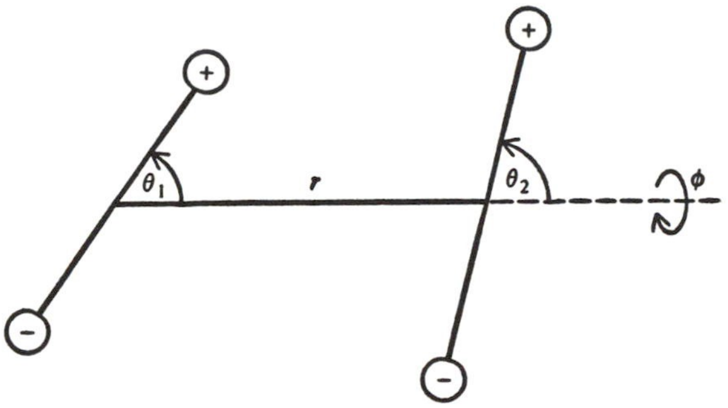

Figure 15–8. **The coordinates describing the mutual orientation of two polar molecules.** (From Hirschfelder, Curtiss, and Bird, *Molecular Theory of Gases and Liquids*. New York: Wiley, 1954.)

15–31. Show that

$$B_2(T) = -2\pi \int_0^\infty \left[\frac{kT}{u_2} e^{-u_1/kT}\left(\frac{\pi^{1/2}}{2} \operatorname{erf} \frac{\sqrt{u_2}}{kT}\right)^2 - 1\right] r^2\, dr$$

for a potential

$$\phi(r, \theta_1, \theta_2, \phi_1 - \phi_2) = u_1(r) + u_2(r)[\cos^2 \theta_1 + \cos^2 \theta_2]$$

15–32. Consider the angle dependent potential

$$u(r, \cos\theta) = u_0(r)[1 + aP_m(\cos\theta)]$$

where $P_m(\cos\theta)$ is a Legendre polynomial of degree m, and $u_0(r) = -\alpha/r^n$. Find an expression for the angular correction factor $f(a)$ in the equation

$$B_2[u(r, \cos\theta)] = f(a) b_2[u_0(r)]$$

Carry out the integration to obtain a "simple" expression for $f(a)$ for a "dipole" interaction $m = 1$ and long-range dispersion forces $n = 6$. As a check of your result, show

$$\lim_{a\to\infty} f(a) = 1$$

For a typical value of $a \sim \frac{1}{5}$, calculate $f(a)$.

15–33. Derive the quantum correction to $B_2(T)$ up through the term in $\hbar^2$; that is, derive Eq. (15–61).

15–34. The hard-sphere virial equation can be written in terms of the variable*

$$y = \sqrt{2}\,\frac{\pi v_0}{6v} = \frac{b_0}{4v}$$

where v_0 is the closest packing volume, and b_0 is the hard-sphere second virial coefficient. In terms of y

$$\frac{pv}{kT} = 1 + 4y + 10y^2 + 18.36y^3 + 28.2y^4 + 39.5y^5 + \cdots$$

We can approximate the terms in this expansion by

$$\frac{pv}{kT} = 1 + 4y + 10y^2 + 18y^3 + 28y^4 + 40y^5 + \cdots$$

Now if you are clever enough to notice that B_n is given by $n^2 + n - 2$, you can write

$$\frac{pv}{kT} = 1 + \sum_{n=2}^{\infty} (n^2 + n - 2) y^{n-1}$$

* See N. F. Carnahan and K. E. Starling, *J. Chem. Phys.*, **51**, p. 635, 1969.

Sum this series exactly, and compare with the result of Thiele, *J. Chem. Phys.*, **39**, p. 474, 1963, as well as the exact molecular dynamics calculation of Alder and Wainwright, *J. Chem. Phys.*, **33**, p. 1439, 1960. In order to make this comparison, calculate the compressibility factor for $v/v_0 = 1.5$, 2.0, 10.0. The Alder and Wainwright values are 12.5, 5.89, 1.36, respectively. Furthermore, calculate the seventh hard-sphere virial coefficient from this scheme and compare to the result $0.0138 \pm 0.0004 b_0{}^6$ calculated from cluster integrals. Would you expect a phase transition for this infinite series virial expansion? Why or why not?

15–35. The virial expansion as originally developed by Ursell* was quite different from the method presented in this chapter. Consider the Boltzmann factor $W_N(\mathbf{r}_1, \mathbf{r}_2, \ldots, \mathbf{r}_N)$, which appears in the configuration integral. Ursell expressed W_N as a sum of products of functions U_l defined by

$$U_1(\mathbf{r}_i) = W_1(\mathbf{r}_i)$$

$$U_2(\mathbf{r}_i, \mathbf{r}_j) = W_2(\mathbf{r}_i, \mathbf{r}_j) - W_1(\mathbf{r}_i)W_1(\mathbf{r}_j)$$

$$U_3(\mathbf{r}_i, \mathbf{r}_j, \mathbf{r}_k) = W_3(\mathbf{r}_i, \mathbf{r}_j, \mathbf{r}_k) - W_2(\mathbf{r}_i \mathbf{r}_j), W_1(\mathbf{r}_k) - W_2(\mathbf{r}_j, \mathbf{r}_k)W_1(\mathbf{r}_i) - W_2(\mathbf{r}_k, \mathbf{r}_i)W_1(\mathbf{r}_j) + 2W_1(\mathbf{r}_i)W_1(\mathbf{r}_j)W_1(\mathbf{r}_k)$$

Show that these *U*-functions, now called Ursell functions, are short-ranged functions in the sense that they vanish unless all of the l molecules in the argument of U_l are at least singly connected.

Invert these equations to write

$$W_1(\mathbf{r}_i) = U_1(\mathbf{r}_i) = 1$$

$$W_2(\mathbf{r}_i, \mathbf{r}_j) = U_2(\mathbf{r}_i, \mathbf{r}_j) + U_1(\mathbf{r}_i)U_1(\mathbf{r}_j)$$

$$W_3(\mathbf{r}_i, \mathbf{r}_j, \mathbf{r}_k) = U_3(\mathbf{r}_i, \mathbf{r}_j, \mathbf{r}_k) + U_2(\mathbf{r}_i, \mathbf{r}_j)U_1(\mathbf{r}_k) + U_2(\mathbf{r}_j, \mathbf{r}_k)U_1(\mathbf{r}_i) + U_2(\mathbf{r}_k, \mathbf{r}_i)U_1(\mathbf{r}_j) + U_1(\mathbf{r}_i)U_1(\mathbf{r}_j)U_1(\mathbf{r}_k)$$

This expresses the integrand of the configuration integral into a sum of products of short-ranged functions. This method can be used to write any property of *N*-molecules in terms of short-ranged functions (*cf.* Problem 15–36). It is easy to show (although we shall not) that the b_j of Section 15–1 are given by

$$b_j = (Vj!)^{-1} \int \cdots \int U_j(\mathbf{r}_1, \mathbf{r}_2, \ldots, \mathbf{r}_j)\, d\mathbf{r}_1 \ldots d\mathbf{r}_j$$

and that the configuration integral can be written as

$$\frac{Z_N}{N!} = \sum_{\{m_j\}}^{*} \prod_{j=1}^{N} \frac{(Vb_j)^{m_j}}{m_j}$$

where the asterisk denotes the condition

$$\sum jm_j = N$$

The Ursell development has the advantage of not being limited to pair-wise additive potentials.

15–36. Consider the polarizability of a system of N polarizable molecules $\alpha_N(\mathbf{r}_1, \ldots, \mathbf{r}_N)$. In general, this quantity cannot be rigorously written as the sum of N isolated-molecule polarizabilities, but can be expressed *exactly* by

$$\alpha_N(\mathbf{r}_1, \ldots, \mathbf{r}_N) = \sum_{i=1}^{N} \alpha_i(\mathbf{r}_i) + \sum_{1 \le i < j \le N}\sum \alpha_{ij}(\mathbf{r}_i, \mathbf{r}_j) + \sum_{1 \le i < j < k \le N}\sum\sum \alpha_{ijk}(\mathbf{r}_i, \mathbf{r}_j, \mathbf{r}_k) + \cdots$$

* See Section 3–2 of Hirschfelder, Curtiss, and Bird in "Additional Reading."

Show that the α_i are polarizabilities of isolated molecules; the α_{ij} are incremental polarizabilities of pairs of molecules, and so on, or, in other words, by inversion determine the α_i, α_{ij}, α_{ijk}, and so on, in terms of the polarizabilities of groups of molecules.*

15–37. Find the set $\{m_j\}$ which maximizes ln $Z_N/N!$, where

$$\frac{Z_N}{N!}=\sum_{\{m_j\}}^{*}\prod_{j=1}^{N}\frac{(Vb_j)^{m_j}}{m_j!}$$

where the asterisk signifies $\Sigma\, jm_j = N$ as a constraint. Call the undetermined multiplier z. The m_j^* (those m_j that yield the maximum term) will come out in terms of z. So will $Z_N/N!$, and hence $Q(N, V, T)$ and p. Solve the equation

$$\sum_{j=1}^{N} jm_j^* = N$$

for z (at least the first few terms of z as a power series in ρ) and derive the first few terms of the virial expansion for p.†

15–38. Find the first few terms of $Z_N/N!$, written in the form (see Problem 15–37)

$$V^N\left(1+\frac{a_1}{V}+\frac{a^2}{V^2}+\cdots+\frac{a_N}{V^N}\right)$$

expand the logarithm and use

$$p=kT\left(\frac{\partial \ln Q}{\partial V}\right)_{T,N}$$

to find the virial expansion.

The techniques of this chapter can also be applied to imperfect gases in electric fields, and so it is possible to extend the treatment of Chapter 13 to include this case. This extension is developed in Problems 15–39 through 15–45, and is based upon a very readable and fundamental paper by Hill.‡ The basic thermodynamic equations can be deduced from Eq. (13–20) by interpreting $\overline{M}_z$ to be the induced moment of a macroscopic sample and replacing $\mathscr{E}_z^*$, the field that the sample sees, by the displacement vector D, assumed to be due to the charges on the plates of a capacitor. D is an intensive thermodynamic quantity that can be controlled by varying the charges on the plates of the capacitor. Thus the basic energy equation for an open system is

$$dE = T\,dS - p\,dV - M\,dD + \sum_i \mu_i\, dN_i$$

15–39. From the above thermodynamic equation, derive

$$U = TS - pV + \sum_i \mu_i N_i$$

$$dA = -S\,dT - p\,dV - M\,dD + \sum_i \mu_i\, dN_i$$

$$A = G - pV$$

$$dG = -S\,dT + V\,dp - M\,dD + \sum_i \mu_i\, dN_i$$

$$d(pV) = S\,dT + p\,dV + M\,dD + \sum_i N_i\, d\mu_i$$

$$\rho\, d\mu + \left(\frac{S}{V}\right) dT - dp + \frac{M}{V}\, dD = 0$$

* See H. B. Levine and D. A. McQuarrie, *J. Chem. Phys.*, **49**, p. 4181, 1968.
† See Chapter 13 of Mayer and Mayer in "Additional Reading."
‡ See T. L. Hill, *J. Chem. Phys.*, **28**, p. 61, 1958.

15–40. The polarization $P = M/V$ satisfies the electrostatic equation $D = \mathscr{E} + 4\pi P$ [Eq. (13–31)]. Show that

$$P_N = \frac{kT}{V}\left(\frac{\partial \ln Q(N, V, T, D)}{\partial D}\right)_{T, V, N}$$

for a closed system containing N molecules and

$$\bar{P} = \frac{kT}{V}\left(\frac{\partial \ln \Xi}{\partial D}\right)_{T, V, \mu}$$

$$= \frac{1}{\Xi}\sum_N P_N Q_N(V, T, D) \exp\left(\frac{N\mu}{kT}\right)$$

for an open system. From a molecular theory point of view, D and P are basic quantities and $\mathscr{E}$, and the dielectric constant ε are subsidiary quantities.

15–41. The configuration integral is [Eq. (15–11)]

$$Z_N(V, T, D) = N!\left(\frac{V}{Q_1}\right)^N Q_N(V, T, D)$$

Define the activity z by

$$z(\mu, T, D) = Q_1(V, T, D)\frac{\lambda}{V}$$

and carry through the development from Eq. (15–10) to Eq. (15–17) to get

$$\frac{p}{kT} = \rho + \sum_{n \geq 2} B_n(T, D)\rho^n$$

where the B_n, are given formally by Eqs. (15–16) and (15–17).

15–42. Show that the polarization $\bar{P}$ has the expansion

$$\bar{P} = P_1 V\rho + kT \sum_{j \geq 1}\left(\frac{\partial b_j}{\partial D}\right) z^j$$

where P_1 is the polarization with $N = 1$. Eliminate z in favor of ρ by using the analog of Eq. (15–15) to get

$$\bar{P} = P_1 V\rho - kT \sum_{n \geq 2}\left(\frac{1}{n-1}\right)\left(\frac{\partial B_n}{\partial D}\right)\rho^n$$

15–43. In general, D, P, and $\mathscr{E}$ are vectors, but for a parallel plate capacitor they are perpendicular to the plates and may be treated as quantities which are simply positive and negative. Convince yourself, then, that by symmetry we may write

$$P_1 V = \alpha_1 D + \alpha_3 D^3 + \cdots$$

$$B_n = B_n^{(0)}(T) + B_n^{(2)}(T)D^2 + \cdots$$

$$\varepsilon = \varepsilon_0(\rho, T) + \varepsilon_2(\rho, T)D^2 + \cdots$$

Using this result and the equation $\bar{P} = [(\varepsilon - 1)/4\pi\varepsilon]D$, show that

$$\frac{\varepsilon_0 - 1}{4\pi\varepsilon_0} = \alpha_1\rho - 2kT \sum_{n \geq 2}\left(\frac{1}{n-1}\right) B_n^{(2)}(T)\rho^n$$

Explain how you could calculate the imperfect gas corrections to Eq. (13–35).*

* Compare D. A. McQuarrie and H. B. Levine, *Physica*, **31**, p. 749, 1965, for a recent detailed calculation along these lines.

15–44. Defining $x = (\varepsilon_0 - 1)/\varepsilon_0$, show that the Clausius-Mossotti function $(\varepsilon_0 - 1)/(\varepsilon_0 + 2)\rho$ is

$$\frac{\varepsilon_0 - 1}{(\varepsilon_0 + 2)\rho} = \frac{1}{3\rho}\left(x + \tfrac{2}{3} + \tfrac{4}{9}x^3 + \cdots\right)$$

and from this derive

$$\frac{\varepsilon_0 - 1}{(\varepsilon_0 + 2)\rho} = \mathscr{A}(T) + \mathscr{B}(T)\rho + \cdots$$

where

$$\mathscr{A} = \frac{4\pi\alpha_1}{3}$$

$$\mathscr{B} = -\frac{8\pi}{3}kTB_2^{(2)} + \frac{32\pi^2\alpha_1^2}{9}$$

and so on.

15–45. Lastly, we shall treat the phenomenon of electrostriction—the change in density that occurs when an electric field is applied across a substance. There are two cases to consider: the change in density in a closed system (at constant T and p), and the change in density for an open system (at constant T and μ). First use the last equation in Problem 15–39 to derive

$$\frac{1}{\rho^2}\left(\frac{\partial \rho}{\partial D}\right)_{T,p} = \left(\frac{\partial (P/\rho)}{\partial p}\right)_{T,D}$$

from which you should be able to obtain

$$\frac{1}{\rho}\left(\frac{\partial \rho}{\partial D}\right)_{T,p} = -\left(\frac{\partial B_2}{\partial D}\right)\rho + \left\{-\left(\frac{\partial B_3}{\partial D}\right) + 2B_2\left(\frac{\partial B_2}{\partial D}\right)\right\}\rho^2 + \cdots$$

$$= 2D[-B_2^{(2)}\rho + (-B_3^{(2)} + 2B_2^{(0)}B_2^{(2)})\rho^2 + \cdots]$$

For electrostriction in an open system, first derive

$$\left(\frac{\partial \rho}{\partial D}\right)_{T,\mu} = \left(\frac{\partial P}{\partial \mu}\right)_{T,D} = \rho\left(\frac{\partial P}{\partial p}\right)_{T,D}$$

and then

$$\frac{1}{\rho}\left(\frac{\partial \rho}{\partial D}\right)_{T,\mu} = D\left\{\frac{\alpha_1}{kT} + \left(-4B_2^{(2)} - 2B_2^{(0)}\frac{\alpha_1}{kT}\right)\rho + \cdots\right\}$$

For an application of these equations for the Lennard-Jones 6–12 potential, see J. F. Ely and D. A. McQuarrie, *J. Phys. Chem.*, **75**, 771, 1971.

15–46. A model "potential" that has been used in the theoretical study of imperfect gases is the so-called Gaussian gas, in which it is assumed that the Mayer f-function $f(r)$ can be approximated by a negative Gaussian, that is, by $-e^{-\alpha r^2}$, where α is a constant. Show that this "potential" simulates a soft repulsive potential. The great advantage of this "potential" is that the many-center integrals involved in the evaluation of the B_n can be done analytically, since it can be readily shown by matrix algebra* that

$$\int_{-\infty}^{\infty} \cdots \int e^{\mathbf{t}\cdot\mathbf{x} - (1/2)\mathbf{x}\cdot\mathbf{A}\cdot\mathbf{x}}\, dx_1\, dx_2 \ldots dx_n = \frac{(2\pi)^{n/2}}{|A|^{1/2}} e^{(1/2)\mathbf{t}\cdot\mathbf{A}^{-1}\cdot\mathbf{t}}$$

* See H. Cramer, *Mathematical Methods of Statistics* (Princeton, N.J.: Princeton University Press, 1946).

where $\mathbf{t} = (t_1, t_2, \ldots, t_n)$, $\mathbf{x} = (x_1, x_2, \ldots, x_n)$, $\mathbf{A}$ is a matrix, and where $|A|$ is the determinant of $\mathbf{A}$. Use this to show that for a Gaussian gas,

$$B_2 = b \equiv \frac{1}{2}\left(\frac{\pi}{\alpha}\right)^{3/2}$$

$$B_3 = 0.257b^2$$

$$B_4 = -0.125b^3$$

and with patience

$$B_5 = 0.013b^4$$

The sixth and seventh virial coefficients are also known for this potential, $B_6 = 0.038b^5$ and $B_7 = -0.030b^6$.*

15–47. Go to the tables in Katsura's paper† to verify the result for B_3 for hard spheres given in Eq. (15–57).

* See *Studies in Statistical Mechanics*, vol. I, ed. by J. DeBoer and G. E. Uhlenbeck (Amsterdam: North-Holland Publishing Co., 1962, p. 182).
† In *Phys. Rev.*, **115**, p. 1417, 1959.

APPENDIX A

VALUES OF SOME PHYSICAL CONSTANTS AND ENERGY CONVERSION FACTORS

Table A–1. **Values of some physical constants**

quantity	symbol	value
Avogadro's number	N_0	6.0222×10^{23}
Planck constant	h	6.6262×10^{-27} erg-sec
	$\hbar$	1.0546×10^{-27} erg-sec
Boltzmann constant	k	1.3806×10^{-16} erg/molecule-deg K
gas constant	R	8.3143×10^{7} ergs/mole-deg K
		1.9872 cal/mole-deg K
speed of light	c	2.9979×10^{10} cm/sec
proton charge	e	4.8032×10^{-10} esu
electron mass	m_e	9.1096×10^{-28} g
atomic mass unit	*amu*	1.6605×10^{-24} g
Bohr magneton	μ_B	9.2741×10^{-21} erg/gauss
nuclear magneton	μ_N	5.0509×10^{-24} erg/gauss

Source: B. N. Taylor, W. H. Parker, and D. N. Langenberg, *Rev. Mod. Phys.*, **41**, p. 375, 1969.

Table A–2. **Energy conversion factors**

	ergs	eV	cm^{-1}	°K	kcal	kcal/mole	atomic units
1 erg	1	6.2420×10^{11}	5.0348×10^{15}	7.2441×10^{15}	2.3901×10^{-11}	1.4394×10^{13}	2.294×10^{10}
1 eV	1.6021×10^{-12}	1	8.0660×10^{3}	1.1605×10^{4}	3.8390×10^{-23}	2.3119×10^{1}	3.675×10^{-2}
1 cm^{-1}	1.9862×10^{-16}	1.2398×10^{-4}	1	1.4388	4.7471×10^{-27}	2.8588×10^{-3}	4.556×10^{-6}
1°K	1.3804×10^{-16}	8.6167×10^{-5}	6.9502×10^{-1}	1	3.2993×10^{-27}	1.9869×10^{-3}	3.116×10^{-6}
1 kcal	4.1840×10^{10}	2.6116×10^{22}	2.1066×10^{26}	3.3009×10^{26}	1	6.0222×10^{23}	9.597×10^{20}
1 kcal/mole	6.9446×10^{-14}	4.3348×10^{-2}	3.4964×10^{2}	5.0307×10^{2}	1.6598×10^{-24}	1	1.594×10^{-3}
1 atomic unit	4.360×10^{-11}	27.21	2.195×10^{5}	3.158×10^{5}	1.042×10^{-21}	6.275×10^{2}	1

APPENDIX B

FOURIER INTEGRALS AND THE DIRAC DELTA FUNCTION

The Fourier transform $g(k)$ of the function $f(x)$ is defined by

$$g(k) = \frac{1}{(2\pi)^{1/2}} \int_{-\infty}^{\infty} f(x) e^{-ikx}\, dx \tag{B–1}$$

The Fourier integral theorem states that if $f(x)$ is single valued and periodic with at most a finite number of finite discontinuities, maxima, and minima, and

$$\int_{-\infty}^{\infty} |f(x)|^2\, dx$$

is finite, then

$$f(x) = \frac{1}{(2\pi)^{1/2}} \int_{-\infty}^{\infty} g(k) e^{ikx}\, dk \tag{B–2}$$

Equations (B–1) and (B–2) are said to be a Fourier transform pair. Other definitions of the Fourier transforms are in use, differing by numerical factors and by the use of complex conjugates of the integrands given here.

In vector notation, if $V(\mathbf{r})$ is a function of the vector $\mathbf{r} = (x, y, z)$ and $\bar{V}(\mathbf{k})$ its Fourier transform, then

$$\bar{V}(\mathbf{k}) = \frac{1}{(2\pi)^{3/2}} \iiint_{-\infty}^{\infty} V(\mathbf{r}) e^{-i\mathbf{k}\cdot\mathbf{r}}\, d\mathbf{r} \tag{B–3}$$

and

$$V(\mathbf{r}) = \frac{1}{(2\pi)^{3/2}} \iiint_{-\infty}^{\infty} \bar{V}(\mathbf{k}) e^{i\mathbf{k}\cdot\mathbf{r}}\, d\mathbf{k} \tag{B–4}$$

Fourier transforms can be used in a formal manner to obtain a useful representation for the Dirac delta function. The Dirac delta function is defined by

$$\int_{-\infty}^{\infty} \delta(x-a)\Phi(x)\,dx = \Phi(a) \tag{B–5}$$

$$\int_{-\infty}^{\infty} \delta(x)\,dx = 1 \tag{B–6}$$

The Fourier transform of $\delta(x)$ is

$$\frac{1}{(2\pi)^{1/2}} \int_{-\infty}^{\infty} \delta(x) e^{-iux}\,dx = \frac{1}{(2\pi)^{1/2}} \tag{B–7}$$

The formal inverse gives the desired relation:

$$\delta(x) = \frac{1}{2\pi} \int_{-\infty}^{\infty} e^{iux}\,du \tag{B–8}$$

More generally, in three dimensions,

$$\frac{1}{(2\pi)^{3/2}} \iiint_{-\infty}^{\infty} \delta(\mathbf{k}) e^{-i\mathbf{k}\cdot\mathbf{r}}\,d\mathbf{k} = \frac{1}{(2\pi)^{3/2}} \tag{B–9}$$

$$\delta(\mathbf{k}) = \frac{1}{(2\pi)^3} \iiint_{-\infty}^{\infty} e^{i\mathbf{k}\cdot\mathbf{r}}\,d\mathbf{r} \tag{B–10}$$

where $\delta(\mathbf{k}) = \delta(k_x)\delta(k_y)\delta(k_z)$.

A triple integral such as

$$\bar{V}(|\mathbf{k}|) = \frac{1}{(2\pi)^{3/2}} \iiint_{-\infty}^{\infty} V(|\mathbf{r}|) e^{-i\mathbf{k}\cdot\mathbf{r}}\,d\mathbf{r} \tag{B–11}$$

may be converted to a single integral, and hence more easily handled, by converting to spherical coordinates:

$$d\mathbf{r} = dx\,dy\,dz \rightarrow r^2 \sin\theta\,dr\,d\theta\,d\phi$$

$$\mathbf{k}\cdot\mathbf{r} \rightarrow kr\cos\theta$$

and so

$$\bar{V}(|\mathbf{k}|) = \frac{1}{(2\pi)^{3/2}} \int_0^{\infty} \int_0^{\pi} \int_0^{2\pi} V(|\mathbf{r}|) e^{-ikr\cos\theta} r^2\,dr \sin\theta\,d\theta\,d\phi$$

The integral over ϕ gives 2π. The integral over θ gives

$$\bar{V}(|\mathbf{k}|) = \left(\frac{2}{\pi}\right)^{1/2} \int_0^{\infty} V(r) \frac{r \sin kr}{k}\,dr \tag{B–12}$$

In order to evaluate Fourier transforms, the identity $e^{iz} = \cos z + i \sin z$ is often useful if one remembers that

$$\int_{-\infty}^{\infty} f(x)\,dx = 2\int_0^{\infty} f(x)\,dx \qquad \text{if } f(x) \text{ is even}$$

$$\int_{-\infty}^{\infty} f(x)\,dx = 0 \qquad \text{if } f(x) \text{ is odd}$$

For example, consider the Fourier transform of $1/(x^2+\alpha^2)$.

$$\phi(k) = (2\pi)^{-1/2} \int_{-\infty}^{\infty} \frac{e^{-ikx}}{x^2+\alpha^2}\, dx$$

$$= 2(2\pi)^{-1/2} \int_0^{\infty} \frac{\cos kx\, dx}{x^2+\alpha^2} = \left(\frac{\pi}{2}\right)^{1/2} \alpha^{-1} e^{-k|\alpha|}$$

You should be able to invert $\phi(k)$ to get $(x^2+\alpha^2)^{-1}$.

PROBLEMS

B–1. Show that the Fourier transform of $f(x) = e^{-\alpha|x|}$ is

$$\left(\frac{2}{\pi}\right)^{1/2} \frac{\alpha}{k^2+\alpha^2}$$

B–2. Show that the Fourier transform of $e^{-\lambda x^2}$ is $(2\lambda)^{-1/2} e^{-k^2/4\lambda}$.

B–3. Solve the equation

$$e^{-|\alpha|} = \int_0^{\infty} \cos \alpha t f(t)\, dt$$

for $f(x)$ given that $f(x)$ is an even function of x.

B–4. Given that $f(\mathbf{r}) = Ne^{-r^2/a^2}$ where $r = |\mathbf{r}|$ (three dimensions), show that its Fourier transform $\phi(\mathbf{k}) = a^3 N e^{-k^2a^2/4}/2^{3/2}$.

B–5. The delta function is not a function in the strict sense, but has meaning only if it occurs under an integral sign. For example, we can assign a meaning to $x\delta(x)$ by multiplying by an arbitrary but continuous function $f(x)$ and integrating to get

$$\int_{-\infty}^{\infty} f(x) x\delta(x)\, dx = 0$$

and thus one often sees $x\delta(x) = 0$. In a similar way, show that

$$x\delta'(x) = -\delta(x)$$

and

$$\delta(ax) = a^{-1}\delta(x)$$

B–6. Show that

$$\int_{-\infty}^{\infty} \delta(x-a) \cos x\, dx = \cos a$$

by first evaluating

$$I(\sigma) = (2\pi\sigma^2)^{-1/2} \int_{-\infty}^{\infty} \cos x e^{-(x-a)^2/2\sigma^2}\, dx$$

and then taking the limit of $\sigma \to 0$. This shows that Eq. (1–72) with $\sigma \to 0$ is one representation of the delta function. When in doubt, this is always a sure way of handling $\delta(x)$.

B–7. Another representation of $\delta(x)$ was used in deriving Eq. (10–131). Let $u_k(x)$ be an orthonormal complete set. Let $\psi(x)$ be a suitably arbitrary function which can be expanded in terms of the $\{u_k(x)\}$:

$$\psi(x) = \sum_k a_k u_k(x)$$

First find the a_k by multiplying by $u_k^*(x)$ and integrating and then substituting the result back into the above equation to get

$$\psi(x) = \int \left[\sum_k u_k^*(x') u_k(x) \right] \psi(x')\, dx'$$

Conclude from this that

$$\delta(x - x') = \sum_k u_k^*(x') u_k(x)$$

Generalize this result to n-dimensions.

APPENDIX C

DEBYE HEAT CAPACITY FUNCTION

Table C–1. **Debye heat capacity function, $C_V/3R$ as a function of Θ_D/T**

Θ_D/T	0.0	0.1	0.2	0.3	0.4	0.5	0.6	0.7	0.8	0.9	1.0
0.0	1.0000	0.9995	0.9980	0.9955	0.9920	0.9876	0.9822	0.9759	0.9687	0.9606	0.9517
1.0	0.9517	0.9420	0.9315	0.9203	0.9085	0.8960	0.8828	0.8692	0.8550	0.8404	0.8254
2.0	0.8254	0.8100	0.7943	0.7784	0.7622	0.7459	0.7294	0.7128	0.6961	0.6794	0.6628
3.0	0.6628	0.6461	0.6296	0.6132	0.5968	0.5807	0.5647	0.5490	0.5334	0.5181	0.5031
4.0	0.5031	0.4883	0.4738	0.4595	0.4456	0.4320	0.4187	0.4057	0.3930	0.3807	0.3686
5.0	0.3686	0.3569	0.3455	0.3345	0.3237	0.3133	0.3031	0.2933	0.2838	0.2745	0.2656
6.0	0.2656	0.2569	0.2486	0.2405	0.2326	0.2251	0.2177	0.2107	0.2038	0.1972	0.1909
7.0	0.1909	0.1847	0.1788	0.1730	0.1675	0.1622	0.1570	0.1521	0.1473	0.1426	0.1382
8.0	0.1382	0.1339	0.1297	0.1257	0.1219	0.1182	0.1146	0.1111	0.1078	0.1046	0.1015
9.0	0.1015	0.09847	0.09558	0.09280	0.09011	0.08751	0.08500	0.08259	0.08025	0.07800	0.07582
10.0	0.07582	0.07372	0.07169	0.06973	0.06783	0.06600	0.06424	0.06253	0.06087	0.05928	0.05773
11.0	0.05773	0.05624	0.05479	0.05339	0.05204	0.05073	0.04946	0.04823	0.04705	0.04590	0.04478
12.0	0.04478	0.04370	0.04265	0.04164	0.04066	0.03970	0.03878	0.03788	0.03701	0.03617	0.03535
13.0	0.03535	0.03455	0.03378	0.03303	0.03230	0.03160	0.03091	0.03024	0.02959	0.02896	0.02835
14.0	0.02835	0.02776	0.02718	0.02661	0.02607	0.02553	0.02501	0.02451	0.02402	0.02354	0.02307
15.0	0.02307	0.02262	0.02218	0.02174	0.02132	0.02092	0.02052	0.02013	0.01975	0.01938	0.01902

Source: K. S. Pitzer, *Quantum Chemistry* (Englewood Cliffs, N.J.: Prentice-Hall, 1953).

INDEX